Compendium of Organic Synthetic Methods

Volume 3

LOUIS S. HEGEDUS

and

LEROY WADE

DEPARTMENT OF CHEMISTRY
COLORADO STATE UNIVERSITY
FORT COLLINS, COLORADO

A Wiley-Interscience Publication

JOHN WILEY & SONS, New York · Chichester · Brisbane · Toronto

PREFACE

By their compilation of Volumes 1 and 2 of this *Compendium*, Ian and Shuyen Harrison filled one of the greatest needs of the synthetic community: a method for rapidly retrieving needed information from the literature by reaction type rather than by the author's name or publication date. We are honored by the opportunity to bring this useful work up to date.

Compendium of Organic Synthetic Methods, Volume 3, presents the functional group transformations and difunctional compound preparations of 1974, 1975, and 1976. We have attempted to follow as closely as possible the classification schemes of the first two volumes; the experienced user of the *Compendium* will require no additional instructions on the use of this volume.

Perhaps it is fitting here to echo the Harrisons' request stated in Volume 2 of the *Compendium*: The synthetic literature would become more easily accessible and more useful if chemists could write well-organized, concise papers with charts and diagrams that allow the reader to assess quickly and easily the scope of the published research. In addition, the reporting of actual, isolated yields and detailed experimental conditions will save a great deal of wasted effort on the part of other chemists hoping to apply the reported reactions to their own synthetic problems.

Anyone who has ever typed a research paper with structures can understand what a Gargantuan project the preparation of the camera-ready copy for this volume has been. Linda Benedict and Jackie Swinehart completed the entire project almost more quickly than our proofreaders, Gary Allen, Joel Slade and Robert Williams, could make corrections. The authors would like to express their thanks to these people for their dedicated work.

Louis S. Hegedus
Leroy Wade

Fort Collins, Colorado
June, 1977

CONTENTS

ABBREVIATIONS

The authors have attempted to use only abbreviations whose meaning will be readily apparent to the reader. Some of those more commonly used are the following:

Ac	acetyl
Bu	butyl
Bz	benzyl
Cp	cyclopentadienyl
DCC	dicyclohexylcarbodiimide
DDQ	2,3-dichloro-5,6-dicyanobenzoquinone
DIBAL	diisobutylaluminum hydride
DME	1,2-dimethoxyethane
DMF	dimethylformamide
DMSO	dimethyl sulfoxide
Et	ethyl
Hex	hexyl
HMPA, HMPT	hexamethylphosphoramide (hexamethylphosphoric triamide)
L	triphenylphosphine ligand (if not specified)
LDA	lithium diisopropylamide
MCPBA	*meta*-chloroperbenzoic acid
Me	methyl
Ms	methanesulfonyl
MVK	methyl vinyl ketone
NBS	*N*-bromosuccinimide
NCS	*N*-chlorosuccinimide
Ni	Raney nickel
Oct	octyl
Ph, ϕ	phenyl
Pr	propyl
Pyr	pyridine

Sia	*secondary*-isoamyl
Tf	trifluoromethane sulfonate
TFA	trifluoroacetic acid
TFAA	trifluoroacetic anhydride
THF	tetrahydrofuran
THP	tetrahydropyranyl
TMS	trimethylsilyl
Ts	*p*-toluenesulfonyl

INDEX, MONOFUNCTIONAL COMPOUNDS

Sections—heavy type
Pages—light type

PREPARATION OF →

FROM ↓

FROM \ PREPARATION OF	Acetylenes	Carboxylic acids, acid halides, anhydrides	Alcohols, phenols	Aldehydes	Alkyls, methylenes, aryls	Amides	Amines	Esters	Ethers, epoxides	Halides, sulfonates	Hydrides (RH)	Ketones	Nitriles	Olefins
Acetylenes	**1** 1	**16** 8	**31** 24	**46** 66	**61** 88	**76** 134		**106** 178		**136** 218		**166** 248		**196** 305
Carboxylic acids, acid halides, anhydrides		**17** 8	**32** 25	**47** 67	**62** 89	**77** 135	**92** 149	**107** 178		**137** 218		**167** 249	**182** 296	**197** 308
Alcohols, phenols		**18** 13	**33** 26	**48** 69	**63** 89	**78** 138	**93** 150	**108** 186	**123** 204	**138** 219	**153** 235	**168** 254	**183** 296	**198** 309
Aldehydes		**19** 13	**34** 26	**49** 74	**64** 90	**79** 138	**94** 150	**109** 188	**124** 208	**139** 222		**169** 261	**184** 296	**199** 312
Alkyls, methylenes, aryls				**50** 76	**65** 91									**200** 315
Amides		**21** 14		**51** 76		**81** 139	**96** 153	**111** 189				**171** 263	**186** 299	
Amines						**82** 143	**97** 154			**142** 223	**157** 239	**172** 264	**187** 299	**202** 316
Esters	**8** 4	**23** 15	**38** 32	**53** 77	**68** 92			**113** 190			**158** 240	**173** 266	**188** 300	**203** 316
Ethers, epoxides			**39** 33	**54** 78	**69** 93		**99** 160	**114** 194	**129** 210	**144** 223		**174** 267		**204** 317
Halides, sulfonates, sulfates	**10** 5	**25** 17	**40** 37	**55** 78	**70** 94	**85** 146	**100** 161	**115** 195	**130** 210	**145** 224	**160** 242	**175** 268	**190** 300	**205** 319
Hydrides (RH)		**26** 19	**41** 38	**56** 82	**71** 105	**86** 146	**101** 163	**116** 197		**146** 227		**176** 272	**191** 301	
Ketones	**12** 6	**27** 19	**42** 40		**72** 107	**87** 147	**102** 163	**117** 198	**132** 212	**147** 231		**177** 273		**207** 324
Nitriles			**43** 51	**58** 83		**88** 147	**103** 165	**118** 201				**178** 285	**193** 301	
Olefins		**29** 20	**44** 51	**59** 83	**74** 111		**104** 166	**119** 201	**134** 215	**149** 232		**179** 286	**194** 303	**209** 331
Miscellaneous compounds	**15** 7	**30** 20	**45** 55	**60** 84	**75** 133		**105** 167			**150** 234		**180** 289	**195** 303	**210** 332

PROTECTION

	Sect.	Pg.
Acetylenes	30A	21
Carboxylic acids	45A	56
Alcohols, phenols	60A	85
Aldehydes	90A	148
Amines	105A	172
Esters		
Ketones	180A	291
Olefins	210A	333

Blanks in the table correspond to sections for which no additional examples were found in the literature.

INDEX, DIFUNCTIONAL COMPOUNDS

Sections—heavy type

Pages—light type

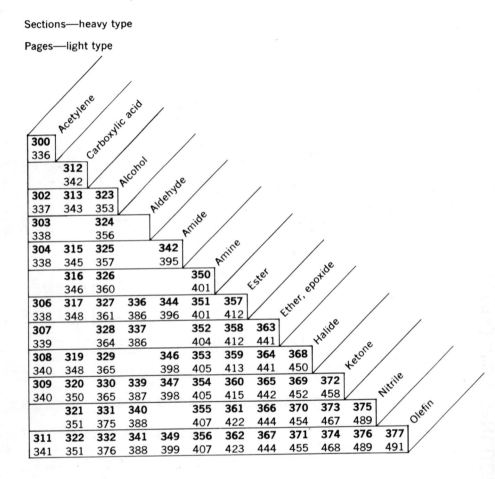

Blanks in the table correspond to sections for which no examples were found in the literature.

INTRODUCTION

Relationship between Volume 3 and Previous Volumes. *Compendium of Organic Synthetic Methods, Volume 3* presents over 1000 examples of published methods for the preparation of monofunctional compounds, updating the 4000 in Volumes 1 and 2. In addition, Volume 3 contains over 1000 additional examples of preparations of difunctional compounds and various functional groups, updating these sections which were initially introduced in Volume 2. The same systems of section and chapter numbering are used in the two volumes.

Classification and Organization of Reactions Forming Monofunctional Compounds. Examples of published chemical transformations are classified according to the reacting functional group of the starting material and the functional group formed. Those reactions that give products with the same functional group form a chapter. The reactions in each chapter are further classified into sections on the basis of the functional group of the starting material. Within each section reactions are listed in a somewhat arbitrary order, although an effort has been made to put chain-lengthening processes before degradations.

The classification is unaffected by allylic, vinylic, or acetylenic unsaturation, which appears in both starting material and product, or increases or decreases in the length of carbon chains; for example, the reactions t-BuOH $\rightarrow$ t-BuCOOH, $PhCH_2OH$ $\rightarrow$ PhCOOH and $PhCH=CHCH_2OH$ $\rightarrow$ $PhCH=CHCOOH$ would all be considered as preparations of carboxylic acids from alcohols. Entries in which conjugate reduction or alkylation of an unsaturated ketone, aldehyde, ester, acid, or nitrile occurs have generally been placed in category 74, Alkyls from Olefins.

The terms hydrides, alkyls, and aryls classify compounds containing reacting hydrogens, alkyl groups, and aryl groups, respectively; for example, RCH_2-H $\rightarrow$ RCH_2COOH (carboxylic acids from hydrides), RMe $\rightarrow$ RCOOH (carboxylic acids from alkyls), RPh $\rightarrow$ RCOOH (carboxylic acids from aryls). Note the distinction between $R_2CO \rightarrow R_2CH_2$ (methylenes from ketones) and RCOR' $\rightarrow$ RH (hydrides from ketones). Alkylations which involve additions across a double bond are found in section 74, Alkyls from Olefins.

The following examples illustrate the application of the classification scheme to some potentially confusing cases:

$RCH{=}CHCOOH \rightarrow RCH{=}CH_2$	(hydrides from carboxylic acids)
$RCH{=}CH_2 \rightarrow RCH{=}CHCOOH$	(carboxylic acids from hydrides)
$ArH \rightarrow ArCOOH$	(carboxylic acids from hydrides)
$ArH \rightarrow ArOAc$	(esters from hydrides)
$RCHO \rightarrow RH$	(hydrides from aldehydes)
$RCH{=}CHCHO \rightarrow RCH{=}CH_2$	(hydrides from aldehydes)
$RCHO \rightarrow RCH_3$	(alkyls from aldehydes)
$R_2CH_2 \rightarrow R_2CO$	(ketones from methylenes)
$RCH_2COR \rightarrow R_2CHCOR$	(ketones from ketones)
$RCH{=}CH_2 \rightarrow RCH_2CH_3$	(alkyls from olefins)
$RBr + RC{\equiv}CH \rightarrow RC{\equiv}CR$	(acetylenes from halides; also acetylenes from acetylenes)
$ROH + RCOOH \rightarrow RCOOR$	(esters from alcohols; also esters from carboxylic acids)
$RCH{=}CHCHO \rightarrow R_2CHCH_2CHO$	(alkyls from olefins)
$RCH{=}CHCN \rightarrow RCH_2CH_2CN$	(alkyls from olefins)

Yields quoted are overall; they are occasionally reduced to allow for incomplete conversion and impurities in the product.

Reactions not described in the given references, but required to complete a sequence, are indicated by a dashed arrow.

Reactions are included even when full experimental details are lacking in the given reference. In some cases the quoted reaction is a minor part of a paper or may have been investigated from a purely mechanistic aspect.

How to Use the Book to Locate Examples of the Preparation or Protection of Monofunctional Compounds. Examples of the preparation of one functional group from another are located via the monofunctional index on p. ix, which lists the corresponding section and page. Thus Section 1 contains examples of the preparation of acetylenes from other acetylenes; Section 2, acetylenes from carboxylic acids; and so forth.

Sections that contain examples of the reactions of a functional group are found in the horizontal rows of the index. Thus Section 1 gives examples of the reactions of acetylenes that form other acetylenes; Section

16, reactions of acetylenes that form carboxylic acids; and Section 31, reactions of acetylenes that form alcohols.

Examples of alkylation, dealkylation, homologation, isomerization, transposition are found in Sections 1, 17, 33, and so on, which lie close to a diagonal of the index. These sections correspond to such topics as the preparation of acetylenes from acetylenes, carboxylic acids from carboxylic acids, and alcohols and phenols from alcohols and phenols. Alkylations which involve conjugate additions across a double bond are found in section 74, Alkyls from Olefins.

Examples of name reactions can be found by first considering the nature of the starting material and product. The Wittig reaction, for instance, is in Section 199 on olefins from aldehydes and Section 207 on olefins from ketones.

Examples of the protection of acetylenes, carboxylic acids, alcohols, phenols, aldehydes, amides, amines, esters, ketones, and olefins are also indexed on p. ix.

The pairs of functional groups alcohol, ester; carboxylic acid, ester; amine, amide; carboxylic acid, amide can be interconverted by quite trivial reactions. When a member of these groups is the desired product or starting material, the other member should, of course, also be consulted in the text.

A few reactions already presented in Volumes 1 and 2 are given again in Volume 3 when significant new publications have appeared. In such cases the starting material and product are shown in a contracted form; for example, ROH instead of $PhCH_2CH_2OH$.

The original literature must be used to determine the generality of reactions. A reaction given in this book for a primary aliphatic substrate may also be applicable to tertiary or aromatic compounds.

The references usually yield a further set of references to previous work. Subsequent publications can be found by consulting the Science Citation Index.

Classification and Organization of Reactions forming Difunctional Compounds. This chapter considers all possible difunctional compounds formed from the groups acetylene, carboxylic acid, alcohol, aldehyde, amide, amine, ester, ether, epoxide, halide, ketone, nitrile, and olefin. Reactions that form difunctional compounds are classified into sections on the basis of the two functional groups of the product. The relative positions of the groups do not affect the classification. Thus preparations of 1,2-aminoalcohols, 1,3-aminoalcohols and 1,4-aminoalcohols are included in a single section. It is recommended that the following illustrative examples of the classification of difunctional compounds be scrutinized closely.

Difunctional Product	Section Title
RC≡C-C≡CR	Acetylene—Acetylene
RCH(OH)COOH	Carboxylic Acid—Alcohol
RCH(COOH)CH₂COOMe	Carboxylic Acid—Ester
RCH(OAc)COOH	Carboxylic Acid—Ester
RCH=CHOMe	Ether—Olefin
RCH(OMe)₂	Ether—Ether
RCHF₂	Halide—Halide
RCH(Br)CH₂F	Halide—Halide
RCH(OAc)CH₂OH	Alcohol—Ester
RCH(OH)COOMe	Alcohol—Ester
RCOCOOEt	Ester—Ketone
RCOCH₂OAc	Ester—Ketone
RCH=CHCH₂COOMe	Ester—Olefin
RCH=CHOAc	Ester—Olefin
RCH(Br)COOEt	Ester—Halide
RCH(Br)CH₂OAc	Ester—Halide
RCH=CHCH₂CH=CH₂	Olefin—Olefin

How to Use the Book to Locate Examples of the Preparation of Difunctional Compounds. The difunctional index on p. x gives the section and page corresponding to each difunctional product. Thus Section 327 (Alcohol—Ester) contains examples of the preparation of hydroxyesters; Section 323 (Alcohol—Alcohol) contains examples of the preparation of diols.

Some preparations of olefinic and acetylenic compounds from olefinic and acetylenic starting materials can, in principle, be classified in either the monofunctional or difunctional sections; for example, RCH=CHBr → RCH=CHCOOH, Carboxylic acids from Halides (monofunctional sections) or Carboxylic acid—Olefin (difunctional sections). In such cases both sections should be consulted.

Reactions applicable to both aldehyde and ketone starting materials are in many cases illustrated by an example that uses only one of them.

Many literature preparations of difunctional compounds are extensions of the methods applicable to monofunctional compounds. Thus the reaction RCl → ROH can clearly be extended to the preparation of diols by using the corresponding dichloro compound as a starting material. Such methods are not fully covered in the difunctional sections.

The user should bear in mind that the pairs of functional groups alcohol, ester; carboxylic acid, ester; amine, amide; carboxylic acid, amide can be interconverted by quite trivial reactions. Compounds of the type RCH(OAc)CH$_2$OAc (Ester—Ester) would thus be of interest to anyone preparing the diol RCH(OH)CH$_2$OH (Alcohol—Alcohol).

Chapter 1   PREPARATION

OF

ACETYLENES

Section 1   <u>Acetylenes from Acetylenes</u>

PhC≡CH $\xrightarrow[\begin{array}{c}\text{2) Bu}_3\text{B}\\\text{3) MeSOCl}\end{array}]{\text{1) BuLi}}$ PhC≡CBu   80%

Tetrahedron (1974) <u>30</u> 2159

HC≡CCH$_2$OH $\xrightarrow[\text{2) }\underline{\text{n}}\text{-C}_6\text{H}_{13}\text{Br}]{\text{1) LiNH}_2\text{, NH}_3}$ $\underline{\text{n}}$-C$_6$H$_{13}$-C≡CCH$_2$OH   60%

JOC USSR (1975) <u>11</u> 517

PhC≡CH + PhI $\xrightarrow[\text{Et}_2\text{NH}]{(\text{Ph}_3\text{P})_2 \text{ PdCl}_2/\text{CuI}}$ PhC≡CPh   90%

Tetr Lett (1975) 4467

PhI + $\underline{\text{n}}$-PrC≡CH $\xrightarrow[\text{NaOMe, DMF}]{\text{Pd(PPh}_3)_4}$ PhC≡C-$\underline{\text{n}}$-Pr   97%

J Organometal Chem (1975) <u>93</u> 253, 259

$[Bu_3BC\equiv CPh]Li \xrightarrow{\quad I_2 \quad} BuC\equiv CPh$ 85%

JOC (1974) <u>39</u> 731

$\underline{i}\text{-}Bu_3B \ + \ LiC\equiv CCl \xrightarrow[\text{2) NaOH/H}_2O_2]{\text{1) I}_2, \ -78°} \ \underline{i}\text{-}BuC\equiv C\text{-}Bu\text{-}\underline{i}$ 85%

Tetr Lett (1975) 1961

+

$LiC\equiv CCHC_5H_{11}$
          |
         OR

$\xrightarrow[\text{2) I}_2]{\quad\quad}$

(cyclohexyl)$-C\equiv C\text{-}CH\text{-}C_5H_{11}$
                          |
                         OR   65%

JOC (1976) <u>41</u> 3947

$Sia_2BC\equiv C\text{-}$(cyclohexyl)

+

$LiC\equiv CCMe_3$

$\xrightarrow[\text{2) I}_2]{\text{1) THF}}$

(cyclohexyl)$-C\equiv C\text{-}C\equiv C\text{-}CMe_3$ 70%

JOC (1976) <u>41</u> 1078

$\underline{t}\text{-}BuCl \ + \ (BuC\equiv C)_3Al \xrightarrow[0°]{\text{CH}_2Cl_2} \underline{t}\text{-}BuC\equiv C\text{-}Bu$ 98%

JACS (1975) <u>97</u> 7385

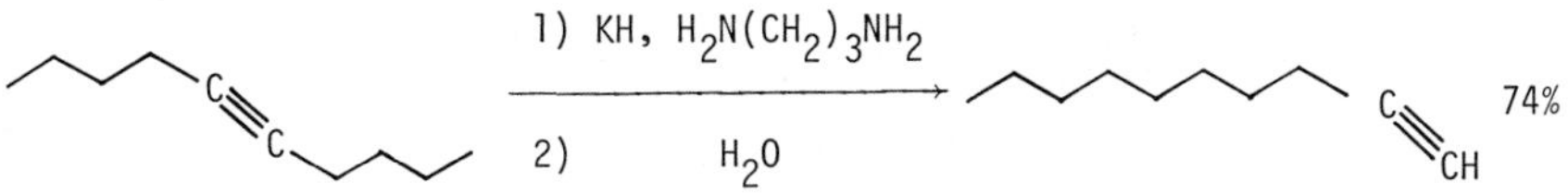

1) KH, $H_2N(CH_2)_3NH_2$

2) H_2O

74%

JACS (1975) 97 891

Section 2   Acetylenes from Carboxylic Acids

No additional examples

Section 3   Acetylenes from Alcohols

No additional examples

Section 4   Acetylenes from Aldehydes

No additional examples

Section 5   Acetylenes from Alkyls, Methylenes and Aryls

No examples

Section 6   Acetylenes from Amides

No additional examples

Section 7   Acetylenes from Amides

No additional examples

Section 8   Acetylenes from Esters

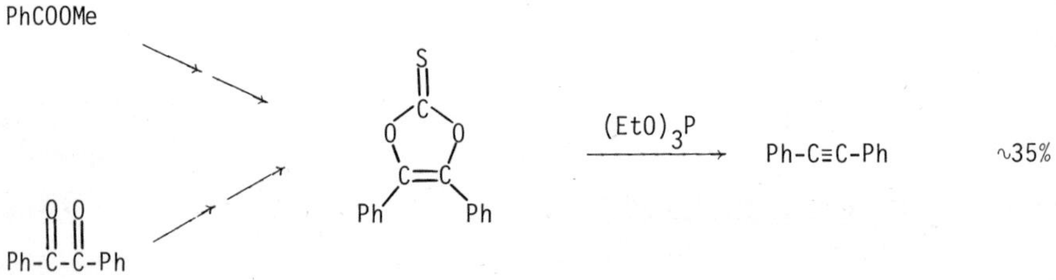

PhCOOMe

$(EtO)_3P$ → Ph-C≡C-Ph   ∿35%

O O
‖ ‖
Ph-C-C-Ph

JOC (1976) 41 2640

Section 9   Acetylenes from Ethers

No examples

Section 10   Acetylenes from Halides

$$HCBr=CBrH \xrightarrow[\text{rfx}]{\text{2 eq. } Ph_2PLi, \text{ THF}} HC\equiv CH \qquad\qquad 80\%$$

Tetr Lett (1975) 4709

$$\text{cyclohexyl}-CH=CCl_2 \xrightarrow[\substack{Et_2NLi \\ Et_2O, \text{ THF}}]{} \text{cyclohexyl}-C\equiv CCl \qquad\qquad 80\%$$

Synthesis (1975) 458

$$R\text{-phenyl}-MgX$$
$$+$$
$$F_2C=CCl_2$$

$$\longrightarrow R\text{-phenyl}-C(F)=C(Cl)Cl \xrightarrow{\text{2 BuLi}} R\text{-phenyl}-C\equiv CLi$$

JOC (1976) 41 1487   ~60-80%

$$\underset{CH_2Cl}{EtOCH(CH_2)_{11}Me} \xrightarrow[Et_2O/\text{hexane}]{BuLi} HC\equiv C(CH_2)_{11}CH_3 \qquad\qquad 81\%$$

Synth Comm (1975) 5 331

Section 11   Acetylenes from Hydrides

No examples

For examples of the reaction $RC{\equiv}CH \rightarrow RC{\equiv}C\text{-}C{\equiv}CR'$ see section 300
(Acetylene - Acetylene)

Section 12   Acetylenes from Ketones

$$\underset{\underset{NNH_2}{\|}}{\overset{\overset{NNH_2}{\|}}{Ph\text{-}C\text{-}C\text{-}Ph}} \xrightarrow[\text{Cu}_2\text{Cl}_2]{\text{O}_2,\ \text{pyr}} \quad Ph\text{-}C{\equiv}C\text{-}Ph \qquad\qquad\qquad 90\%$$

Chem Lett (1976) 147

Section 13   Acetylenes from Nitriles

No examples

Section 14   Acetylenes from Olefins

No additional examples

Section 15   Acetylenes from Miscellaneous Compounds

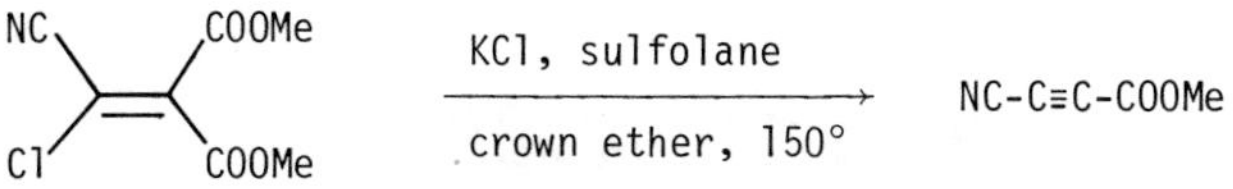

NC   COOMe   KCl, sulfolane
  \   /   ─────────────────────→   NC-C≡C-COOMe   16%
   C＝C   crown ether, 150°
  /   \
Cl   COOMe

Tetr Lett (1975) 2429

Section 15A   Protection of Acetylenes

No additional examples

Chapter 2   PREPARATION

OF

CARBOXYLIC ACIDS

ACID HALIDES

AND ANHYDRIDES

Section 16   <u>Carboxylic Acids from Acetylenes</u>

H-C≡C-OEt

$+$ $\xrightarrow[\text{ZnCl}_2]{\text{Hg(OAc)}_2}$ $H_3C-\underset{\underset{\text{OPh}}{|}}{\overset{\overset{\text{OEt}}{|}}{C}}-OPh$ 75%

PhOH

Rec Trav Chim (1975) <u>94</u> 209

Also via: Esters - Section 106, Amides - Section 76. Also see any relevant Difunctional Compounds.

Section 17   <u>Carboxylic Acids, Acid Halides and Anhydrides</u>
                          <u>from Carboxylic Acids</u>

$CH_2=\overset{\overset{\text{CH}_3}{|}}{C}-CO_2Li$

$+$ $\xrightarrow[\text{2. H}_2\text{O}]{}$ $CH_2(\underset{\underset{\text{CH}_3}{|}}{C}HCO_2Li)_2$ 66%

$CH_3CH(Li)CO_2Li$

Gazz Chim Ital (1976) <u>106</u> 201

8

$$\text{(3-methylbut-2-enoic acid)} \xrightarrow[\text{2) MeI}]{\text{1) 2 LDA, THF}} \text{(2-methyl-3-butenoic acid derivative)} \quad 99\%$$

Gazz Chim Ital (1974) <u>104</u> 625

$$Me_3SiCH_2COOH \xrightarrow[\text{2) } \underline{n}\text{-BuI}]{\text{1) 2 LDA}} Me_3SiCHCOOH \quad 87\%$$
$$\qquad\qquad\qquad\qquad\qquad\qquad | $$
$$\qquad\qquad\qquad\qquad\qquad\quad Bu$$

JCS Chem Comm (1975) 537

$$PhSCH_2COOH \xrightarrow[\text{2) } \underline{i}\text{-PrI}]{\text{1) 2 LDA}} PhSCH-COOH \quad 99\%$$
$$\qquad\qquad\qquad\qquad\qquad\qquad | $$
$$\qquad\qquad\qquad\qquad\qquad \underline{i}\text{-Pr}$$

JCS Chem Comm (1975) 714

$$MeSCH_2COOH \xrightarrow[\text{2) } \underline{i}\text{-BuBr}]{\text{1) LDA}} MeSCHCOOH \quad 80\%$$
$$\qquad\qquad\qquad\qquad\qquad\qquad | $$
$$\qquad\qquad\qquad\qquad\qquad \underline{i}\text{-Bu}$$

Tetr Lett (1975) 3797

$$PhCH=CH-\text{(oxazoline)} \xrightarrow[\substack{\text{2) MeOH} \\ \text{3) } H_3O^+}]{\text{1) EtLi, THF, } -78°} Ph-\overset{*}{C}HCH_2COOH \quad 66\%$$
$$\qquad\qquad\qquad\qquad\qquad\qquad\qquad\qquad\qquad | $$
$$\qquad\qquad\qquad\qquad\qquad\qquad\qquad\qquad Et$$

(high optical yield)

JACS (1975) <u>97</u> 6266

CH_3CH_2 — (oxazoline with Ph, MeO)

(-)

1) LDA
2) BuI
   -78°
3) H_3O^+

→

$Bu \overset{*}{\underset{H_3C}{C}} \begin{smallmatrix} COOH \\ H \end{smallmatrix}$

70% yield
40-80% optical yields

JACS (1974) 96 6508
JOC (1974) 39 2778
Tetr Lett (1974) 3495

$MeCH_2COOH$ - - - - - → $MeCH_2$ — (oxazoline with Ph, MeO)

1) LDA
2) BuI
3) H^+

→

$\overset{Bu}{\underset{Me}{\overset{H}{\diagdown}}} C - COOH$

42% yield
75% optical purity

JACS (1974) 96 268, 6508
JOC (1974) 39 618, 2778

1) PhLi
2) H_3O^+

→

(aromatic ring with COOH, Ph, OMe)

70%

JACS (1975) 97 7383

1) BuLi

2) H_3O^+

65%

S 81% ee

Tetr Lett (1976) 1947

1) LDA

2) BuI

3) H_3O^+

92%

66% ee

JACS (1976) 98 567

1)

Li←OMe

2) H_3O^+

86%

34% ee

JACS (1976) 98 2290

Carboxylic Acids may be alkylated and homologated via ketoacid, ketoester
and olefinic acid intermediates. See section 320 (Carboxylic Acid - Ketone),
section 360 (Ester - Ketone) and section 322 (Carboxylic Acid - Olefin).
Conjugate reductions of unsaturated acids are listed in Section 74 (Alkyls
from Olefins).

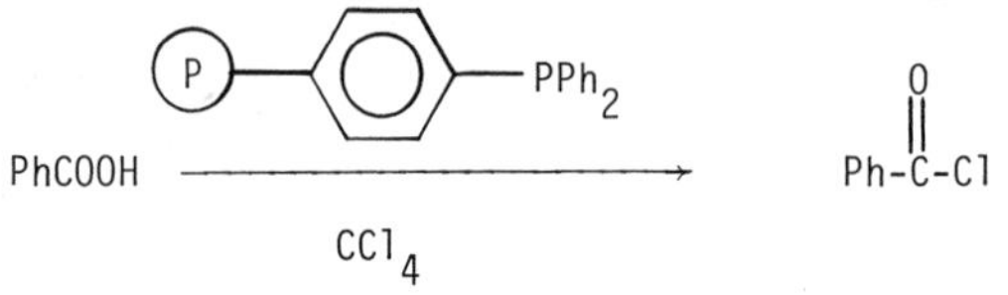

PhCOOH $\xrightarrow[\text{CCl}_4]{\text{(P)}\!-\!\text{C}_6\text{H}_4\!-\!\text{PPh}_2}$ Ph-C-Cl   90%

JCS Chem Comm (1975) 622

RCOOH $\xrightarrow{\text{Polymer-PCl}_2}$ RCOCl

JACS (1974) **96** 6469

$\text{CH}_3\text{-C-C-OH} \xrightarrow{\text{Cl}_2\text{CHOCH}_3} \text{CH}_3\text{-C-C-Cl}$ 54%

Synthesis (1975) 163

RCOOH or
                + SeF$_4$ $\longrightarrow$ RC-F   80-95%
(RCO)$_2$O

JACS (1974) **96** 925

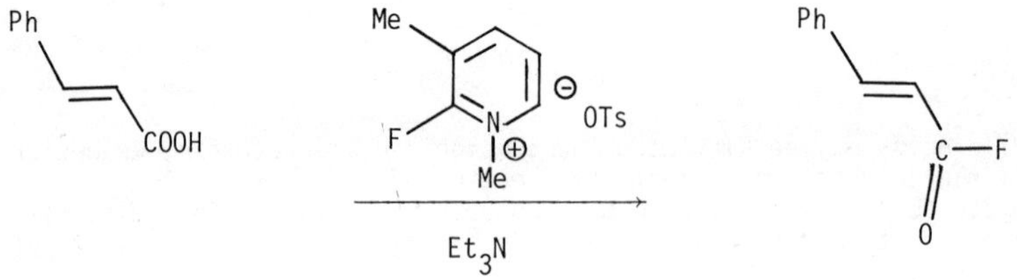

Chem Lett (1976) 303

Ph-C-Cl $\xrightarrow{\text{[morpholine N-SF}_3\text{]}}$ Ph-C-F   70%

Synthesis (1975) 801

Section 18   Carboxylic Acids from Alcohols

PhH + aq. KMnO$_4$ + Bu$_4$NBr $\longrightarrow$ "Purple Benzene" $\xrightarrow{\text{PhCH}_2\text{OH}}$ PhCOOH

Tetr Lett (1974) 1511

Section 19   Carboxylic Acids and Anhydrides from Aldehydes

[cyclohexenyl]–CH=NOH $\xrightarrow{\text{Na}_2\text{O}_2}$ [cyclohexenyl]–COO$^{\ominus}$ Na$^{\oplus}$ 74%

Synthesis (1976) 807

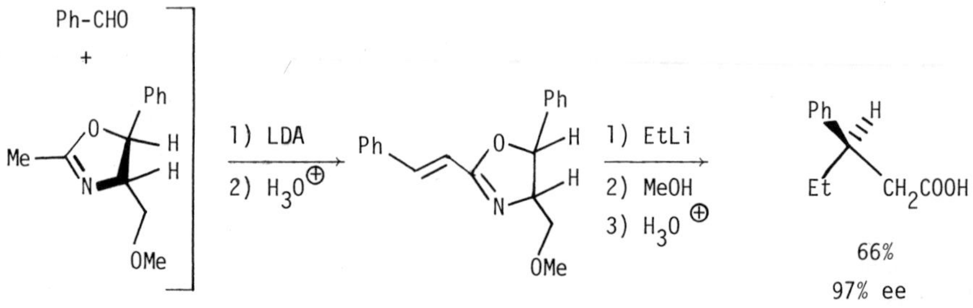

JACS (1975) 97 6266

Related methods: Carboxylic Acids from Ketones (Section 27). Also via: Esters - Section 105, Amides - Section 79. Also see any relevant Difunctional Compounds.

Section 20   Carboxylic Acids from Alkyls

No additional examples

Section 21   Carboxylic Acids from Amides

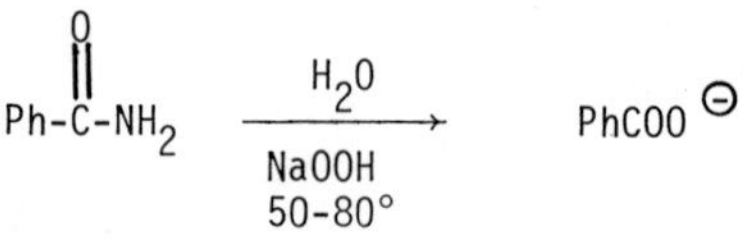

JOC (1975) 40 1187

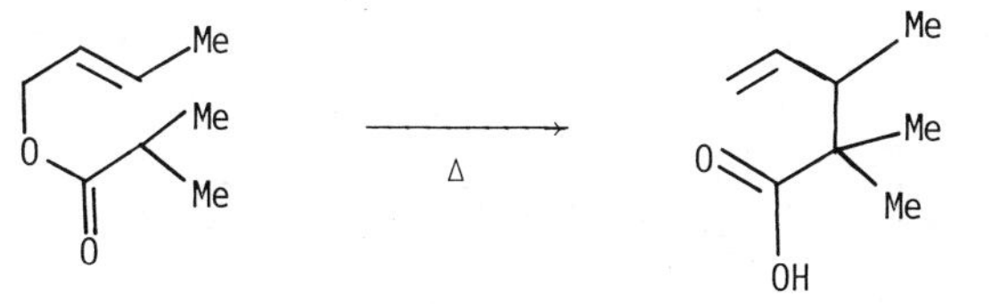

JACS (1976) <u>98</u> 2868

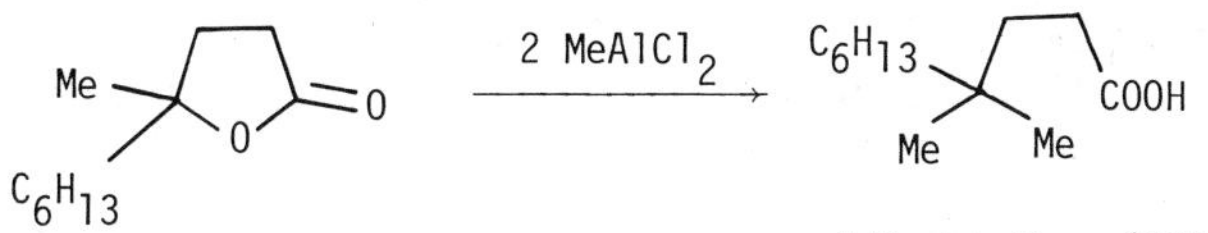

80%

2 MeAlCl$_2$

J Prakt Chem (1974) <u>316</u> 215

91%

Section 24   Carboxylic Acids from Ethers

No additional examples

Section 25   Carboxylic Acids from Halides

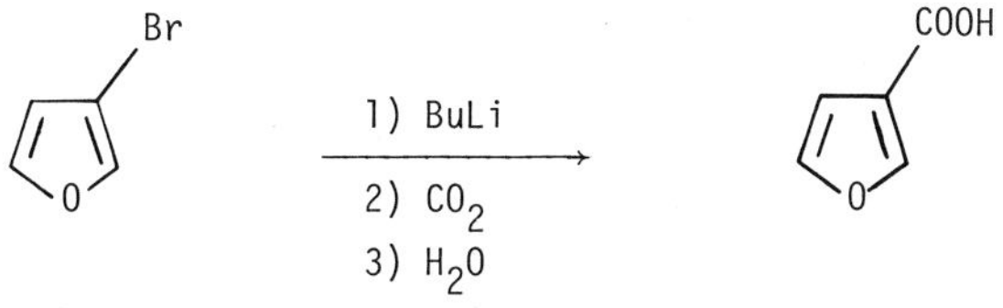

1) BuLi

2) CO$_2$

3) H$_2$O

97%

Synthesis (1974) 443

$$PhCuMgX_2 \quad \xrightarrow[\begin{array}{l} 2)\ CO_2 \\ 3)\ H^{\oplus} \end{array}]{1)\ (EtO)_3P,\ Et_2O/HMPA} \quad PhCOOH \qquad 95\%$$

J Organometal Chem (1975) <u>94</u> 463

$$RX \ + \ CO \ + \ 2\ NaOH \quad \xrightarrow[NR_4^{\oplus}\ X^{\ominus}]{Pd(PPh_3)_4} \quad RCOONa$$

R = Bz, Ar, vinyl, heterocyclic

J Organometal Chem (1976) <u>121</u> C55

$$\xrightarrow[2)\ CrO_3/H_2SO_4]{1)\ B_2H_6} \quad Ph-\overset{\overset{\textstyle Me}{|}}{C}H-COOH \qquad 45\%$$

Synth Comm (1976) <u>6</u> 349

$$\xrightarrow[\begin{array}{l} 2)\ ^{\ominus}OH,\ H_2O \\ 3)\ H_3O^{\oplus} \end{array}]{1)\ CO_2} \qquad \overset{\textstyle R}{\underset{\textstyle R'}{\diagdown}}{-}COOH \qquad 80\text{-}90\%$$

Tetr Lett (1974) 1275

Also via: Esters - Section 115, Amides - Section 85. Also see any relevant Difunctional Compounds.

Section 26   Carboxylic Acids from Hydrides

Ph-OCH$_3$ $\xrightarrow[\text{PdCl}_2, \text{ AcOH}]{\text{CH}_2(\text{COO}^\ominus)_2\text{Na}_2}$

55%

JOC (1976) 41 2049

(CH$_2$)$_3$CHCOOH
| |
NHR   NHR

1) Cl-COOEt
2) CH$_2$N$_2$
3) AgOCOPh
4) $^\ominus$OH

$\longrightarrow$

(CH$_2$)$_3$CHCH$_2$COOH
| |
NHR   NHR

$$R = PhCH_2O\overset{O}{\overset{\|}{C}}-$$

or t-BuOC-
        ‖
        O

Bull Chem Soc Japan (1975) 48 2401

Also via: Esters (Section 116)

Section 27   Carboxylic Acids from Ketones

$\xrightarrow{^\ominus OH}$

97%

JCS Perkin I (1974) 927

Also via: Esters - Section 117. See also relevant Difunctional Compounds.

Section 28   Carboxylic Acids from Nitriles

No additional examples

Section 29   Carboxylic Acids from Olefins

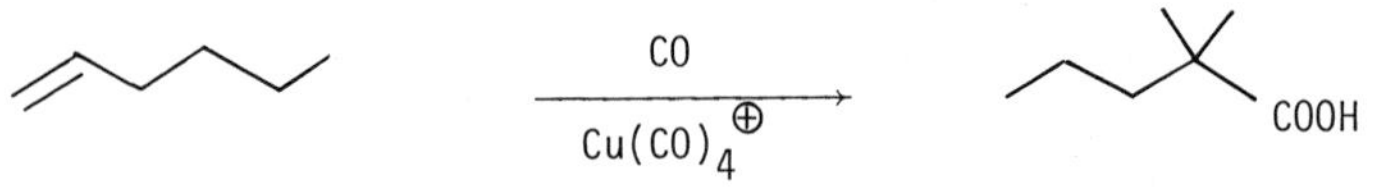

$$\xrightarrow[\text{Cu(CO)}_4^{\oplus}]{\text{CO}}$$

60%

Bull Chem Soc Japan (1976) 49 3296

Section 30   Carboxylic Acids from Miscellaneous Compounds

Review: "Syntheses of Tetracarboxylic Acids"

Russ Chem Rev (1973) 42 939

Section 30A   <u>Protection of Carboxylic Acids</u>

$$R\text{-}COOH \xrightarrow[\text{I}_2,\ \text{AcOH}]{\text{Ph}_2\text{C=NNH}_2} R\text{-}\overset{\overset{\textstyle O}{\|}}{C}\text{-OCHPh}_2 \qquad >90\%$$

Useful for protecting amino acids.

JCS Perkin I (1975) 2019

1) CuSO$_4$
2) EDTA

$\sim$74%

JOC (1975) <u>40</u> 3287

$$R\text{-}\overset{\overset{\textstyle O}{\|}}{C}\text{-ONa}$$

$$R\text{-}COOH$$

PhCH$_2$OCH$_2$Cl

HMPA

HCl, THF

<u>or</u> H$_2$, Pd/C

$$R\text{-}\overset{\overset{\textstyle O}{\|}}{C}\text{-OCH}_2\text{OCH}_2\text{Ph}$$

JOC (1975) <u>40</u> 2962

1) Cu/Quinoline, 200°
2) CH$_3$OH

70%

Synthesis (1976) 40

$$Ph\text{-}\overset{O}{\overset{\|}{C}}NHCH_2\overset{O}{\overset{\|}{C}}\text{-}OCH_2CCl_3 \xrightarrow[\text{pH } 5.5]{\text{Zn, THF/H}_2O} Ph\text{-}\overset{O}{\overset{\|}{C}}NHCH_2COOH \qquad 83\%$$

Synthesis (1976) 457

$$Ph\text{-}\overset{O}{\overset{\|}{C}}\text{-}OCH_2CH_2Cl \xrightarrow{\text{Co(I)phthalocyanine}} PhCOOH \qquad 66\%$$

Angew Int Ed (1976) 15 681

Use of the trimethylsilyl group to protect the carboxyl function of peni-
cillin sulfoxides during their conversion to deacetoxycephalosporins.

JOC (1975) 40 1346

(stable to carbanions)

JOC (1974) 39 2787

JACS (1974) <u>96</u> 590

Use of TlOR to cleave protected peptides from Merrifield resin.

Can J Chem (1974) <u>52</u> 2832

Other reactions useful for the protection of carboxylic acids are included
in Section 107 (Esters from Carboxylic Acids and Acid Halides) and Section 23
(Carboxylic Acids from Esters).

Chapter 3   PREPARATION

OF

ALCOHOLS

AND

PHENOLS

Section 31   <u>Alcohols from Acetylenes</u>

Bu_3B $\xrightarrow{\begin{array}{l}\text{1) } LiC\equiv CH \\ \text{2) aq. HCl, } \Delta \\ \text{3) NaOH, } H_2O_2\end{array}}$ $CH_3C(Bu)_2OH$ 82%

JOC (1975) <u>40</u> 2845

$MeC\equiv CEt$ $\xrightarrow{\begin{array}{l}\text{1) } Bu_2BH \\ \text{2) MeLi} \\ \text{3) HCl} \\ \text{4) NaOH/}H_2O_2\end{array}}$ $MeCH_2\overset{\overset{\displaystyle Bu}{|}}{\underset{\underset{\displaystyle OH}{|}}{C}}\text{-Et}$ 70%

Synthesis (1974) 339

Section 32   <u>Alcohols from Carboxylic Acids</u>

$$\text{PhCOCl} \xrightarrow{\overset{\oplus\ominus}{\text{Bu}_4\text{N BH}_4}} \text{PhCH}_2\text{OH} \qquad\qquad 100\%$$

JOC (1976) <u>41</u> 690

$$\text{CH}_2\text{CHCOOH} \xrightarrow[\text{THF}]{\text{B}_2\text{H}_6} \text{Ar-CH}_2\text{CHCH}_2\text{OH} \qquad 80\%$$

JCS Perkin I (1974) 191

$$\text{PhCOOH} + \text{BH}_3\cdot\text{MeSH} + \text{B(OMe)}_3 \longrightarrow \text{PhCH}_2\text{OH}$$

JOC (1974) <u>39</u> 3052

$$\text{RCOOH} + \text{Ph} \longrightarrow \xrightarrow[\text{2) NaBH}_4]{\text{1) Et}_3\text{N}} \text{RCH}_2\text{OH} \qquad 80\text{-}100\%$$

JOC (1974) <u>39</u> 111

$$CH_2Cl_2 \quad \xrightarrow[\substack{\text{2) PhCOCl, } -78° \\ \text{3) MeOH/HOAc}}]{\text{1) BuLi, } -100°} \quad \underset{\overset{|}{OH}}{PhC-(CHCl_2)_2}$$

Chem Ber (1975) <u>108</u> 328

Also via: Esters (Section 38)

Section 33   <u>Alcohols from Alcohols and Phenols</u>

1) BuLi, $(CH_2NMe_2)_2$

2) H_2O

Org Synth (1976) <u>55</u> 1

Section 34   <u>Alcohols and Phenols from Aldehydes</u>

2 Ph-OMgBr

+

PhCHO

$\xrightarrow{\text{benzene}}$

JCS Chem Comm (1976) 309

$$RCHO \quad + \quad Me_2CuLi \quad \xrightarrow[\text{(MeO)}_3\text{P}]{\text{Et}_2\text{O/HMPA}} \quad RCHOHMe \qquad\qquad >90\%$$

Tetr Lett (1975) 2353

$$CH_3(CH_2)_2CHO \quad \xrightarrow[\text{TiCl}_4, \ \text{CH}_2\text{Cl}_2]{\text{Me}_3\text{SiCH}_2\text{CH=CH}_2}$$

87%

OH

Tetr Lett (1976) 1295

$$PhCH_2SH \quad \xrightarrow[\text{2) PhCHO}]{\text{1) BuLi}} \quad$$

PhCHSH
|
CH(OH)   70%
|
Ph

Angew Int Ed (1974) 13 202

$$Me_2NCH_2CO_2\text{-}\underline{t}\text{-Bu} \quad \xrightarrow[\text{2) CH}_3\text{CHO}]{\text{1) LDA, -78}°} \quad CH_3CH(OH)CHCO_2\text{-}\underline{t}\text{-Bu} \qquad 85\%$$

NMe$_2$

Tetr Lett (1975) 1477

PhCHO + $\quad \xrightarrow[\text{2) MeOH}]{\text{1) TiCl}_4, \ \text{CH}_2\text{Cl}_2} \quad$ PhCH(OH)CH$_2$COCH$_2$ 91%

COOMe

Chem Lett (1975) 161

$$\text{MeCOCCl}_3 \xrightarrow[\text{2) EtCHO}]{\text{1) LDA, THF, } -75°}$$

$$\underset{\underset{\text{CH(OH)Et}}{|}}{\text{MeCOCCl}_2} \qquad 73\%$$

Bull Soc Chim France (1975) 1876

$$\text{MeN(NO)Me} \xrightarrow[\text{2) PhCHO}]{\text{1) LDA, THF, } -80°} \text{MeN(NO)CH}_2\text{CH(OH)Ph} \qquad 85\%$$

Chem Ber (1975) 108 1293

1) LDA
2) PhCHO

81%

Synthesis (1975) 512

$$\text{PhCHO} + \underline{n}\text{-Bu}_3\text{SnCCl}_3 \xrightarrow[\text{2) H}_2\text{O}]{\text{1) 80°}} \text{PhCHOHCCl}_3 \qquad 63\%$$

J Organometal Chem (1975) 102 423

$$\text{Cl}_3\text{CCO}_2\text{-}\underline{i}\text{-Pr} \xrightarrow[\text{2) Me}_2\text{CHCHO}]{\text{1) Et}_2\text{NLi, THF-ØH-HMPA}} \underset{\underset{\text{OH}}{|}}{\text{Me}_2\text{CHCHCCl}_2\text{CO}_2\text{-}\underline{i}\text{-Pr}} \qquad 71\%$$

Synthesis (1975) 524, 533

CH_2Br_2 $\xrightarrow[\text{2) PrCHO}]{\text{1) LDA, THF/Et}_2\text{O, -90°}}$ PhCH-CHBr$_2$ 63%
$\qquad\qquad\qquad\qquad\qquad\qquad\qquad\qquad$ |
$\qquad\qquad\qquad\qquad\qquad\qquad\qquad\qquad$ OH

$\qquad\qquad\qquad\qquad\qquad\qquad$ Bull Soc Chim France (1975) 1797

PhCHO + CHCl$_3$ $\xrightarrow{\text{t-BuOK, NH}_3\text{, -75°}}$ PhCH-CCl$_3$ 84%
$\qquad\qquad\qquad\qquad\qquad\qquad\qquad\qquad\qquad$ |
$\qquad\qquad\qquad\qquad\qquad\qquad\qquad\qquad\qquad$ OH

$\qquad\qquad\qquad\qquad\qquad\qquad$ J Gen Chem USSR (1974) $\underline{44}$ 2590

PhSCH$_2$SePh $\xrightarrow[\text{2) PhCHO}]{\text{1) BuLi}}$ PhSCH$_2$-CHPh   94%
$\qquad\qquad\qquad\qquad\qquad\qquad\qquad\qquad\qquad$ |
$\qquad\qquad\qquad\qquad\qquad\qquad\qquad\qquad\qquad$ OH

$\qquad\qquad\qquad\qquad\qquad\qquad$ Tetr Lett (1975) 1617
$\qquad\qquad\qquad\qquad\qquad\qquad$ Angew Int Ed (1975) $\underline{14}$ 350, 700

$\qquad\quad$ Li
$\qquad\quad$ |
PhSe(O)CHCH$_2$CH$_2$Ph + EtCHO $\xrightarrow{\text{NaI}}$ PhSe(O)CHCH$_2$CH$_2$Ph   87%
$\qquad\qquad\qquad\qquad\qquad\qquad\qquad\qquad\qquad\qquad\qquad$ |
$\qquad\qquad\qquad\qquad\qquad\qquad\qquad\qquad\qquad\qquad\qquad$ HC(OH)Et

$\qquad\qquad\qquad\qquad\qquad\qquad$ JCS Chem Comm (1975) 790

$(\underline{n}\text{-C}_6\text{H}_{13})_2$BCl + BuC=N-$\underline{t}$-Bu $\xrightarrow{\begin{array}{l}\text{1) Et}_2\text{O, -78°}\\\text{2) HSCH}_2\text{COOH}\\\text{3) NaOH, diglyme}\\\text{4) NaOH, H}_2\text{O}_2\end{array}}$ $(\underline{n}\text{-C}_6\text{H}_{13})_2$C-OH   87%
$\qquad\qquad\qquad\qquad$ |
$\qquad\qquad\qquad\qquad$ Li $\qquad\qquad\qquad\qquad\qquad\qquad\qquad\qquad\qquad\qquad\qquad\qquad$ Bu

$\qquad\qquad\qquad\qquad\qquad\qquad$ JOC (1975) $\underline{40}$ 3644
$\qquad\qquad\qquad\qquad\qquad\qquad$ Tetr Lett ($\overline{1975}$) 2689

$$\xrightarrow[\text{Et}_3\text{N, KOH, MeOH}]{\text{Ra-Ni (Cr-promoted)}}$$

88%

Tetr Lett (1976) 4681

$$\xrightarrow[\text{alumina}]{\text{Me}_2\text{CHOH}}$$

82%

Tetr Lett (1975) 3601

$$\xrightarrow[\text{THF}]{\text{FeCl}_2\cdot\text{NaH}}$$

Chem Lett (1976) 581

$$\text{RCHO} + \text{Et}_3\text{SiH} \xrightarrow{\text{H}^{\oplus}} \text{RCH}_2\text{OH}$$

JOC (1974) <u>39</u> 2740

Benzaldehyde is reduced in preference to acetophenone by NaBH(OAc)$_3$.

JCS Chem Comm (1975) 535

$LiAlH(O\text{-}\underline{t}\text{-}Bu)_3$
$NaBH_4$
$LiBH_4$

reduce benzaldehyde in the presence of acetophenone, and butanal in the presence of 2-butanone.

Aust J Chem (1975) 28 1383

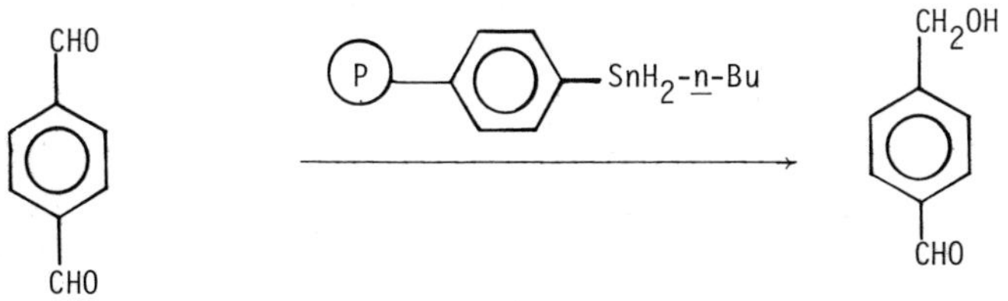

CHO
CHO
P
SnH2-n-Bu
CH2OH
CHO

78%

JOC (1975) 40 1966

CHO
MeO
+
MCPBA
CH2Cl2
OH
MeO

60-90%

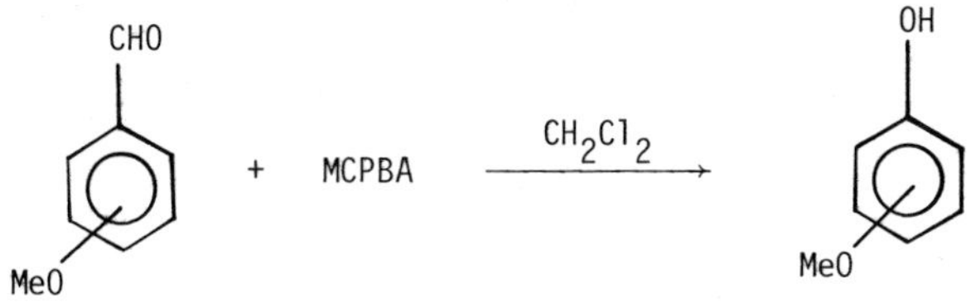

JCS Perkin I (1974) 1353

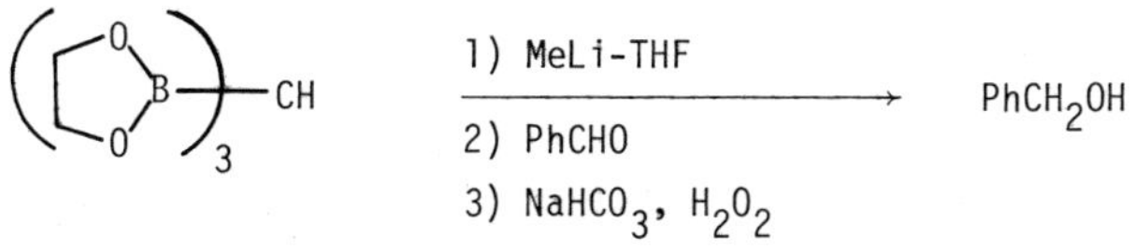

O
B
O
3
CH
1) MeLi-THF
2) PhCHO
3) NaHCO3, H2O2
PhCH2OH

97%

JACS (1975) 97 5608

Related methods: Alcohols from Ketones (Section 42)

Section 35   Alcohols and Phenols from Alkyls, Methylenes and Aryls

No examples of the reaction RR' → ROH (R'=alkyl, aryl, etc.) occur in the literature. For reactions of the type RH → ROH (R=alkyl or aryl) see Section 41 (Alcohols and Phenols from Hydrides).

Section 36   Alcohols and Phenols from Amides

No additional examples

Section 37   Alcohols and Phenols from Amines

No additional examples

Section 38   Alcohols from Esters

$$
\underset{\text{Ph-C-OMe}}{\overset{\overset{\displaystyle O}{\|}}{}} \quad \xrightarrow[\text{THF, rfx 20 hrs.}]{\text{NaBH}_4,\ \text{HSCH}_2\text{CH}_2\text{SH}} \quad \text{PhCH}_2\text{OH} \qquad 100\%
$$

Tetr Lett (1975) 3295

$$PhCH_2CH_2COOEt \xrightarrow[CH_2Cl_2]{Na^{\oplus}[OBH_3C(CH_3)NPh]^{\ominus}} Ph(CH_2)_3OH \qquad 97\%$$

Chem Pharm Bull (1976) <u>24</u> 1059

1) 3 eq R'MgX

2) H_2O

"excellent yield"

Comptes Rendus <u>C</u> (1975) <u>280</u> 123

Related methods: Carboxylic Acids from Esters - Section 23,
                   Protection of Alcohols - Section 45A

Section 39   <u>Alcohols and Phenols from Ethers and Epoxides</u>

HI, NaI

CH_3OCHCl_2

84%

JOC (1976) <u>41</u> 367

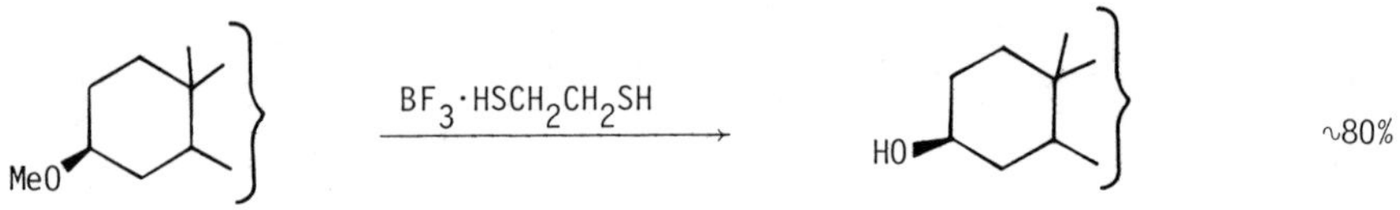

$$BF_3 \cdot HSCH_2CH_2SH$$

~80%

MeO   HO

JCS Perkin I (1976) 2237

Additional examples of ether cleavages may be found in Section 45A
(Protection of Alcohols and Phenols).

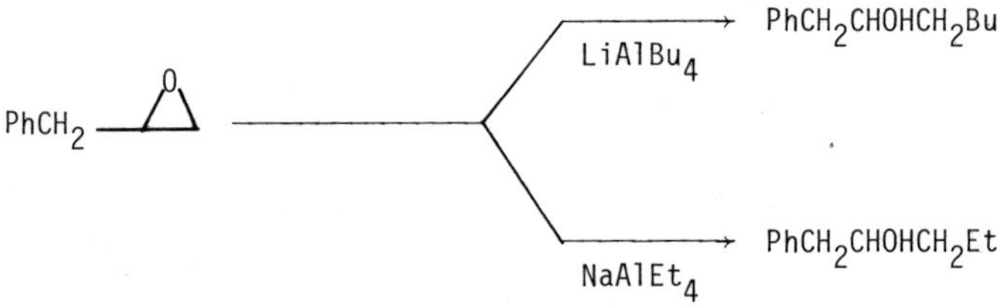

$PhCH_2$ —

$LiAlBu_4$ → $PhCH_2CHOHCH_2Bu$

$NaAlEt_4$ → $PhCH_2CHOHCH_2Et$

Tetr Lett (1975) 2521

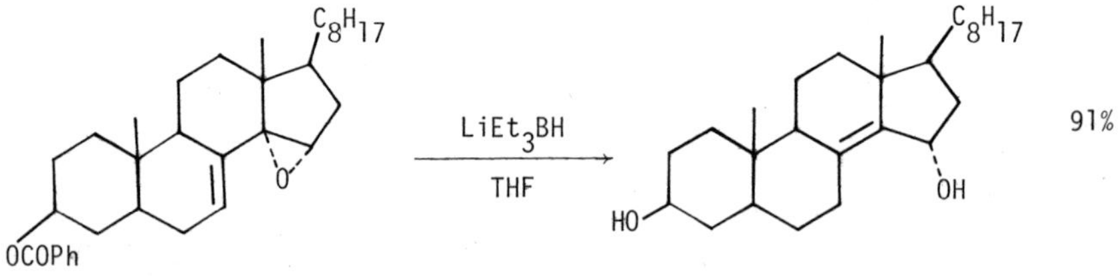

C_8H_{17}

$LiEt_3BH$

THF

91%

OCOPh   HO   OH

Tetr Lett (1976) 3775

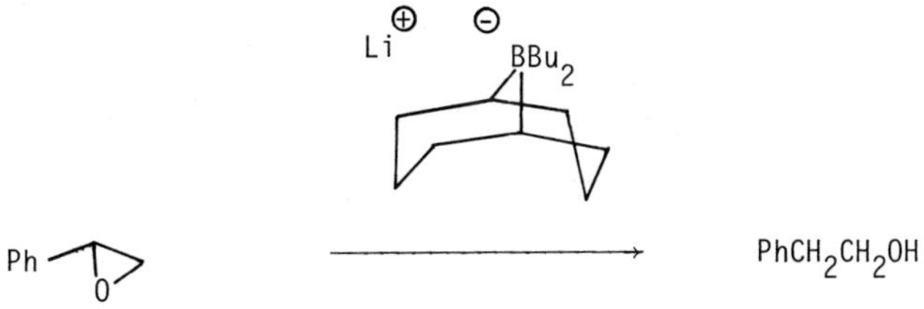

Li
BBu2
Ph
O
PhCH2CH2OH

99%

JCS Chem Comm (1976) 672

S
S
Li
O
O
S
S
O
OH

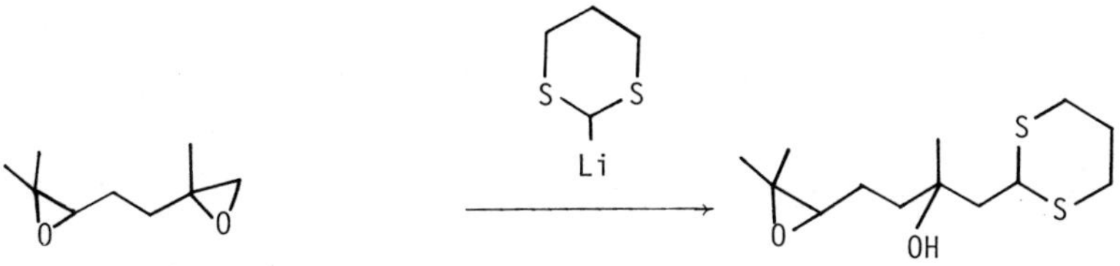

98%

Used in juvenile hormone synthesis.

JOC (1974) 39 3645

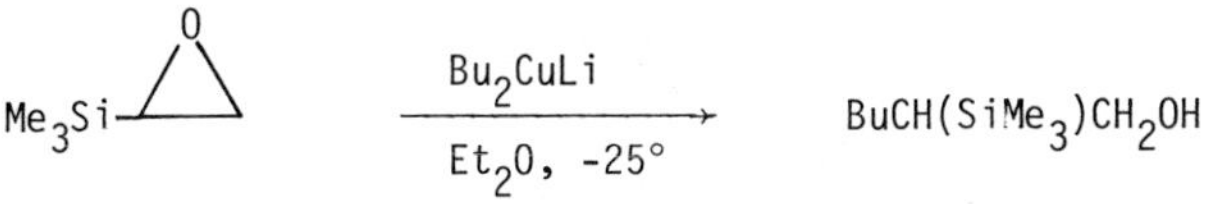

O
Me3Si
Bu2CuLi
Et2O, -25°
BuCH(SiMe3)CH2OH

88%

JOC (1975) 40 2263

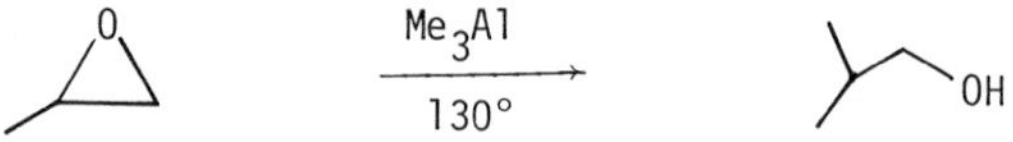

$$\xrightarrow[130°]{Me_3Al}$$

90%

J Organometal Chem (1974) 73 187

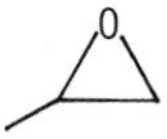

+ $(CH_3)_2CuLi$ $\longrightarrow$ $CH_3CHOHCH_2CH_3$ 89%

Org React (1975) 22 253

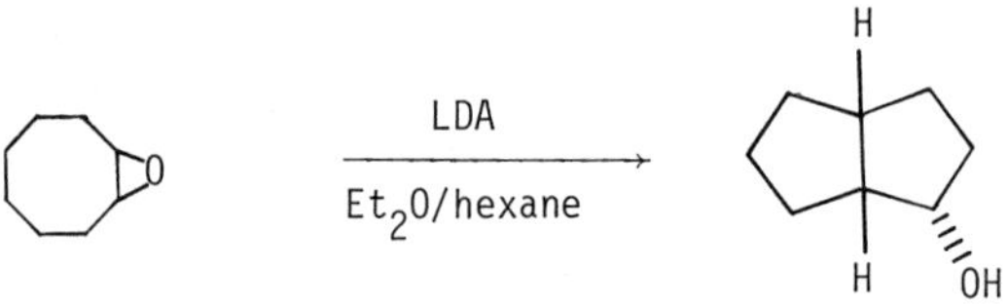

$$\xrightarrow[Et_2O/hexane]{LDA}$$

80%

Synthesis (1975) 602

1) Et_2NH, Li,
   HMPA/benzene

2)

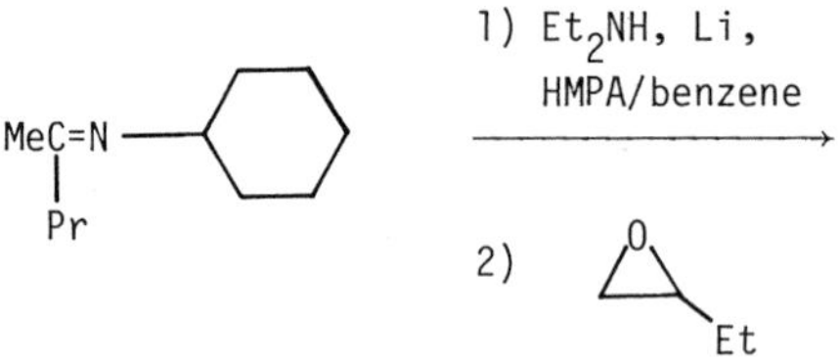

$\longrightarrow$ $\underline{n}$-PrCOCH$_2$CH$_2$CH(OH)Et   95%

Synthesis (1975) 256

Section 40   Alcohols and Phenols from Halides and Sulfonates

$$\underset{CH_3SO_2O \quad \overset{|}{H}}{R \diagdown \diagup \diagdown C_5H_{11}} \quad \xrightarrow[\text{DMSO/DMF/DME}]{KO_2, \text{ crown ether}} \quad \underset{H \quad \overset{|}{OH}}{R \diagdown \diagup \diagdown C_5H_{11}} \qquad 75\%$$

Tetr Lett (1975) 3183

$$\underset{\overset{|}{OTs}}{\overset{*}{C_6H_{13}-CHCH_3}} \quad \xrightarrow[\text{2) } H_2O]{\text{1) } KO_2, \text{ DMSO, crown ether}} \quad \underset{\overset{|}{OH}}{\overset{*}{C_6H_{13}-CHCH_3}} \qquad 75\%$$

(S)   (R)

JOC (1975) 40 1678

$$RBr \ + \ HgClO_4 \ + \ H_2O \ \longrightarrow \ ROH \qquad 64\text{-}98\%$$

21 cases

Tetrahedron (1974) 30 2467

$$\underline{n}\text{-BuSnCH}_2OH \quad \xrightarrow[\text{2) } \underline{n}\text{-Oct Cl}]{\text{1) 2 } \underline{n}\text{-BuLi}} \quad \underline{n}\text{-Oct-CH}_2OH$$

Angew Int Ed (1976) 15 438

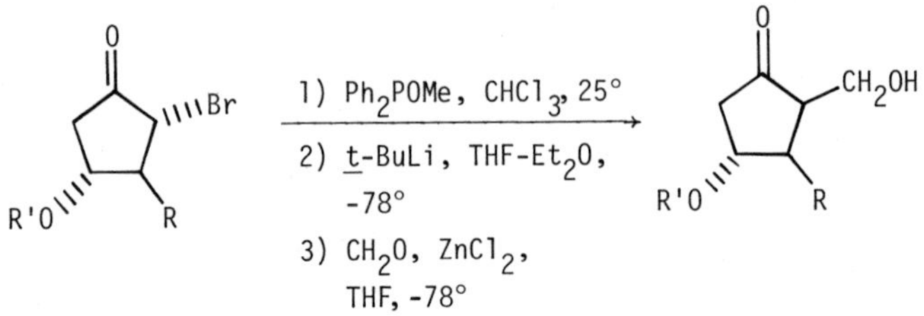

O
Br
R'O
R
1) Ph2POMe, CHCl3, 25°
2) t-BuLi, THF-Et2O, -78°
3) CH2O, ZnCl2, THF, -78°
O
CH2OH
R'O
R

80-90%

JACS (1975) 97 4745, 6260

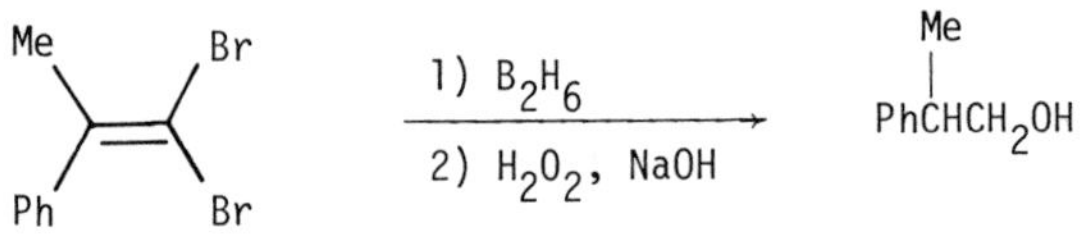

Me
Br
Ph
Br
1) B2H6
2) H2O2, NaOH
Me
PhCHCH2OH

97%

Synth Comm (1976) 6 349

Section 41 Alcohols and Phenols from Hydrides

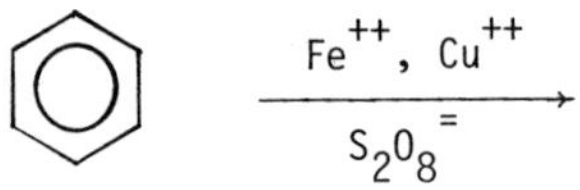

Fe++, Cu++
S2O8=

Ph-OH 64%

JACS (1975) 97 1603

$PhH + Tl(OCOCF_3)_3 \longrightarrow PhTl(OCOCF_3)_2 \xrightarrow[2)\ H_2O]{1)\ B_2H_6} PhOH$ 58%

JCS Chem Comm (1975) 36

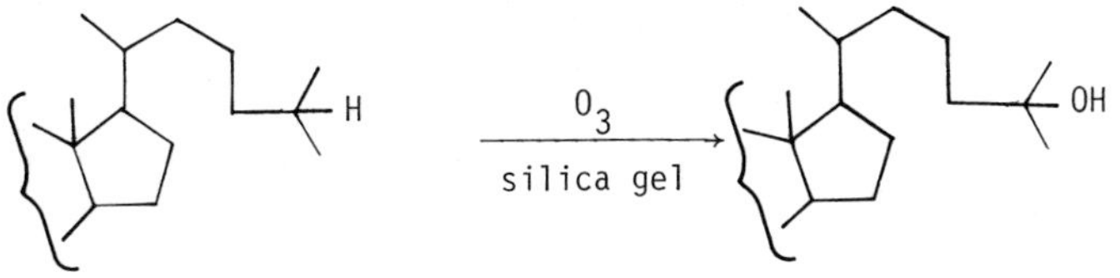

51% yield
11% conversion

JOC (1976) 41 2651

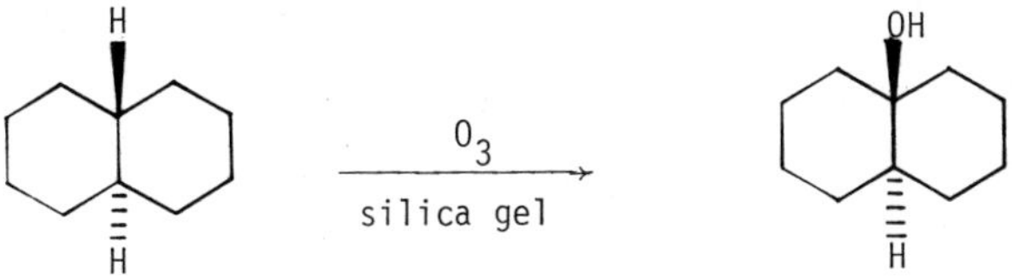

72%

JOC (1975) 40 2141

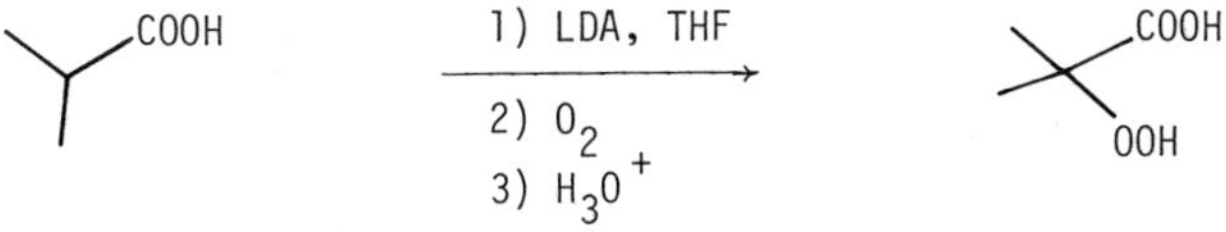

1) LDA, THF

2) O$_2$

3) H$_3$O$^+$

~80%

JOC (1976) 41 370

hν

H$_2$O, O$_2$

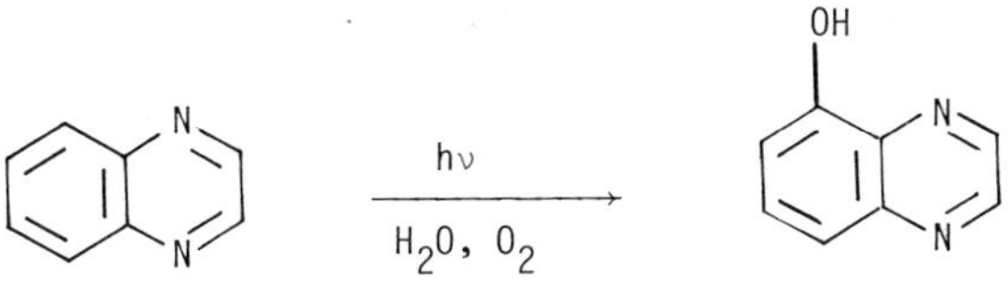

~80%

Rec Trav Chim (1976) 95 285

Review: Photochemical Hydroxylation of Aromatic Compounds

Synthesis (1974) 173

Section 42   Alcohols from Ketones

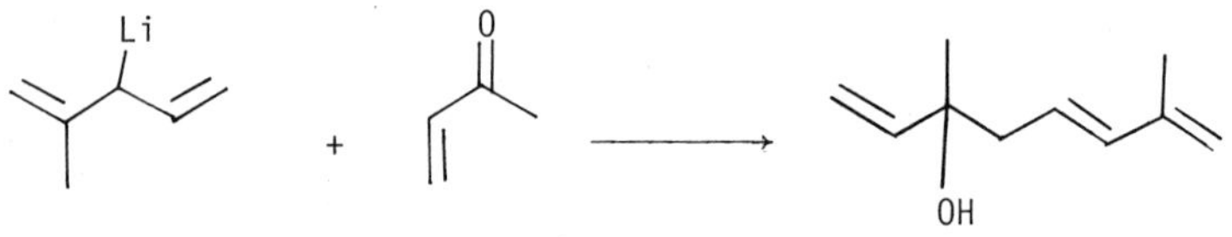

JOC (1976) 41 3209

Me_2CO + $BrCH_2CMe=CHCOOEt$ $\xrightarrow[\substack{2)~\Delta \\ 3)~H^{\oplus}}]{1)~Zn,~HgI_2}$ $Me_2C(OH)CHCMe=CH_2$ 70%

$\overset{\underset{|}{COOEt}}{}$

J Organometal Chem (1975) 96 149

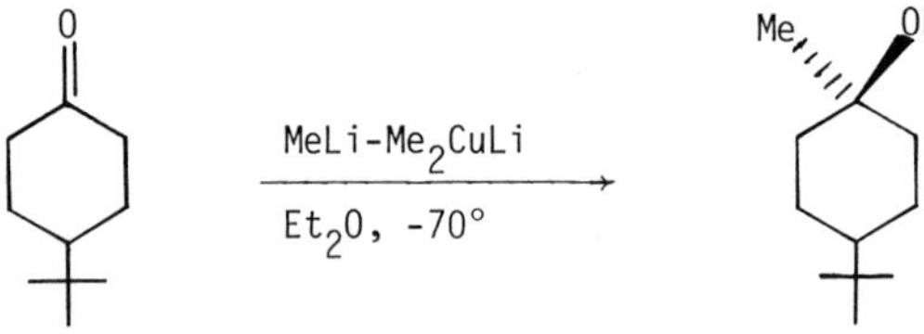

(high yield axial OH)   91%

JACS (1975) 97 5280

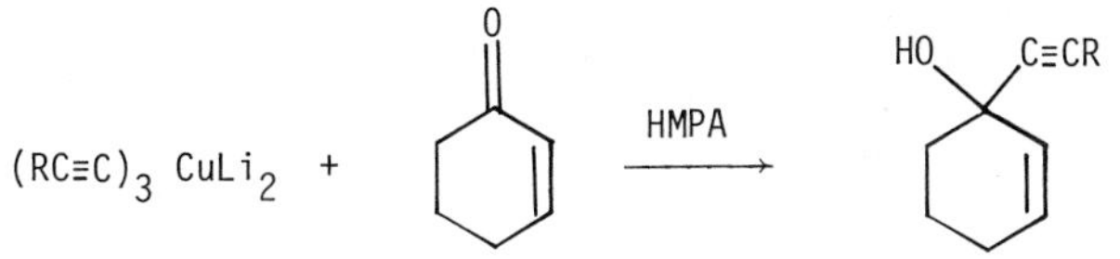

$(RC \equiv C)_3$ CuLi$_2$ + $\xrightarrow{\text{HMPA}}$ ~85%

JCS Chem Comm (1975) 892

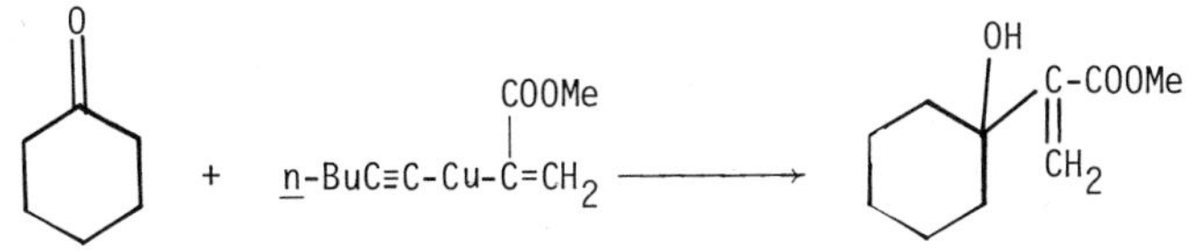

+ $\underline{n}$-BuC≡C-Cu-C=CH$_2$ with COOMe ⟶   100%

Tetr Lett (1975) 3897

Review: Stereochemistry of Organometallic Compound Addition to Ketones

Chem Rev (1975) <u>75</u> 521

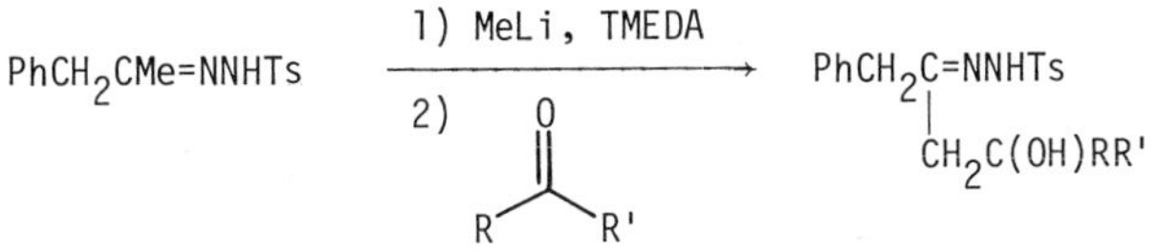

PhCH$_2$CMe=NNHTs $\xrightarrow[\text{2)}]{\text{1) MeLi, TMEDA}}$ PhCH$_2$C=NNHTs

$\underset{R \quad R'}{\overset{O}{\|}}$

CH$_2$C(OH)RR'

Tetr Lett (1975) 1811

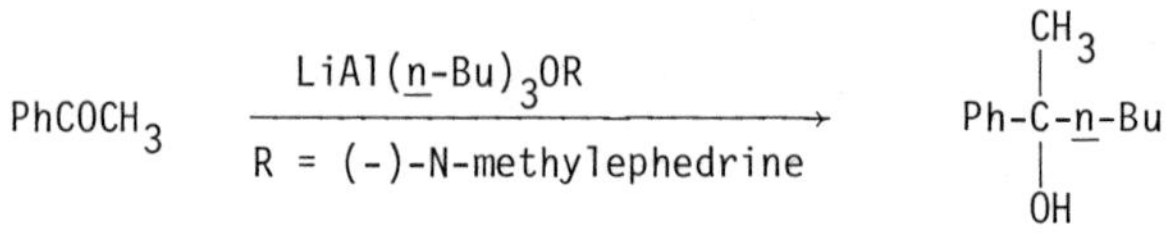

PhCOCH$_3$ $\xrightarrow[\text{R = (-)-N-methylephedrine}]{\text{LiAl(}\underline{\text{n}}\text{-Bu)}_3\text{OR}}$

$$\underset{\text{OH}}{\overset{\text{CH}_3}{\text{Ph-C-}\underline{\text{n}}\text{-Bu}}}$$

56%

31% ee

Tetr Lett (1976) 4781

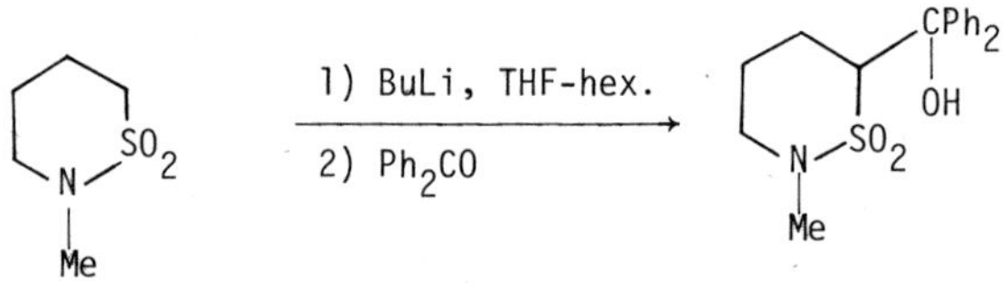

1) BuLi, THF-hex.

2) Ph$_2$CO

91%

JOC (1975) 40 1342
Synthesis (1975) 333

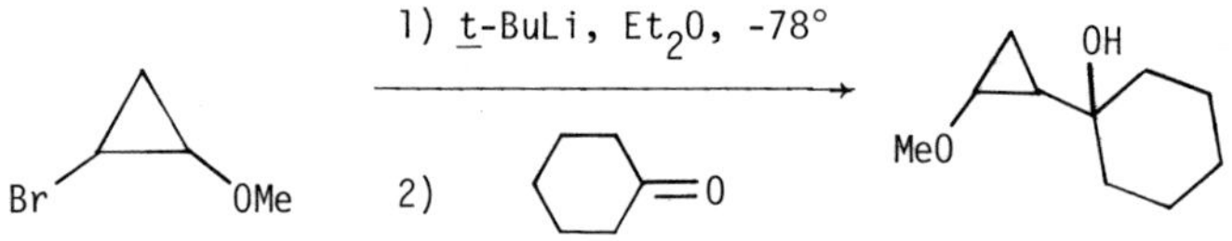

1) $\underline{\text{t}}$-BuLi, Et$_2$O, -78°

2) $\quad$ cyclohexanone =O

92%

Tetr Lett (1975) 3685

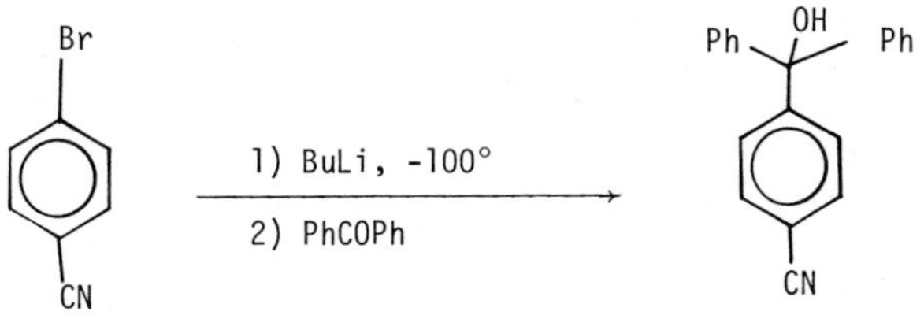

Br
1) BuLi, -100°
2) PhCOPh
CN
OH
Ph
Ph
CN

83%

JOC (1976) 41 1187

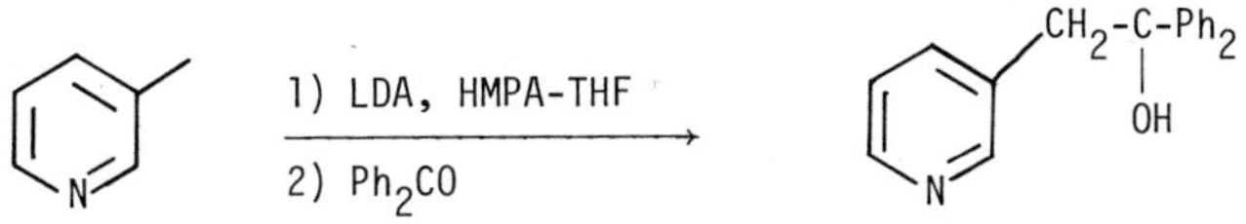

1) LDA, HMPA-THF
2) Ph2CO
N
CH2-C-Ph2
OH
N

79%

Synthesis (1975) 705

MeCOCH=CH2
NaC≡CH, MgSO4
liq NH3/EtO
OH
Me-C-C≡CH
CH=CH2

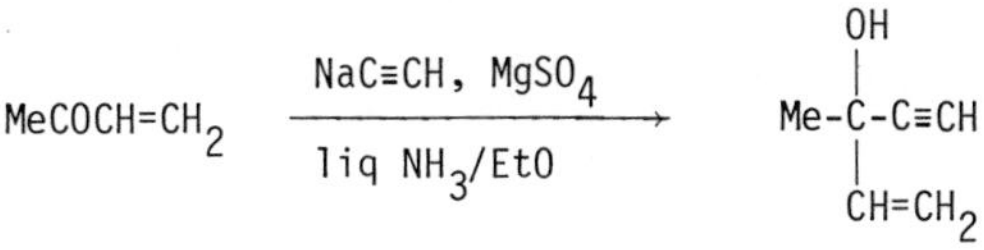

70%

Can J Chem (1975) 53 2157

CH_3COS-t-Bu —(1) LDA, -78°; 2) Ph_2CO)→ $Ph_2C(OH)CH_2COS$-t-Bu 80%

Tetr Lett (1975) 3255

$$\underset{\substack{| \\ Cl}}{Et-C(Me)CO_2\text{-}\underline{i}\text{-}Pr} \quad \xrightarrow[\text{2) } (CH_3)_2CO]{\text{1) Li, THF}} \quad \underset{\substack{| \\ Me_2COH}}{\overset{\substack{Me \\ |}}{Et\overset{|}{C}\text{-}CO_2\text{-}\underline{i}\text{-}Pr}} \qquad 71\%$$

J Organometal Chem (1975) <u>102</u> 129

$$NCCH_2CH_2COOH \quad \xrightarrow[\text{2) } Ph_2CO]{\text{1) LiNH}_2, \text{ liq NH}_3} \quad \underset{\substack{| \ | \\ OH \ CN}}{Ph_2C\text{—}CHCH_2COOH} \qquad 61\%$$

J Organometal Chem (1975) <u>92</u> 125

$$\underset{PhCCH_3}{\overset{O}{\|}} \quad \xrightarrow[\text{2) } NH_4Cl]{\text{1) NaAlEt}_4, \text{ NiCl}_2} \quad \underset{\substack{| \\ OH}}{\overset{\substack{Et \\ |}}{Ph\text{-}\overset{|}{C}\text{-}CH_3}} \qquad 84\%$$

Tetr Lett (1976) 993

$$\xrightarrow[\substack{P(\text{-O-Ph-OMe})_3 \\ AlEt_3}]{Ni(acac)_2} \qquad 62\%$$

Bull Chem Soc Japan (1976) <u>49</u> 3646

$$Ph_2CO \quad \xrightarrow[Ni(acac)_2]{Me_3Al} \quad \underset{\substack{| \\ Me}}{Ph_2COH} \qquad 58\%$$

Aust J Chem (1974) <u>27</u> 2569

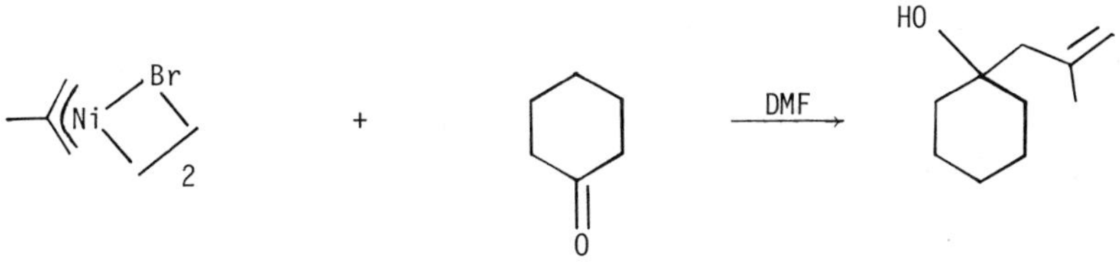

JOC (1975) <u>40</u> 593

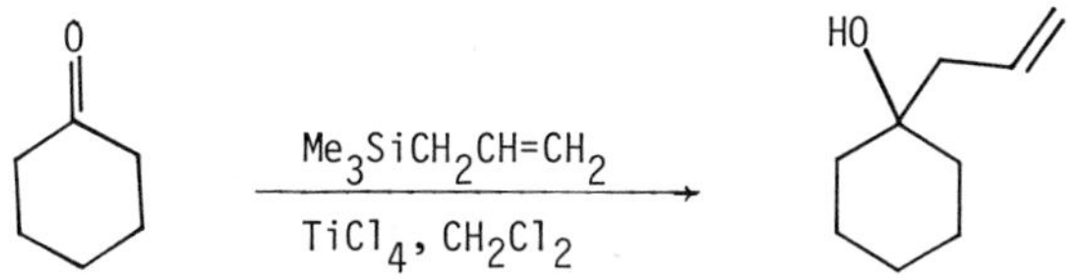

70%

Tetr Lett (1976) 1295

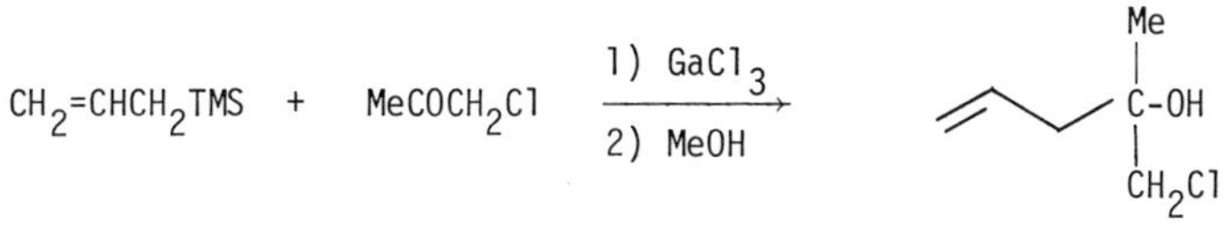

$CH_2=CHCH_2TMS$ + $MeCOCH_2Cl$ $\xrightarrow[\text{2) MeOH}]{\text{1) GaCl}_3}$

60%

J Organometal Chem (1975) <u>93</u> 43

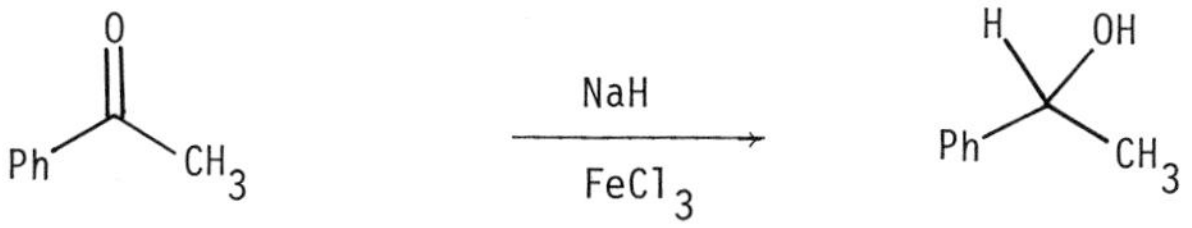

72%

JOC (1976) <u>41</u> 1667

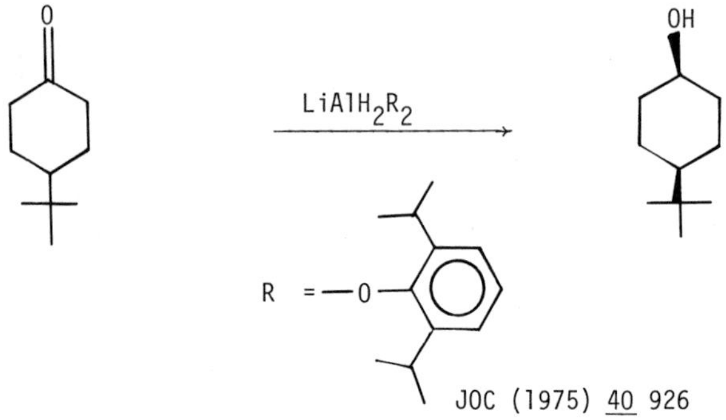

$$LiAlH_2R_2$$

$$R = -O-$$

70%

JOC (1975) <u>40</u> 926

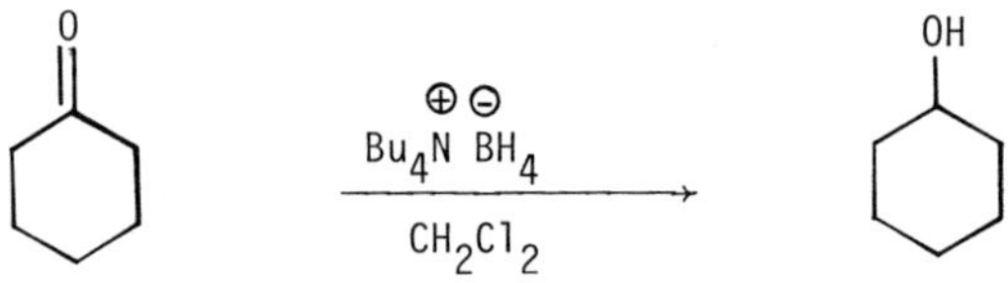

$$Bu_4N \overset{\oplus}{} \overset{\ominus}{BH_4}$$

$$CH_2Cl_2$$

98%

JOC (1976) <u>41</u> 690

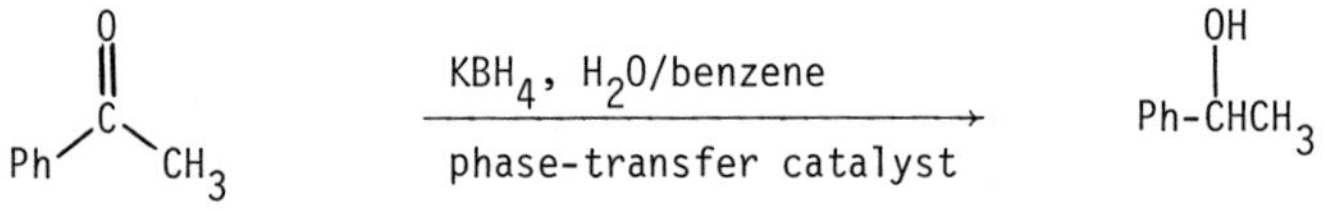

Ph–C(=O)–CH₃

$$\xrightarrow[\text{phase-transfer catalyst}]{KBH_4, H_2O/benzene}$$

$$Ph-\underset{OH}{\overset{}{C}}HCH_3$$

97%

Synthesis (1975) 531

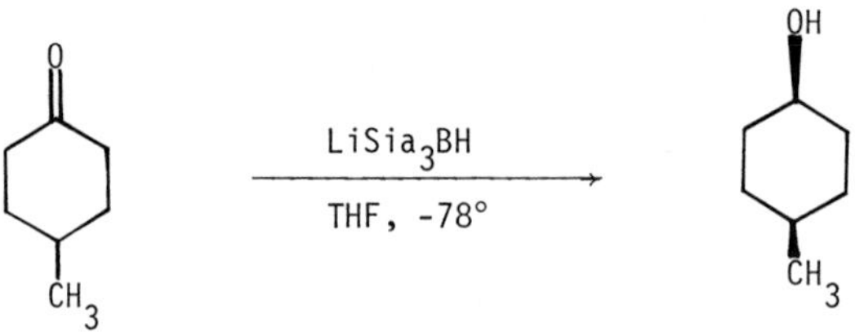

$$LiSia_3BH$$

$$THF, -78°$$

99%

JACS (1976) <u>98</u> 3383

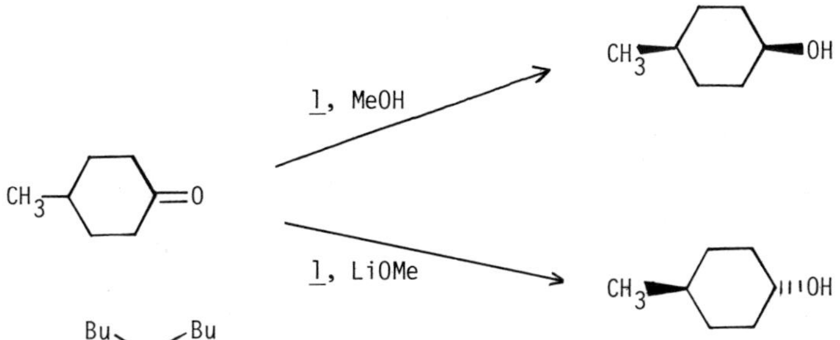

84%

1, MeOH

CH₃ — cyclohexane — OH

1, LiOMe

CH₃ — cyclohexane — ⅠⅠOH   90%

Bu   Bu
 \   /
  B ⊖ Li⊕
1 =

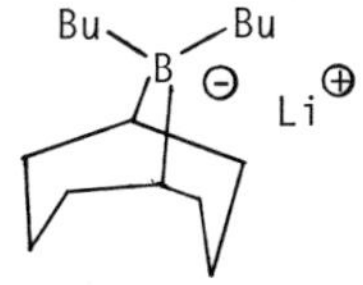

JACS (1976) 98 1965

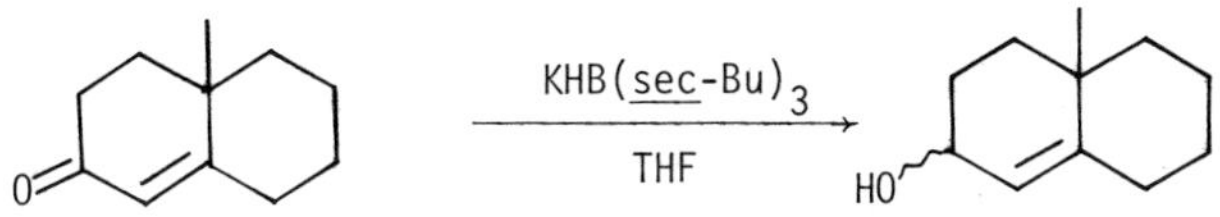

KHB(sec-Bu)₃
————————
THF

JOC (1975) 40 146

O
‖
Ph-C-CH₃

amino-ester borane
————————————
BF₃·Et₂O

OH
|
Ph-CHCH₃   48%
*

17% ee

Tetr Lett (1976) 295

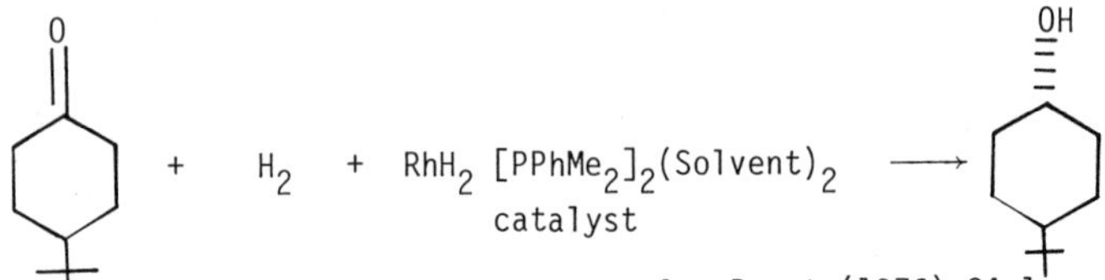

86%

Org React (1976) 24 1

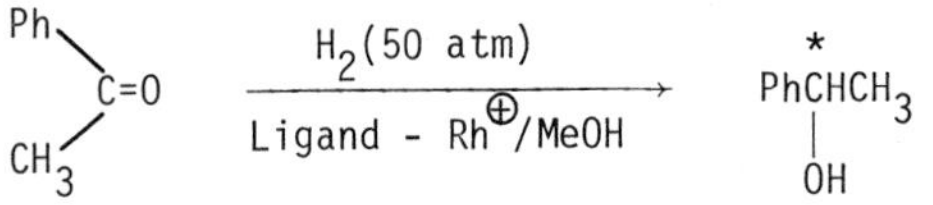

R 43% ee

96%

Ligand =

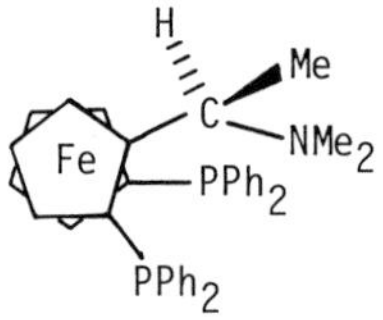

Tetr Lett (1976) 4351

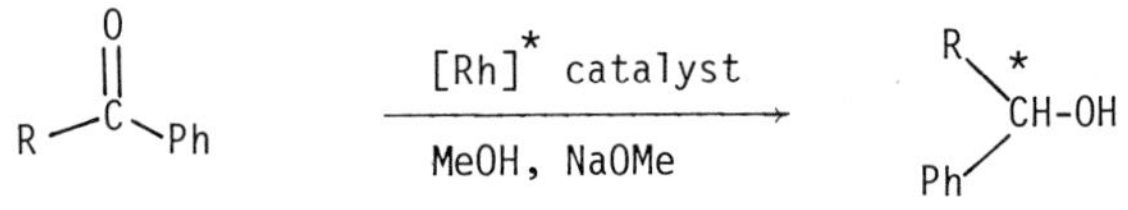

R = 1°,2°,3° alkyl

up to 58% ee

>95%

J Organometal Chem (1976) 122 83

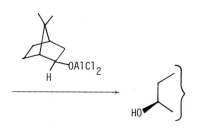

O
‖
Ph-C-CF₃

$$Ph-\overset{OH}{\underset{|}{C}}HCF_3 \qquad 88\%$$

(S) 68% ee

Synthesis (1975) 701

O꞊

(various 3-oxo
triterpenoids)

HO⟋

~70%

75-100% axial alcohol

JCS Perkin I (1976) 321

NMe

MeO

O꞊

formamidinesulfinic
acid

⟍OH

63%

JOC (1976) **41** 3624

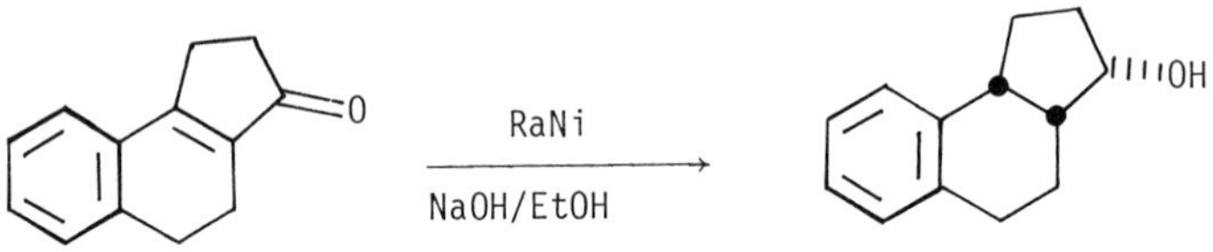

RaNi

NaOH/EtOH

95%

Synthesis (1975) 702

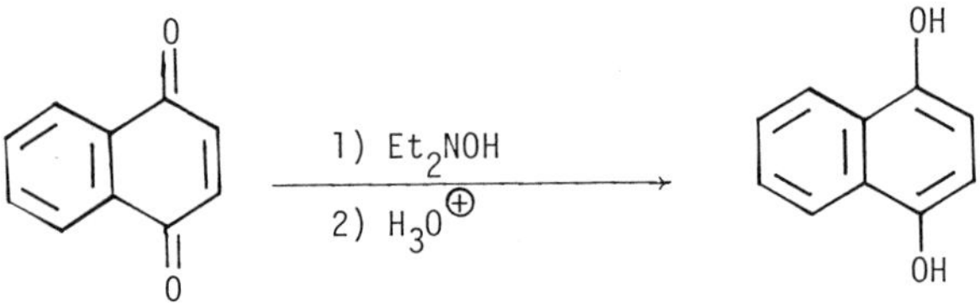

1) Et$_2$NOH

2) H$_3$O$^{\oplus}$

83%

Tetr Lett (1975) 1695

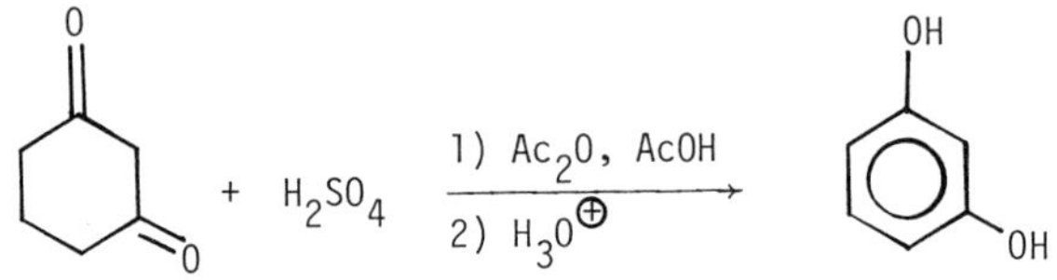

+ H$_2$SO$_4$

1) Ac$_2$O, AcOH

2) H$_3$O$^{\oplus}$

95%

same
conditions

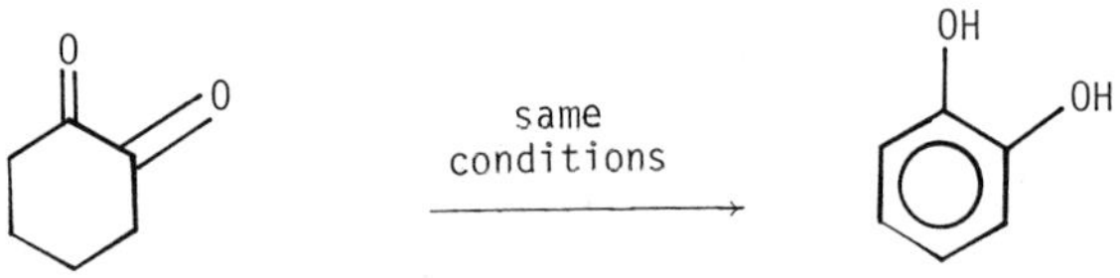

95%

JOC (1974) <u>39</u> 3697

Related methods: Alcohols from Aldehydes (Section 34)

Section 43   Alcohols and Phenols from Nitriles

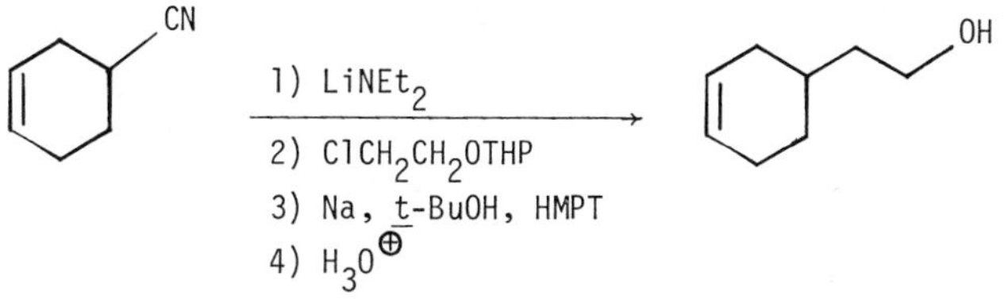

1) LiNEt$_2$

2) ClCH$_2$CH$_2$OTHP

3) Na, t-BuOH, HMPT

4) H$_3$O$^{\oplus}$

90%

Synthesis (1976) 391

Section 44   Alcohols from Olefins

For the preparation of diols from olefins see Section 323 (Alcohol-Alcohol)

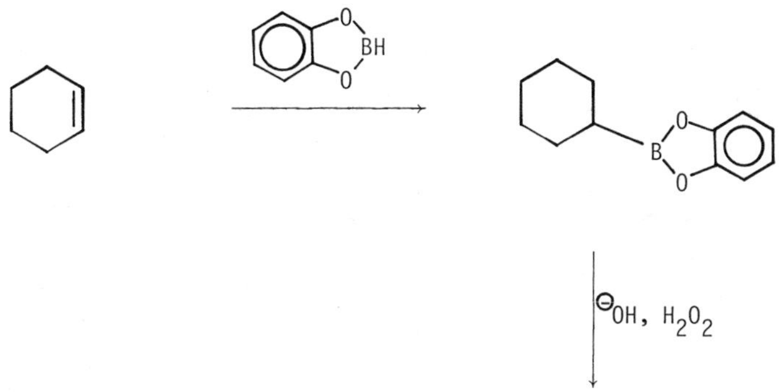

$^{\ominus}$OH, H$_2$O$_2$

cyclohexanol   90%

JACS (1975) 97 5249

$$\text{CH}_3\text{CH}_2\text{CH}_2\text{CH}_2\text{CH=CH}_2 \xrightarrow[\text{2) }^{\ominus}\text{CN}]{\text{1) BH}_3} \text{Hex}_3\text{BCN}^{\ominus} \xrightarrow[\text{2) }^{\ominus}\text{OH, H}_2\text{O}_2]{\text{1) TFAA}} \text{Hex}_3\text{COH} \qquad 96\%$$

JCS Perkin I (1975) 138

$$(\underline{n}\text{-hexyl})_3\text{B} + \underset{(-)}{\text{R'C(SR'')}_2} \xrightarrow[\text{2) H}_2\text{O}_2,\ ^{\ominus}\text{OH}]{\text{1) HgCl}_2} (\underline{n}\text{-hexyl})_2\underset{\underset{\text{R'}}{|}}{\text{C-OH}} \qquad \sim 80\%$$

JCS Chem Comm (1974) 863

Use of $\text{BH}_3\cdot\text{Me}_2\text{S}$ as a hydroboration agent.

JOC (1974) <u>39</u> 1437

$$\xrightarrow[\substack{\text{2) MeI}\\ \text{3) NaOH, H}_2\text{O}_2}]{\text{1) LiCH}_2\text{SMe}}$$

(cyclopentyl)–CH$_2$OH   97%

JOC (1975) <u>40</u> 814

$$\text{R}_3\text{B} \xrightarrow[\substack{\text{2) HCl, -78}°\\ \text{3) NaOH, H}_2\text{O}_2}]{\text{1) LiCH=CH}_2} \underset{\underset{\text{OH}}{|}}{\text{R}_2\text{C-Me}} \qquad 87\text{-}94\%$$

JACS (1975) <u>97</u> 5017

$$(C_6H_{13})_3B \quad + \quad Me_3\overset{\oplus}{N}\text{-}\overset{\ominus}{O} \quad \longrightarrow \quad 3\ C_6H_{13}OH \qquad\qquad 95\%$$

JOC (1975) $\underline{40}$ 1776

1) NaBH$_4$, AcOH/THF

2) NaOH, H$_2$O, H$_2$O$_2$

Simplified hydroboration procedure

Synthesis (1974) 340

Review: Thexylborane, A Highly Versatile Reagent for Organic Synthesis
via Hydroboration

Synthesis (1974) 77

Review: Organoboranes as Reagents for Organic Synthesis

Chem Soc Rev (1974) $\underline{3}$ 443

Full experimental details for preparation and use of 9-BBN
as a highly selective reagent for hydroboration of olefins.

JACS (1974) $\underline{96}$ 7765

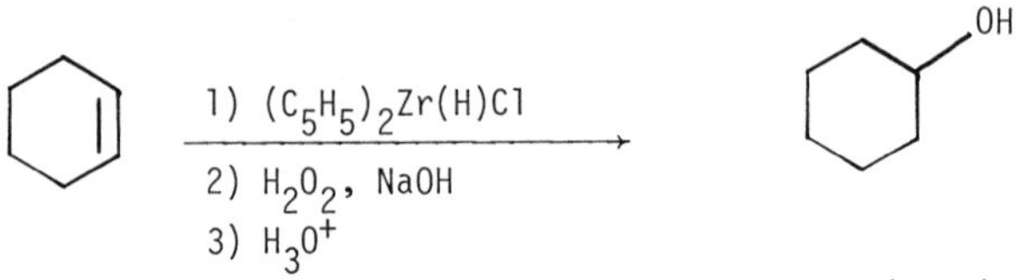

76%

Tetr Lett (1975) 3041

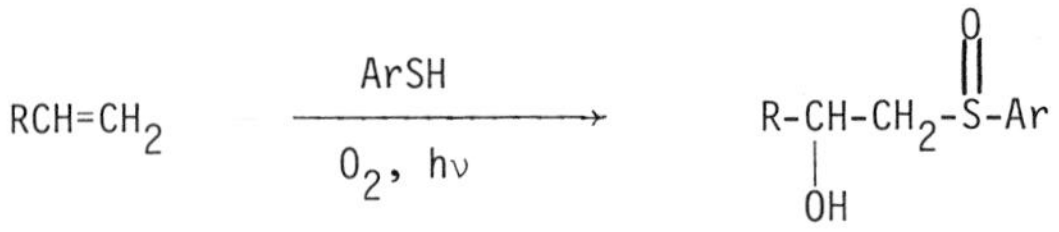

$RCH=CH_2$ $\xrightarrow[\text{O}_2, \text{ h}\nu]{\text{ArSH}}$ R-CH-CH$_2$-S-Ar

54-92%

JOC (1974) **39** 1170

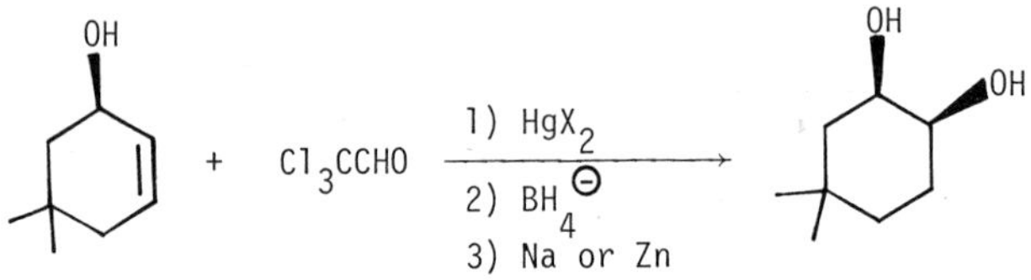

moderate yield

JOC (1974) **39** 1474

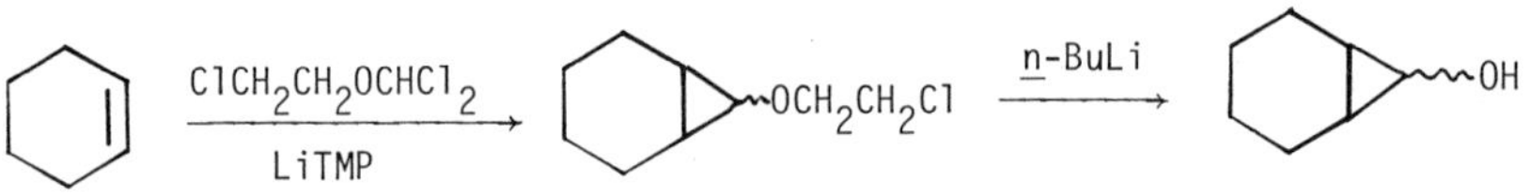

58%   no yield

Tetr Lett (1976) 3783

Section 45 Alcohols from Miscellaneous Compounds

NO2
PhCH=N
O
X
OH
X

20-94%

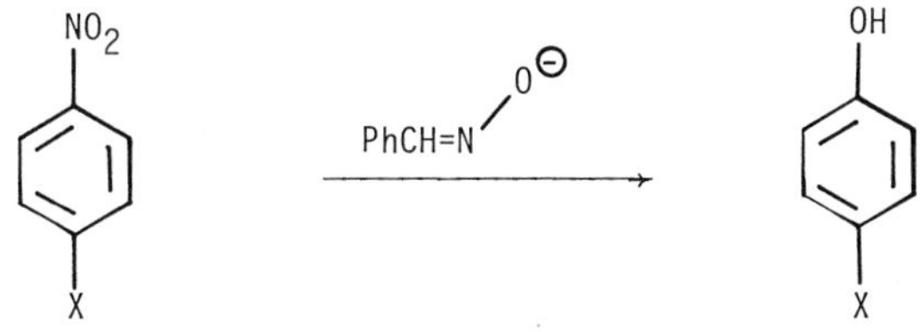

X = COOR, NO_2, CN, $COCH_3$, CHO, $CONR_2$

JOC (1974) 39 3343

Ph
NO2
1) 2 n-BuLi
2) PhCHO
3) AcOH
HO
Ph
Ph
NO2

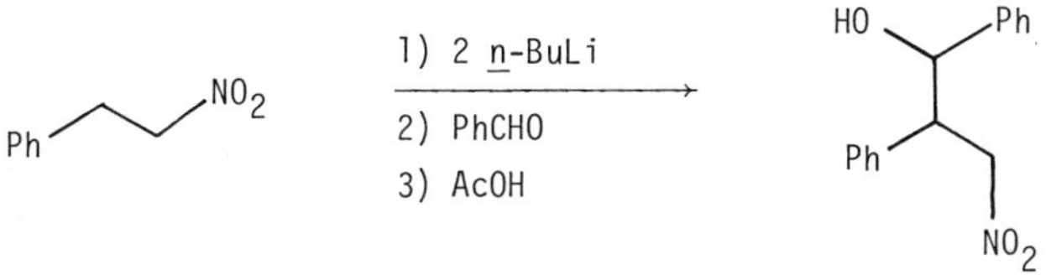

80%

Helv Chim Acta (1976) 59 2213

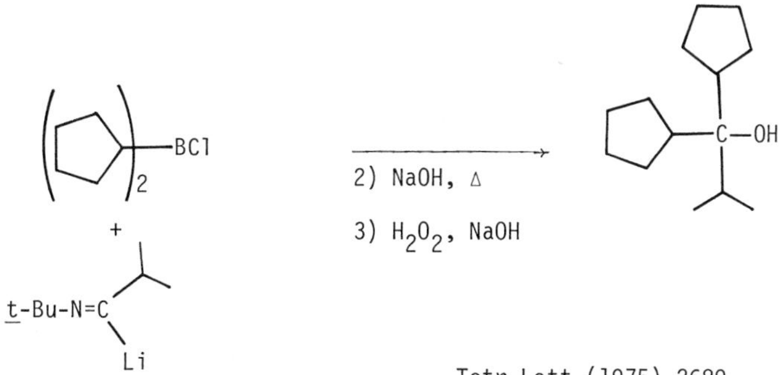

BCl
2
+
t-Bu-N=C
Li
2) NaOH, Δ
3) H2O2, NaOH
C—OH

77%

Tetr Lett (1975) 2689

$(\underline{n}\text{-Hex})_3B$

$+$

$\underline{t}\text{-Bu-N=C} \overset{\text{n-Bu}}{\underset{\text{Li}}{\diagup}}$

$\xrightarrow{\begin{array}{l} 1)\ HSCH_2CO_2H \\ \hline 2)\ H_2O_2,\ NaOH \end{array}}$

$(\underline{n}\text{-Hex})_2\underset{\underline{n}\text{-Bu}}{\overset{\mid}{C}}\text{-OH}$ 87%

JOC (1975) <u>40</u> 3644

$\left(\bigcirc \right)_3 \!\!- B$

$\xrightarrow{\begin{array}{l} 1)\ Li\text{-}\overset{OMe}{\overset{\mid}{C}}\text{=}CH_2 \\ \hline 2)\ HCl \\ 3)\ H_2O_2,\ NaOH \end{array}}$

$\left(\bigcirc \right)_2 \!\!- \underset{}{\overset{OH}{\overset{\mid}{C}}}\text{-}CH_3$ 94%

Tetr Lett (1976) 2201

Section 45A   Protection of Alcohols and Phenols

ROH   + $\left(\overset{O}{\overset{\|}{C}}\!\!-\!\!CH_2CH_2\!\!-\!\!\overset{O}{\underset{\|}{C}} \right)_2 O$ $\longrightarrow$ $R\text{-O-}\overset{O}{\overset{\|}{C}}\!\!-\!\!CH_2CH_2\!\!-\!\!\overset{O}{\underset{\|}{C}}$

NaBH$_4$

JACS (1975) <u>97</u> 1614

Synth Comm (1975) 5 91

$$Ar-(CH_2)_4-\underset{\underset{Me}{|}}{\overset{\overset{Me}{|}}{C}}-OH \quad \xrightarrow[\text{HgCl}_2, \text{ CH}_3\text{CN/H}_2\text{O}]{\text{Ac}_2\text{O, DMSO}} \quad Ar-(CH_2)_4-\underset{\underset{Me}{|}}{\overset{\overset{Me}{|}}{C}}-OCH_2SCH_3 \qquad 90\%$$

Tetr Lett (1976) 65

$$RO^{\ominus} \; + \; ClCH_2SCH_3 \quad \xrightarrow{\text{NaI}} \quad ROCH_2SCH_3$$

Stable to base and mild acid.

$$ROCH_2SCH_3 \quad \xrightarrow[\text{or Ag}^+]{\text{Hg}^{++}} \quad ROH$$

Tetr Lett (1975) 2643, 3269

$$\text{ROH} \xrightarrow{\text{DMSO, HOAc, Ac}_2\text{O}} \text{ROCH}_2\text{SCH}_3$$

$$\text{CH}_3\text{I, NaHCO}_3, \text{H}_2\text{O}$$

R = 1°,2°,3° alkyl

Tetr Lett (1976) 3067

Use of the β-methoxymethyl (MEM) group for the protection of alcohols.

$$\text{ROH} + \text{MEM-Cl} + (\underline{i}\text{-Pr})_2\text{NEt} \longrightarrow \text{RO-MEM}$$

$$\text{ROH} + \text{MEM-N}\overset{\oplus}{\text{Et}}_3 \overset{\ominus}{\text{Cl}} \xrightarrow[\text{rfx}]{\text{MeCN}} \text{RO-MEM}$$

$$\text{RO}^{\ominus} + \text{MEM-Cl} \longrightarrow \text{RO-MEM}$$

$$\text{MEM} = -\text{CH}_2\text{OCH}_2\text{CH}_2\text{OCH}_3$$

Stable to strong bases, reducing agents, some oxidizing agents, and mild acids. Removed by ZnBr_2 or TiCl_4 in CH_2Cl_2.

Tetr Lett (1976) 809

$$\text{R-OH} + \text{CH}_2(\text{OMe})_2 \xrightarrow{\text{P}_2\text{O}_5} \text{ROCH}_2\text{OCH}_3 \qquad\qquad 94\text{-}99\%$$

R = 1°,2° alkyl

Synthesis (1975) 276

ROH $\xrightarrow[\text{Et}_3\text{N}]{\text{THF, SOCl}_2}$

Removed with mild acid.

Tetr Lett (1976) 1725

ROH $\xrightarrow[\text{AgNO}_3,\ \text{MeCN}]{\text{PhSeCH}_2\text{CH}_2\text{Br}}$ ROCH$_2$CH$_2$SePh

$\xleftarrow{\text{1) H}_2\text{O}_2}$

2) HCl   Synth Comm (1975) $\underline{5}$ 367

$\xrightarrow[\text{mol. sieves, CH}_2\text{Cl}_2]{\text{H}_2\text{C(OMe)}_2,\ \text{TsOH}}$

~70%

Synthesis (1976) 244

$\xrightarrow{\text{NaAlH}_2(\text{OCH}_2\text{CH}_2\text{OCH}_3)_2}$

~50%

JOC (1976) $\underline{41}$ 2545

R-O-CH$_2$—⟨O⟩—OMe $\xrightarrow{\text{electrochemical oxidation}}$ ROH + anisaldehyde

74-98%

R = alkyl, benzyl, propargyl, etc.

JOC (1975) <u>40</u> 1356

n-C$_8$H$_{17}$OBz $\xrightarrow{\text{UF}_6}$ n-C$_8$H$_{17}$OH   69%

JACS (1976) <u>98</u> 6717

MeO—⟨O⟩—CHO (with MeO, MeO substituents) $\xrightarrow[\text{HMPT/Toluene}]{\text{NaS—⟨O⟩—CH}_3}$ HO—⟨O⟩—CHO (with MeO, MeO substituents)   90%

Synthesis (1976) 191

Further examples of ether cleavages are included in Section 39 (Alcohols and Phenols from Ethers and Epoxides)

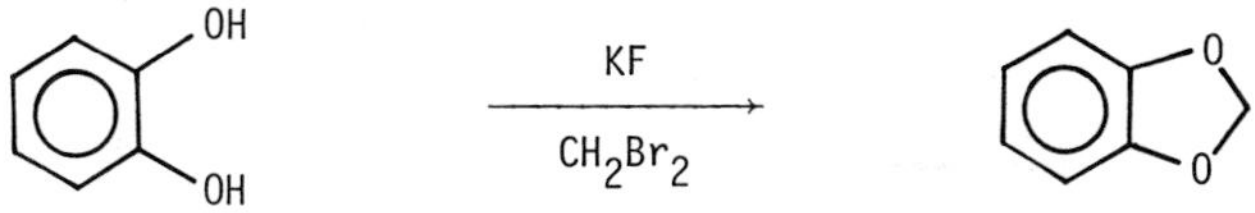

Tetr Lett (1976) 3361

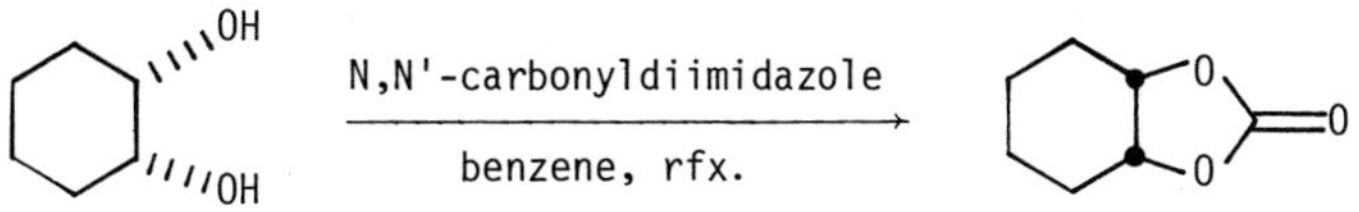

Synth Comm (1975) 5 47

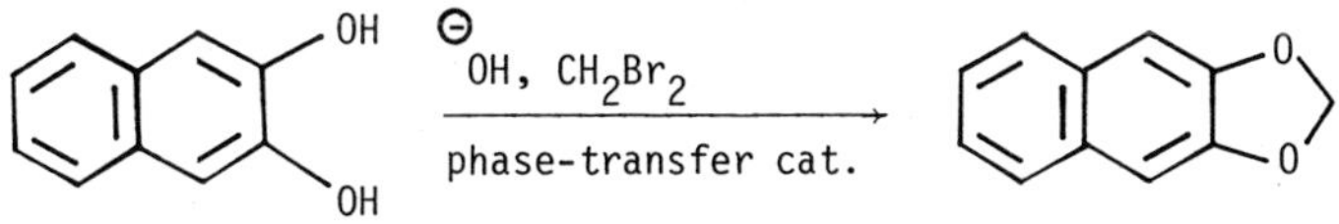

Tetr Lett (1975) 3489

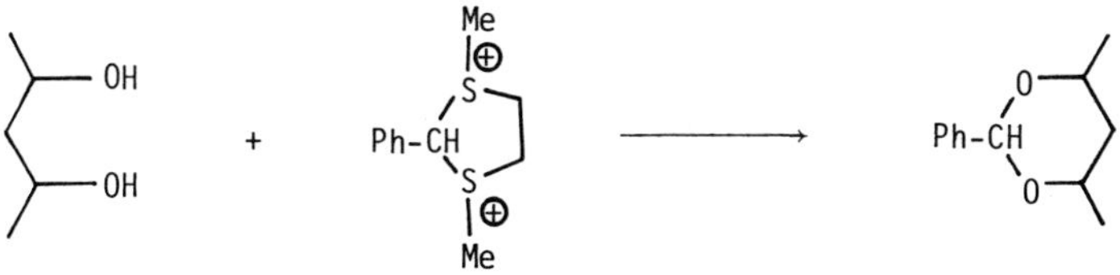

Tetr Lett (1975) 4543

Can J Chem (1974) $\underline{52}$ 187

Me
|
t-Bu-Si
|
Me   cannot be used for protecting hydroxy groups in molecules where there is an undesired opportunity for acyl migration, e.g. acyl glycerols.

JCS Chem Comm (1975) 249

Use of polymer-bound trityl chloride residues to block selectively one primary OH group of a polyhydroxy alcohol

Can J Chem (1976) $\underline{54}$ 926, 935

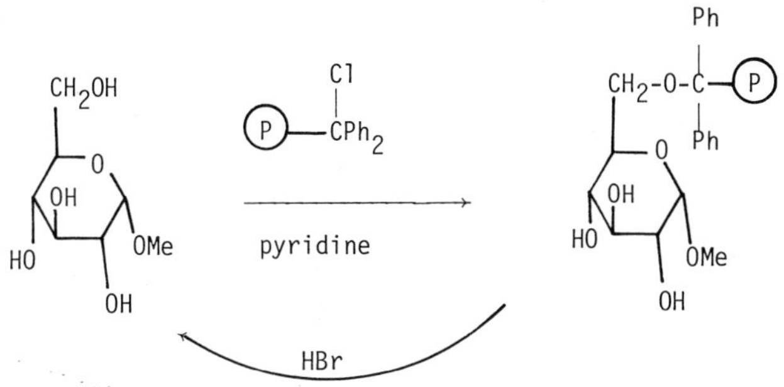

Tetr Lett (1975) 3055

Use of a polymer to protect glucose OH groups:

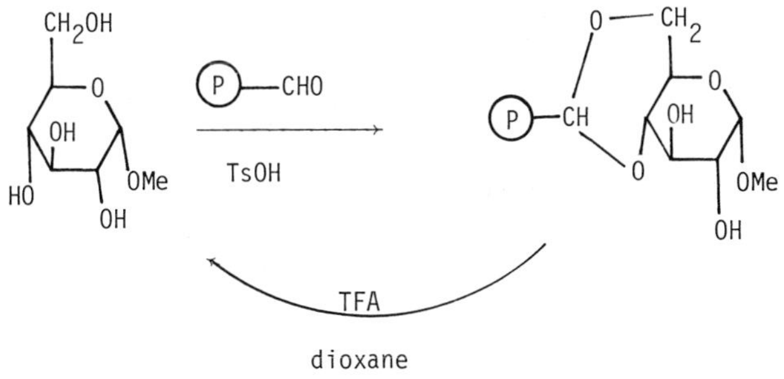

JCS Chem Comm (1975) 225

$$R-O \diagup\hspace{-0.3em}\diagup\hspace{-0.3em}\backslash\backslash \quad \xrightarrow[\text{H}_2\text{O or MeOH}]{\text{Pd/C}} \quad \text{ROH} \qquad 78\text{-}95\%$$

R = alkyl, aryl

Angew Int Ed (1976) <u>15</u> 558

Allyl ether as a protecting group in carbohydrate chemistry.

JCS Perkin I (1974) 1446

5'-nucleotide + [structure]$_2$ POCl $\underset{\text{H}_2/\text{Pd}}{\overset{\longrightarrow}{\longleftarrow}}$ 5'-phosphate

Highly selective for 1° OH over 2° OH

Stable to dil. base, acid

JOC (1974) <u>39</u> 3767

Use of (<u>t</u>-Bu)SiMe$_2$ and (<u>i</u>-PrO)$_3$Si as protecting groups for 2', and 2',5'- positions of ribonucleosides.

Tetr Lett (1974) 2861, 2865

Use of $HgCl_2/H_2S$ for removal of the S-trityl group from α-SR β lactams.

JCS Chem Comm (1974) 12

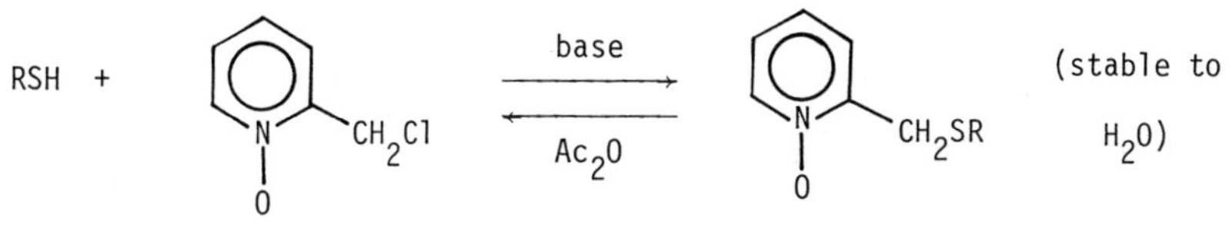

RSH +

(stable to

H_2O)

RSH = cysteine, glutathione, thiouridine.

Chem Pharm Bull (1974) 22 2889

Review: "Photosensitive Protecting Groups"

Israel J Chem (1974) 12 103

Review: "Electro-Deprotection--Electrochemical Removal of Protecting Groups"

Angew Int Ed (1976) 15 281

Chapter 4   PREPARATION

OF

ALDEHYDES

R-C≡CH

1) [benzodioxaborole structure] BH

⟶

2) H_2O_2, $^{\ominus}OH$

RCH_2CHO

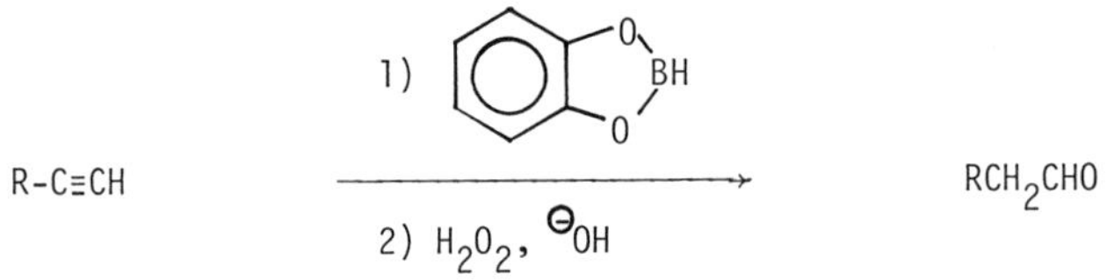

JACS (1975) <u>97</u> 5249

$$H-C≡C-\overset{\overset{\displaystyle O}{||}}{C}-OCH_3$$

$\xrightarrow[\text{NaCN}]{\text{MeOH}}$

$(MeO)_2CHCH_2COOCH_3$

75%

JOC (1976) <u>41</u> 3765

66

Section 47   <u>Aldehydes from Carboxylic Acids and Acid Halides</u>

RCOOH   +   [benzene ring with two SH groups]   1) $POCl_3$, $HClO_4$
                                                 2) $LiAlH_4$
                                                 3) H_2O RCHO   50-80%

J Het Chem (1974) <u>11</u> 943

[structure: Ph-C-Cl with double bond O]
+
[benzene ring with OH and SH groups] $\longrightarrow$ [benzodioxathiole cation structure with Ph]   1) $NaBH_4$
                                                                                                          2) hydrolysis   PhCHO   87%

JCS Perkin I (1976) 323

PhCH=CHCOCl   +   PhNHN=C(SMe)NHPh   1) $NaBH_4$
                                     2) $H_3O^{\oplus}$ PhCH=CHCHO   ∼65%

Tetr Lett (1974) 2649

[structure: R-C-Cl with double bond O] $\longrightarrow$ $\longrightarrow$ [heterocyclic structure with Ph, N, SMe, N-N, Ph, R]   HOAc
                                                                                                                                   H_2O RCHO   60-90%

R = alkyl, aryl, cinnamoyl

Tetrahedron (1976) <u>32</u> 2549

58%

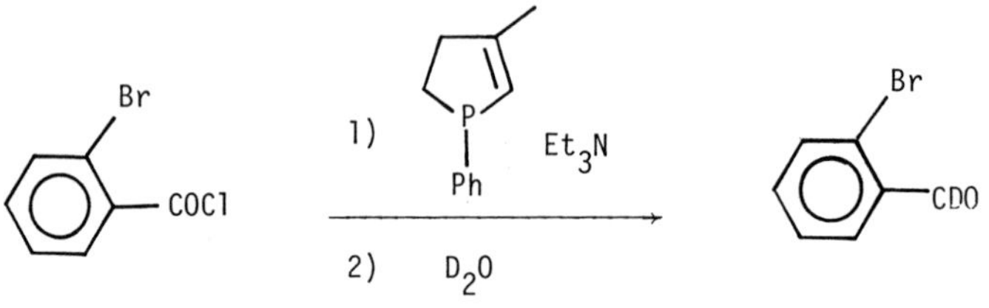

Synth Comm (1976) 6 135

$$\underline{n}\text{-}C_{11}H_{23}\overset{O}{\overset{\|}{C}}\text{-}Cl \xrightarrow[\text{2,6-lutidine}]{H_2/Pd\text{-}C,\ THF} \underline{n}\text{-}C_{11}H_{23}\overset{O}{\overset{\|}{C}}\text{-}H$$

95%

Synthesis (1976) 767

80%

JCS Chem Comm (1975) 459

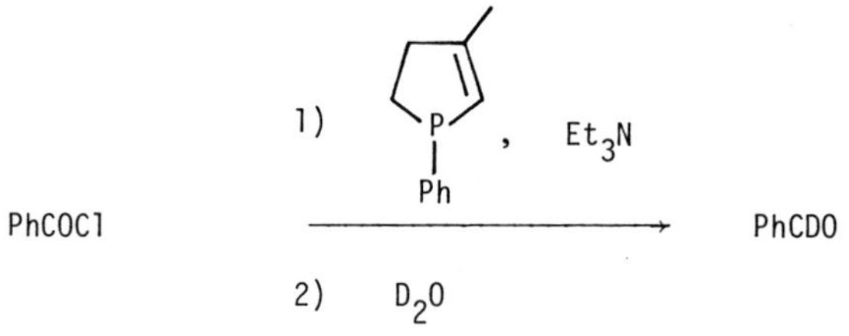

Section 48   Aldehydes from Alcohols and Phenols

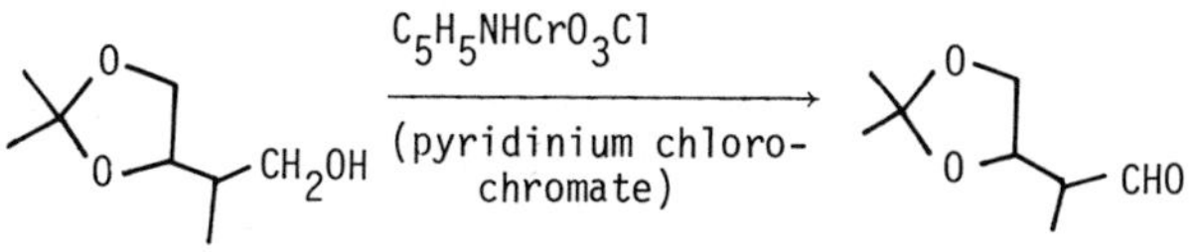

Tetr Lett (1975) 2647

$R\text{-}CH_2OH \xrightarrow{\quad CrO_2Cl_2 \quad} R\text{-}CHO$ ~80%

May isomerize double bonds.

JACS (1975) $\underline{97}$ 5927

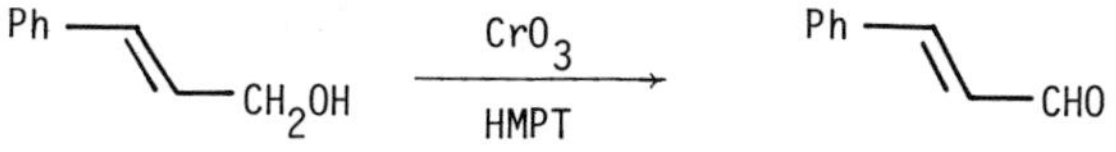

Synthesis (1976) 394

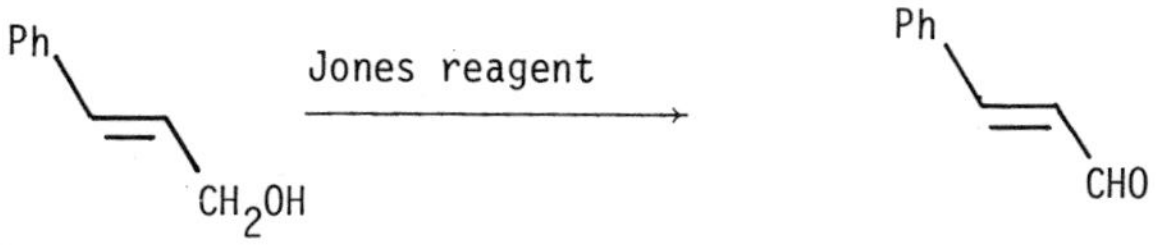

JOC (1975) $\underline{40}$ 1664

Ph—CH=CH—CH₂OH $\xrightarrow[\text{resin}]{\text{CrO}_3}$ Ph—CH=CH—CHO   96%

JACS (1976) <u>98</u> 6737

Ph—CH=CH—CH₂OH $\xrightarrow[\text{3) H}_2\text{O}]{\substack{\text{1) TMSCl} \\ \text{2) Ph}_3\text{CBF}_4}}$ Ph—CH=CH—CHO   100%

JOC (1976) <u>41</u> 1479

R-CH₂OH $\xrightarrow[\text{celite}]{\text{Ag}_2\text{CO}_3}$ R-CHO   50-97%

R = pyridine, pyrrole, indole, furyl, etc.

J Het Chem (1976) <u>13</u> 525

PhCH₂OH $\xrightarrow[\text{MCPBA}]{}$ PhCHO   76%

JOC (1975) <u>40</u> 1860

Ph-CH=CH-CH$_2$OH $\quad\xrightarrow[\text{Et}_3\text{N}]{\begin{array}{c}\text{DMSO}\\\text{TFAA}\end{array}}\quad$ Ph-CH=CH-CHO   83%

JOC (1976) <u>41</u> 957

Me$_3$C-CH$_2$OH $\quad\xrightarrow[\text{TFAA}]{\text{DMSO}}\quad$ Me$_3$C-CHO   81%

JOC (1976) <u>41</u> 3329

Ph-CH$_2$OH $\quad\xrightarrow{\text{Br}_2/\text{HMPT}}\quad$ Ph-CHO   86%

Synthesis (1976) 811

PhCHO $\quad\xrightarrow[\text{CH}_2\text{Cl}_2]{\overset{\oplus}{\text{Bu}_4\text{N}}\;\overset{\ominus}{\text{BH}_4}}\quad$ PhCH$_2$OH   92%

JOC (1976) <u>41</u> 690

Ph-CH$_2$OH $\quad\xrightarrow[\text{CH}_2\text{Cl}_2]{\overset{\oplus}{\text{NO}}\;\overset{\ominus}{\text{BF}_4}}\quad$ PhCHO   62%

Synthesis (1976) 609

PhCH$_2$OH $\quad\xrightarrow[\begin{array}{c}\text{tetrabutylammonium}\\\text{bisulfate}\end{array}]{\text{NaOCl}}\quad$ PhCHO   76%

Tetr Lett (1976) 1641

$PhCH_2OSnBu_3$ $\xrightarrow[CCl_4]{NBS}$ PhCHO   86%

JACS (1976) <u>98</u> 1629

$PhCH_2OH$ $\longrightarrow$ $PhCH_2O-\overset{O}{\overset{||}{C}}-\overset{O}{\overset{||}{C}}-CH_3$ $\xrightarrow[C_6H_6]{h\nu}$ PhCHO   95%

Synth Comm (1976) <u>6</u> 281

$CH_3(CH_2)_5CH_2OH$ $\xrightarrow[\text{2)} \quad h\nu]{\text{1) } CH_3-\overset{O}{\overset{||}{C}}-\overset{O}{\overset{||}{C}}-Cl, \text{ pyridine}}$ $CH_3(CH_2)_5CHO$ 77%

JOC (1976) <u>41</u> 3030

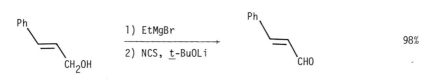

1) EtMgBr

2) NCS, <u>t</u>-BuOLi   98%

Chem Lett (1975) 691

1) Et_3SnOCH_3
2) Br_2
3) $Na_2S_2O_3$, KOH   92%

Chem Lett (1975) 145

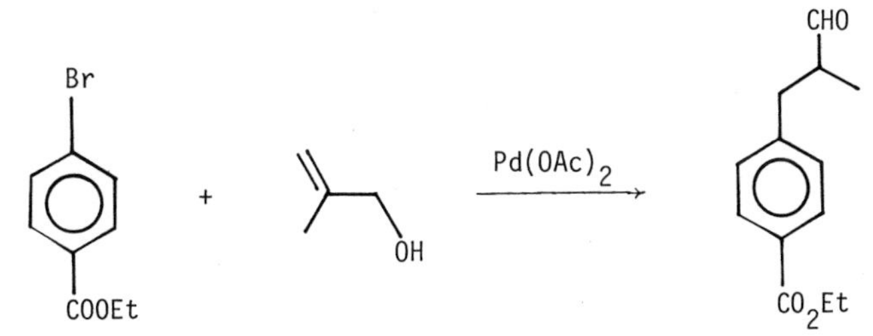

JOC (1976) 41 1206

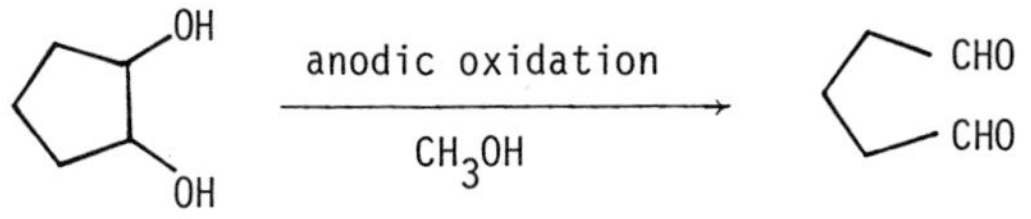

dialdehyde 56%

acetals   40%

JACS (1975) 97 2546

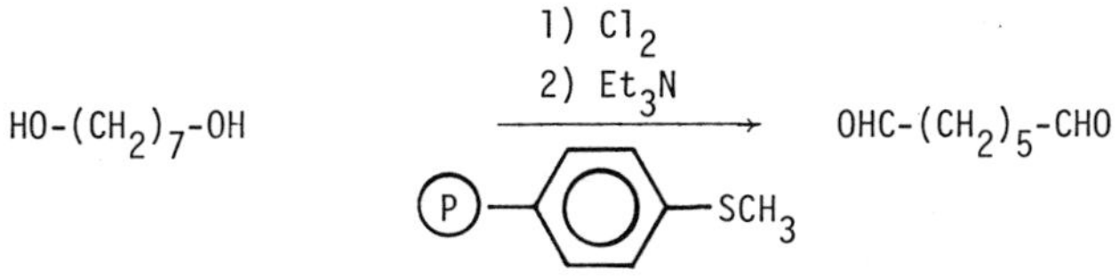

HO-(CH$_2$)$_7$-OH

1) Cl$_2$

2) Et$_3$N

OHC-(CH$_2$)$_5$-CHO   40%

JACS (1975) 97 2232

Related methods: Ketones from Alcohols and Phenols (Section 168)

Section 49   Aldehydes from Aldehydes

Me_2CHCHO -------→ $Me_2CHCH=N-\underline{t}-Bu$
$\xrightarrow{\begin{array}{l}1)\ LDA,\ DME\\2)\ PhCH_2Br\\3)\ H^{\oplus}\end{array}}$
$PhCH_2\underset{\underset{Me_2}{|}}{C}CHO$ 76%

JOC (1974) 39 3102

Me_2CHCHO
$\xrightarrow{\begin{array}{l}1)\ Me_3CNH_2,\ K_2CO_3\\2)\ EtMgBr\\3)\ BzCl\end{array}}$
$Bz-\underset{\underset{Me}{|}}{\overset{\overset{Me}{|}}{C}}-CHO$ 30%

Org Synth (1974) 54 46

1) PhMgBr
2) $H_3O^{\oplus}$

41%

63% ee

JACS (1976) 98 7450

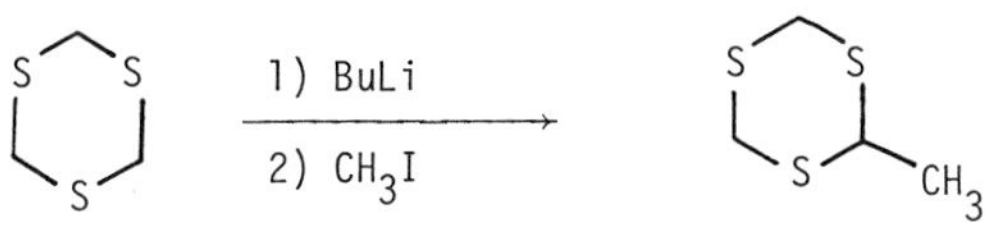

1) BuLi
2) CH_3I
S
S
S
CH_3

92%

Chem Ber (1974) 107 367

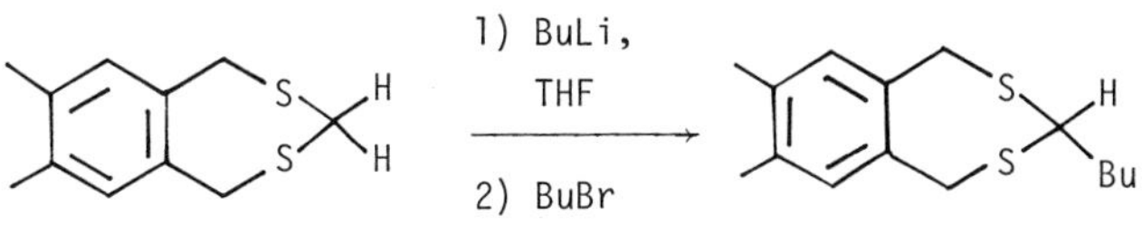

1) BuLi,
THF
2) BuBr
S
H
Bu

92%

Synthesis (1975) 720

N-Hex
1) LDA
2)
Br
N-Hex

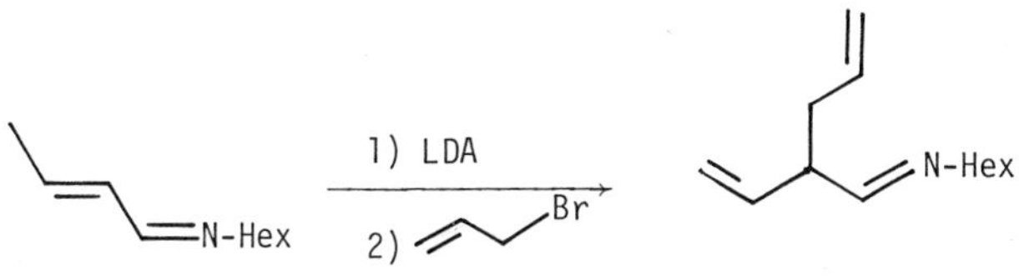

100%

Tetr Lett (1976) 597

Related methods: Aldehydes from Ketones (Section 57), Ketones from Ketones (Section 177). Also via: Olefinic aldehydes (Section 341).

Section 50   Aldehydes from Alkyls

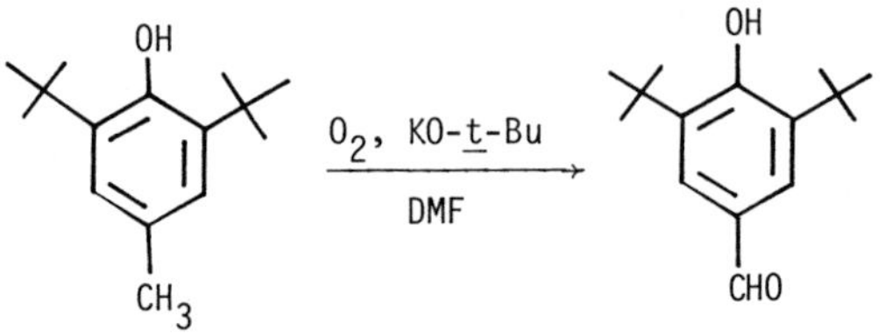

O_2, KO-t-Bu

DMF

90%

Angew Int Ed (1975) 14 356

Section 51   Aldehydes from Amides

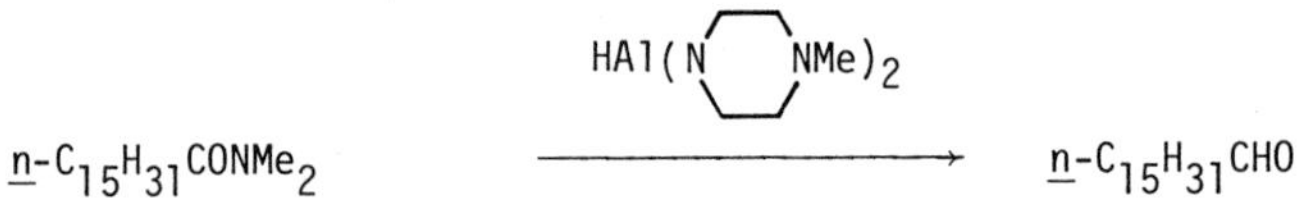

n-$C_{15}H_{31}CONMe_2$

HAl(N͡NMe)$_2$

n-$C_{15}H_{31}CHO$ 71%

Chem Lett (1975) 875

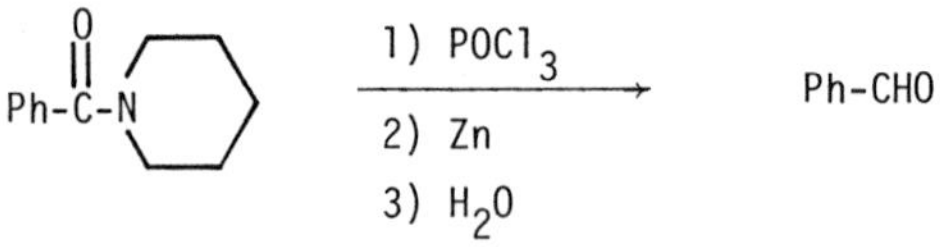

Ph-C-N

1) $POCl_3$
2) Zn
3) H_2O

Ph-CHO   95%

JCS Chem Comm (1976) 594

Section 52 Aldehydes from Amines

No additional examples

Section 53 Aldehydes from Esters

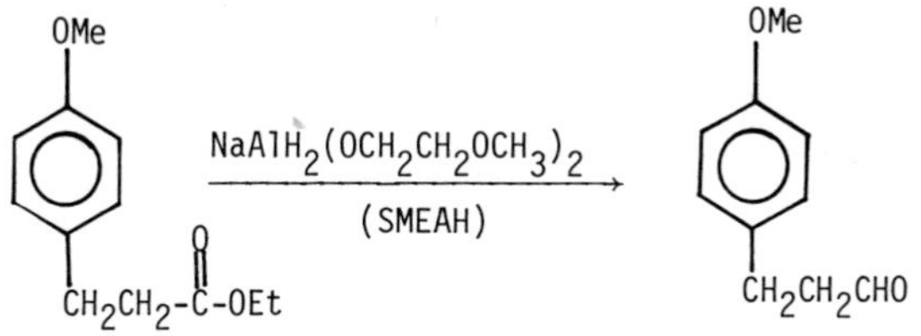

OMe
$NaAlH_2(OCH_2CH_2OCH_3)_2$
(SMEAH)
CH_2CH_2-C(=O)-OEt
OMe
CH_2CH_2CHO

88%

Synthesis (1976) 526

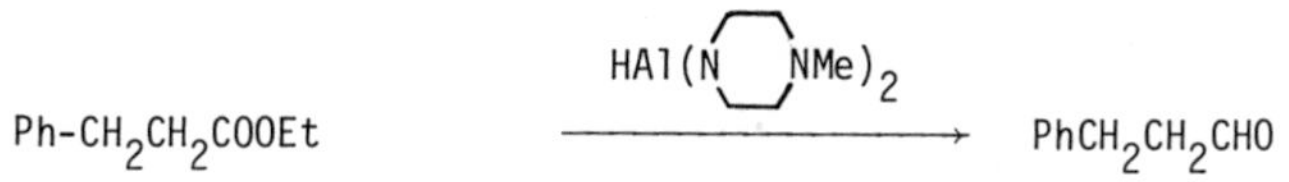

$Ph-CH_2CH_2COOEt$
HAl(N NMe)$_2$
$PhCH_2CH_2CHO$

76%

Chem Lett (1975) 215

Cbz-NH-CH(Bz)-COOMe —DIBALH→ Cbz-NH-CH(Bz)CHO 55%

Chem Pharm Bull (1975) 23 3081

$$\underset{\substack{|| \\ RC-OCH_2CH_2OMe}}{O} \xrightarrow{\overset{\oplus \ominus}{Et_3O\ BF_4}} \underset{H}{\overset{R}{>}}\!\!\!\!<\!\!\!\overset{O}{\underset{O}{\rfloor}} \qquad 76\text{-}91\%$$

Synthesis (1974) 808

Section 54   Aldehydes from Ethers and Epoxides

$$PhCH_2OCH_3 \xrightarrow{UF_6} PhCHO \qquad 78\%$$

JACS (1976) 98 6717

Section 55   Aldehydes from Halides

$$Ph(CH_2)_3Br \xrightarrow[\substack{2) \quad H_2O_2}]{\substack{Li \\ | \\ 1)\ Me_3SiCHSePh}} Ph(CH_2)_3CHO \qquad 66\%$$

Tetr Lett (1976) 4223

70%

JCS Chem Comm (1974) 410

$$\text{(1,3-dithiane)} \xrightarrow[\text{3) } H_3O^{\oplus}]{\substack{\text{1) BuLi} \\ \text{2) } \underline{n}\text{-}C_5H_{11}Br}} C_5H_{11}CHO \qquad \sim 90\%$$

JOC (1975) $\underline{40}$ 231

$$\xrightarrow{HC(OEt)_2OPh} \qquad 68\%$$

Chem Ber (1974) $\underline{107}$ 2295

Org Synth (1974) $\underline{54}$ 42

$$RCH_2Br \; + \; AgBF_4 \xrightarrow[\text{2) } Et_3N]{\text{1) DMSO}} RCHO \qquad 60\text{-}85\%$$

Tetr Lett (1974) 917

$$\underline{n}\text{-C}_5\text{H}_{11}\text{I} \quad + \quad \underset{\overset{|}{\underset{\|}{\text{S}}}}{\overset{\text{SMe}}{\text{LiCHSCNMe}_2}} \quad \longrightarrow \quad \underset{\text{MeOH}}{\overset{\text{Hg}^{++}}{\longrightarrow}} \quad \underline{n}\text{-C}_5\text{H}_{11}\text{CH(OMe)}_2 \qquad 70\%$$

Synthesis (1974) 705

$$(\text{Me}_2\text{N})_2\text{PON(Me)CH}_2\underset{\text{H}}{\text{C}}\text{=CH}_2 \quad \xrightarrow[\substack{\text{2) PhCH}_2\text{Cl} \\ \text{3) H}_3\text{O}^{\oplus}}]{\text{1) BuLi}} \quad$$

70%

Comptes Rendus (1974) <u>279</u> 609

$$\xrightarrow[\substack{\text{H}_2\text{, CO, R}_3\text{N} \\ 80\text{-}150°}]{\text{L}_2\text{PdX}_2}$$

60-90%

JACS (1974) <u>96</u> 7761

$$+ \quad \text{Li}^{\oplus}\text{PhSCu}^{\ominus} \underset{\|}{\overset{}{\diagup}}\text{CH(OEt)}_2 \quad \longrightarrow \quad$$

74%

JOC (1976) <u>41</u> 726

$$\xrightarrow[\text{HMPT, crown ether}]{\text{K}_2\text{CrO}_4}$$

78%

JCS Chem Comm (1976) 190

84%

JOC (1976) $\underline{41}$ 265
JOC (1976) $\underline{41}$ 273

PhCH$_2$Br

PhCHO   98%

Tetr Lett (1976) 3985

PhCH$_2$Br + Hg$_2$(NO$_3$)$_2$ $\xrightarrow[\ominus\text{OH}]{\text{aq. EtOH}}$ PhCHO   70-95%

(OMe, NO$_2$, CN, COOR groups are incompatible)

Synth Comm (1974) 45

CH$_3$SCH$_2$S-C-NMe$_2$ $\xrightarrow{\begin{array}{l}1)\ \text{BuLi}\\2)\ \underline{n}\text{-C}_5\text{H}_{11}\text{Br}\\3)\ \text{Hg}^{++},\ \text{MeOH}\end{array}}$ $\underline{n}$-C$_5$H$_{11}$-CH$\begin{array}{l}\diagup\text{OMe}\\\diagdown\text{OMe}\end{array}$ 70%

Synthesis (1974) 705

Section 56   <u>Aldehydes from Hydrides</u>

Tetr Lett (1974) 3463   20-40%

JCS Perkin I (1976) 540

COOMe
SeO2
glyme
COOMe
CHO

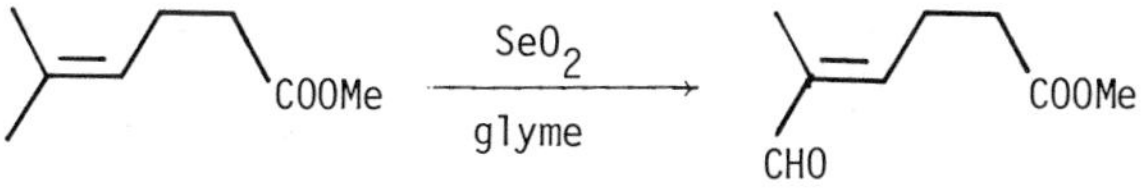

41%

Org React (1976) 24 261

Section 57 Aldehydes from Ketones

No additional examples

Section 58 Aldehydes from Nitriles

$n\text{-BuCN} + Et_3O^{\oplus}\ BF_4^{\ominus} \xrightarrow[\text{2) } H_3O^{\oplus}]{\text{1) } Et_3SiH} n\text{-BuCHO}$ 71%

JCS Chem Comm (1974) 45

Section 59 Aldehydes from Olefins

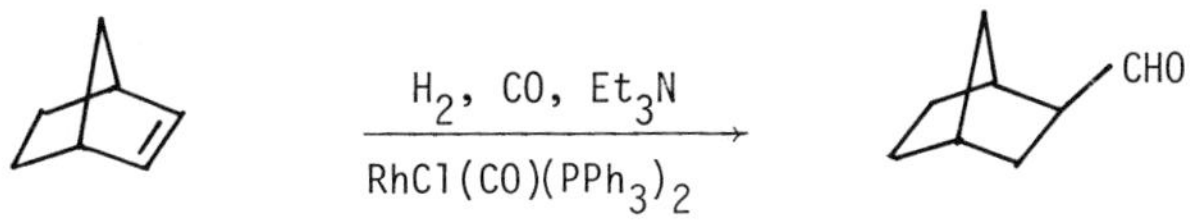

H2, CO, Et3N
RhCl(CO)(PPh3)2
CHO

82%

Synth Comm (1976) 6 199

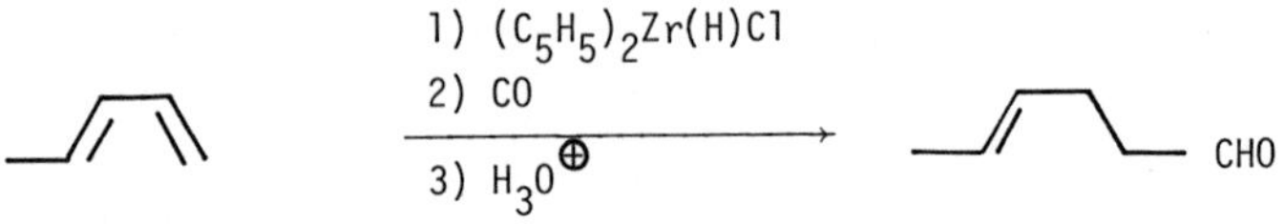

$$\text{1) } (C_5H_5)_2Zr(H)Cl$$

2) CO

3) $H_3O^{\oplus}$ → CHO   81%

JACS (1976) 98 262

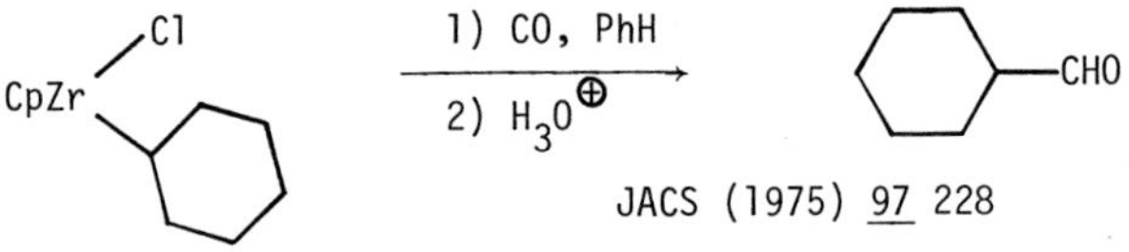

CpZr—Cl

1) CO, PhH

2) $H_3O^{\oplus}$ → CHO   97%

JACS (1975) 97 228

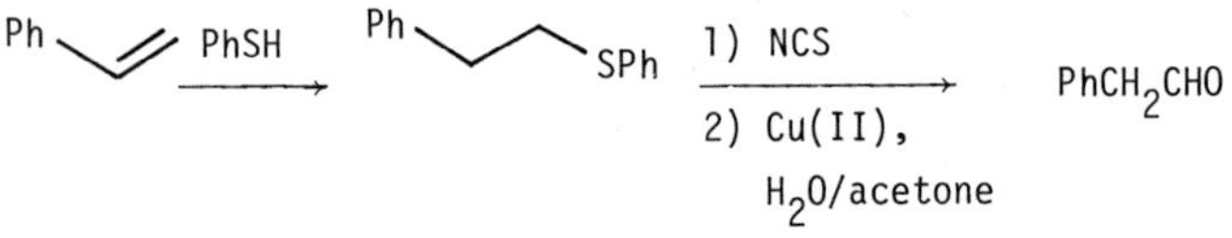

Ph⟍⟍ PhSH → Ph⟍⟍SPh

1) NCS

2) Cu(II), H_2O/acetone → $PhCH_2CHO$ 60%

JOC (1976) 41 2769

Section 60   Aldehydes from Miscellaneous Compounds

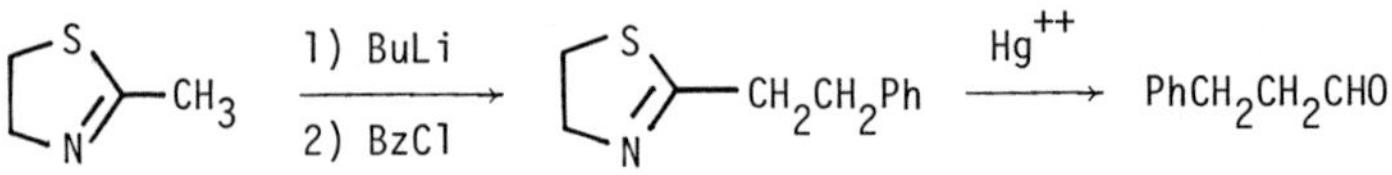

1) BuLi

2) BzCl → CH₂CH₂Ph

Hg^{++} → $PhCH_2CH_2CHO$ 56%

JOC (1975) 40 2021

Section 60A   Protection of Aldehydes

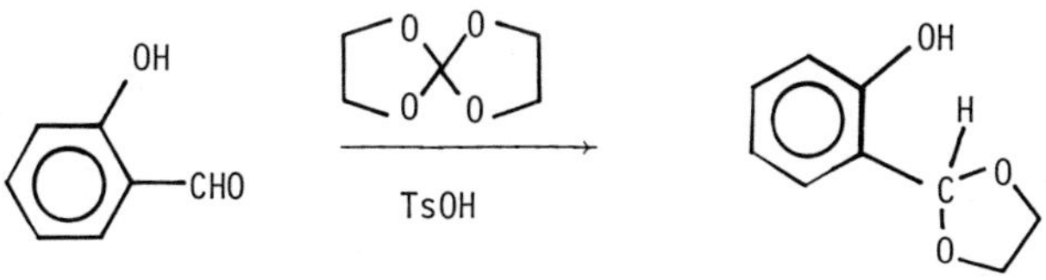

JCS Chem Comm (1975) 432

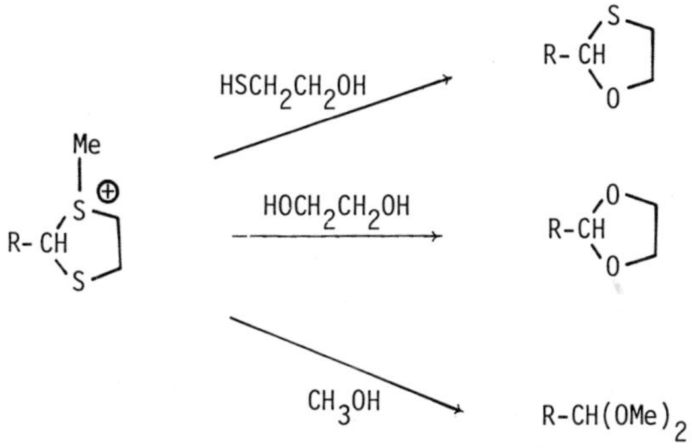

Tetr Lett (1975) 3267

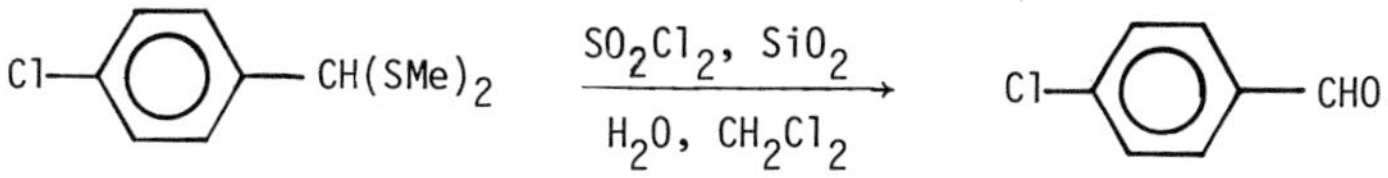

Synthesis (1976) 678

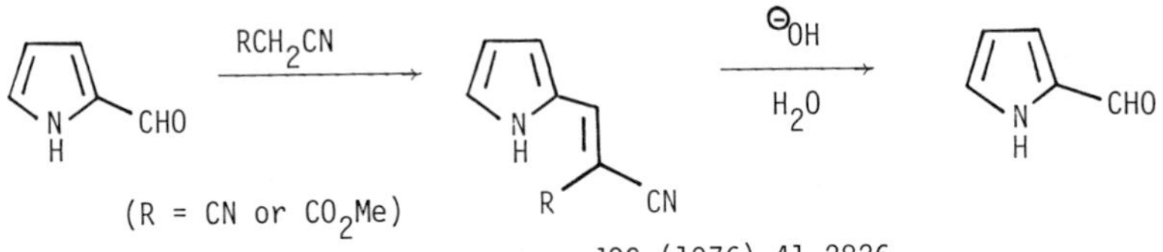

RCH2CN
CHO
(R = CN or CO2Me)
R
CN
OH
H2O
CHO

JOC (1976) 41 2826

O2N
CHO
P
AlCl3
n-butanol
O2N
CH(OBu)2

48%

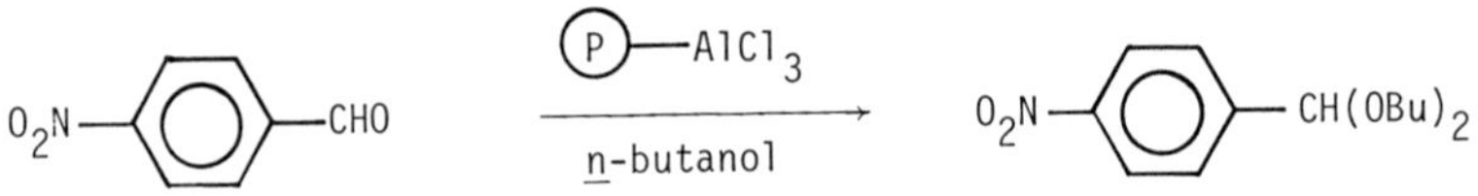

JOC (1975) 40 959

PhCHO
MeS-TMS
MeS
SMe
C
Ph
H

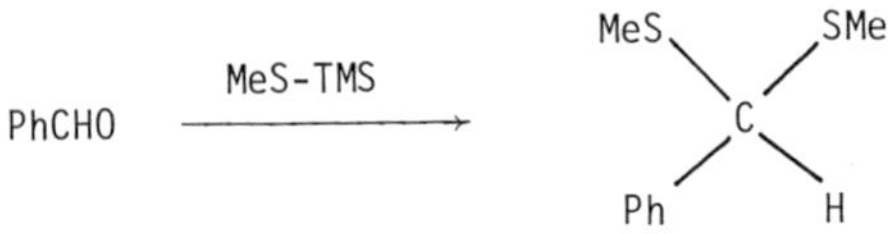

90%

No acid catalyst required.

JACS (1975) 97 3229

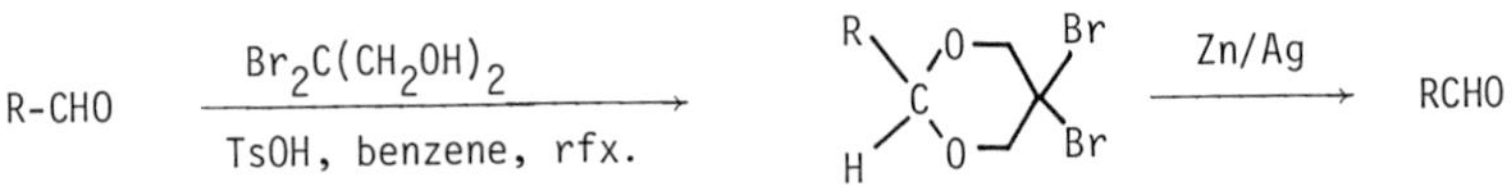

R-CHO
Br2C(CH2OH)2
TsOH, benzene, rfx.
R
O
Br
C
H
O
Br
Zn/Ag
RCHO

>90%

Tetr Lett (1976) 4577

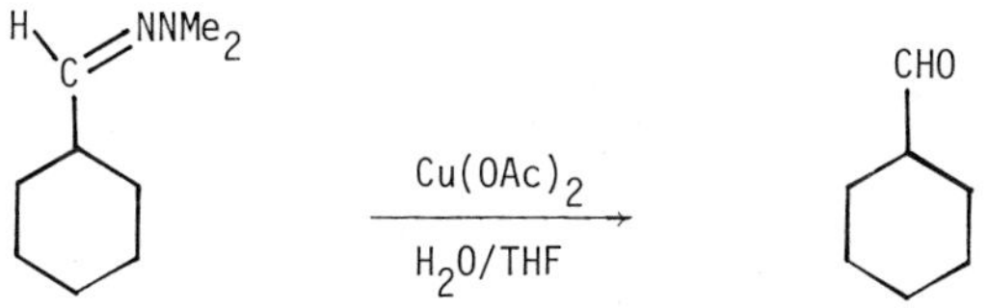

Tetr Lett (1976) 3667

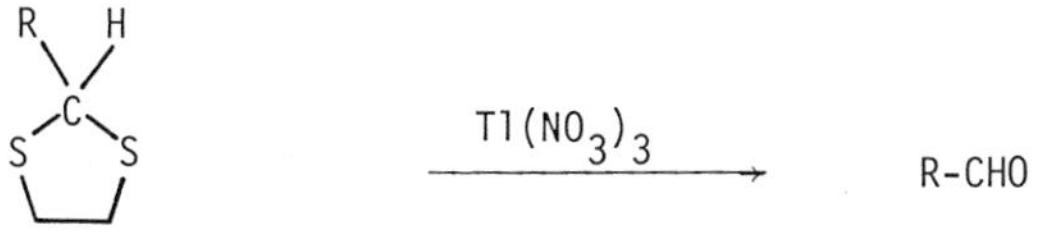

R = H, alkyl, aryl

Chem Pharm Bull (1976) <u>24</u> 1115

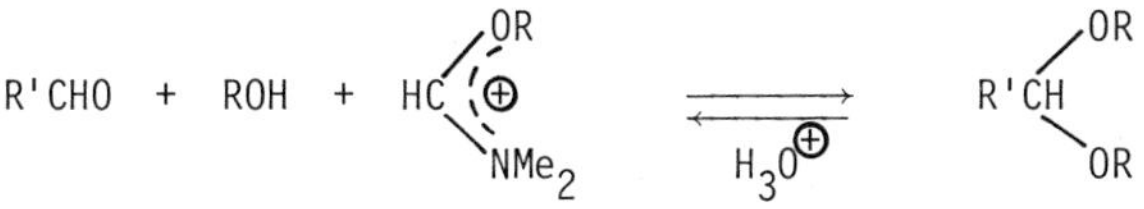

Liebigs Ann Chem (1974) 690

Related methods: Protection of Ketones - Section 180A, Enol Ethers -
                   Section 367

Chapter 5 PREPARATION OF ALKYLS METHYLENES AND ARYLS

This chapter lists the conversion of functional groups into Me, Et...,
CH_2, Ph, etc.

Section 61   Alkyls, Methylenes and Aryls from Acetylenes

$$R-C \equiv CH \xrightarrow{\text{FeCl}_2\text{-NaH}} RCH_2CH_3$$

60-90%

R = alkyl, aryl

Chem Lett (1976) 581

$$HC \equiv CCH_2CH_2C \equiv CH \xrightarrow[\substack{CpCo(CO)_2 \\ octane,\ rfx}]{Me_3SiC \equiv CSiMe_3}$$

60%

JACS (1975) 97 5600
Angew Int Ed (1975) 14 712

$BrCH_2C \equiv CH$ + $CH_2 = CCH_2ZnBr$ $\longrightarrow$ 68%
                            |
                            Me   Me

Comptes Rendus C (1975) 280 1389, 1473

Section 62   Alkyls and Aryls from Carboxylic Acids

$$CH_3(CH_2)_{14}COOH \xrightarrow[\Delta]{Me_3Al} CH_3(CH_2)_{14}CMe_3 \qquad 91\%$$

Aust J Chem (1974) 27 1665

Section 63   Alkyls from Alcohols

Reactions in which hydroxyl groups are replaced by alkyl, e.g., ROH → RMe, are included in this section. For the conversion ROH → RH see Section 153 (Hydrides from Alcohols and Phenols)

$$R_3COH \xrightarrow[100-200°]{\text{excess } Me_3Al} R_3CCH_3$$

Aust J Chem (1974) 27 1639

$$PhCH_2OH \xrightarrow[\text{THF}]{TiCl_3-LiAlH_4} PhCH_2CH_2Ph \qquad 78\%$$

JOC (1975) 40 2687

Section 64   Alkyls from Aldehydes

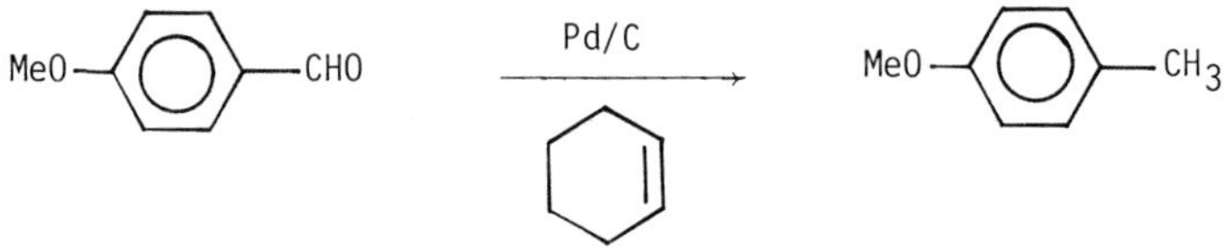

MeO—⟨benzene⟩—CHO $\xrightarrow{\text{Pd/C}}$ MeO—⟨benzene⟩—CH_3 80%

JCS Chem Comm (1976) 757

Cp_2TiCl_2 +   Na $\xrightarrow{\text{PhH}}$ $[Cp_2Ti]_{1-2}$ $\xrightarrow{\text{RCHO}}$ RCH_3

JACS (1974) 96 5290

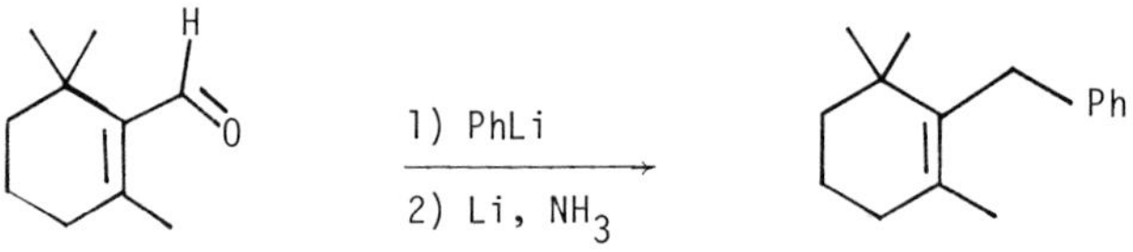

$\xrightarrow[\text{2) Li, NH}_3]{\text{1) PhLi}}$ 99%

JOC (1976) 41 3465

PhCHO   +   [cyclohexene structure] $\xrightarrow[\text{BF}_3\cdot\text{Et}_2\text{O}]{\text{Zn}}$ [bicyclic cyclopropyl structure]—Ph   43%

Tetrahedron (1975) 31 2785

Review: "New Alkylation Methods Using Cyclopropyl Ylides"

Accts Chem Res (1974) 7 85

Related methods: Alkyls and Methylenes from Ketones (Section 72)

Section 65   Alkyls and Aryls from Alkyls and Aryls

Review: "Photosubstitution Reactions of Aromatic Compounds"

Chem Rev (1975) 75 353

Section 66   Alkyls, Methylenes and Aryls from Amides

No additional examples

Section 67   Alkyls, Methylenes and Aryls from Amines

No additional examples

Section 68   Alkyls, Methylenes and Aryls from Esters

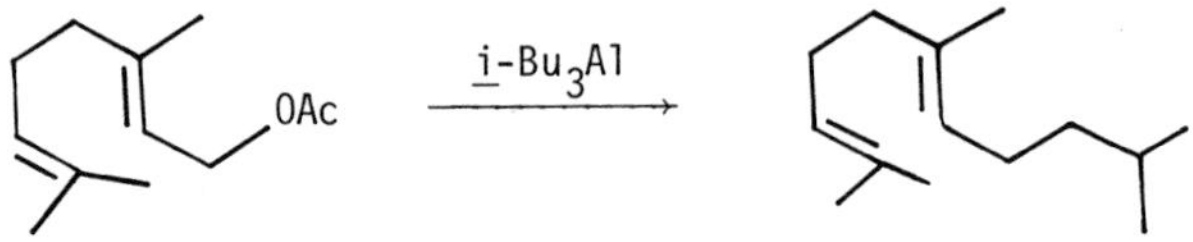

Tetr Lett (1976) 2615

66%

$$Cp_2TiCl_2 \ + \ Na \ \xrightarrow{\text{PhH}} \ [Cp_2Ti]_{1-2} \ \xrightarrow{\text{RCOOR'}} \ RCH_3$$

JACS (1974) 96 5290

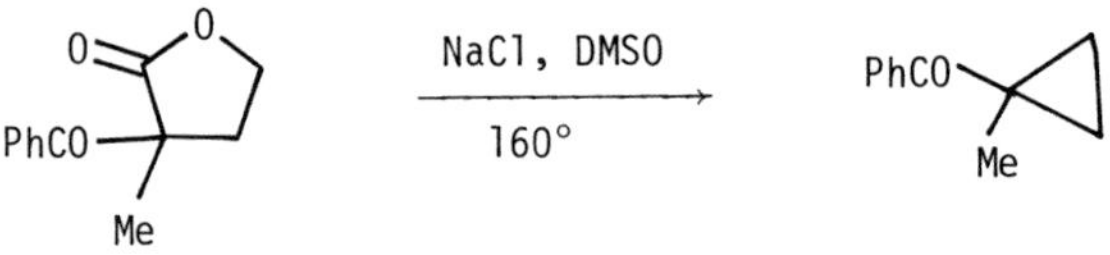

Tetr Lett (1975) 4389
Chem Lett (1975) 1149

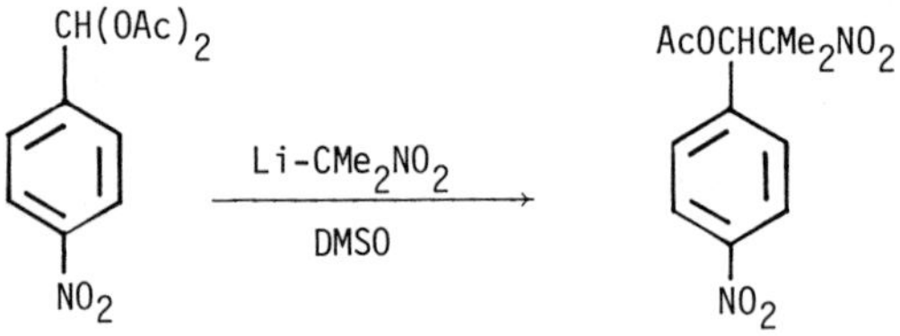

82%

Tetr Lett (1975) 431

80%

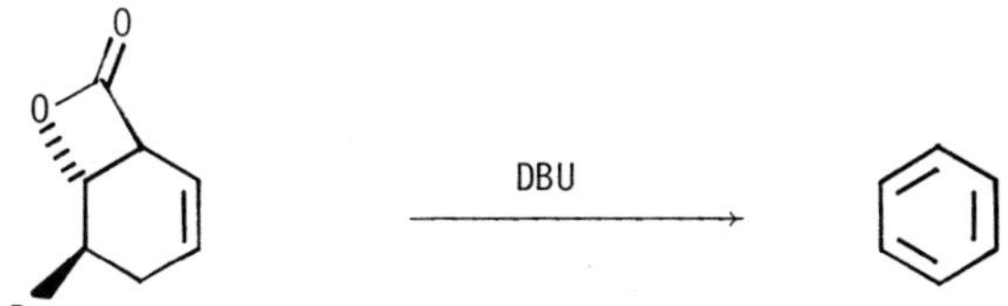

DBU

~100%

Tetr Lett (1976) 4435

Section 69   Alkyls and Aryls from Ethers

The conversion ROR → RR' (R'=alkyl, aryl) is included in this section.

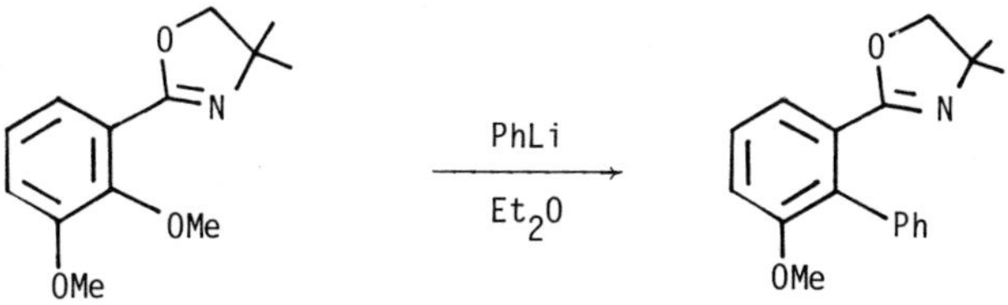

PhLi

Et$_2$O

100%

JACS (1975) 97 7383

$Me_2C=CHCH_2OEt$ $\xrightarrow[\text{.5\% CuBr}]{\text{n-}C_7H_{15}MgCl}$ $Me_2C=CHCH_2C_7H_{15}$ 80%

Tetr Lett (1975) 3837

Section 70   Alkyls and Aryls from Halides

The replacement of halogen by alkyl or aryl groups is included in this
section. For the conversion RX → RH (X=halo) see Section 160 (Hydrides
from Halides and Sulfonates)

$$\xrightarrow[\text{Et}_2\text{O}]{(\underline{n}\text{-Bu})_2\text{CuLi}}$$

80%

Tetr Lett (1976) 1161

$$\underline{n}\text{-C}_5\text{H}_{11}\text{Br} \xrightarrow{(\underline{\text{sec}}\text{-Bu})_2\text{CuLi}} \text{3-Methyloctane} \qquad 94\%$$

(alkenyl, alkynyl, aryl, benzyl, allyl, and propargyl halides also work well)

Org React (1975) 22 253

$$\underline{n}\text{-C}_8\text{H}_{17}\text{CH=C=CHBr} \xrightarrow[\text{2) MeI}]{\text{1) BuLi}} \underline{n}\text{-C}_8\text{H}_{17}\text{CH=C=CHCH}_3 \qquad 90\%$$

JCS Chem Comm (1975) 561

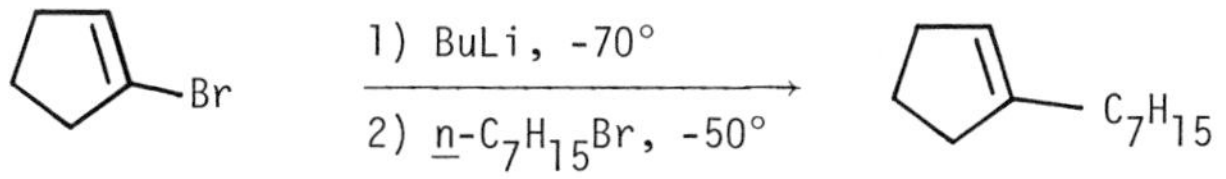

Br
1) BuLi, -70°
2) n-C7H15Br, -50°
C7H15

75%

Synthesis (1975) 434

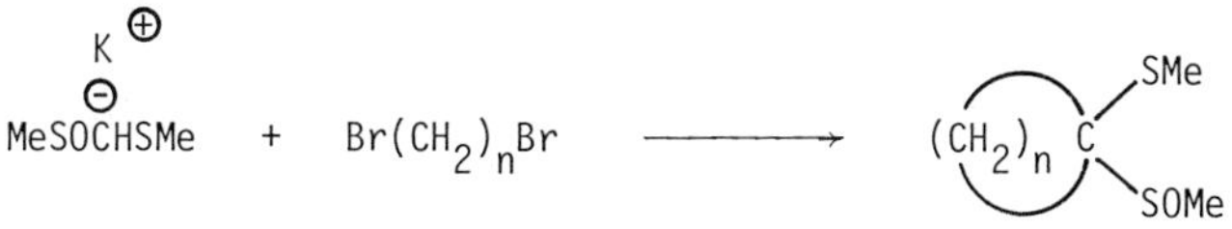

K⊕
⊖
MeSOCHSMe
+
Br(CH2)nBr
(CH2)n
C
SMe
SOMe

~80%

n = 3-5

Tetr Lett (1975) 2767

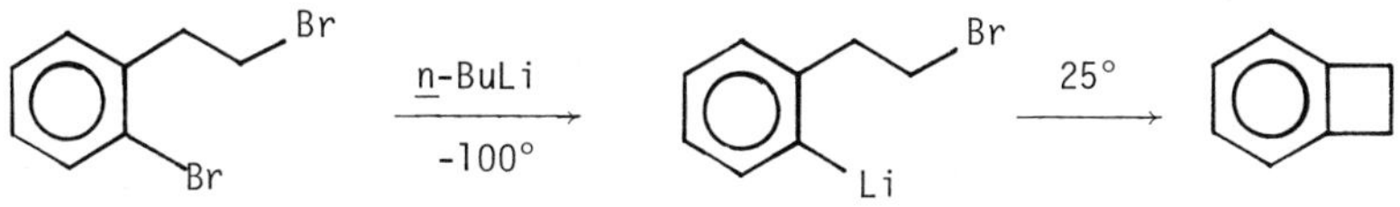

Br
Br
n-BuLi
-100°
Br
Li
25°

68%

JOC (1976) 41 1184

$Me_3SiC{\equiv}CCH_2N{=}CHPh$ —— 1) BuLi, THF; 2) $HC{\equiv}CCH_2Br$ ——→ $Me_3SiC{\equiv}CCH(CH_2C{\equiv}CH){-}N{=}CHPh$ 62%

Tetr Lett (1975) 3337

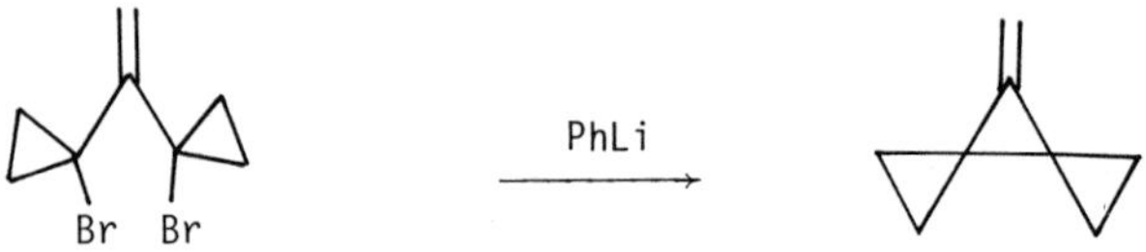

PhLi
Br Br

52%

Angew Int Ed (1976) 15 762

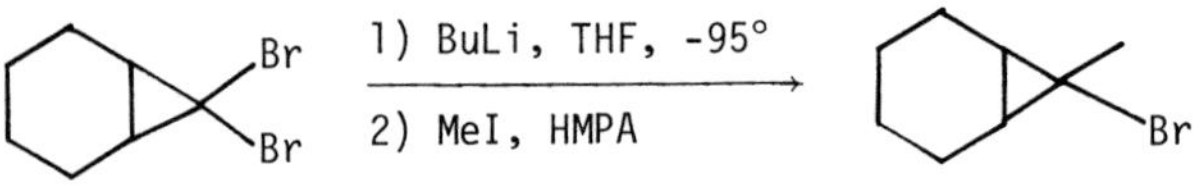

1) BuLi, THF, -95°
2) MeI, HMPA
Br
Br
Br

86%

JACS (1975) 97 949

Cl
BuLi
THF
-10°

62%

Tetr Lett (1974) 1207

Bu
Et
I
RLi
Mn2+
Bu
Et
R

85-90%

J Organometal Chem (1976) 113 99

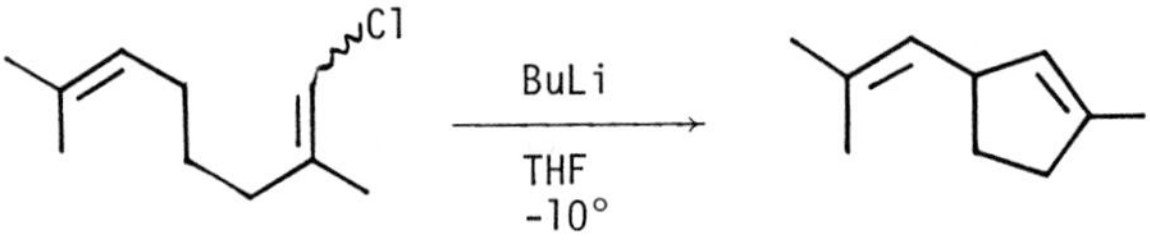

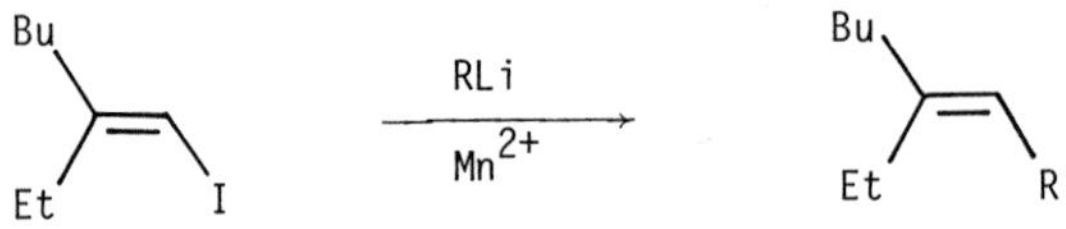

$$\text{EtMgBr} \; + \; C_6H_{11}\text{OTs} \quad \xrightarrow[\text{-78° to 25°}]{\text{Li}_2\text{CuCl}_4} \quad \text{Et-}C_6H_{11} \qquad\qquad 98\%$$

Angew Int Ed (1974) <u>13</u> 82

$\text{PhCH}_2\text{Li} \; +$ $\xrightarrow{\text{Et}_2\text{O}}$ 86%

JOC (1974) <u>39</u> 1168

$$\underline{t}\text{-Bu-CH-CHCl} \; + \; \underline{n}\text{-BuLi} \quad \longrightarrow \quad \underline{t}\text{-Bu-CH-CH-}\underline{n}\text{-Bu}$$

Tetr Lett (1974) 951

$\xrightarrow{\text{TiCl}_3/\text{LiAlH}_4}$ 85%

Synthesis (1976) 607

$$\underline{t}\text{-BuN-Me} \quad \xrightarrow[\text{2) PhCH}_2\text{Br}]{\text{1) LDA, THF, -80°}} \quad \underline{t}\text{-BuNCH}_2\text{CH}_2\text{-Ph} \qquad 95\%$$
$$\qquad\;\; | \qquad\qquad\qquad\qquad\qquad\qquad\qquad\qquad\;\; |$$
$$\qquad \text{NO} \qquad\qquad\qquad\qquad\qquad\qquad\qquad\qquad \text{NO}$$

Chem Ber (1975) <u>108</u> 1293

$(PhS)_2CH_2$ +   n-BuBr $\xrightarrow[\text{THF, HMPA}]{\text{NaNH}_2}$ $(PhS)_2CBu_2$ 84%

Synthesis (1975) 387

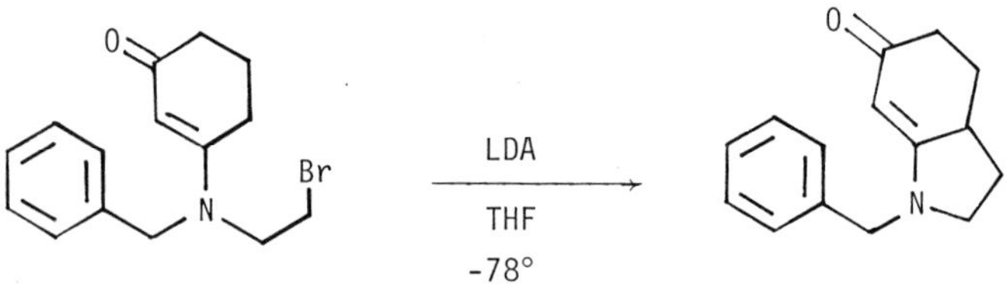

LDA
THF
-78°

35%

Tetr Lett (1974) 3963

$Bu_3\overset{\ominus}{B}CH_3 \overset{\oplus}{Li}$

+ $\xrightarrow{\text{CuCN}}$ $PhCH_2Bu$ 68%

$PhCH_2Br$

Synthesis (1976) 618

1) Mg, THF
2) AgTf

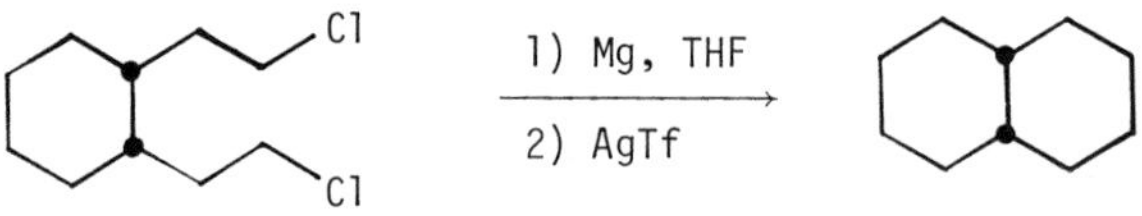

57%

JOC (1976) 41 2882

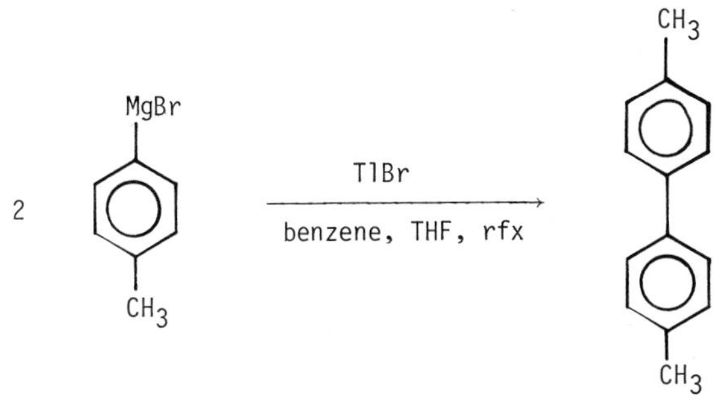

Org Synth (1976) <u>55</u> 48

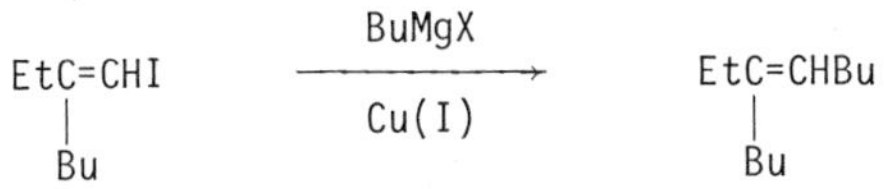

Comptes Rendus (1974) <u>278</u> 967

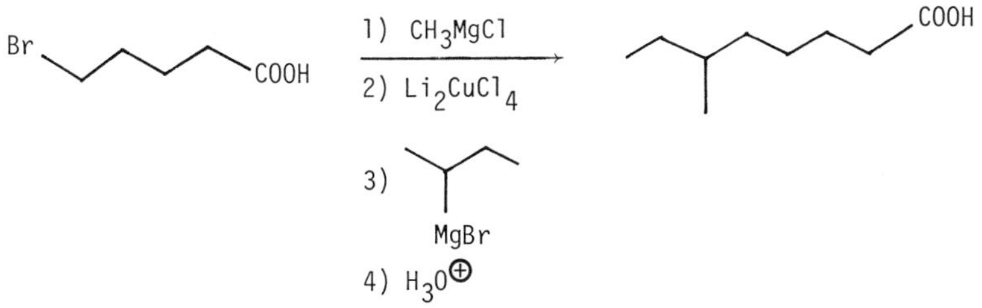

Tetr Lett (1976) 4697

1) Li, Et$_2$O

2) CuI

3) n-octyl-I, HMPA

~90%

Org Synth (1976) 55 103

1) Cu/quinoline, 200°

2) CH$_3$OH

55%

Synthesis (1976) 40

$$\underset{\underset{Br}{|}}{\overset{\overset{NNHTs}{||}}{PhCCH_2}} \quad \xrightarrow[Et_2O/THF]{PhCu,\ -60°} \quad \overset{\overset{NNHTs}{||}}{PhC\text{-}CH_2Ph}$$

81%

JACS (1975) 97 7372

$$t\text{-BuCO-C}(t\text{-Bu})Br_2 \quad + \quad Me_2CuLi \quad \longrightarrow \quad \xrightarrow{H_2O} \quad \underset{\underset{t\text{-Bu}}{|}}{t\text{-BuCOCH(Me)}}$$

90%

Comptes Rendus (1975) 280 217

t-BuCHBrCOCHBr-t-Bu + Me$_2$CuLi

$\xrightarrow{\text{H}_2\text{O}}$ t-Bu(Me)CHCOCH$_2$ 85%
                                             |
                                           t-Bu

$\xrightarrow{\text{MeI}}$ t-Bu(Me)CHCOCH(Me)t-Bu   80%

Tetrahedron (1975) 31 1223, 1227

TsO～～～～ $\xrightarrow{\underline{n}\text{-Bu}_2\text{CuLi}}$ n-Bu ～～～～   94%

Tetrahedron (1976) 32 2281

Cl
|
Ph-C=CH$_2$ +   t-BuLi $\longrightarrow$ Ph$\diagdown$ $\diagup$H
                                                       C=C
                                                  H$\diagup$ $\diagdown$t-Bu   45%

Tetr Lett (1974) 2935

▷-Li, PhSCu

93%

Tetr Lett (1976) 3233, 3241, 3245

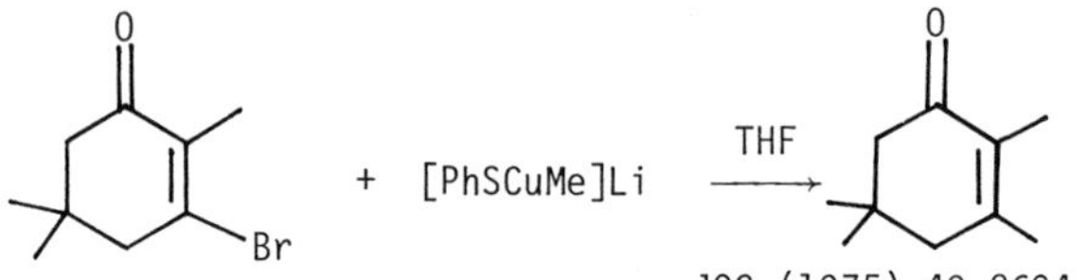

+ [PhSCuMe]Li $\xrightarrow{\text{THF}}$

82%

JOC (1975) <u>40</u> 2694

BzBr $\xrightarrow[\text{benzene}]{\text{AgClO}_4}$ Ph-CH$_2$-Ph

52%

Coll Czech (1976) <u>41</u> 1777

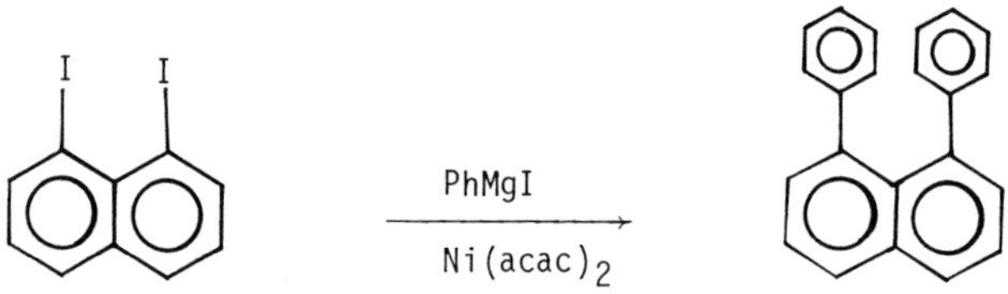

$\xrightarrow[\text{Ni(acac)}_2]{\text{PhMgI}}$

70%

JOC (1976) <u>41</u> 2252

$(Ph_3P)_2NiCl_2$ $\xrightarrow[\text{2) ArX}]{\text{1) } Ph_3P, \text{ Zn, DMF}}$ Ar-Ar   50-80%

Tetr Lett (1975) 3375
JCS Perkin I (1975) 121
      [Pd(OAc)$_2$ catalyst]
JACS (1975) <u>97</u> 3873
      (intramolecular biaryl formation)

PhMgX + Mesityl Br $\xrightarrow{Ni(dpp)_2Cl_2}$ Ph-mesityl   78%

Chem Lett (1975) 133

1) PdL$_4$, PhH
2) MeLi, Et$_2$O   88%

J Organometal Chem (1975) <u>91</u> C39

$(Ph_3P)_2Pd(Ph)I$ 51%

J Organometal Chem (1976) <u>118</u> 349

$PhSO_2CHBr_2$ + $PhCH_2Cl$ $\xrightarrow[\text{TEBA}]{\text{50\% aq. NaOH}}$ $PhSO_2\underset{\underset{CH_2Ph}{|}}{C}Br_2$ 75%

JOC (1975) 40 266

$\xrightarrow[\substack{Et_2O \\ 0°}]{t\text{-BuOK}}$ 80%

JACS (1974) 96 7355

$(allyl\text{-}SPh)^{\ominus}$ +   allyl'Cl $\longrightarrow$ biallyl

Synthesis (1974) 129
Tetrahedron (1974) 30 715

Review: "Oxidative Coupling via Organocopper Compounds"

Angew Int Ed (1974) 13 291

Review: "The Ullmann Synthesis of Biaryls"

Synthesis (1974) 9

Review: "Activation of Grignard Reagents by Transition Metal Compounds"

Tetrahedron (1975) $\underline{31}$ 2735

Section 71   Alkyls and Aryls from Hydrides

This section lists examples of the reaction RH → RR' (R,R'=alkyl or aryl).
For the reaction C=CH → C=CR (R=alkyl or aryl) see Section 209 (Olefins
from Olefins).

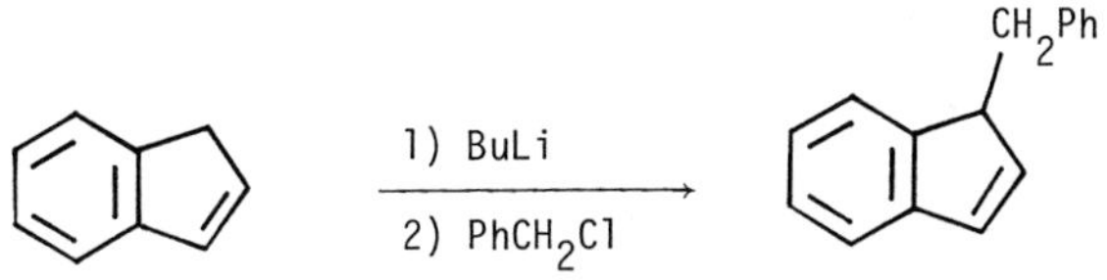

1) BuLi
2) $PhCH_2Cl$

84%

Acta Chem Scand (1974) $\underline{B28}$ 295

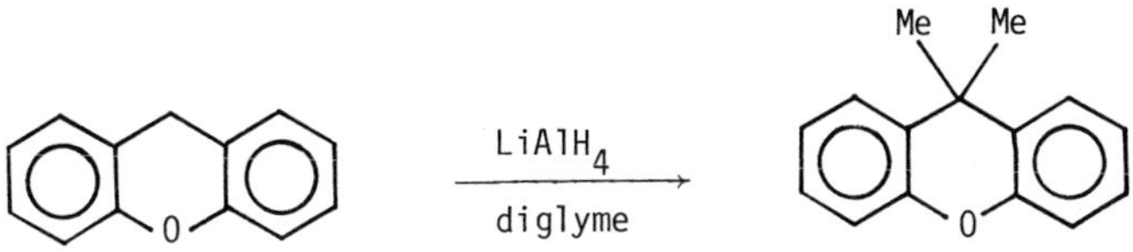

$LiAlH_4$
diglyme

50%

JCS Perkin I (1976) 2380

1) LDA, THF
   HMPA, -78°
2) CH$_2$=C(Me)(CH$_2$)$_2$-I

60%

JACS (1975) <u>97</u> 7152

1) BuLi
2) RX

80-95%

JOC (1975) <u>40</u> 2021

Ph$_2$TlOCOCF$_3$ $\xrightarrow[\text{benzene}]{h\nu}$ Ph-Ph

95%

JOC (1975) <u>40</u> 2351

PhCr(CO)$_3$ $\xrightarrow[\text{2) I}_2]{\text{1) }\underline{t}\text{-BuLi}}$ Ph-C(CH$_3$)$_3$

97%

JACS (1975) <u>97</u> 1247

2 PhHgOAc + Cu $\xrightarrow[\text{PdCl}_2,\ 115°C]{\text{pyridine}}$ Ph-Ph

86%

JOC (1976) <u>41</u> 2661

$$PhCH_2\overset{\overset{S}{\|}}{S}CNMe_2 \quad \xrightarrow[\text{THF, } -60°]{\text{LDA}} \quad \xrightarrow[\text{THF}]{\text{MeI}} \quad PhCHMeS\overset{\overset{S}{\|}}{-}C-NMe_2 \qquad 99\%$$

$$\xrightarrow[\text{HMPA/THF}]{\text{MeI}} \quad MeSCH\overset{\overset{S}{\|}}{-}CNMe_2 \qquad 90\%$$
$$\underset{Ph}{|}$$

JACS (1975) 97 1608

Review: "Alkylation and Arylation of Unsaturated Compounds with
         the Aid of Transition Metal Complexes"

Russ Chem Rev (1975) 44 552

Section 72   Alkyls and Methylenes from Ketones

The conversions $R_2CO \rightarrow RR$, R_2CH_2, R_2CHR', etc. are listed in this section.

1) n-BuLi
2) HexBr
3) Li, Et_3N

65%

Tetr Lett (1976) 2643

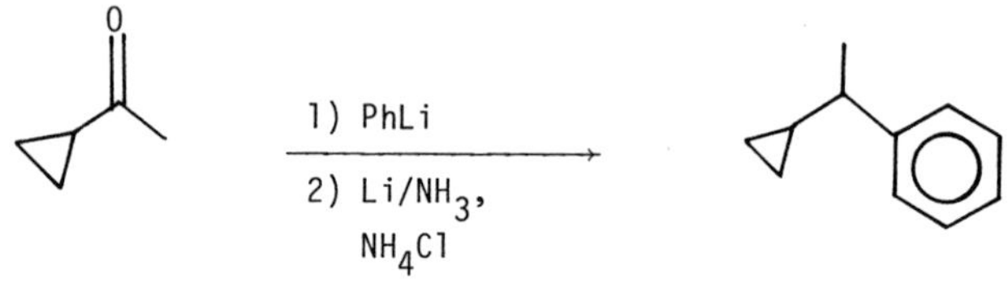

1) PhLi

2) Li/NH$_3$, NH$_4$Cl

91%

JOC (1975) <u>40</u> 271

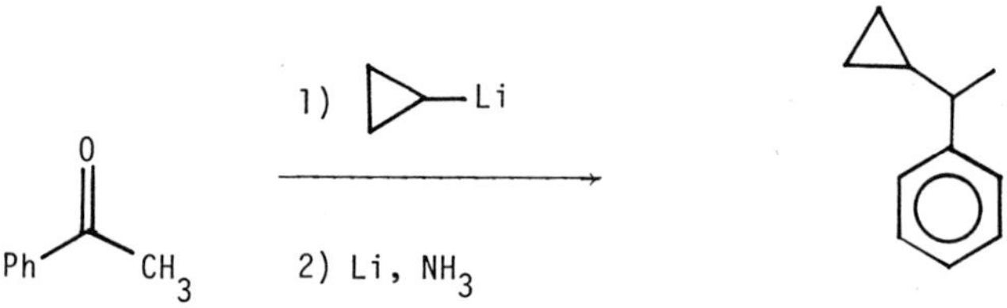

1) [cyclopropyl]-Li

2) Li, NH$_3$

82%

JOC (1976) <u>41</u> 1494

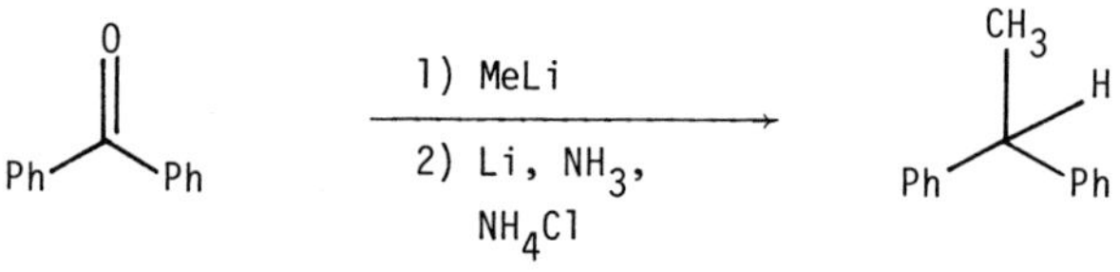

1) MeLi

2) Li, NH$_3$, NH$_4$Cl

Org Synth (1976) <u>55</u> 7

RCOR'

1) PhLi, Et$_2$O

2) Li, liq. NH$_3$, NH$_4$Cl

RR'CHPh   (20 examples)

JOC (1975) <u>40</u> 271
Synth Comm (1975) <u>5</u> 441

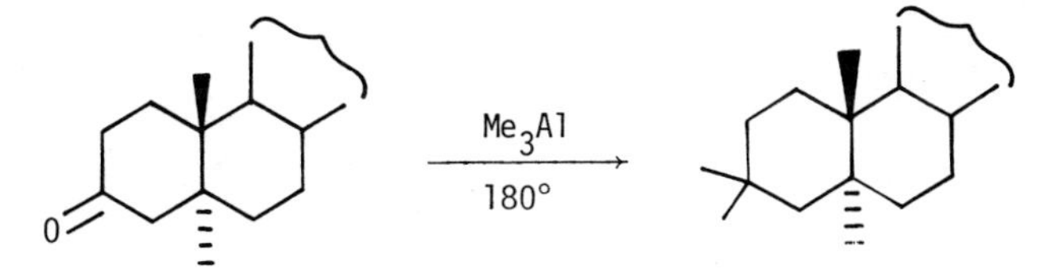

$$\xrightarrow[180°]{Me_3Al}$$

78%

Aust J Chem (1974) 27 1655

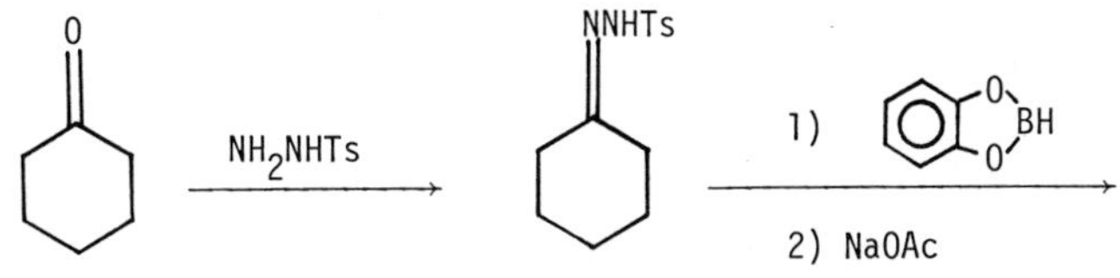

$$\xrightarrow{NH_2NHTs}$$

$$\xrightarrow[\text{2) NaOAc}]{\text{1)}}$$

92%

JOC (1975) 40 1834

$(CH_2)_5\ (CHMe)_n$

$$\xrightarrow{CH_3CHN_2}$$

$(CH_2)_5\quad (CHMe)_{n+1}$

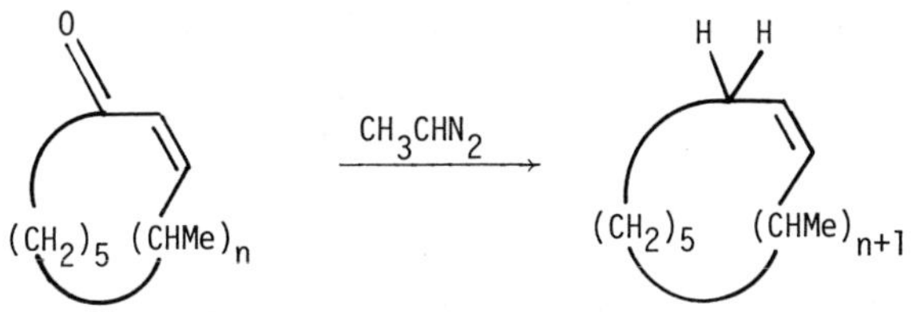

~60%

JCS Chem Comm (1975) 142

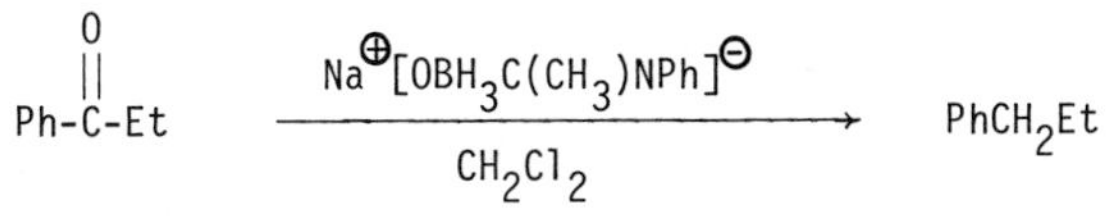

$$\underset{Ph-\overset{O}{\overset{||}{C}}-Et}{} \xrightarrow[CH_2Cl_2]{Na^{\oplus}[OBH_3C(CH_3)NPh]^{\ominus}} PhCH_2Et$$

78%

Chem Pharm Bull (1976) 24 1059

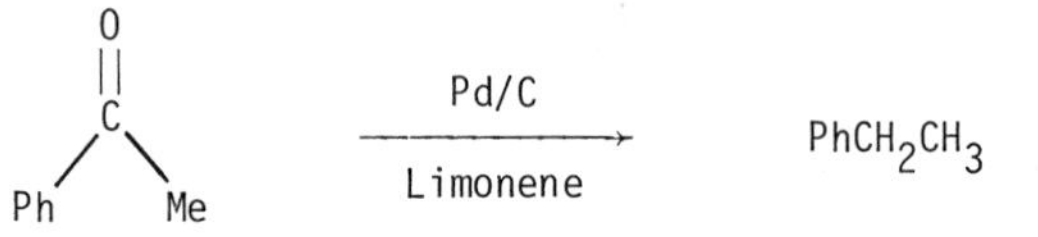

O
C
Ph
Me
Pd/C
Limonene
$PhCH_2CH_3$

100%

JCS Chem Comm (1976) 757

R
C=O
R'
R
C(SeMe)$_2$
R'
Li
Et$_3$N
R
CH$_2$
R'

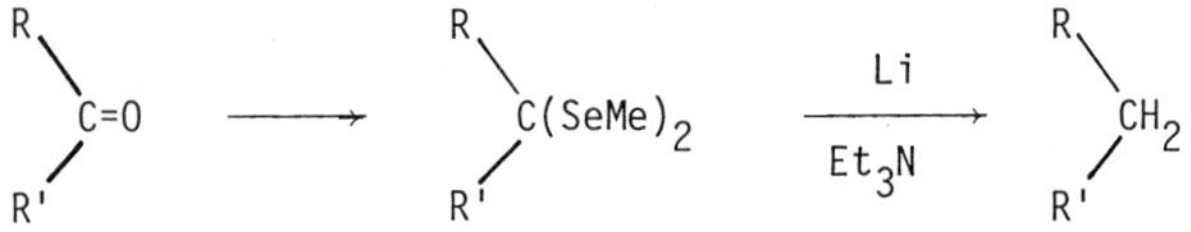

~65%

R,R' = n-alkyl

Tetr Lett (1976) 2643

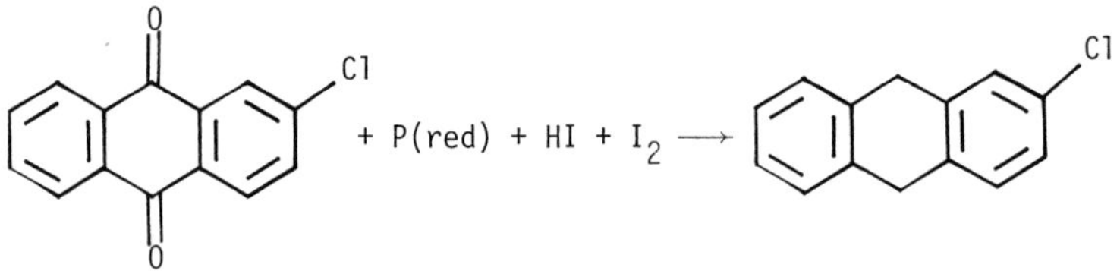

O
Cl
O
+ P(red) + HI + I$_2$
Cl

90%

Can J Chem (1974) 52 1229

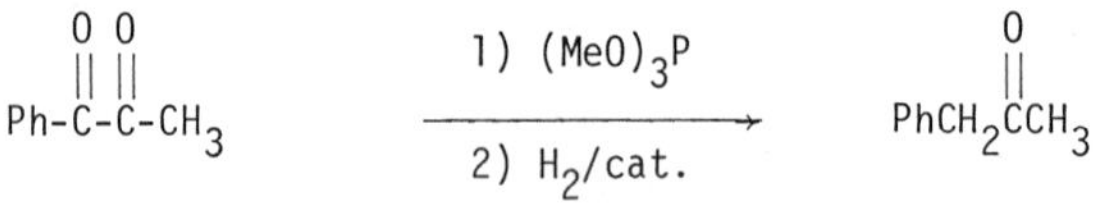

O O
Ph-C-C-CH$_3$
1) (MeO)$_3$P
2) H$_2$/cat.
O
PhCH$_2$CCH$_3$

100%

JOC (1976) 41 2928

Related methods: Alkyls from Aldehydes (Section 64)

Section 73   Alkyls, Methylenes and Aryls from Nitriles

No additional examples

Section 74   Alkyls, Methylenes and Aryls from Olefins

The following reaction types are included in this section:

1. Hydrogenation of olefins (and aryls).
2. Dehydrogenations to form aryls.
3. Alkylations and arylations of olefins.
4. Conjugate reductions of conjugated aldehydes, ketones,
    acids, esters and nitriles.
5. Conjugate alkylations.
6. Cyclopropanations, including halocyclopropanations.

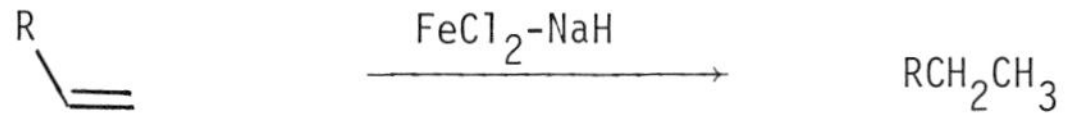

$$\text{R}\diagdown \!=\quad \xrightarrow{\text{FeCl}_2\text{-NaH}}\quad RCH_2CH_3 \qquad 60\text{-}90\%$$

R = alkyl, aryl   Chem Lett (1976) 581

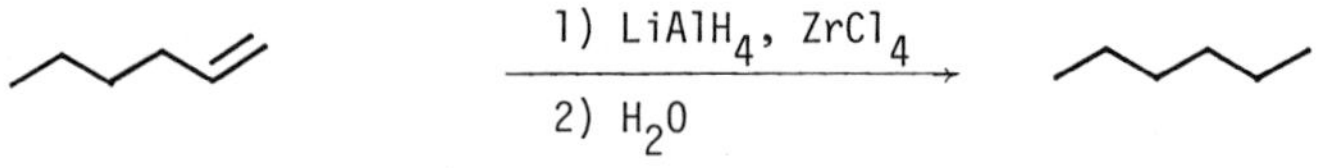

1) LiAlH$_4$, ZrCl$_4$

2) H$_2$O

~90%

J Organometal Chem (1976) 122 C25

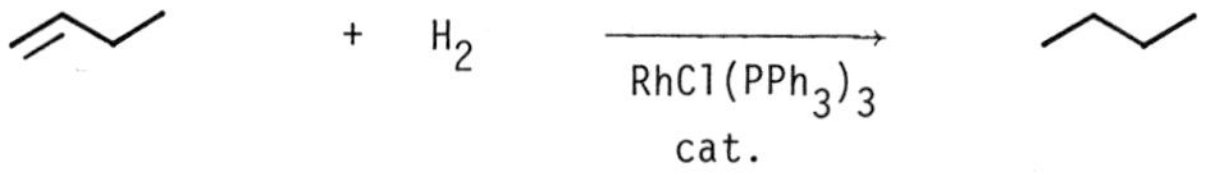

+ H$_2$

RhCl(PPh$_3$)$_3$
cat.

100%

Org React (1976) 24 1

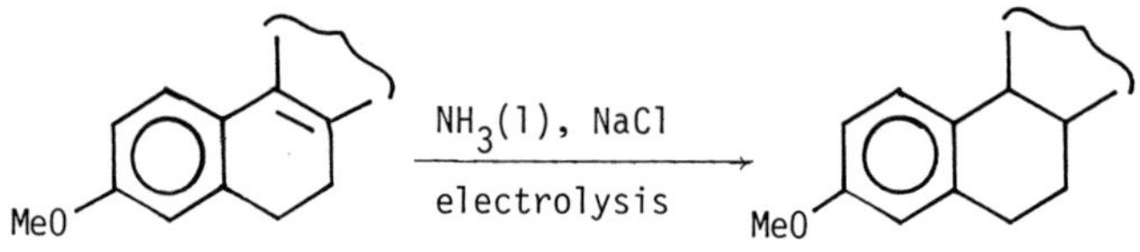

NH$_3$(1), NaCl

electrolysis

Chem Ber (1976) 109 395

H$_2$, NaBH$_4$

NiCl$_2$

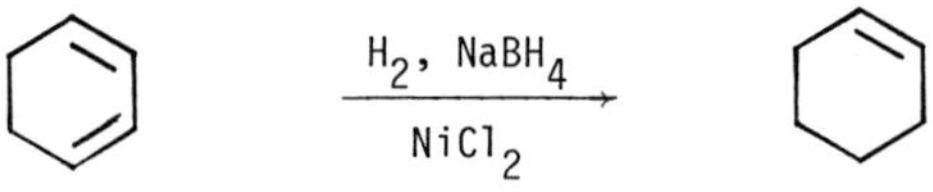

95%

Z Naturforsch B (1975) 30 643

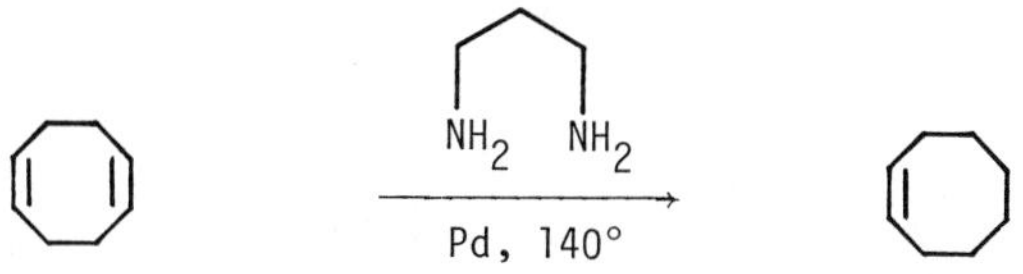

60%

Tetr Lett (1975) 4235

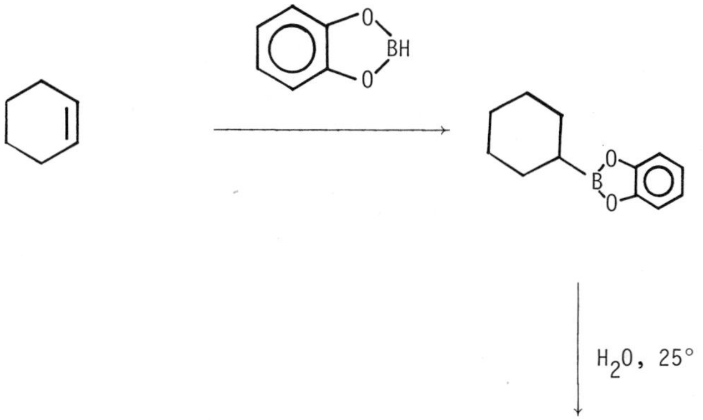

H$_2$O, 25°

cyclohexane

JACS (1975) <u>97</u> 5249

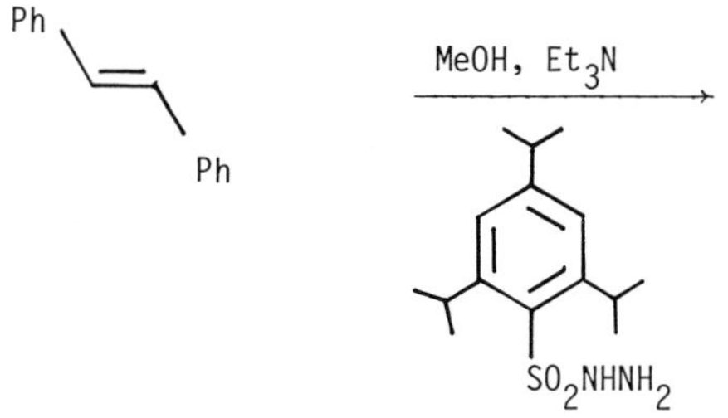

MeOH, Et$_3$N

PhCH$_2$CH$_2$Ph   90%

Tetrahedron (1976) <u>32</u> 2157

$$\text{LiAlH}_4/\text{TiCl}_4$$

$\underline{n}\text{-octane}$

Tetr Lett (1976) 15

P —⟨◯⟩— iminodiacetate + PdCl$_2$

Catalyst for reduction of conjugated diolefins to monoolefins at 30° and 1 atm. H$_2$.

Chem Lett (1976) 165

Use of for catalytic hydrogenation of olefins. Rate is 20-30 times greater than that for the free catalyst.

CH$_2$

TiCl$_n$ J Organometal Chem (1976) 120 49

+ H$_2$ + π-allylCo[P(OMe)$_3$]$_3$ $\xrightarrow[\;25°\;]{\;1\ \text{atm.}\;}$ ◯

catalyst

Org React (1976) 24 1

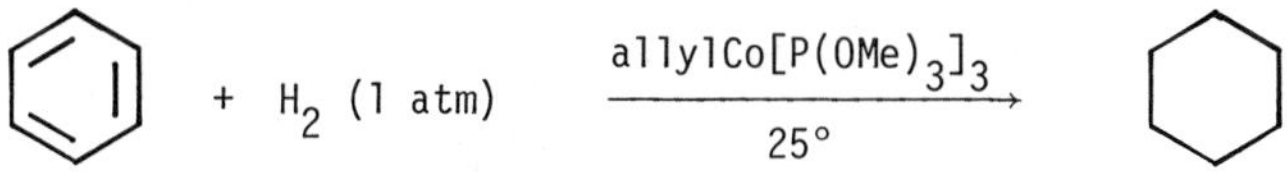

+ H$_2$ (1 atm) $\xrightarrow[25°]{\text{allylCo[P(OMe)}_3]_3}$

JACS (1974) 96 4063

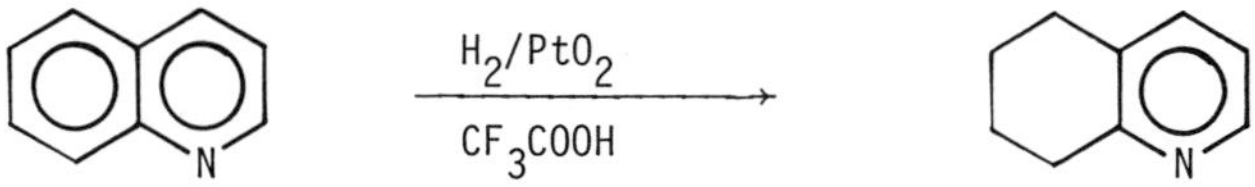

$\xrightarrow[\text{CF}_3\text{COOH}]{\text{H}_2/\text{PtO}_2}$

70%

JOC (1975) 40 2729

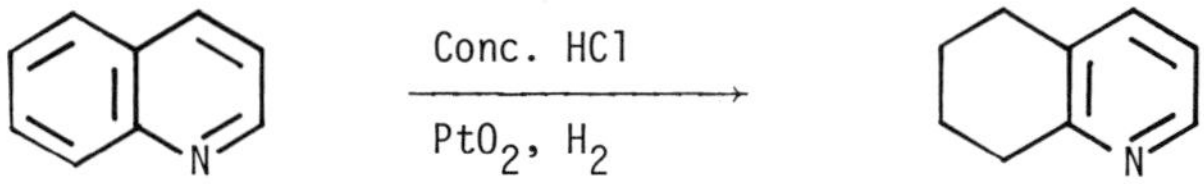

$\xrightarrow[\text{PtO}_2, \text{H}_2]{\text{Conc. HCl}}$

70%

JACS (1974) 96 2256

$\xrightarrow[\substack{\text{MeOH} \\ \text{varying HCl}}]{\text{H}_2/\text{Pt}}$

+

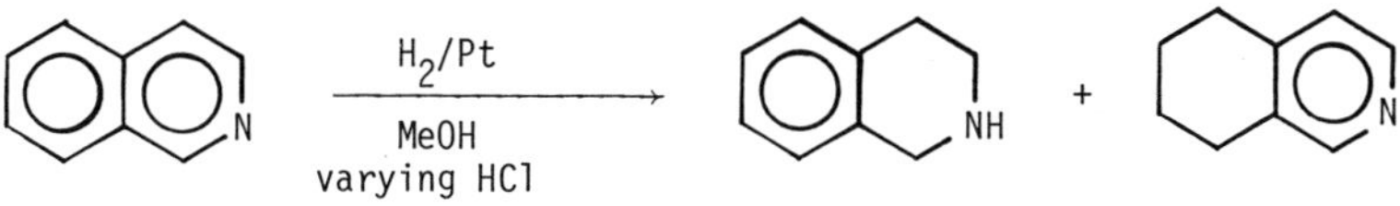

87:13 without HCl
13:87 with 4N HCl

JOC (1975) 40 1191

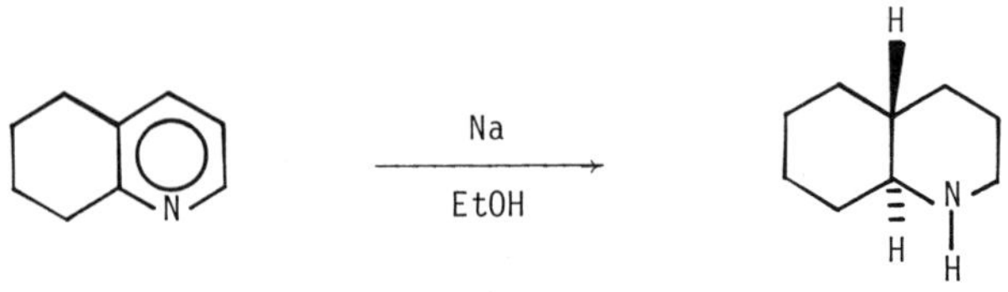

90%

JOC (1975) 40 2734

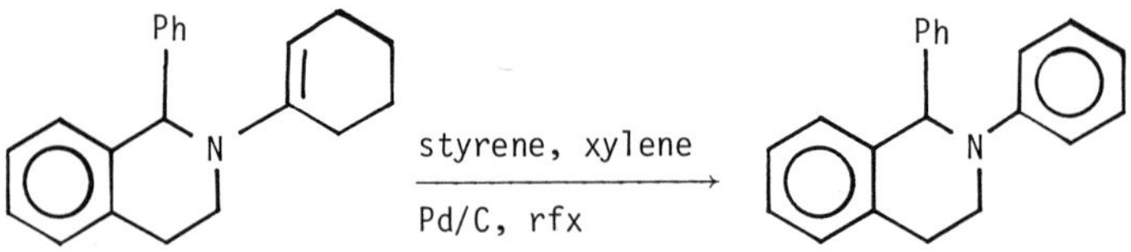

styrene, xylene
Pd/C, rfx

35%

Bull Soc Chim Belges (1976) 85 1

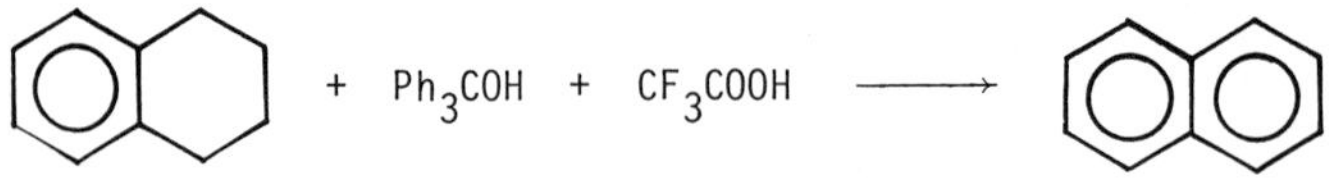

+ Ph₃COH + CF₃COOH ⟶

Tetr Lett (1974) 3217

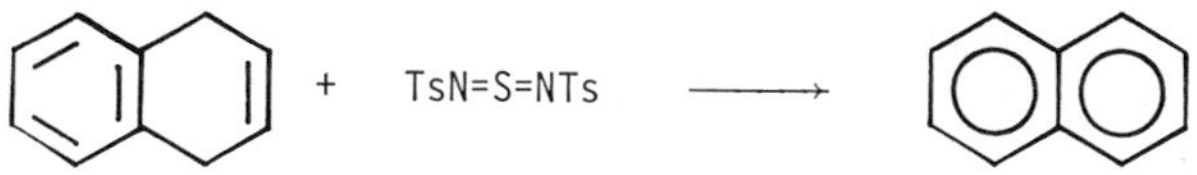

Liebigs Ann Chem (1974) 847

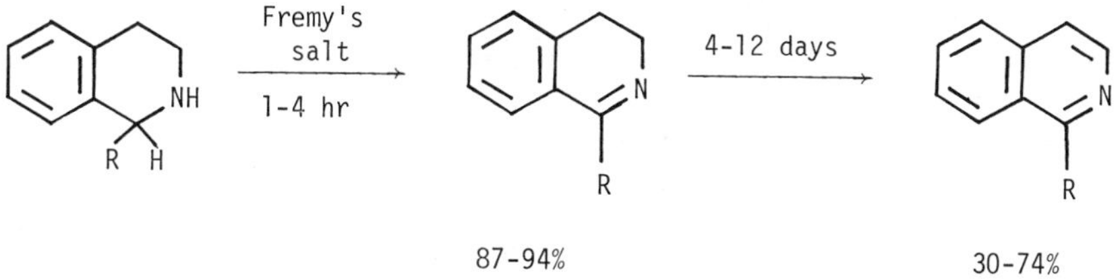

87-94%   30-74%

Synthesis (1974) 288

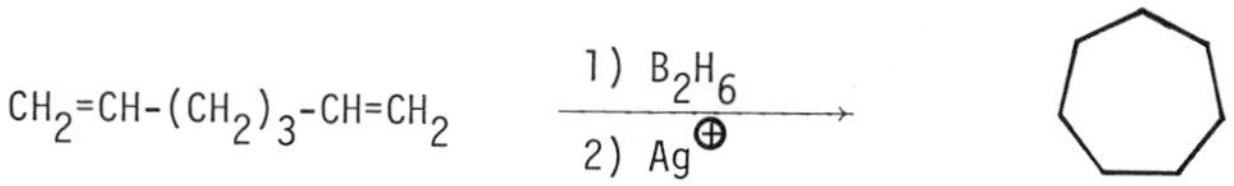

$CH_2=CH-(CH_2)_3-CH=CH_2$ $\xrightarrow{\begin{array}{c}1)\ B_2H_6\\ \hline 2)\ Ag^{\oplus}\end{array}}$ 67%

Tetr Lett (1976) 463

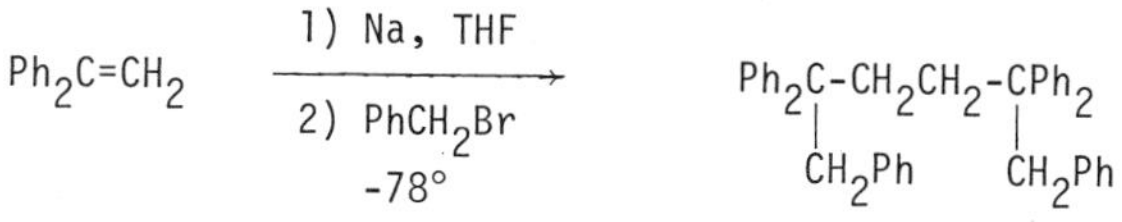

$Ph_2C=CH_2$ $\xrightarrow{\begin{array}{c}1)\ Na,\ THF\\ \hline 2)\ PhCH_2Br\\ -78°\end{array}}$ $Ph_2\underset{\underset{CH_2Ph}{|}}{C}-CH_2CH_2-\underset{\underset{CH_2Ph}{|}}{C}Ph_2$ 95%

JCS Perkin I (1975) 1474

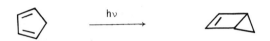

Org Synth (1976) <u>55</u> 15

Review: "Cyclobutanes from Dimerization of Olefins"

Synthesis (1974) 539

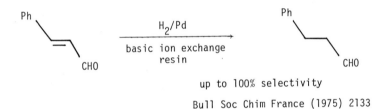

up to 100% selectivity

Bull Soc Chim France (1975) 2133

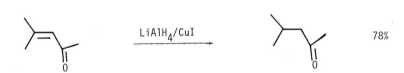

·78%

Tetr Lett (1975) 4453

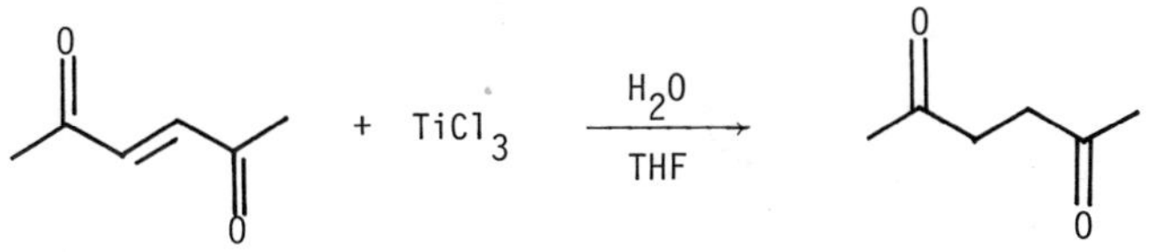

+ TiCl$_3$ $\xrightarrow[\text{THF}]{\text{H}_2\text{O}}$ 86%

JOC (1974) <u>39</u> 258

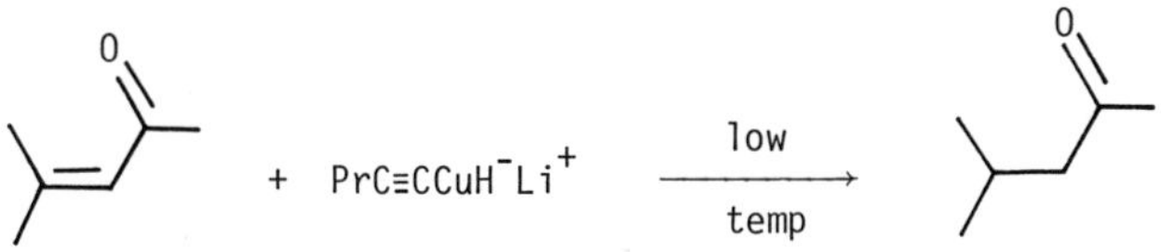

+ PrC≡CCuH$^-$Li$^+$ $\xrightarrow[\text{temp}]{\text{low}}$ 80%

JACS (1974) <u>96</u> 1625

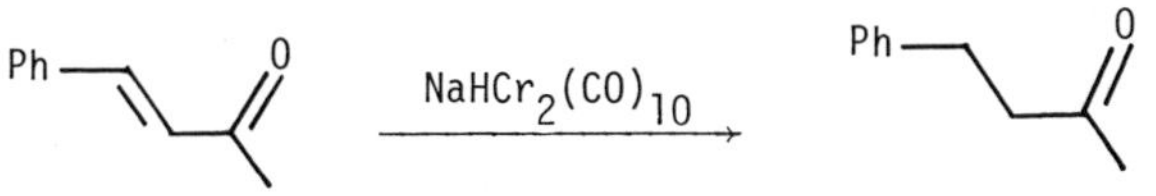

$\xrightarrow{\text{NaHCr}_2\text{(CO)}_{10}}$ 80%

Synthesis (1976) 596

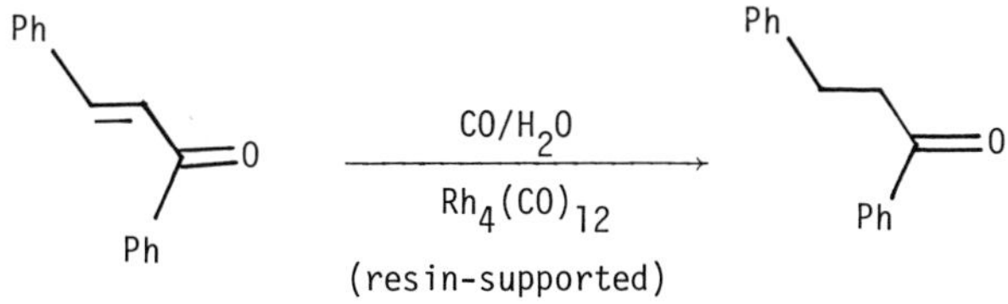

$\xrightarrow[\text{Rh}_4\text{(CO)}_{12}]{\text{CO/H}_2\text{O}}$ 83%

(resin-supported)

Chem Lett (1975) 203

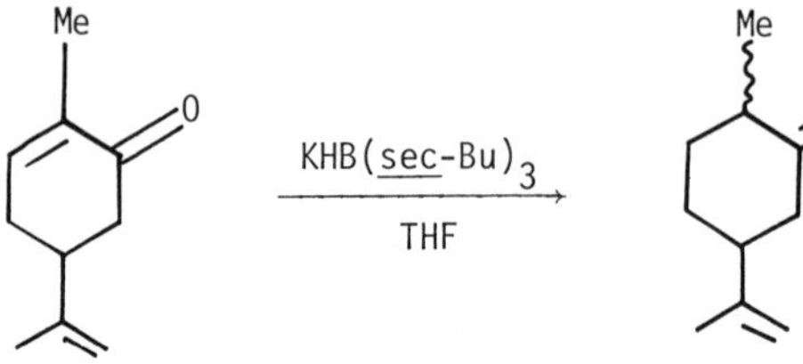

JOC (1975) 40 146

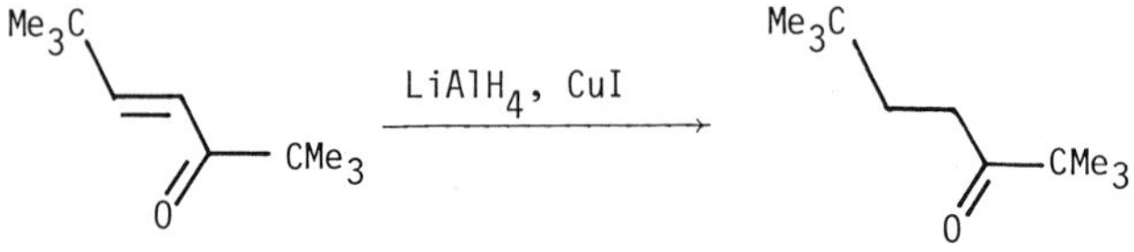

JOC (1976) 41 1939

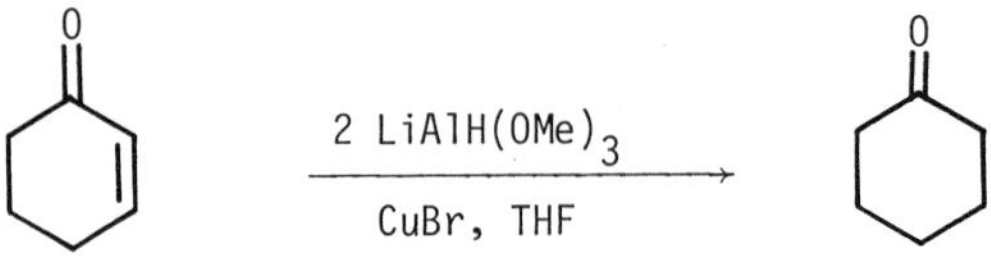

JOC (1975) 40 3619

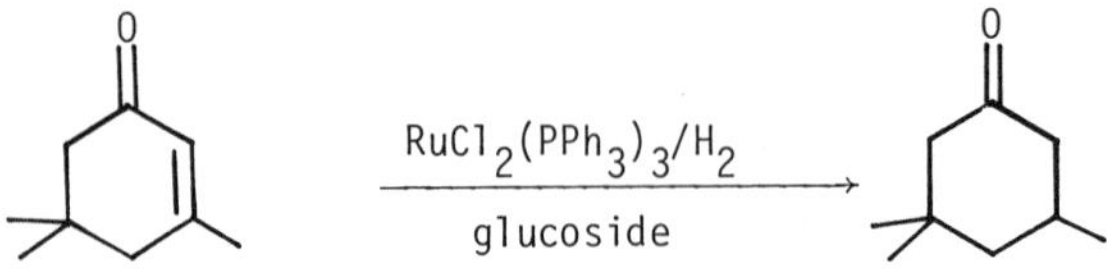

S 34% ee

Tetr Lett (1976) 4083

98%

99%

84%

40%

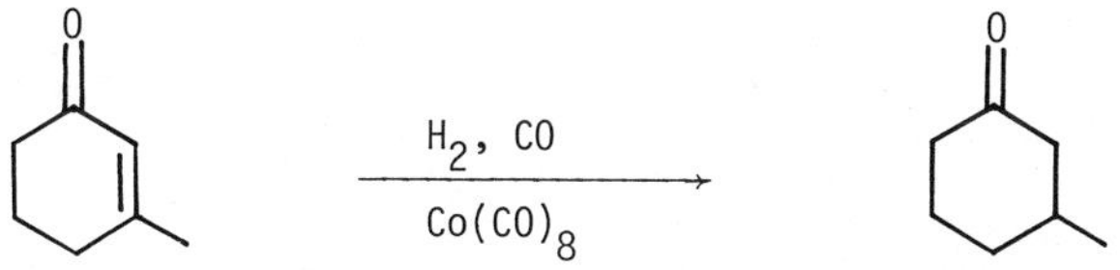

$$\text{H}_2, \text{CO} \quad \xrightarrow{} \quad \text{Co(CO)}_8$$

100%

Comptes Rendus C (1975) 281 877

PhCH=CHCOPh $\xrightarrow[\text{Et}_2\text{O}]{\text{K/NH}_3}$ PhCH$_2$CH$_2$COPh   98%

Org React (1976) 23 1

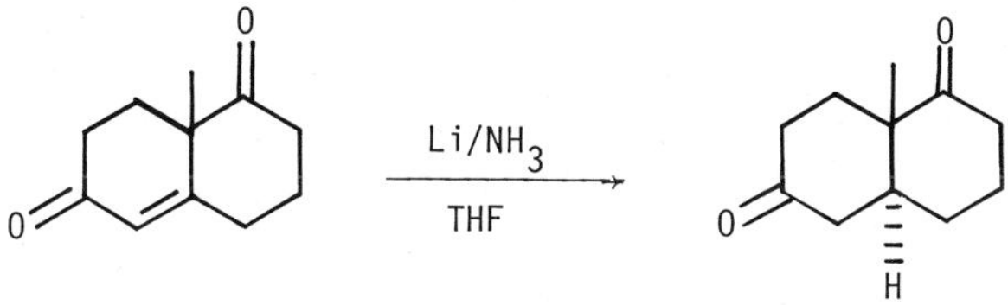

$\xrightarrow[\text{THF}]{\text{Li/NH}_3}$

95%

Org React (1976) 23 1

$\xrightarrow[\text{EtOH}]{\text{Zn/ZnCl}_2}$

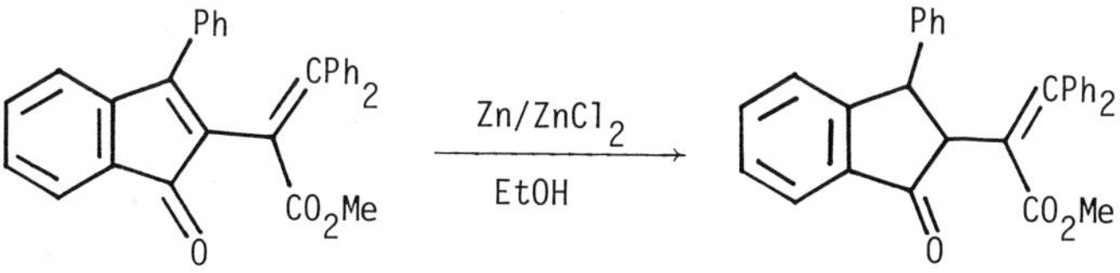

71%

Chem Lett (1976) 695

56%

Synth Comm (1976) 6 113

$$\underline{i}\text{-PrCH=CCOOH} \quad + \quad \text{RhCl (chiral diphos.)} \xrightarrow{\text{H}_2} \quad \underline{i}\text{-PrCH}_2\text{CHCO}_2\text{H}$$

with NHCOPh above the reactant carbon and NHCOPh above the product carbon

cat.

optically active

Org React (1976) 24 1

$$\text{PhCH=C-COOH} \xrightarrow{\text{H}_2, \; \underline{I}} \text{PhCH}_2\text{-CHCOOH}$$

NHCOMe below reactant, NHCOMe below product, * above product

(S 93% ee)

~90%

$$\underline{I} \quad = \quad$$

Tetr Lett (1976) 1133

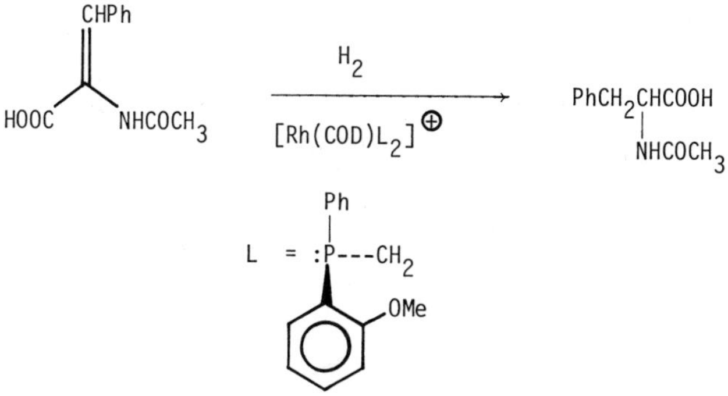

$$PhCH_2CHCOOH \quad >90\% \text{ ee}$$
$$| \atop NHCOCH_3$$

JACS (1975) 97 2567

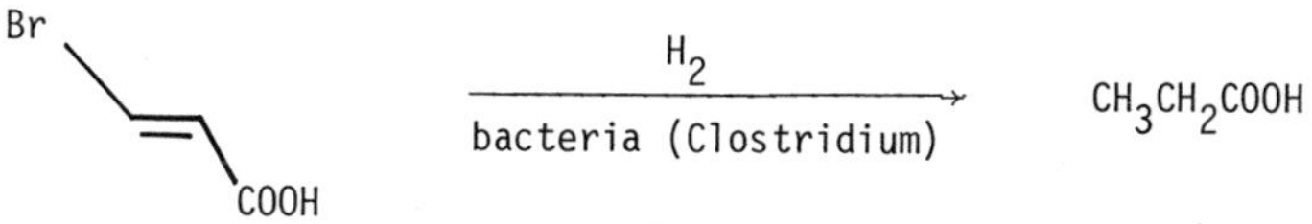

$$CH_3CH_2COOH \quad 100\%$$

Angew Int Ed (1975) 14 106

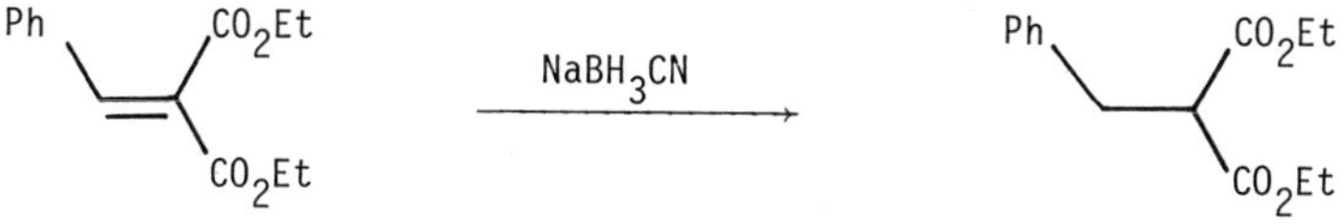

90%

JOC (1976) 41 3279

$$C_6H_{13}CH=CHCOOMe \xrightarrow[\underline{t}\text{-BuOH}]{LiBu_3BH} C_6H_{13}CH_2CH_2COOMe \quad 92\%$$

JOC (1975) 40 2846

$$\underset{\substack{Ph \\ }}{\overset{COCH_3}{\diagup}}\diagdown \overset{COCH_3}{\diagdown} \quad \xrightarrow{\text{KHFe(CO)}_4} \quad PhCH_2CH_2COCH_3 \qquad 76\%$$

Tetr Lett (1975) 1867

$$\underset{\text{Me}}{\overset{\text{CN}}{\diagup}} \quad \xrightarrow[\text{2) HCl}]{\text{1) Mg, MeOH}} \quad \underset{\text{Me}}{\overset{\text{CN}}{\diagdown}} \qquad 77\%$$

JOC (1975) 40 127

$$\underset{NC}{\overset{Ph}{\diagup}}\diagdown\underset{H}{\overset{Ph}{\diagdown}} \quad \xrightarrow{\text{NaBH}_4} \quad \underset{NC}{\overset{Ph}{H}}\overset{Ph}{\underset{H}{C}}-\overset{Ph}{\underset{H}{C}} \qquad 100\%$$

Bull Chem Soc Japan (1976) 49 2643

$$CH_3-(CH_2)_{10}-CHO$$
+
$$\underset{CH_2=CH-\overset{O}{\overset{||}{C}}-CH_3}{}$$

$$\xrightarrow[\text{thiazolium salts}]{\text{Et}_3N} \quad CH_3-(CH_2)_{10}-\overset{O}{\overset{||}{C}}CH_2CH_2\overset{O}{\overset{||}{C}}CH_3 \qquad 76\%$$

Chem Ber (1976) 109 3426

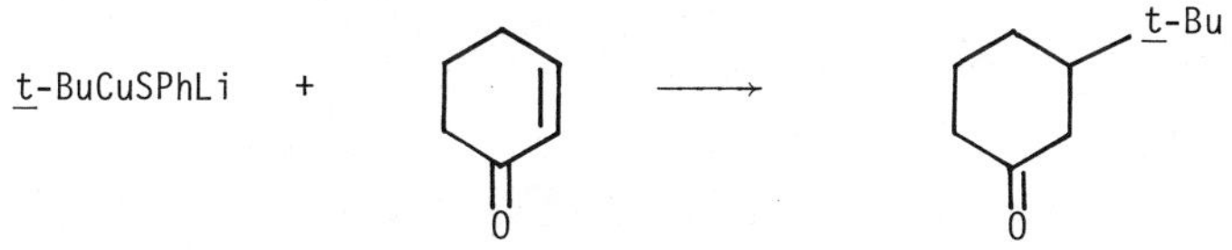

t-BuCuSPhLi +   ⟶   91%

Synthesis (1974) 662

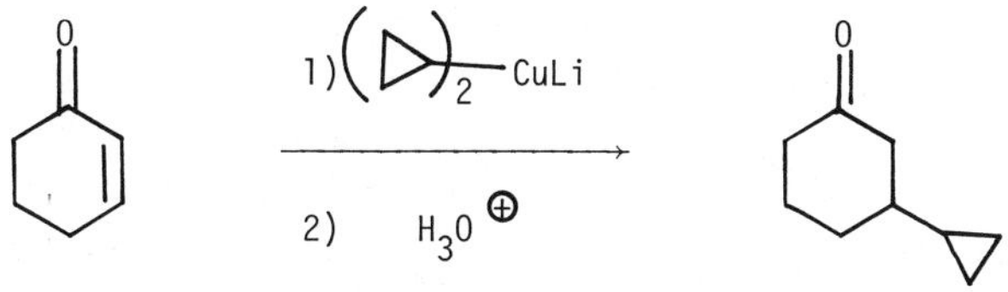

83%

JOC (1976) 41 3629

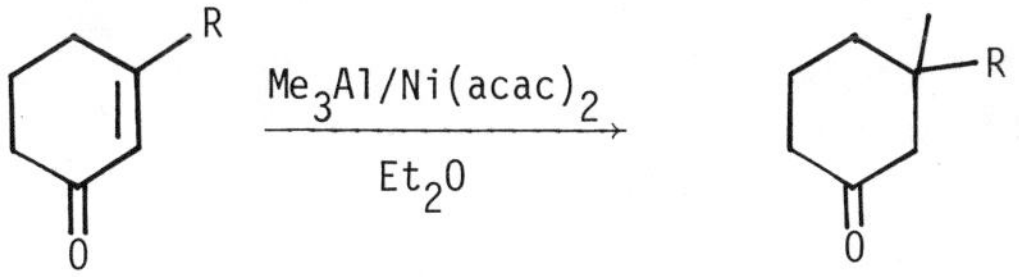

68%

Aust J Chem (1975) 28 801

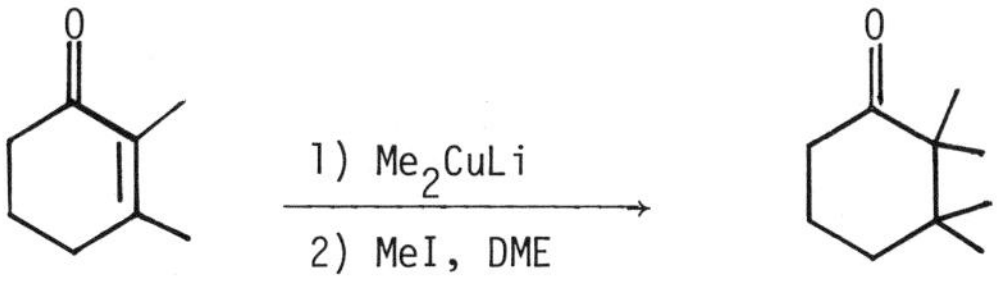

86%

JOC (1974) 39 275

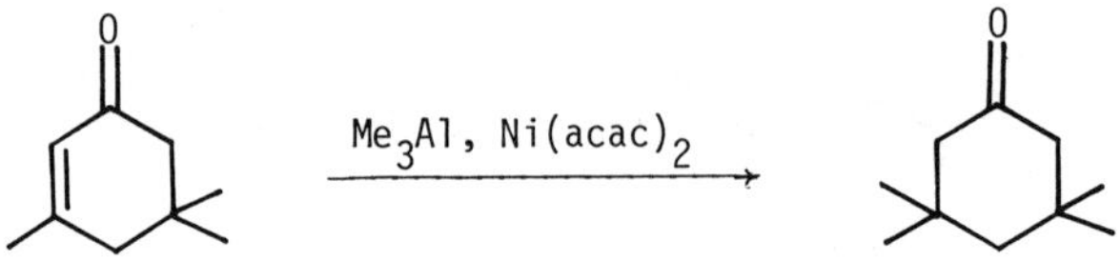

JOC (1974) 39 3297   85%

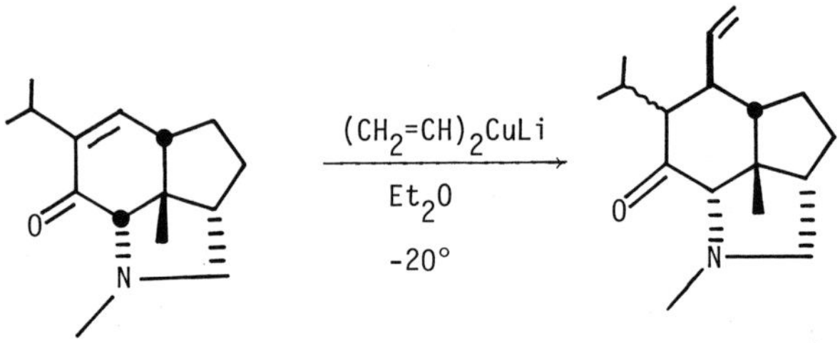

JACS (1974) 96 4334   80%

PhCH=CHNO₂   +   Me₂CuLi ⟶   PhCHMeCH₂NO₂   54%

Tetr Lett (1975) 3591

Review: "Catalytic Transfer Hydrogenation"

Chem Rev (1974) 74 567

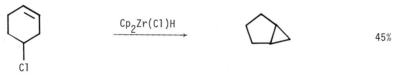

JCS Perkin I (1974) 1440

Tetr Lett (1974) 3329, 3333, 3327

Org Synth (1976) <u>55</u> 12

J Organometal Chem (1976) <u>108</u> Cl

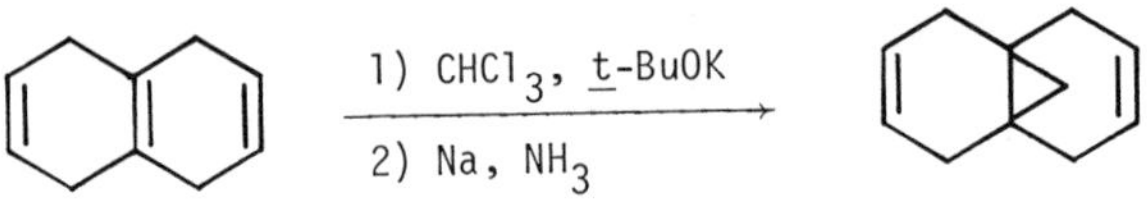

1) $CHCl_3$, t-BuOK
2) Na, NH_3

~35%

Org Synth (1974) 54 11

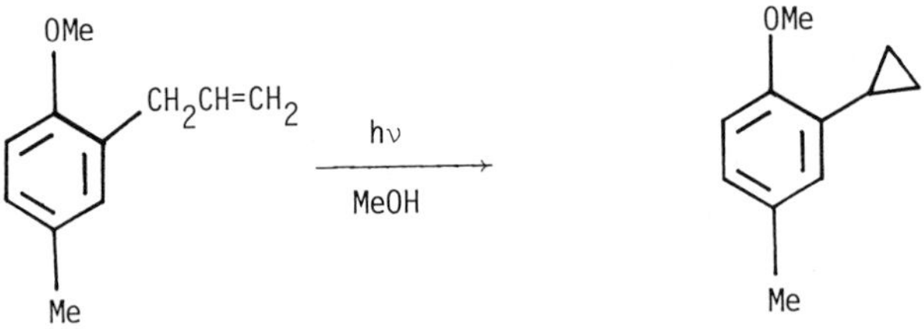

OMe
$CH_2CH=CH_2$
Me
hν
MeOH
OMe
Me

70%

Helv Chim Acta (1975) 58 178

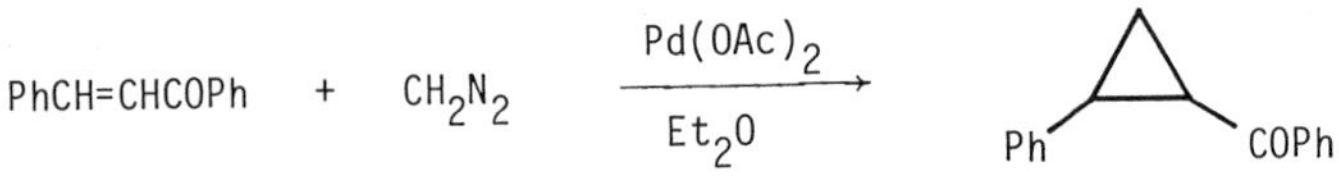

PhCH=CHCOPh + CH_2N_2
$Pd(OAc)_2$
Et_2O
Ph
COPh

98%

Tetr Lett (1975) 629
Synthesis (1975) 636

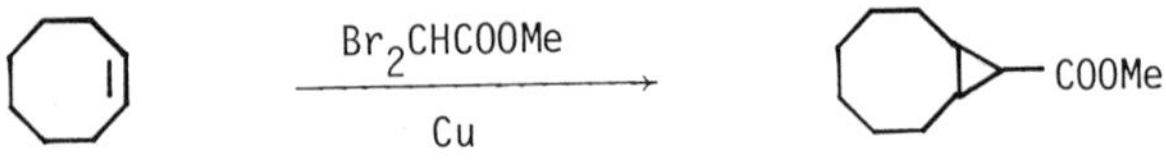

$Br_2CHCOOMe$
Cu
COOMe

71%

JACS (1976) 98 2676

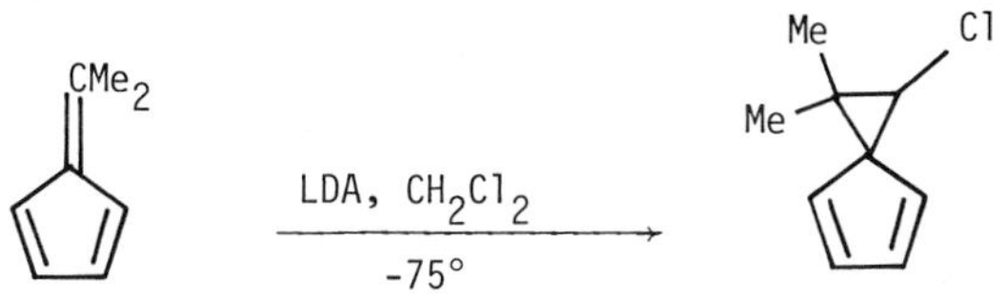

JACS (1975) <u>97</u> 3830

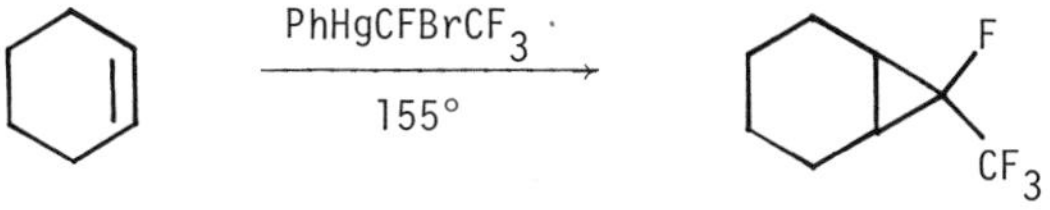

80%

J Organometal Chem (1975) <u>92</u> 7

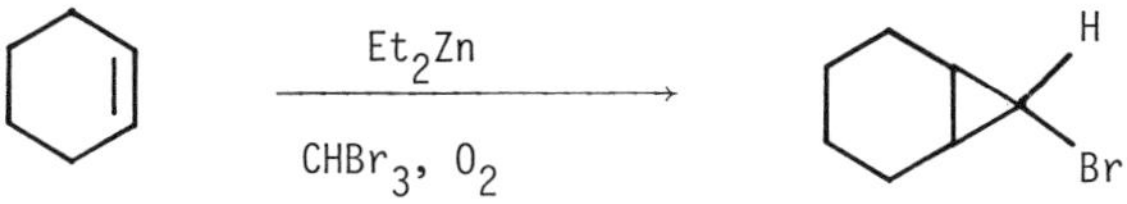

87%

JCS Chem Comm (1975) 364
Bull Chem Soc Japan (1975) <u>48</u> 3665

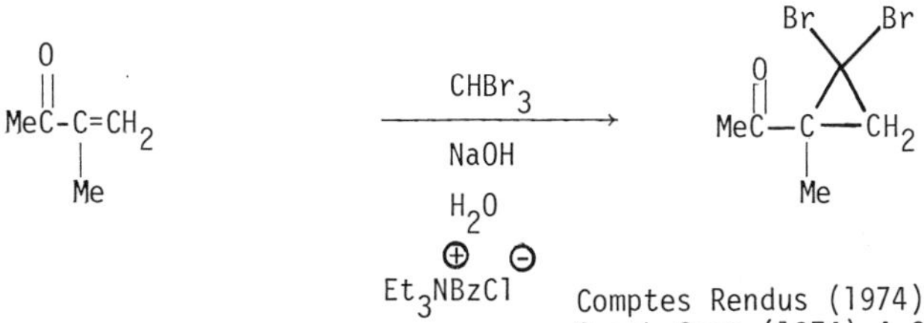

84%

Comptes Rendus (1974) <u>278</u> 621
Synth Comm (1974) <u>4</u> 341
Synthesis (1974) 736

85%

Ph—C(=CH$_2$)—CH$_3$

+

CHCl$_3$

$\xrightarrow{\text{NaOH, H}_2\text{O}}$

(P)—⟨C$_6$H$_4$⟩—CH$_2$NMe$_2$Bu $\overset{\oplus}{}$ Cl $\overset{\ominus}{}$

Ph / Me — cyclopropane — Cl, Cl   99%

JACS (1975) 97 5956

cyclohexene

$\xrightarrow[\substack{\text{CHCl}_3, \\ \text{phase-transfer catalyst}}]{\text{Cl}_3\text{C-CO}_2\text{Na}}$

bicyclic Cl, Cl   57%

Tetr Lett (1976) 91

cyclohexene

$\xrightarrow{\text{PhHgCClBrI}}$

bicyclic Cl, Br

JOC (1975) 40 1620

Ph_3PbCCl_3
125°, octane
Cl
Cl

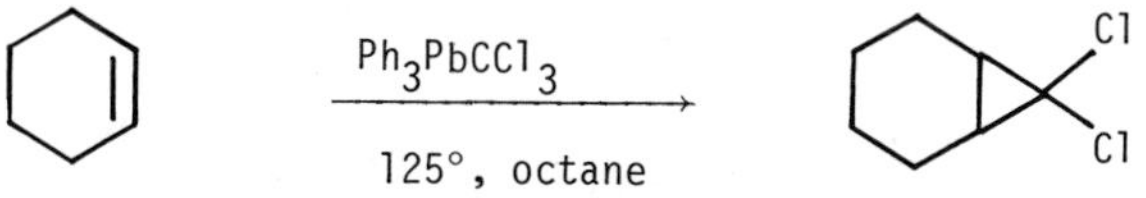

98%

J Organometal Chem (1975) 90 173

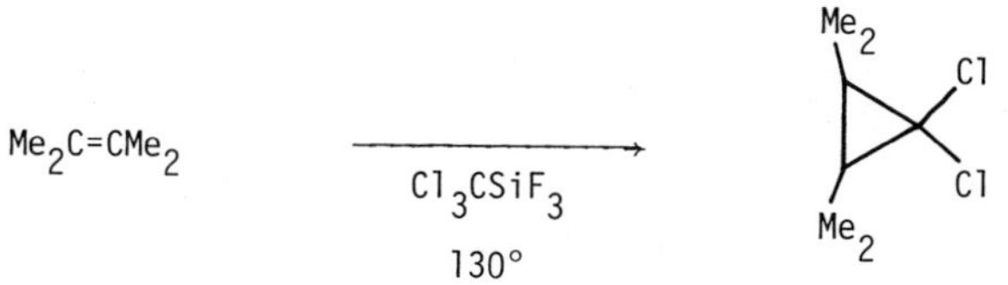

$Me_2C{=}CMe_2$
Cl_3CSiF_3
130°
Me_2
Cl
Cl
Me_2

98%

JCS Perkin II (1975) 1051

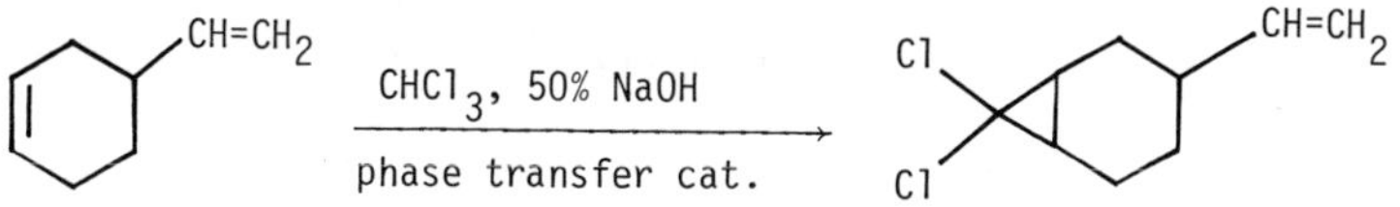

$CH{=}CH_2$
$CHCl_3$, 50% NaOH
phase transfer cat.
Cl
Cl
$CH{=}CH_2$

93%

Tetr Lett (1975) 3013

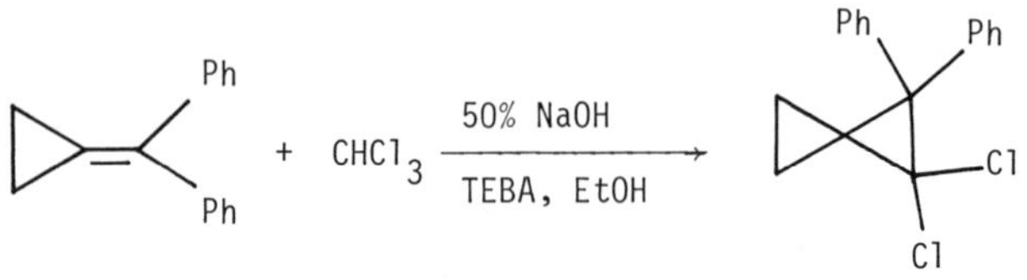

95%

Synthesis (1975) 323

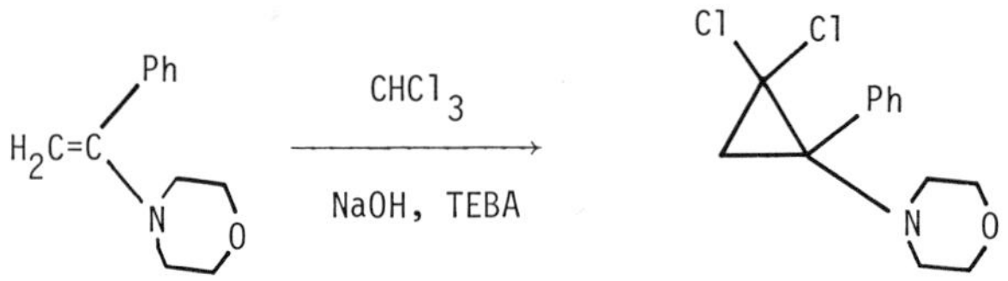

90%

Z Chem (1975) 15 14

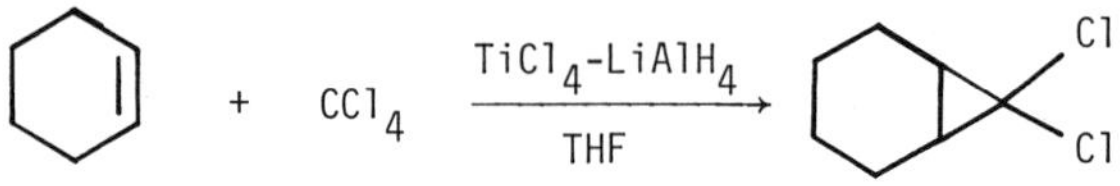

79%

Chem Lett (1975) 711

Review: "Cyclopropanation of Silyl Enol Ethers. A Powerful Synthetic Tool"

Pure & Appl Chem (1975) 43 317

Review: "Chemistry of Phosphorylcarbenes. Cyclopropanation of Various
Substrates"

Angew Int Ed (1975) 14 222

Review: "Two-Phase Reactions in the Chemistry of Carbanions and
           Halocarbenes - A Useful Tool in Organic Synthesis"

                                      Pure & Appl Chem (1975) $\underline{43}$ 439

Section 75   Alkyls, Methylenes and Aryls from Miscellaneous Compounds

$$\text{Ph} \overset{S}{\underset{}{\diagup \diagdown}} \text{Ph} \quad \xrightarrow[\text{DME}]{\text{HFe(CO)}_4{}^{\ominus}} \quad \text{PhCH}_2\text{Ph} \qquad 60\%$$

$$\text{Me-C-NHPh} \quad \xrightarrow[\text{DME}]{\text{HFe(CO)}_4{}^{\ominus}} \quad \text{MeCH}_2\text{NHPh} \qquad 51\%$$

                                      JOC (1975) $\underline{40}$ 2694

                                      JCS Perkin I (1976) 1257

Review: "Novel Approaches to Alkylations"

                                      Pure & Appl Chem (1975) $\underline{43}$ 563

Chapter 6   PREPARATION
OF
AMIDES

Section 76   Amides from Acetylenes

RC≡CR'   + $\xrightarrow{AlCl_3}$ 30-40%

Bull Soc Chim France (1974) 1949

Ph-C≡C-MgBr $\xrightarrow[\text{2) H}_2\text{O}]{\text{1) Me}_3\text{SiNCO}}$ Ph-C≡C-C-NH$_2$ 48%

Tetr Lett (1975) 981

Section 77   Amides from Carboxylic Acids and Acid Halides

CH_3COOH $\xrightarrow[\text{2) PhNH}_2]{\text{1) Ph}_3\text{P(OTf)}_2}$

$$\underset{\underset{O}{\parallel}}{CH_3C}\text{-NHPh}$$

80%

Tetr Lett (1975) 277

$PhCH_2COOH$ $\xrightarrow[\text{2) Bu}_2\text{NH}]{\text{1) Cl}}$

$$PhCH_2\underset{\underset{O}{\parallel}}{C}NBu_2$$

95%

Chem Lett (1975) 1163

$Ph(CH_2)_nCOOH$
+
HNR_2

$\xrightarrow{\text{betaine}}$

$$Ph(CH_2)_n\underset{\underset{O}{\parallel}}{C}NR_2$$

n = 0,1,2
R = Bu, Ph, Bz

Chem Lett (1976) 57

PhCOOH

+

(piperidine)

$\xrightarrow[\text{Et}_3\text{N, DBU}]{\text{BF}_3\cdot\text{Et}_2\text{O}}$

Ph-C(=O)-N(piperidine)

63%

Synthesis (1975) 715

MeOCH$_2$CO$_2$H

+

Ph-N=CHPh

$\xrightarrow[\text{Et}_3\text{N}]{\text{ClPO(OPh)}_2}$

(β-lactam: MeO, Ph substituents, N-Ph, C=O)

70%

Synth Comm (1976) 6 435

RCOOH   +   N$_3$PO(OPh)$_2$ $\xrightarrow[\text{R'OH}]{\text{Et}_3\text{N}}$ RNHCO$_2$R'

moderate to high yield

Tetrahedron (1974) 30 2151

Ar-C(=O)-Cl

+

TMS$_2$NOLi

$\xrightarrow{\text{2) H}_3\text{O}^{\oplus}}$ ArCONH$_2$

~50%

Synthesis (1975) 788

PhCHCOCl + PhCH=NPh $\xrightarrow{\text{Et}_3\text{N}}$
|
OAc

90%

Tetr Lett (1974) 2633
Tetr Lett (1974) 3135
JOC (1974) <u>39</u> 312

CH_3 ——⁄⁄—— COOH ⟶ CH_3 ——⁄⁄—— CN3 $\xrightarrow[110°]{\text{EtOH}}$ CH_3 ——⁄⁄—— NHCOEt
| || ||
                          O   O

80%

Tetr Lett (1976) 3089

Review: "The Azide Method in Peptide Synthesis"

Synthesis (1974) 549

Related methods: Amides from Amines (Section 82)

Section 78   Amides from Alcohols

$$\text{ROH} + \text{Ph}_2\text{CClSbCl}_6 \xrightarrow{\text{R'CN}} \xrightarrow{\text{H}_2\text{O}} \underset{\substack{\text{O} \\ \|}}{\text{RNH-C-R'}} \qquad 40\text{-}86\%$$

(used with steroidal alcohols)

JCS Perkin I (1974) 2101

72%

JCS Perkin I (1976) 2205

Section 79   Amides from Aldehydes

$$\text{RCHNOH} + \text{silica gel} \xrightarrow{\text{Xylene, rfx}} \text{RCONH}_2 \qquad 61\text{-}92\%$$

Tetrahedron (1974) $\underline{30}$ 2899

JOC (1974) <u>39</u> 3924

Section 80   Amides from Alkyls, Methylenes and Aryls

No additional examples

Section 81   Amides from Amides

Me$_2$CHCONMe$_2$ $\xrightarrow[\text{2) BuBr}]{\text{1) LDA, HMPA}}$ Me$_2$C-CONMe$_2$ 86%
                                                                   |
                                                                   Bu

Comptes Rendus (1974) <u>278</u> 1105

JOC (1974) <u>39</u> 2475

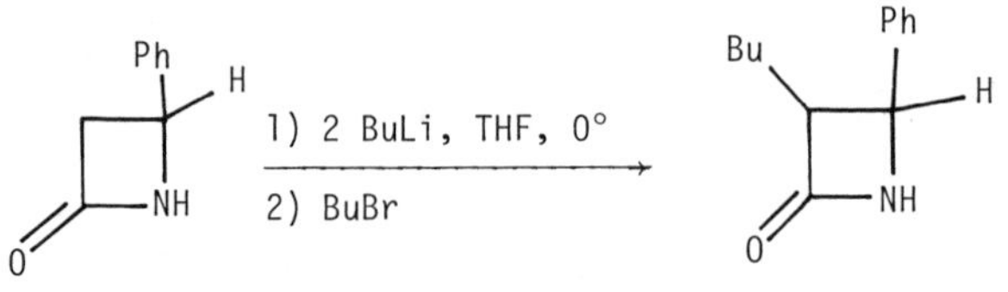

66%

Can J Chem (1974) 52 3206

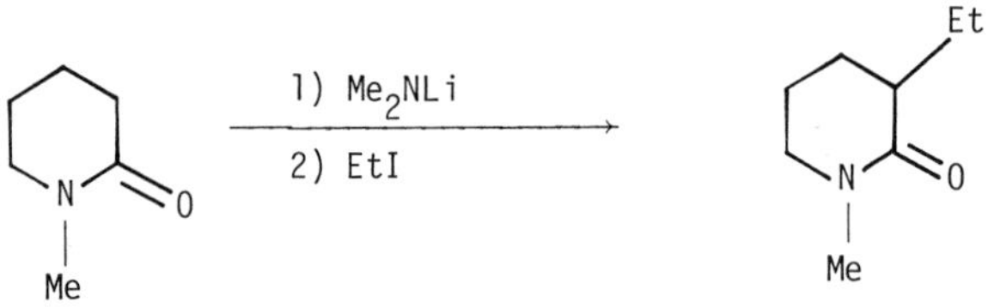

85%

Can J Chem (1975) 53 1682

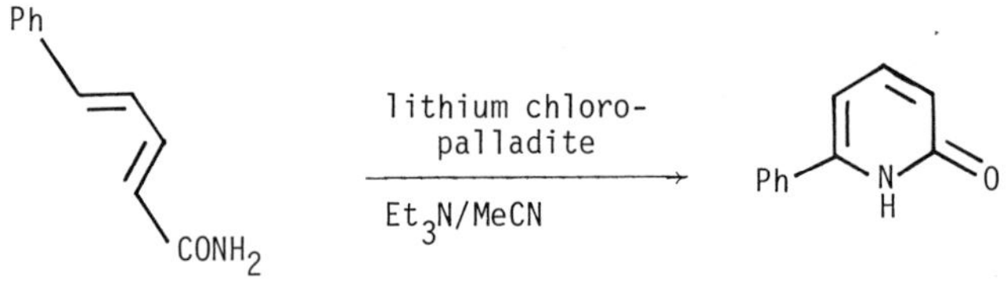

65%

Chem and Ind (1975) 745

$RCONH_2$ $\xrightarrow[\text{-40 to -15°}]{\text{NaOMe, MeOH, Br}_2}$ $RNHCOCH_3$ 68-90%

(improved Hoffmann)

Synthesis (1974) 291

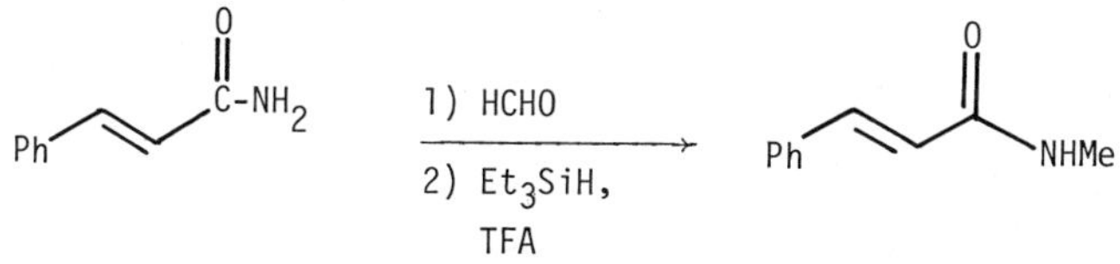

JOC (1976) <u>41</u> 725

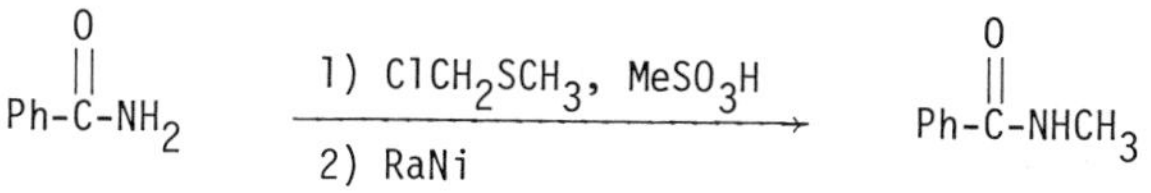

83%

JCS Chem Comm (1975) 320

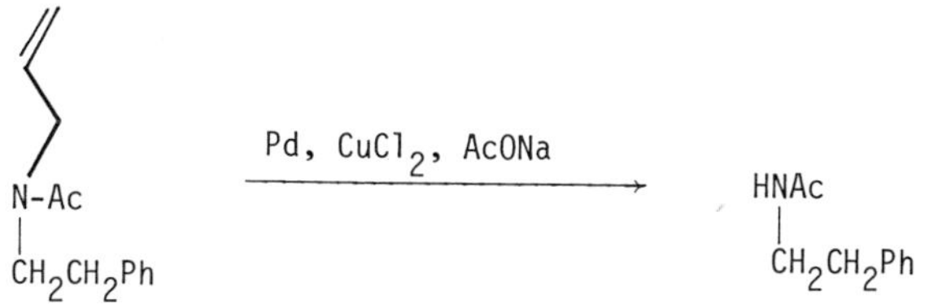

Chem Pharm Bull (1976) <u>24</u> 1992

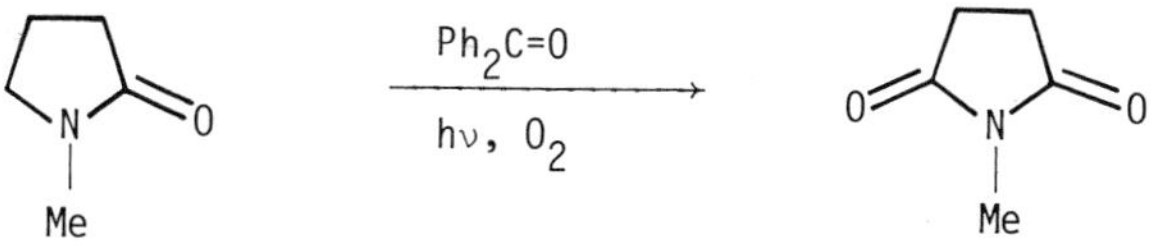

60%

JCS Chem Comm (1976) 194

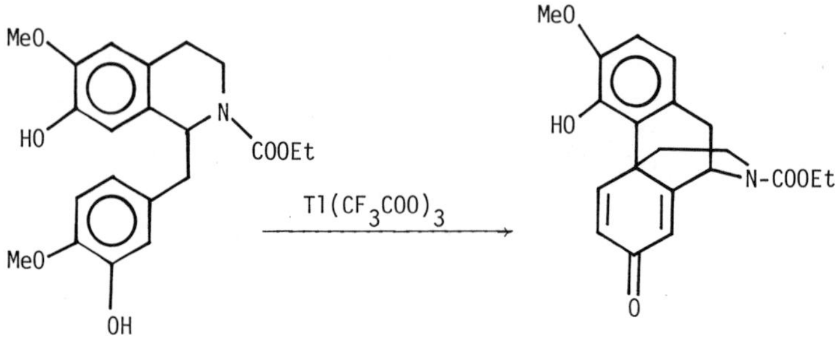

JACS (1975) <u>97</u> 1239

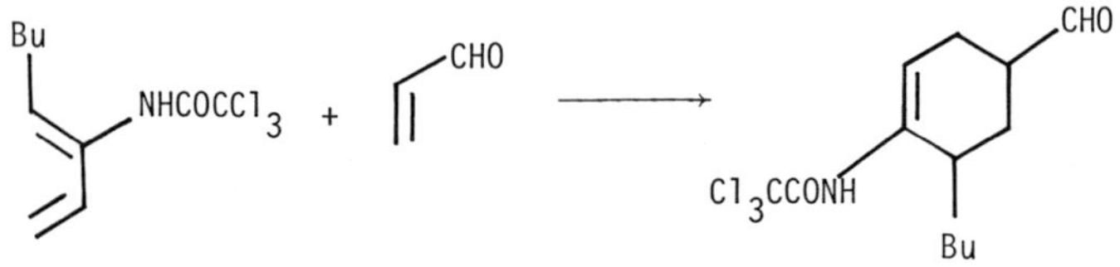

23%

JACS (1976) <u>98</u> 2352

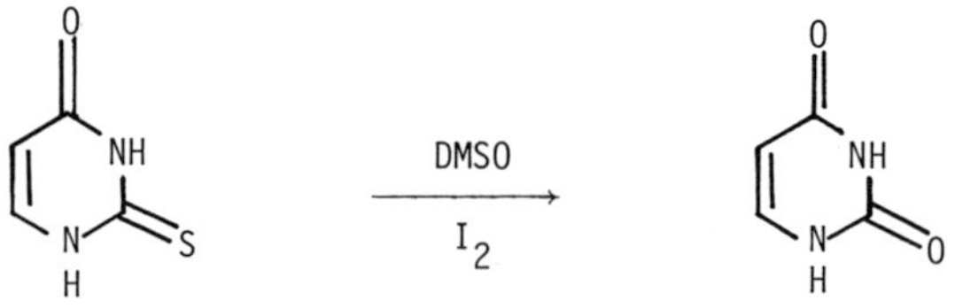

70%

Synthesis (1975) 114

80%

Review: "Alkenylation of Imides and Activated Amides"

Synthesis (1975) 685

Conjugate reductions of unsaturated amides are listed in Section 74
(Alkyls from Olefins).
Related methods: Amides from Halides (Section 85)

Section 82   Amides from Amines

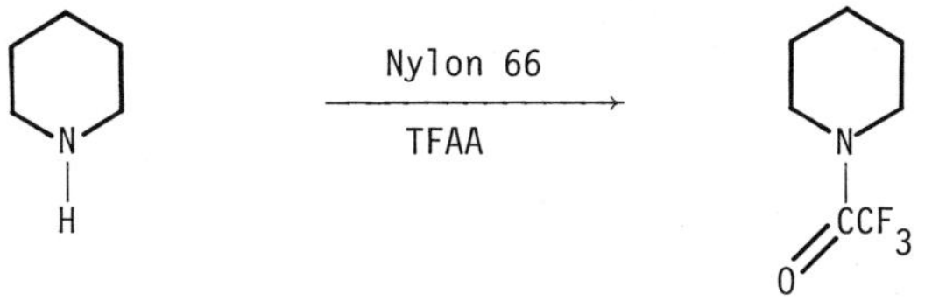

95%

Angew Int Ed (1976) 15 777

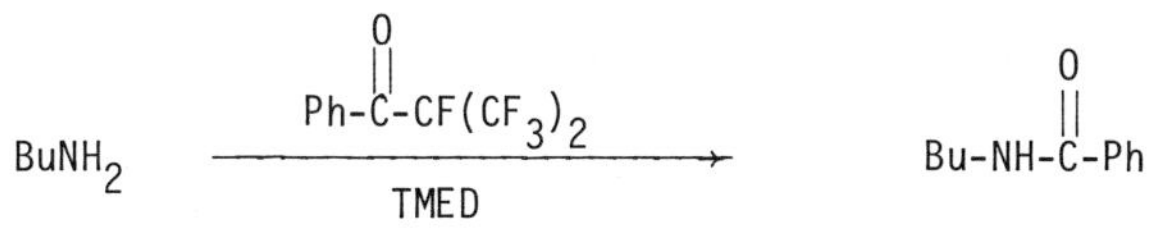

100%

Chem Lett (1976) 673

$$Et_2NH \xrightarrow[\text{phase-transfer cat.}]{\text{CHCl}_3, \text{ NaOH}} Et_2NCHO \qquad\qquad 78\%$$

Z Chem (1974) $\underline{14}$ 434

$$Ac_2O \;+\; RCN \xrightarrow[\text{PhH}]{\Delta} \underset{\overset{|}{CN}}{\overset{\overset{CN}{|}}{AcO-C-H}} \;+\; RNH_2 \longrightarrow RNH-\overset{\overset{O}{\|}}{C}-CH_3 \qquad 85\text{-}90\%$$

specific for R=alkyl: $ArNH_2$, ArOH unreactive

Synth Comm (1974) $\underline{4}$ 351

Synthesis (1975) 109

JOC (1976) $\underline{41}$ 2780

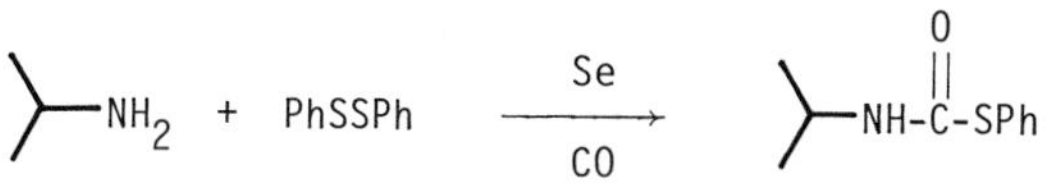

$$\text{>}-NH_2 \ + \ PhSSPh \ \xrightarrow[CO]{Se} \ \text{>}-NH-\overset{\overset{\displaystyle O}{\parallel}}{C}-SPh \qquad 85\%$$

Tetr Lett (1975) 2087

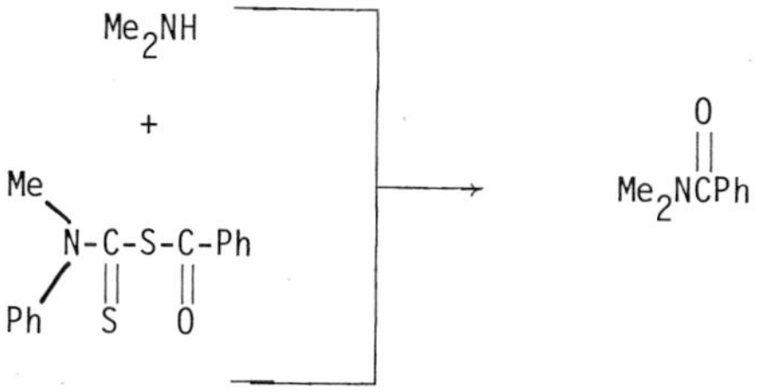

$$Me_2NCPh \qquad \sim 60\%$$

Indian J Chem (1975) **13** 35

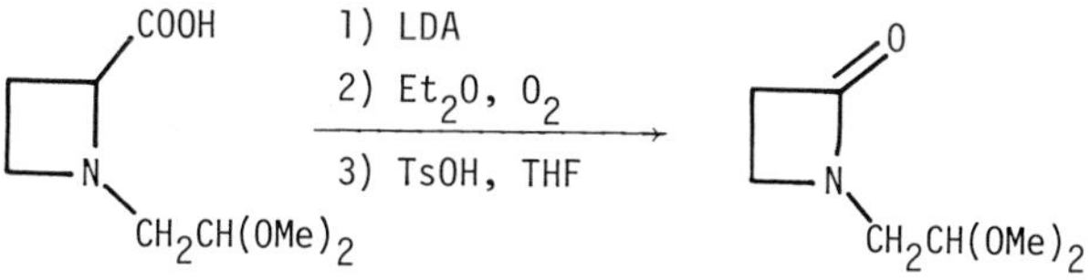

$$\sim 50\%$$

Tetr Lett (1976) 4613

Related methods: Amides from Carboxylic Acids and Acid Halides (Section 77)
                   Protection of Amines (Section 105A)

Section 83   Amides from Esters

No additional examples

Section 84   Amides from Ethers and Epoxides

No additional examples

Section 85   Amides from Halides

PhMgBr $\xrightarrow[\text{2) NH}_4\text{Cl, H}_2\text{O}]{\text{1) TMSNCO}}$ PhCONH$_2$ 49%

Tetr Lett (1975) 981

Section 86   Amides from Hydrides

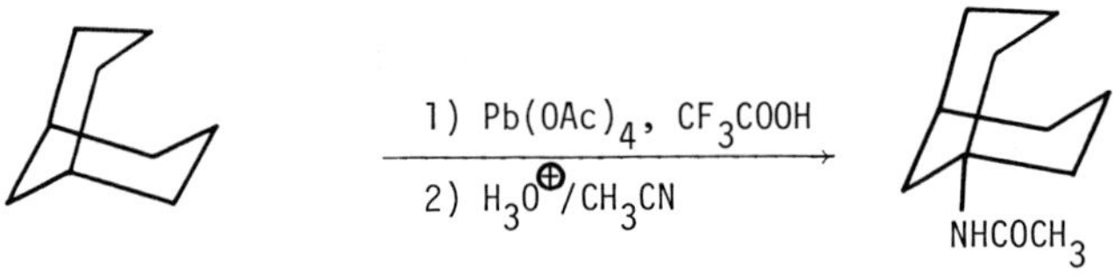

$\xrightarrow[\text{2) H}_3\text{O}^{\oplus}\text{/CH}_3\text{CN}]{\text{1) Pb(OAc)}_4\text{, CF}_3\text{COOH}}$ 78%

NHCOCH$_3$

Synthesis (1976) 32

Section 87   Amides from Ketones

Ph-C(=O)-NHt-Bu   95%

JCS Perkin I (1976) 1511

Review: "The Willgerodt Reaction"

$$Ar-\overset{O}{\overset{||}{C}}-(CH_2)_nH \xrightarrow{S, NH_4OH} Ar(CH_2)_nCONH_2$$

Synthesis (1975) 358

Section 88   Amides from Nitriles

$$RCN + PdCl_2 + H_2O \xrightarrow{\Delta} RCONH_2 \qquad 60\text{-}80\%$$

Synthesis (1974) 574

$$Ph-C\equiv N \xrightarrow[\underline{t}\text{-BuOH}]{KOH} Ph-\overset{O}{\overset{||}{C}}-NH_2 \qquad 90\%$$

JOC (1976) 41 3769

Section 89   Amides from Olefins

No additional examples

Section 90   Amides from Miscellaneous Compounds

No additional examples

Section 90A   Protection of Amides

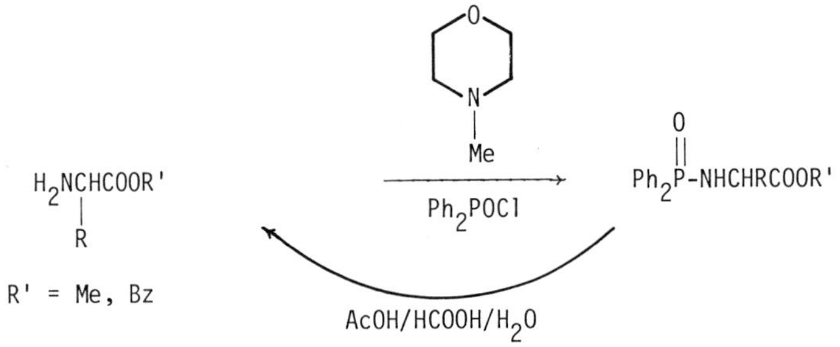

H$_2$NCHCOOR'
|
R

R' = Me, Bz

Ph$_2$POCl

Ph$_2$P-NHCHRCOOR'

AcOH/HCOOH/H$_2$O

Tetr Lett (1976) 3623

Chapter 7   PREPARATION

OF

AMINES

Section 91   Amines from Acetylenes

No additional examples

Section 92   Amines from Carboxylic Acids and Acid Halides

$ArNH_2$ + RCOOH + $NaBH_4$ $\longrightarrow$ $ArNHCH_2R$ 60-90%

JACS (1974) <u>96</u> 7814

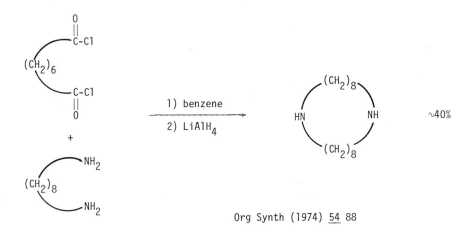

Org Synth (1974) <u>54</u> 88

Section 93   <u>Amines from Alcohols and Phenols</u>

92%

Tetr Lett (1975) 471

Section 94   <u>Amines from Aldehydes</u>

R_2NH + R'CHO + $NaHFe(CO)_4$ ⟶ R_2NCH_2R' 60-80%

Synthesis (1974) 733

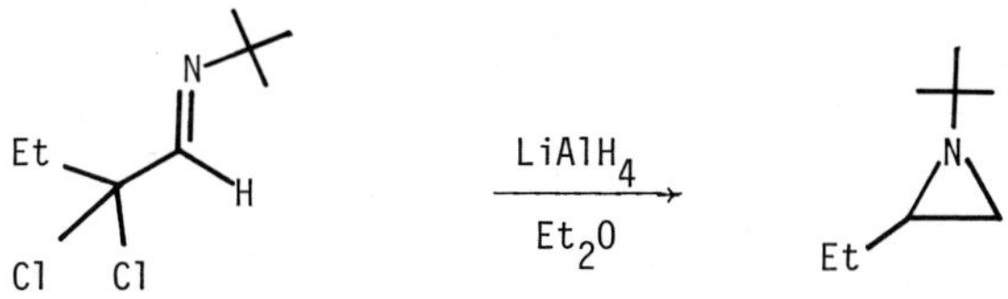

Synth Comm (1975) 5 269

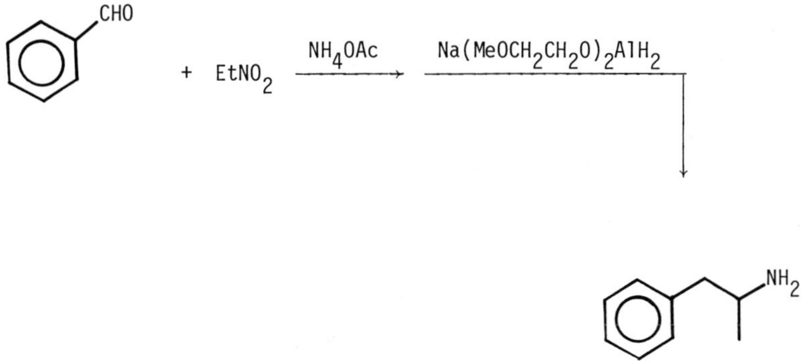

JCS Chem Comm (1974) 307

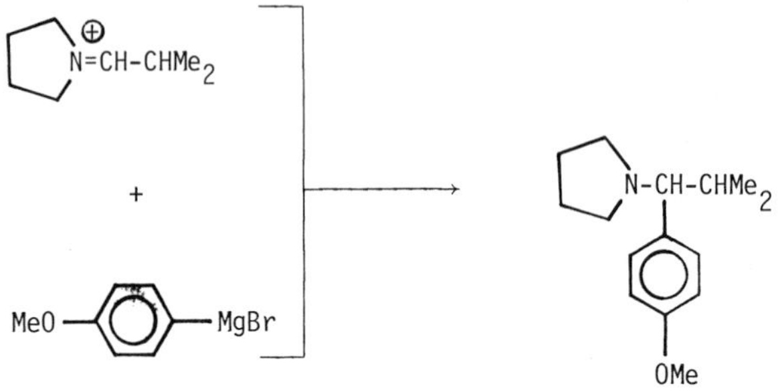

59%

Chem Ber (1975) <u>108</u> 2827

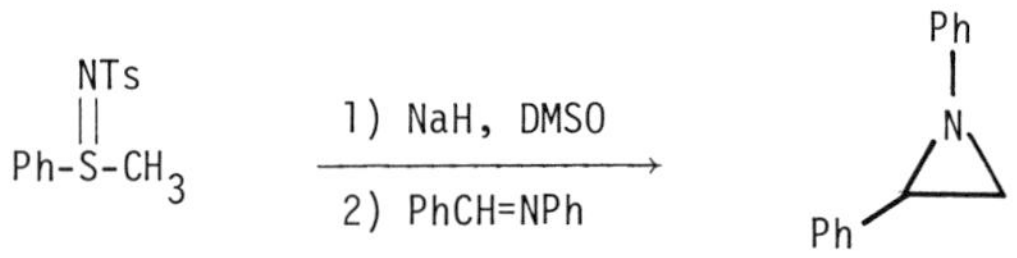

$$Ph-\overset{\overset{\displaystyle NTs}{\|}}{S}-CH_3 \quad \xrightarrow[\text{2) PhCH=NPh}]{\text{1) NaH, DMSO}}$$

47%

Synthesis (1976) 35

Related methods: Amines from Ketones (Section 102)

Section 95   <u>Amines from Alkyls, Methylenes and Aryls</u>

No examples

Section 96   Amines from Amides

$$\underset{\text{PhCNHCH}_3}{\overset{\text{O}}{\overset{||}{}}} \xrightarrow[\text{AcOH}]{\text{NaBH}_4} \text{PhCH}_2\text{NHCH}_3 \qquad\qquad 82\%$$

Tetr Lett (1976) 763

$\xrightarrow[\text{HSCH}_2\text{CH}_2\text{SH}]{\text{NaBH}_4, \text{ THF}}$

~100%

same conditions

~100%

Chem and Ind (1976) 322

$$\underset{\text{Ph-C-N}}{\overset{\text{O}}{\overset{||}{}}}\diagup\diagdown \xrightarrow[\text{2) NaBH}_4]{\text{1) PCl}_5} \text{Ph-CH}_2\text{-N}\diagup\diagdown \qquad\qquad 90\%$$

Tetr Lett (1976) 219

$$\underset{\overset{|}{\text{COCH}_3}}{\text{TsNHCH}_2\text{CH}_2\text{-N-CH}_2\text{CO}_2\text{C}_6\text{Cl}_5} \xrightarrow[\text{2) H}_3\text{O}^{\oplus}]{\text{1) BH}_3 \cdot \text{THF}} \underset{\overset{|}{\text{CH}_2\text{CH}_3}}{\text{TsNHCH}_2\text{CH}_2\text{-N-CH}_2\text{COOH}}$$

JOC (1976) **41** 149

Tetr Lett (1976) 3217

Section 97   <u>Amines from Amines</u>

Tetrahedron (1975) <u>31</u> 2517

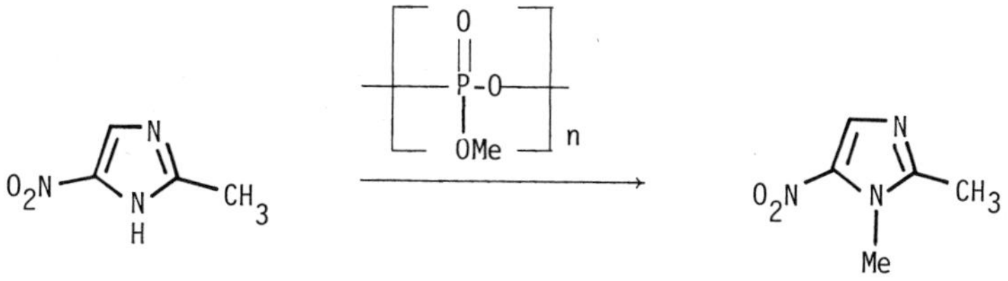

Synthesis (1975) 596

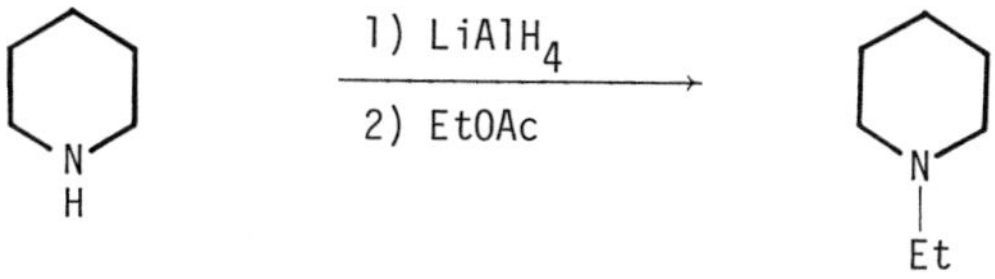

Synthesis (1975) 607

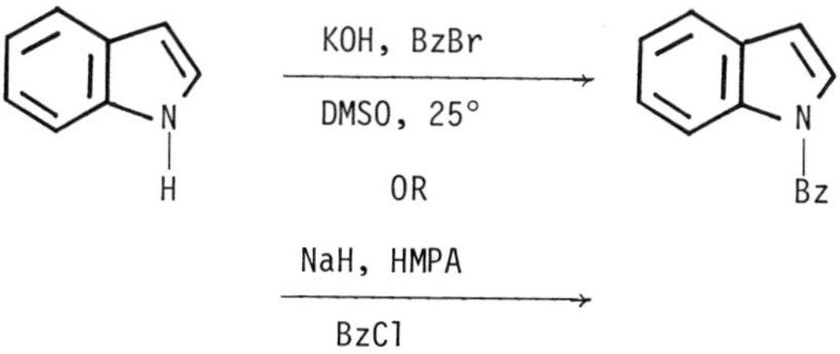

Org Synth (1974) $\underline{54}$ 58, 60

~100%

Bull Chem Soc Japan (1976) <u>49</u> 1378

$BzNH_2$ $\xrightarrow[\text{Et}_3\text{N}]{\text{Ph}_2\text{POCl}}$ $Ph_2PONHBz$ $\xrightarrow[\substack{\text{2) MeI} \\ \text{3) MeOH, TsOH}}]{\text{1) NaH}}$ $BzNH_2\overset{\oplus}{Me}\ \overset{\ominus}{OTs}$ 56%

Tetr Lett (1976) 4005

$R-NH_2$ $\xrightarrow[\text{MeOH}]{\text{NaHCO}_3,\ \text{MeI}}$ $R\overset{\oplus}{N}Me_3\overset{\ominus}{I}$ 75-96%

R = amino acid

Also works with secondary amines.

Can J Chem (1976) <u>54</u> 3310

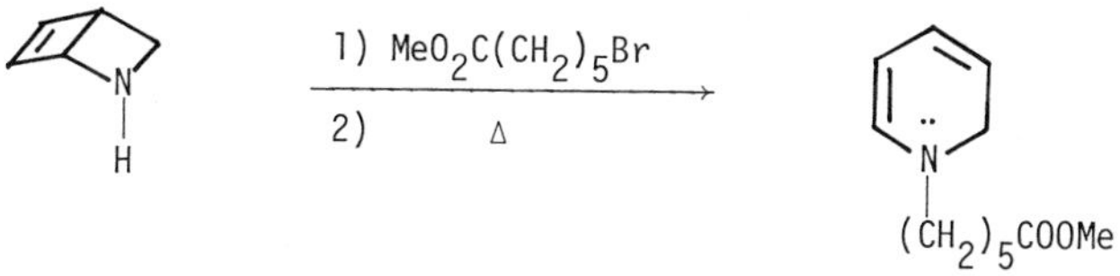

1) $MeO_2C(CH_2)_5Br$

2) Δ

$(CH_2)_5COOMe$

90%

JACS (1976) <u>98</u> 2344

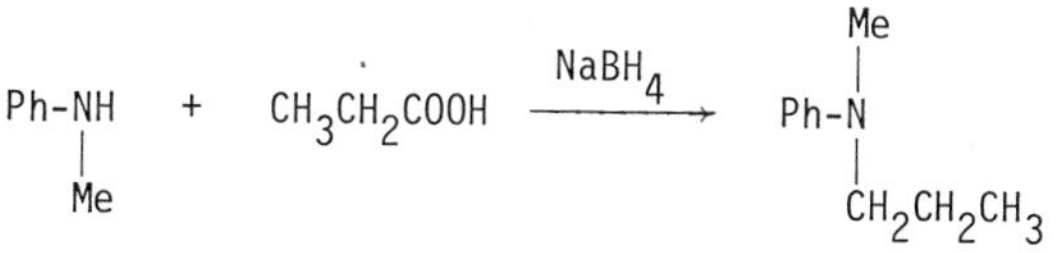

Ph-NH + CH_3CH_2COOH $\xrightarrow{NaBH_4}$ Ph-N
| |
Me

Me
|
Ph-N
|
$CH_2CH_2CH_3$

65%

JOC (1975) <u>40</u> 3453

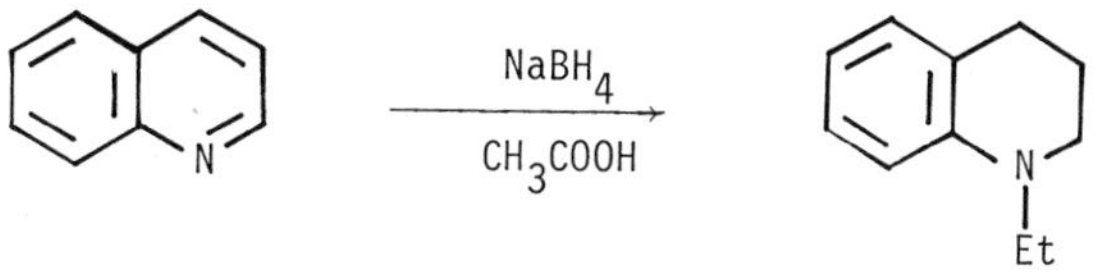

$\xrightarrow[CH_3COOH]{NaBH_4}$

Et

68%

Synthesis (1975) 650

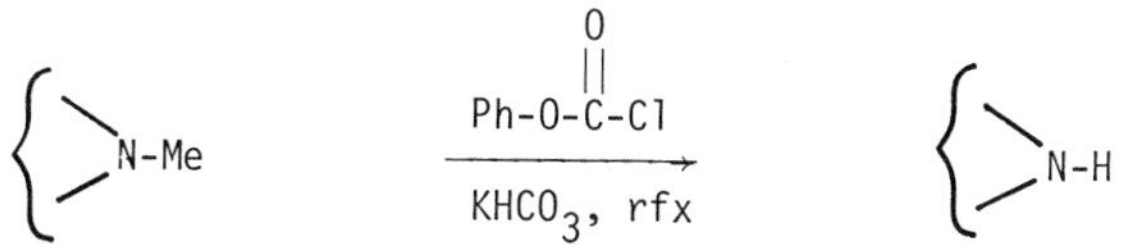

N-Me

$\overset{O}{\overset{||}{Ph-O-C-Cl}}$

$\xrightarrow{KHCO_3, \ rfx}$

N-H

84-89%

morphine, codeine

JOC (1975) <u>40</u> 1850

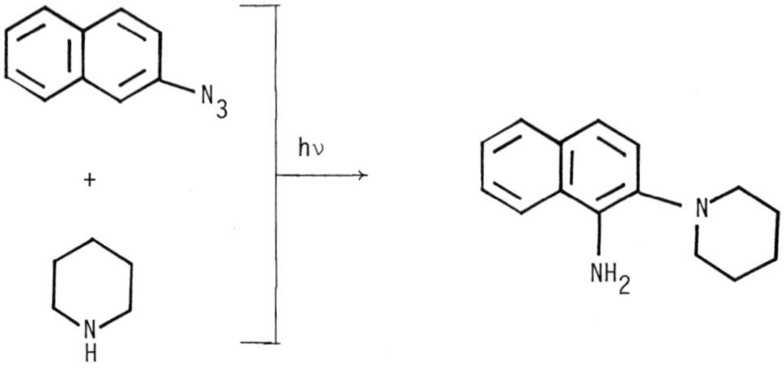

75%

Synthesis (1975) 710

1) Cl_3CCH_2OCOCl
2) Zn/HOAc

CH_3NR_2 $\xrightarrow{\hspace{2cm}}$ R_2NH >70%

(proceeds in 75% with morphine)

Tetr Lett (1974) 1325

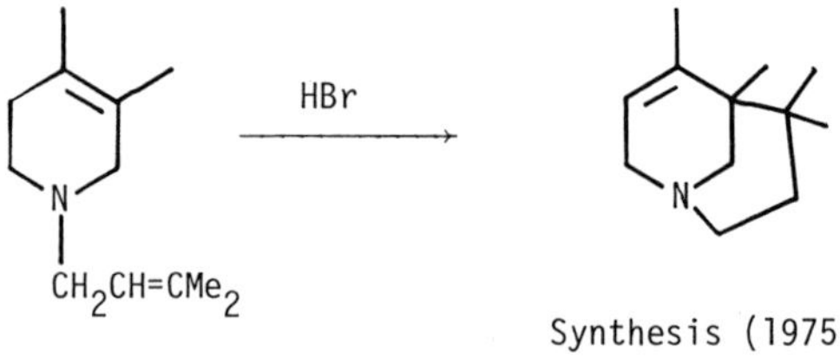

HBr

83%

$CH_2CH=CMe_2$

Synthesis (1975) 440

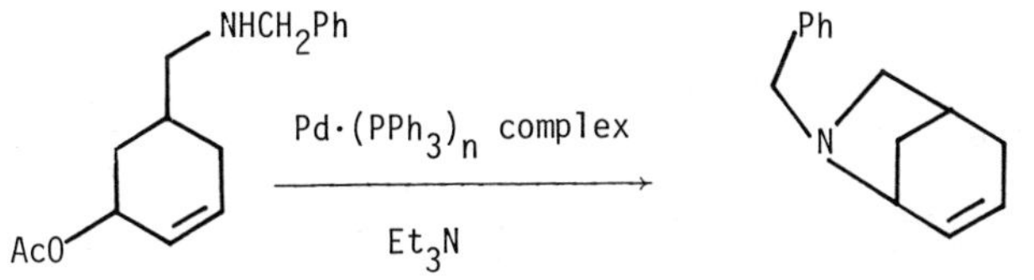

$$Pd \cdot (PPh_3)_n \text{ complex}$$

$$Et_3N$$

67%

JACS (1976) 98 8516

$$Ph_3P \quad + \quad CCl_4 \quad + \quad CH_3CH(OH)CH_2NHCH_2Ph \quad + \quad Et_3N \longrightarrow$$

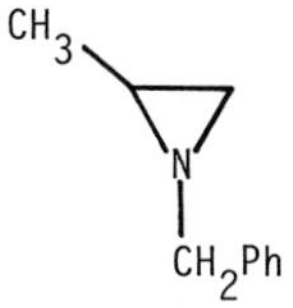

80%

Chem Ber (1974) 107 5

$$RNHCH_2CH_2\overset{\overset{\displaystyle CH_3}{|}}{C}HOH$$

$$\xrightarrow[Et_3N]{Ph_3PBr_2}$$

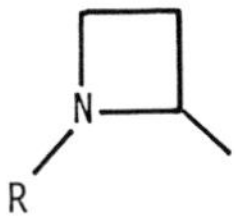

44-65%

R = alkyl, Bz

Synthesis (1974) 894

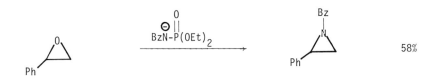

Ph⟍   ⟍COOMe
  ⟍   ⟍COOMe $\xrightarrow{Me_2S=CHCOPh}$ [structure with Ph, COPh, COOMe, COOMe, N, Ph]   85%
    N
    Ph

JOC (1975) 40 2990

Review: "Umpolung of Amine Reactivity"

Angew Int Ed (1975) 14 15

Section 98   Amines from Esters

No additional examples

Section 99   Amines from Epoxides

[epoxide structure Ph] $\xrightarrow{\underset{BzN-P(OEt)_2}{\overset{O}{\ominus\,\|}}}$ [aziridine structure with Bz, N, Ph]   58%

Tetr Lett (1976) 4003

Section 100   Amines from Halides

$$\text{(2-chlorocyclopentanone)} + \text{ArNHMe} \xrightarrow[\text{K}_2\text{CO}_3]{80°,\ \text{dioxane}} \text{(2-(N-Ar-N-Me-amino)cyclopentanone)} \qquad \sim 50\%$$

Monatshefte (1976) 107 401

$$\text{ArI} + (\text{Me}_3\text{Si})_2\text{NCu} \longrightarrow \xrightarrow{\text{MeOH}} \text{ArNH}_2 \qquad 30\text{-}60\%$$

JCS Chem Comm (1974) 256

$$\text{(1,5-dichloropentane)} \xrightarrow{(\text{Bu}_3\text{Sn})_2\text{NEt}} \text{(N-ethylpiperidine)} \qquad 60\%$$

Comptes Rendus C (1975) 281 47

PhLi   +   HMPA $\xrightarrow[-30°]{\text{THF}}$ [complex] $\xrightarrow{\text{PhLi}}$ PhCH$_2$NHMe   75%

JOC (1974) <u>39</u> 3042

NH$_2$NH$_2$ +   CHCl$_3$ +   KOH $\xrightarrow[\text{18-crown-6}]{\text{ether}}$ CH$_2$N$_2$ 40%

Tetr Lett (1974) 2983

$$\underset{\text{HC}\equiv\text{C-C=C-CH}_2\text{Cl}}{\overset{\text{R R'}}{|\ |}}\quad \xrightarrow[\text{2) NaOH}]{\text{1) NHMe}_2}$$

55-60%

Comptes Rendus (1974) <u>278</u> 801

$$\underset{\text{Ph-C(CH}_2)_3\text{CH}_2\text{Cl}}{\overset{\text{NMe}}{||}}\quad \xrightarrow{\text{rfx}}$$

Org Synth (1974) <u>54</u> 93

$$\text{Me}_2\text{NNHCCH}_3$$

O
||

1) MeI
2) EtI

3) $^{\ominus}$OH, H$_2$O

$\longrightarrow$

H$_2$N-NMeEt

~80%

JOC USSR (1975) <u>11</u> 56

Section 101 <u>Amines from Hydrides</u>

 $\xrightarrow{\text{Se(N-}\underline{t}\text{-Bu)}_2}$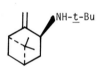

62%

JACS (1976) <u>98</u> 269

Section 102 <u>Amines from Ketones</u>

 $\xrightarrow[\text{NaAlH}_2(\text{OCH}_2\text{CH}_2\text{OCH}_3)_2]{\text{LiAlH}_4}$

54%

Can J Chem (1975) <u>53</u> 3227

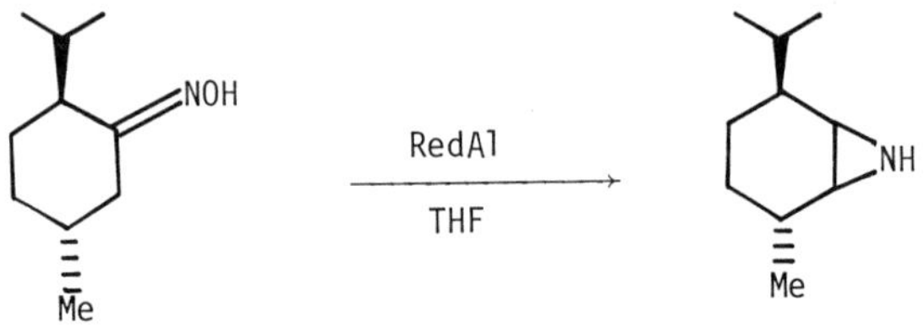

Tetr Lett (1976) 1175

Tetr Lett (1975) 3263

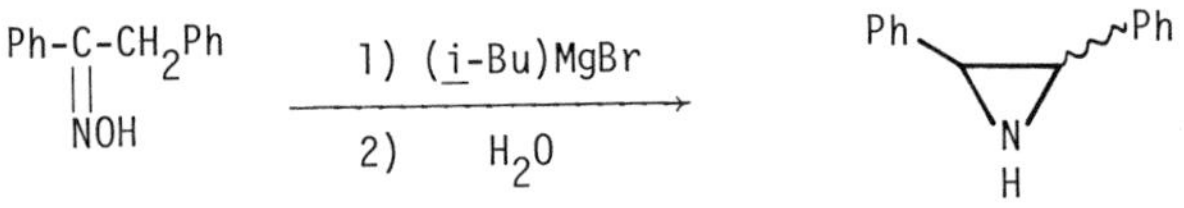

Chem Pharm Bull (1976) <u>24</u> 1083

Ph-C-CH$_2$Me

PhMgBr

$\longrightarrow$

Ph, Ph, Me aziridine (N-H)   80%

Tetr Lett (1975) 355

Related methods: Amines from Aldehydes (Section 94)

Section 103   Amines from Nitriles

PhCH$_2$CN $\xrightarrow[\text{THF}]{\text{NaBH}_3(\text{OCOCF}_3)}$ PhCH$_2$CH$_2$NH$_2$ 70%

Tetr Lett (1976) 2875

MeO, MeO, Br ring with CH-CN / OH $\xrightarrow{\text{B}_2\text{H}_6}$ MeO, MeO, Br ring with CHCH$_2$NH$_2$ / OH

Does not hydrogenolyze halides.

JCS Perkin I (1974) 1015

Section 104   <u>Amines from Olefins</u>

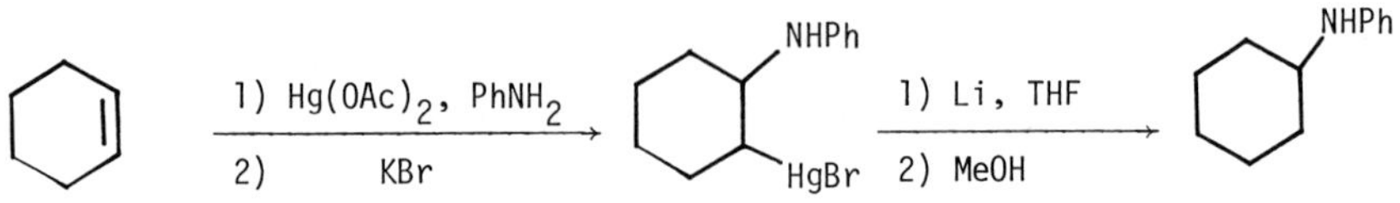

Synthesis (1975) 116

R_3B or R_2BH + $MesSO_3NH_2$ ————→ RNH_2 24-50%

Synthesis (1974) 196

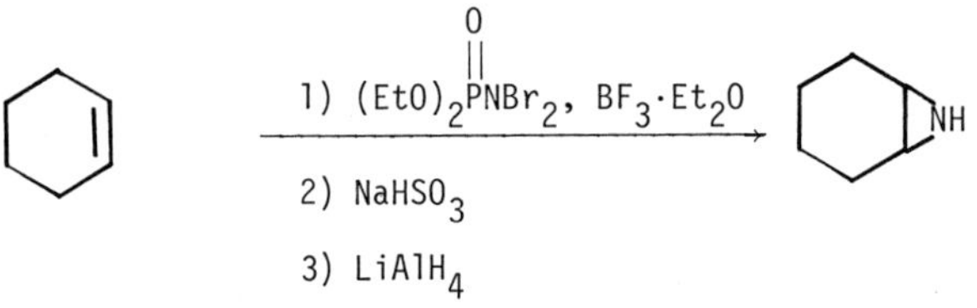

Angew Int Ed (1976) <u>15</u> 302

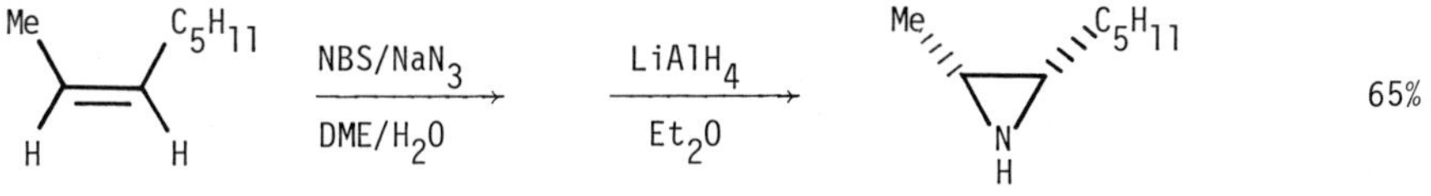

Angew Int Ed (1974) 13 279

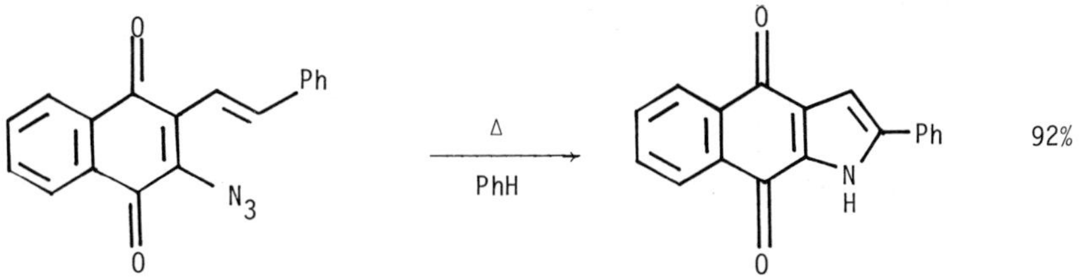

JOC (1974) 39, 775

Section 105   Amines from Miscellaneous Compounds

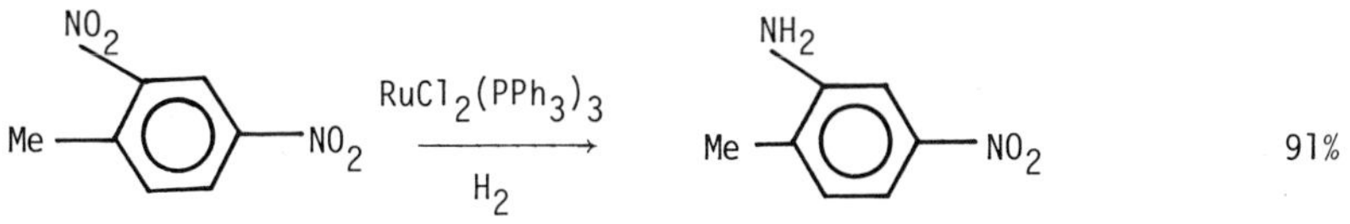

Tetr Lett (1975) 2163

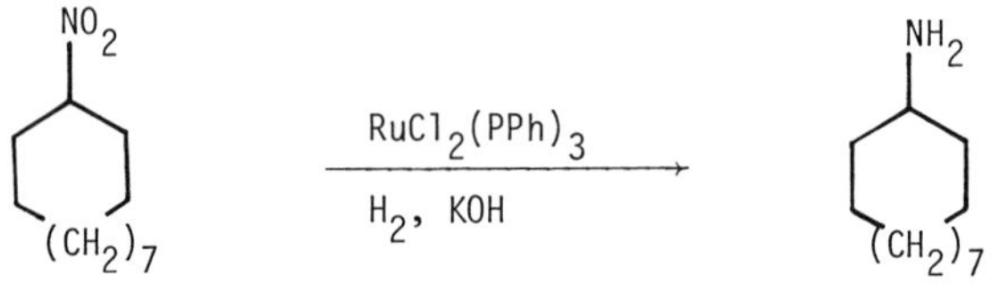

JOC (1975) 40 519

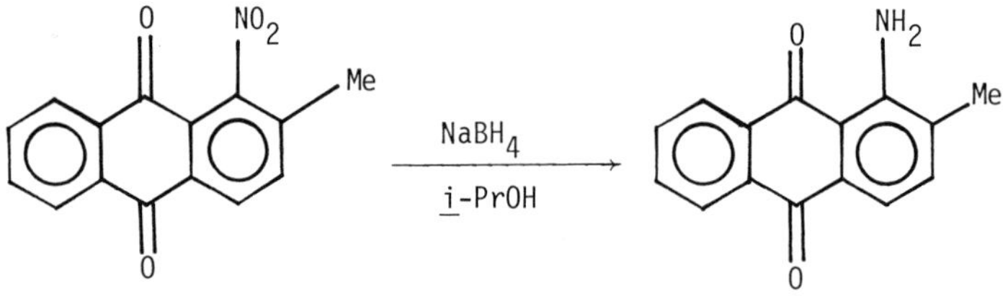

Synthesis (1976) 528

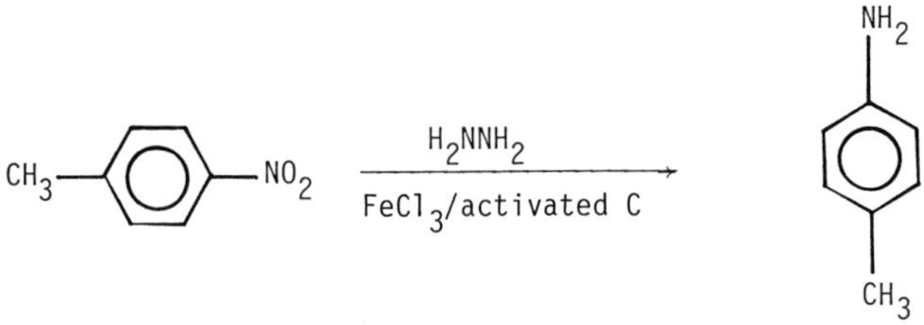

Chem Lett (1975) 259

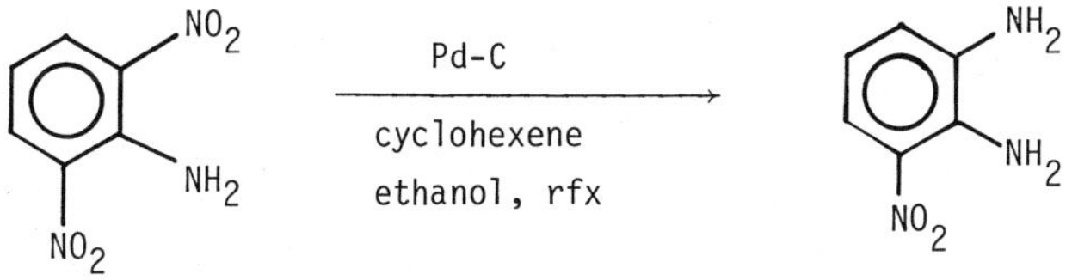

JCS Perkin I (1975) 1300

$PhNO_2$ $\xrightarrow[\text{2) \quad HF}]{\text{1) \quad (NH}_4)_2\text{S}}$ $\underline{p}$-F-PhNH$_2$ 30%

JOC (1974) $\underline{39}$ 1758

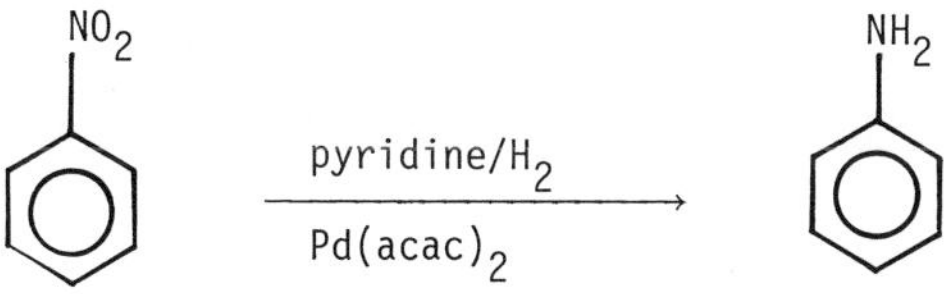

Chem and Ind (1975) 1057

$Ph-N_3$ $\xrightarrow[\text{H}_2\text{O/THF}]{\text{VCl}_2}$ $PhNH_2$ 70%

Synthesis (1976) 815

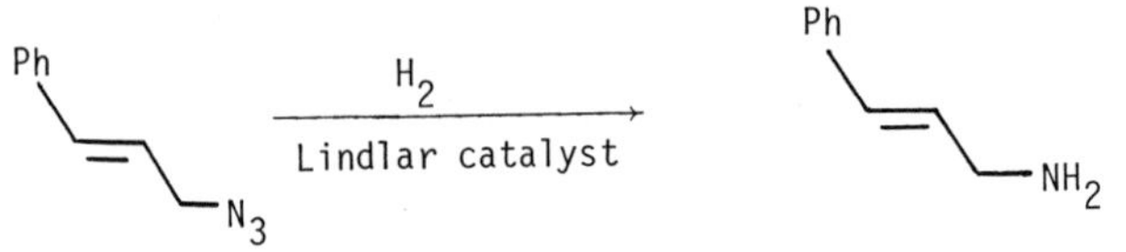

93%

Synthesis (1975) 590

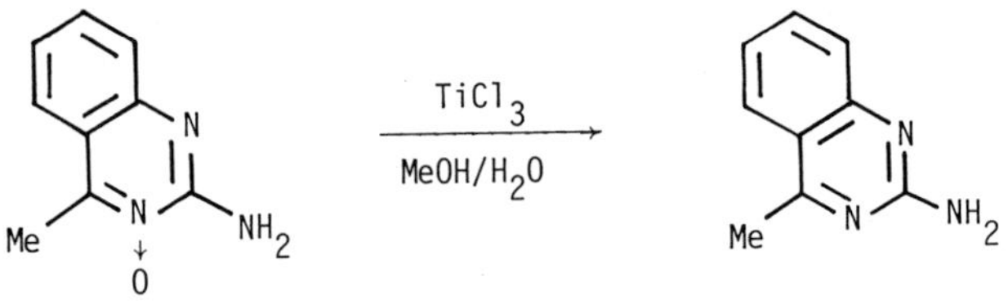

83%

Synthesis (1975) 335

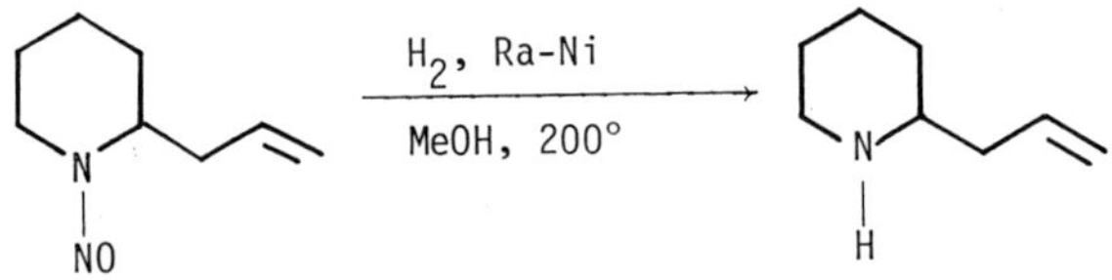

95%

Synthesis (1976) 548

$$\text{Synthesis (1976) 540}$$

Chem Pharm Bull (1976) 24 369

Bull Chem Soc Japan (1975) 48 2397

Section 105A   Protection of Amines

"Bic" protecting group for amine functions in peptide synthesis.
Removed by Et$_3$N/DMF followed by hydrolysis or by hydrogenolysis.

"Bic"   =

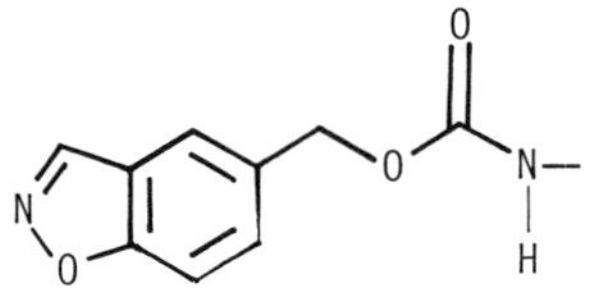

Tetr Lett (1975) 4625

A photolabile protecting group for histidine:

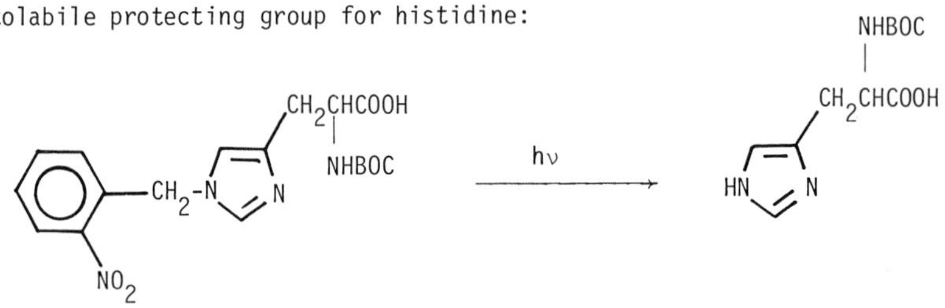

JACS (1975) 97 440

Use of the "Dobz" amine-protecting group to vary solubility properties of
reactants and products in peptide synthesis.

"Dobz"   =

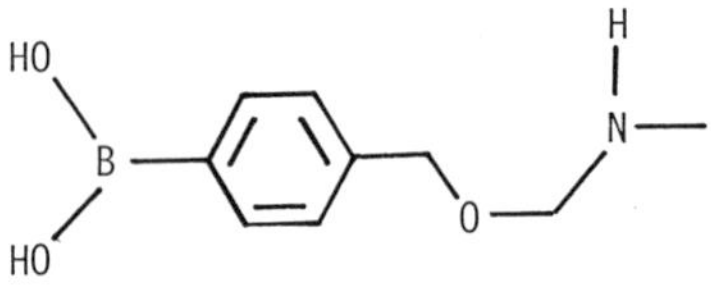

Tetr Lett (1975) 4629

The introduction of tertiary amino substituents into urethane-type amino protecting groups increases their acid-stability and provides a functional group which can be used to alter solubility characteristics.

example:

$$\text{N} \overset{\displaystyle\bigcirc}{} -\text{CMe}_2\text{O-}\overset{\displaystyle\underset{\|}{O}}{\text{C}}\text{-N} \overset{\diagup}{\diagdown}$$

JCS Chem Comm (1975) 939

$$\text{CH}_3\text{-CH-COOH} \quad + \quad \overset{\displaystyle\underset{\|}{O}}{\text{O-CN}_3} \quad \xrightarrow{\quad\quad\quad\quad} \quad \text{CH}_3\text{CH-COOH}$$
$$\underset{\text{NH}_2}{|} \qquad\qquad\qquad\qquad \underset{H_2/Pd}{} \qquad \underset{\text{HN-}\underline{t}\text{-Boc}}{|}$$

liq. NH$_3$

JACS (1974) $\underline{96}$ 4978

Cleavage of $\underline{t}$-Boc, Z(OMe), and NPS protecting groups by CF_3SO_3H in polypeptide synthesis.

JCS Chem Comm (1974) 107

Use of the 2-(triphenylphosphonio)ethoxycarbonyl group as an amino protective function in peptide chemistry. Stable to acids, removed with base.

Chem Ber (1976) $\underline{109}$ 2670

RCH-COO$^{\ominus}$ +

|

NH$_3$

$\oplus$

Stable to DCC

JOC (1974) $\underline{39}$ 3351

Cleavage of N$^{\alpha}$-benzaloxycarbonyl from methionine via H$_2$/Pd in liquid NH$_3$.

$$\text{PhCH}_2\text{O-}\overset{\text{O}}{\overset{||}{\text{C}}}\text{-R, PhCH}_2\text{OR, PhCH}_2\text{OC(O)-R are cleaved.}$$

$$\underline{t}\text{-BuO-}\overset{\text{O}}{\overset{||}{\text{C}}}\text{-R, }\underline{t}\text{-BuO-R, etc. are }\underline{\text{not}}\text{ cleaved.}$$

Tetr Lett (1974) 3259

Me$_3$SiClO$_4$ cleaves the $\underline{t}$-Butoxycarbonyl (BOC) group selectively in the presence of benzyloxycarbonyl (Z) groups and $\underline{t}$-butyl esters.

Angew Int Ed (1975) $\underline{14}$ 818

$$
\begin{array}{c}
R \\
| \\
H_2N-CH-COOH \cdot Et_3N
\end{array}
$$

+

$$
\begin{array}{c}
t\text{-}BuO \\
\diagdown \\
C \\
\| \\
O
\end{array}
\begin{array}{c}
CN \\
\diagdown \\
O-N=\!\!\!< \\
\diagup \\
Ph
\end{array}
$$

$\xrightarrow{\text{H}_2\text{O/Dioxane}}$

$$
\begin{array}{c}
O \quad R \\
\| \quad | \\
t\text{-}BuO \diagdown C \diagup NH-CH-COOH \cdot Et_3N
\end{array}
$$

80-90%

Tetr Lett (1975) 4393

$$
\begin{array}{c}
Bz \\
| \\
H_2N-CH-COOH
\end{array}
\xrightarrow[\text{Ni, THF}]{\text{Me}_2\text{PhSiH}}
\begin{array}{c}
Bz \\
| \\
Me_2PhSi-NH-CH-COOSiMe_2Ph
\end{array}
$$

75%

Tetr Lett (1975) 3207

$$
\begin{array}{c}
O \\
\| \\
Ph-NHC-OCH_2CH_2Cl
\end{array}
\xrightarrow{\text{Co(I)phthalocyanine}}
PhNH_2
$$

83%

Angew Int Ed (1976) 15 681

Use of $Cl-\!\!\bigcirc\!\!-OCH_2O-C(CF_3)_2$ to block imidazole in histidine

(with N-heterocycle attached below)

1) stable to saponification and H_2
2) removed by mild H^+
3) derivatives soluble in organic solvents

Tetr Lett (1974) 2637

Use of 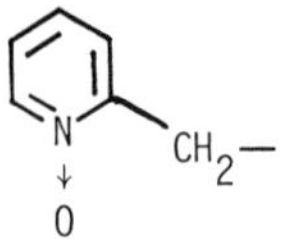 as NH_2 or OH protecting group in nucleotides and
nucleosides. Removed by Ac_2O followed by $NH_3/MeOH$.

JOC (1974) **39** 1250

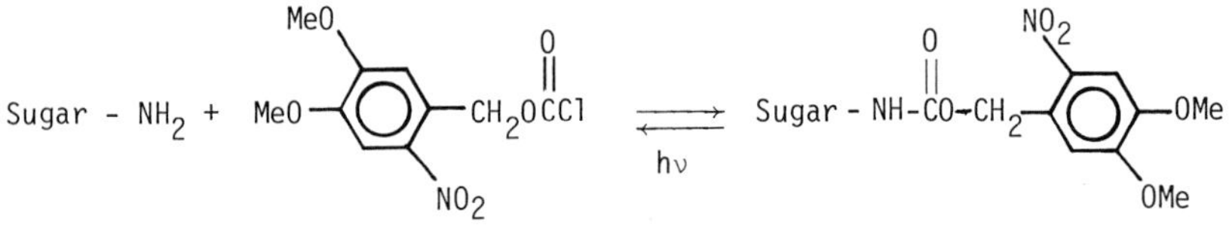

Sugar - NH_2 + $\underset{h\nu}{\rightleftharpoons}$

JOC (1974) **39** 192

Use of $PhCH_2$- to protect basic N in alkaloids.   Removed by n-PrSLi/HMPA at 0°.

Synth Comm (1974) **4** 183

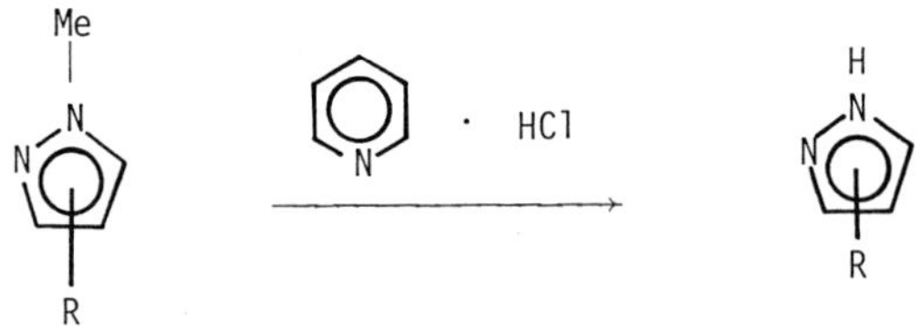

R = alkyl

50-70%

JOC (1975) **40** 1353

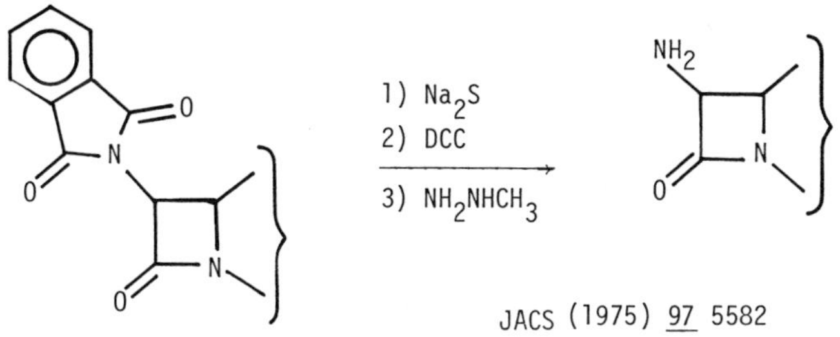

1) Na$_2$S
2) DCC
3) NH$_2$NHCH$_3$

JACS (1975) <u>97</u> 5582

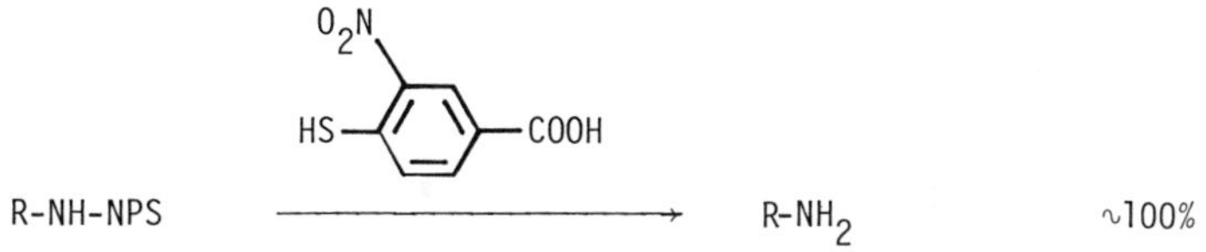

R-NH-NPS $\longrightarrow$ R-NH$_2$ ~100%

NPS = <u>o</u>-nitrophenylsulfenyl

Helv Chim Acta (1976) <u>59</u> 855

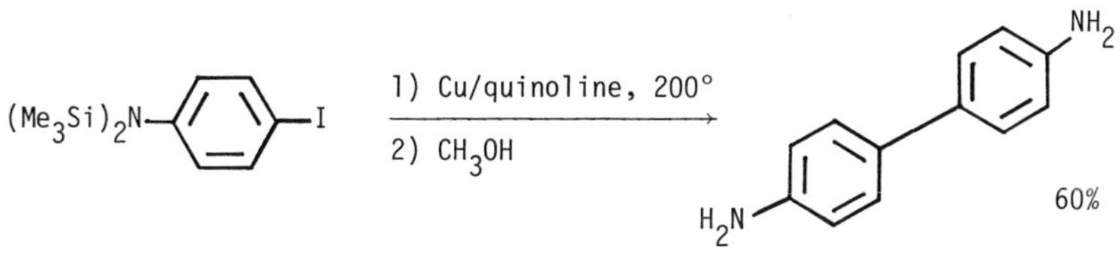

(Me$_3$Si)$_2$N—⟨benzene⟩—I

1) Cu/quinoline, 200°
2) CH$_3$OH

60%

Synthesis (1976) 40

Review: "Stability of Side Chain Protecting Groups in Solid-Phase
         Peptide Synthesis"

Israel J Chem (1974) <u>12</u> 79

Chapter 8   PREPARATION

OF

ESTERS

Section 106   Esters from Acetylenes

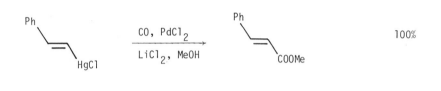

$$\text{JOC (1975) } \underline{40} \text{ 3237}$$

Section 107   Esters from Carboxylic Acids and Acid Halides

The following types of reactions are found in this section:

1. Esters from carboxylic acids (and acid halides) and alcohols.
2. Lactones from hydroxy acids.
3. Esters from carboxlic acids and halides, sulfates, and miscellaneous
    compounds.

Use of "graphite bisulfate" to catalyze esterification

$$RCOOH + R'OH \longrightarrow RCOOR'$$

Fast, high yield; excess R'OH is not required.

3° alcohols undergo elimination.

JACS (1974) <u>96</u> 8113

Use of $BF_3 \cdot Et_2O \cdot ROH$ to effect esterification of carboxylic acids.

Synth Comm (1974) <u>4</u> 167

$$\xrightarrow[\text{2) EtOH}]{\text{1) } Ph_3P(OTf)_2}$$

78%

Tetr Lett (1975) 277

$$R\text{-}COOH \xrightarrow[\text{BuOH, pyridine}]{\text{BzOH, DCC}} R\text{-}COOBz \qquad \sim 85\%$$

R = N-protected amino acid

Z Naturforsch <u>B</u> (1976) <u>31</u> 1157

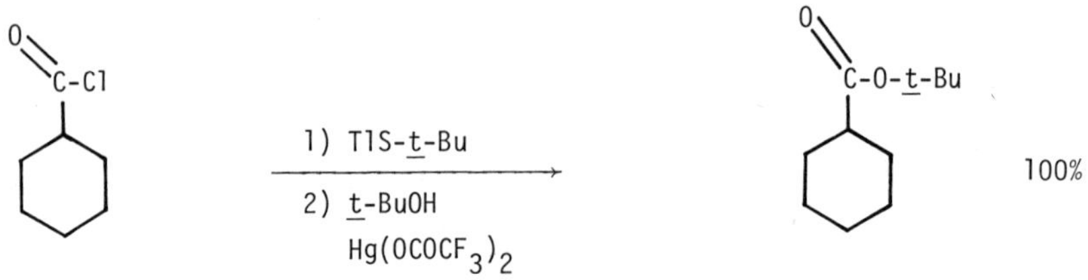

JACS (1975) <u>97</u> 3515

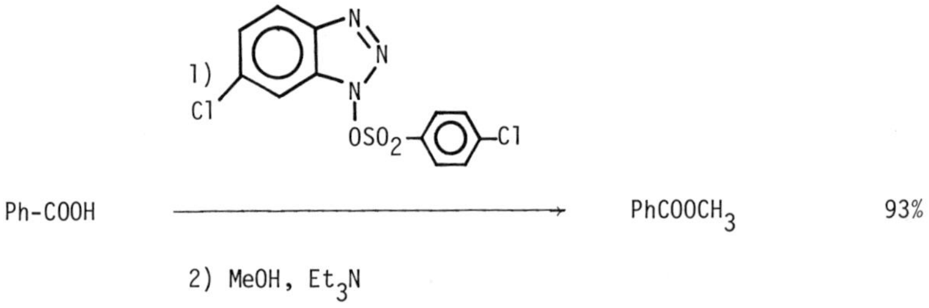

Ph-COOH ⟶ PhCOOCH₃   93%

2) MeOH, Et₃N

Synthesis (1975) 456

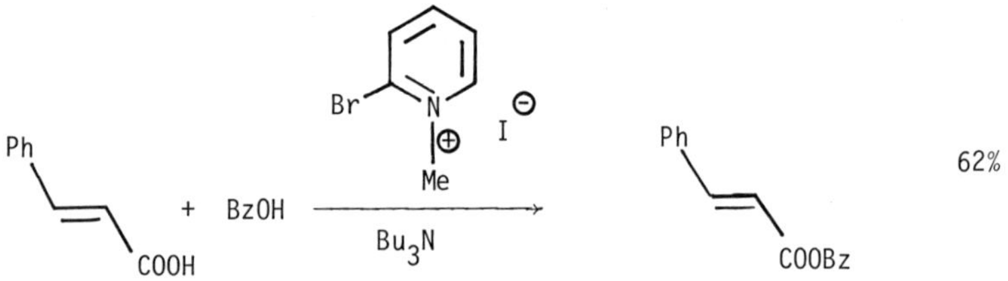

Chem Lett (1975) 1045

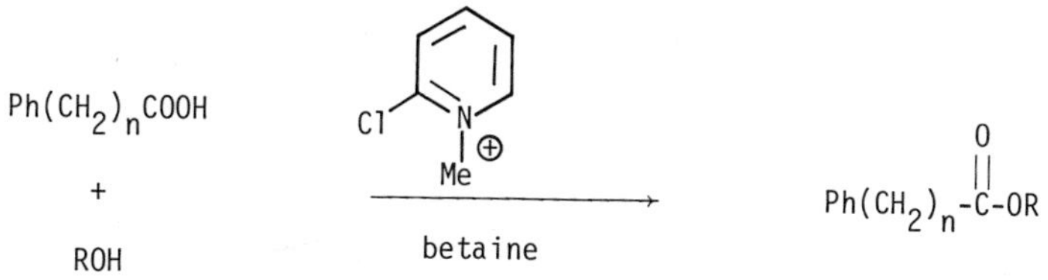

Ph(CH2)nCOOH
+
ROH
Cl
N
Me
betaine
O
Ph(CH2)n-C-OR

n = 1, 2

R = Ph, alkyl, cinnamyl, etc.

Chem Lett (1976) 13

HO-(CH2)nCOOH
Cl
N
Me
(CH2)n
O-C
O

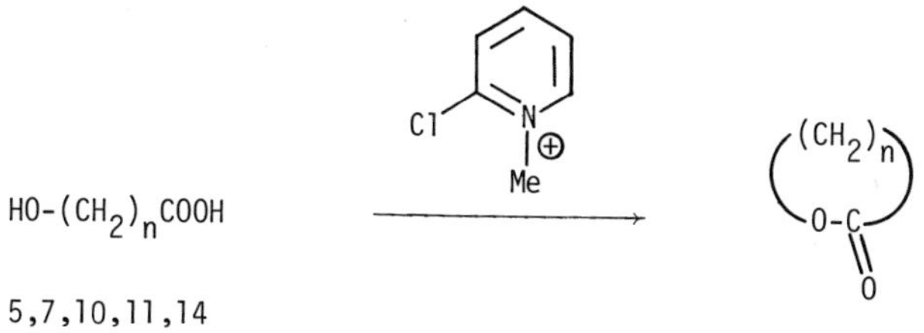

n = 5,7,10,11,14

Chem Lett (1976) 49

HO-(CH2)15-COOH
+
i-Pr
N
S
N
t-Bu
2
Ph3P
O
C
O
(CH2)15

83%

Tetr Lett (1976) 3409

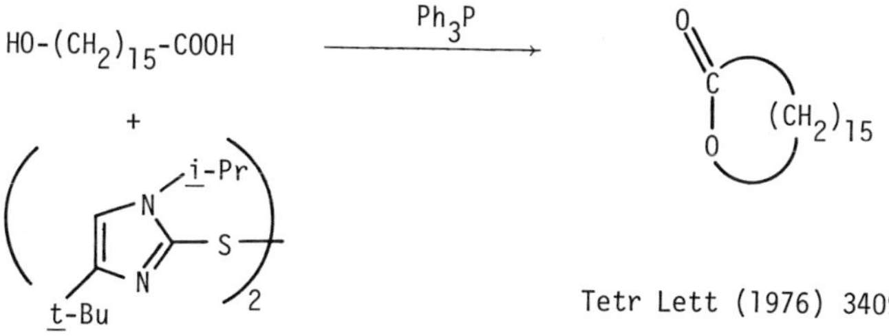

OH
COOH
2,2'-dipyridyl disulfide
Ph3P
OR
C5H11
OR
O
O
C5H11
OR
OR
92%

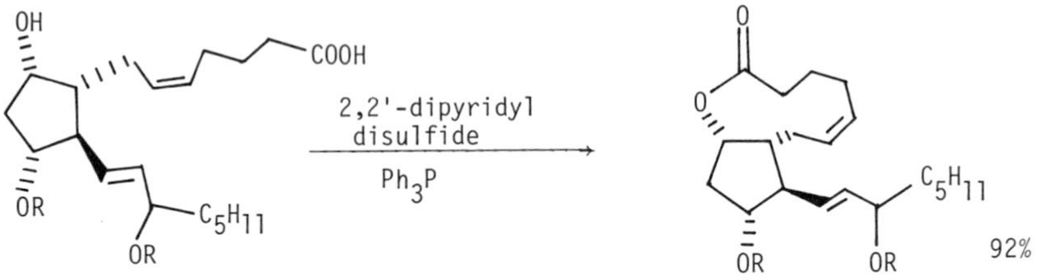

JACS (1975) 97 653, 654

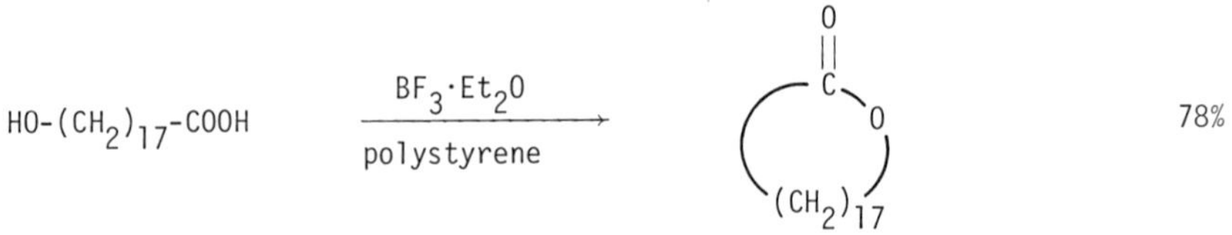

HO-(CH2)17-COOH
BF3·Et2O
polystyrene
O
C
O
(CH2)17
78%

Synthesis (1976) 738

HO-(CH2)11-COOH
+
EtOOC-N=N-COOEt
Ph3P
O
C
(CH2)11
O
63%

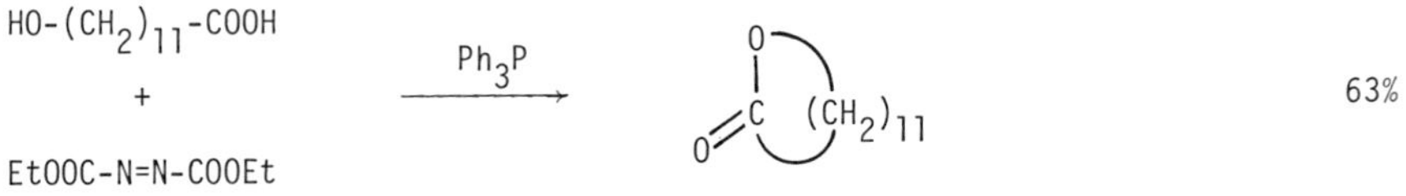

Tetr Lett (1976) 2455

$$HO-(CH_2)_{10}COOH \quad + \quad \left(\!\!\left(\bigcirc\!\!\bigcirc_{N}\!\!\right)\!\!-S\right)_2 \quad \longrightarrow \quad \xrightarrow[xylene]{\Delta} \quad$$

(cyclic ester structure) ~70%

JACS (1974) <u>96</u> 5614

Further examples of the reaction RCOOH + ROH → RCOOR are included in Section 108 (Esters from Alcohols and Phenols) and Section 10A (Protection of Carboxylic Acids).

$$R-COOK \quad + \quad BzCl \quad \xrightarrow[TMEDA]{MeCN} \quad RCOOBz \qquad \sim 80\%$$

R = hindered alkyl

Comptes Rend <u>C</u> (1976) <u>283</u> 483

(acrylic acid derivative) $\xrightarrow[\underline{i}\text{-PrBr}]{\text{Resin-NMe}_3^{\oplus}OH^{\ominus}}$ (isopropyl ester) 59%

Synthesis (1975) 723

$$Ph-COOH \quad + \quad CH_2Cl_2 \quad \xrightarrow{Bu_4N^{\oplus} OH^{\ominus}} \quad$$

(bis-benzoate of formaldehyde) 88%

Tetr Lett (1975) 2303

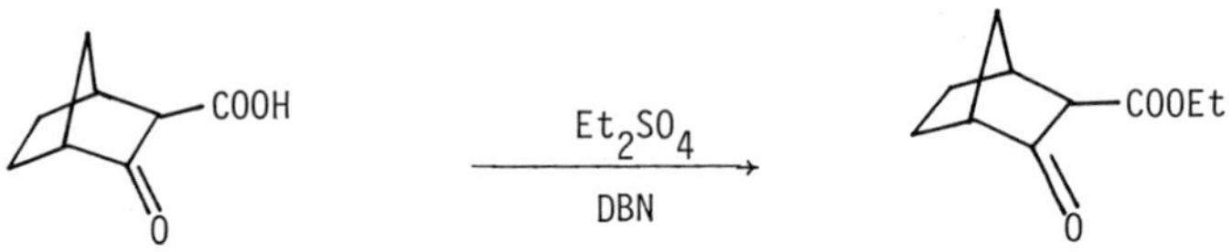

76%

Synth Comm (1976) <u>6</u> 89

$RCO_2^{\ominus}$ + $PhCH_2\overset{\oplus}{N}(Me)_2Ph$ $\xrightarrow{\Delta}$ RCO_2CH_2Ph 65-80%

Synthesis (1974) 727

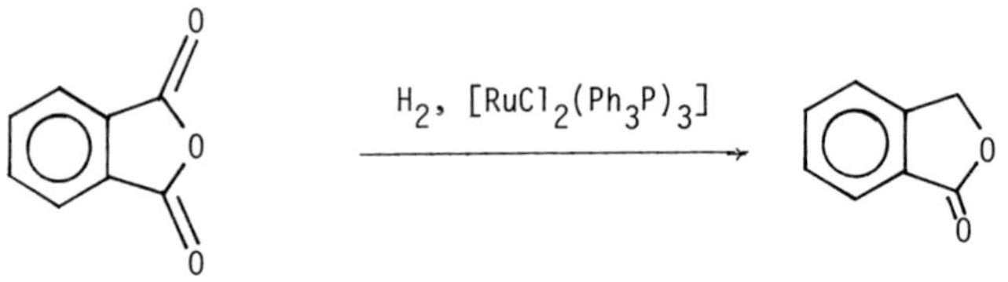

H_2, $[RuCl_2(Ph_3P)_3]$

100%

JCS Chem Comm (1975) 412

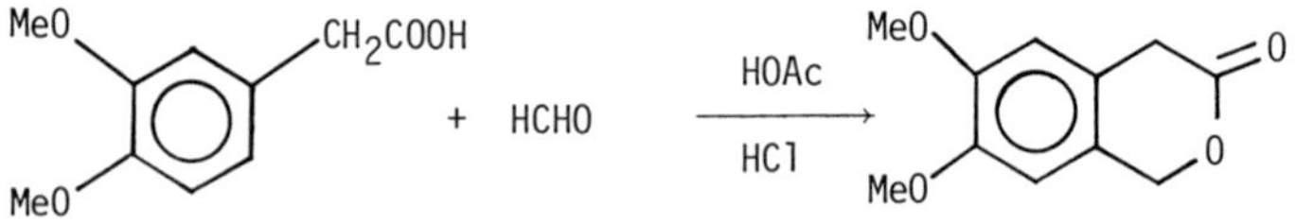

+ HCHO

HOAc
—————→
HCl

56%

Org Synth (1976) <u>55</u> 45

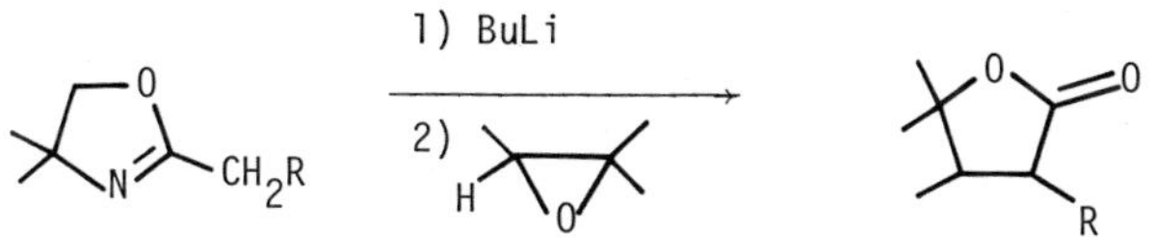

1) BuLi

2)

high yield

JOC (1974) 39 2783

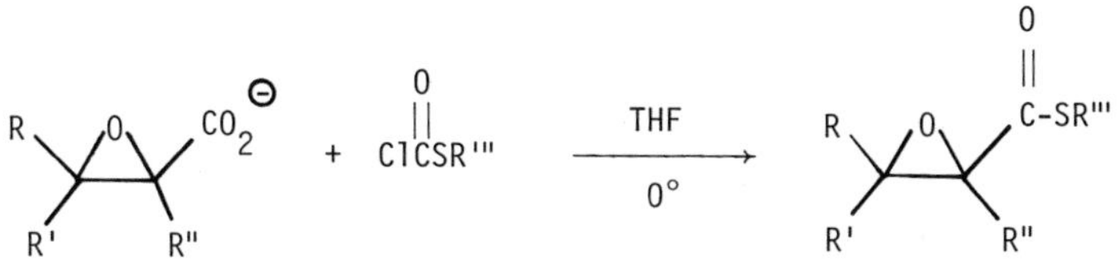

+ ClCSR'''

THF
0°

40-80%

Synthesis (1974) 811

RCOOH + R'SH $\xrightarrow[\text{Et}_3\text{N in DMF}]{\text{NCPO(OEt)}_2}$ RC-SR'   70-95%

JOC (1974) 39 3302

Section 108   Esters from Alcohols and Phenols

$\underline{n}\text{-}C_{16}H_{33}OTHP$ $\xrightarrow[\text{CH}_2\text{Cl}_2]{\text{TFAA}}$ $\underline{n}\text{-}C_{16}H_{33}OCOCF_3$ 89%

Synth Comm (1976) <u>6</u> 21

MeOH $\xrightarrow[\text{TMED}]{\overset{\displaystyle O}{\overset{\displaystyle \|}{\text{Ph-C-CF(CF}_3)_2}}}$ $\overset{\displaystyle O}{\overset{\displaystyle \|}{\text{Me-O-C-Ph}}}$ 89%

Chem Lett (1976) 673

JOC (1976) <u>41</u> 165

Further examples of the reaction ROH → R'COOR are included in Section 107
(Esters from Carboxylic Acids and Acid Halides) and Section 45A (Protection
of Alcohols and Phenols).

BuONa + CO $\xrightarrow[\text{BuOH, THF}]{\text{Se}}$ $(\text{BuO})_2\text{CO}$ 80%

Bull Chem Soc Japan (1975) <u>48</u> 108

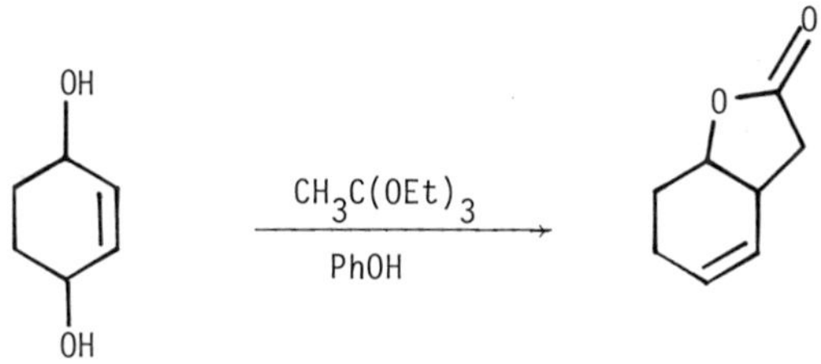

OH
OH
$CH_3C(OEt)_3$
PhOH
O
O

85%

Angew Int Ed (1975) 14 103

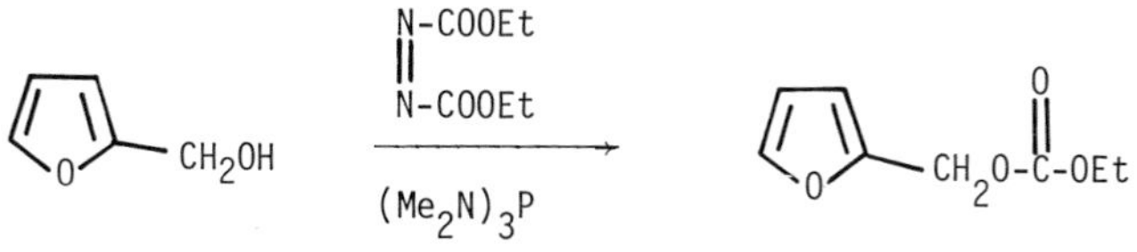

N-COOEt
N-COOEt
$(Me_2N)_3P$
CH_2OH
O
CH_2O-C-OEt
O

71%

Tetrahedron (1975) 31 1411

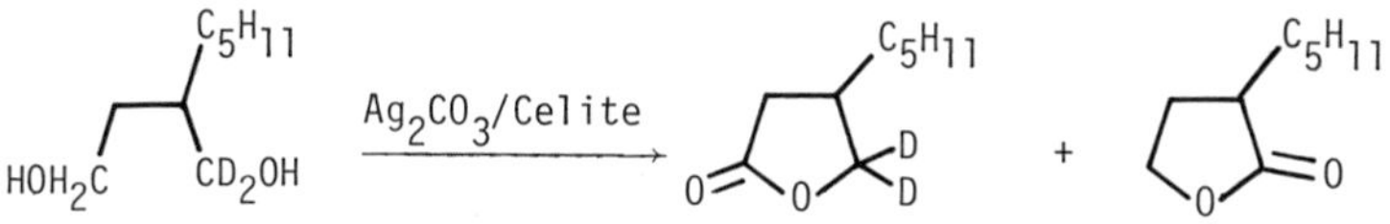

C_5H_{11}
HOH_2C
CD_2OH
Ag_2CO_3/Celite
D
D
O
+
O

90% 10%

Tetrahedron (1975) 31 171

OH
H_2O_2/PhCN
O
O

~80%

Tetr Lett (1976) 3305

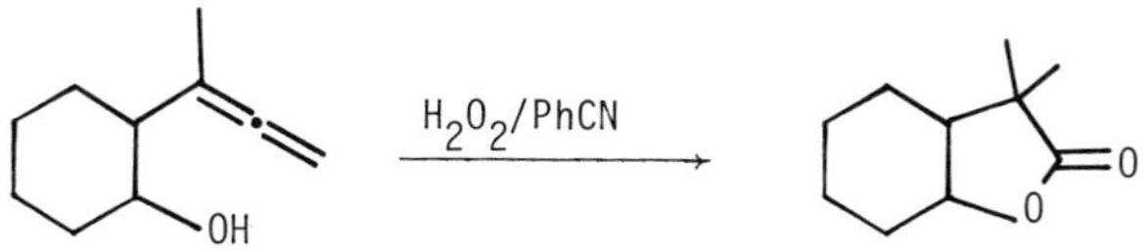

Section 109   Esters from Aldehydes

ROSnBu$_3$

    + $\xrightarrow[\text{CCl}_4]{\text{NBS}}$ R'-C-OR   ~90%

R'CHO

R = alkyl
R'= alkyl, Ph

JACS (1976) 98 1629

Ph$_2$COH

        $\xrightarrow[\substack{\text{2. Ag}_2\text{CO}_3\text{-celite} \\ \text{xylene, rfx}}]{\text{1. 10\% H}_2\text{SO}_4}$ 71%

JOC (1975) 40 892

PhCH(OCH$_2$CH=CH$_2$)$_2$ $\xrightarrow{\Delta}$ Ph-C-OCH$_2$CH=CH$_2$ 72%

Synth Comm (1975) 5 213

Related methods: Esters from Ketones (Section 117)

Section 110   Esters from Alkyls, Methylenes and Aryls

No examples of the reaction RR → RCOOR' or R'COOR (R,R'=alkyl, aryl, etc.)
occur in the literature. For the reaction RH → RCOOR' or R'COOR see
Section 116 (Esters from Hyrides).

Section 111   Esters from Amides

HO~~~~C(=O)NMe$_2$ $\xrightarrow{\text{H}^{\oplus}(\text{ion-exchange resin})}$ (lactone) 86%

Chem Ber (1975) 108 48

RCONHR' $\xrightarrow{\text{PhS(OR}_f)_2}$ RCO$_2$R$_f$ + R'N=SPh$_2$

R$_f$ = PhC(CF$_3$)$_2$

JACS (1975) 97 6137

Section 112   Esters from Amines

No additional information

Section 113   Esters from Esters

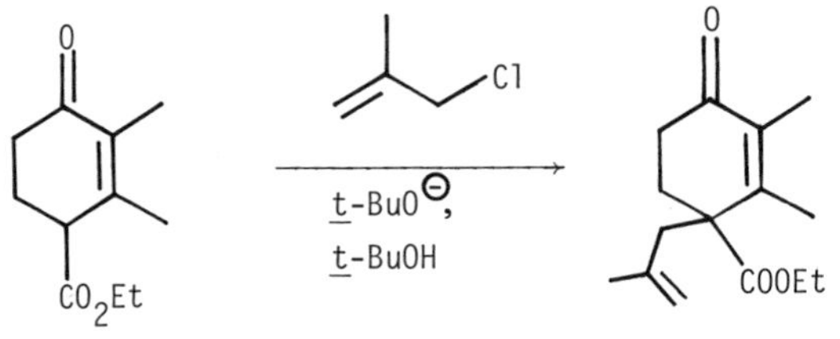

JOC (1974) 39 2323

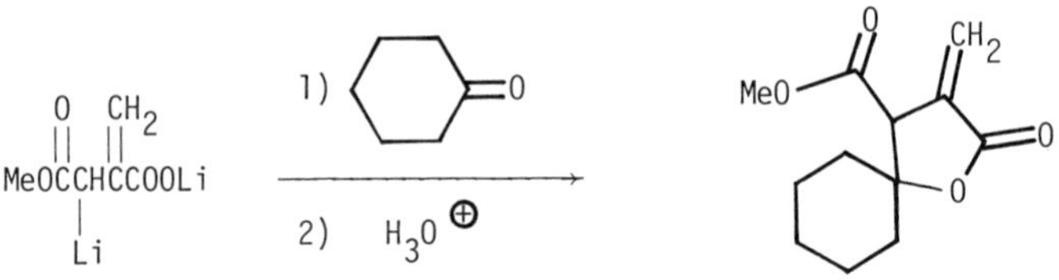

71%

JOC (1976) 41 4065

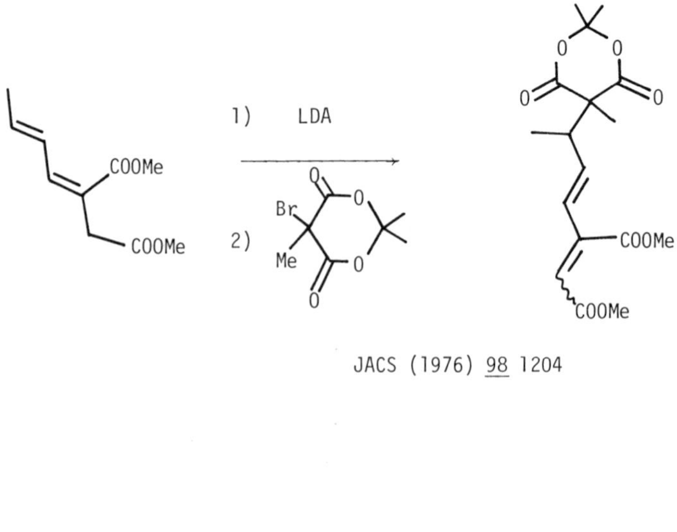

66%

JACS (1976) 98 1204

$C_6H_{13}CH=CHCOOMe$　$\xrightarrow[\text{2) PhCH}_2\text{Br}]{\text{1) L-Selectride}}$　$C_7H_{15}CH(CH_2Ph)CO_2Me$　　50%

JOC (1975) <u>40</u> 2846

Et_3SiCH_2COOMe　$\xrightarrow[\text{2) MeI}]{\text{1) (Me}_3\text{Si)}_2\text{NNa}}$　$Et_3SiCHMeCO_2Me$　　86%

J Gen Chem USSR (1975) <u>45</u> 78

$EtCCl_2COO\text{-}\underline{i}\text{-}Pr$　$\xrightarrow[\text{2) MeI}]{\text{1) Li, THF}}$　$EtCCl(Me)CO_2\text{-}\underline{i}\text{-}Pr$　　85%

J Organometal Chem (1975) <u>102</u> 129

$MeSCH_2CH=CHCOOMe$　$\xrightarrow[\text{2) MeI, HMPA}]{\text{1) LDA, THF}}$　$MeSCH=CHCHMeCOOMe$　　92%

Tetr Lett (1975) 405

$EtOOCCH_2N=C(SMe)_2$ + $\underline{t}$-BuOK + MeI $\longrightarrow$ $EtOOC\text{-}CH\text{-}N=C(SMe)_2$　　70%

$\underset{\displaystyle Me}{\overset{\displaystyle |}{}}$

Angew Chem Int Ed (1975) <u>14</u> 424

MeC≡CCOOH $\xrightarrow[\substack{\text{THF, HMPA}}]{\text{1) Li, TMP}}$ Me$_2$C=CH(CH$_2$)$_2$C≡CCOOMe   50%

2) Me$_2$C=CHCH$_2$Br

3) MeI

JOC (1975) <u>40</u> 269

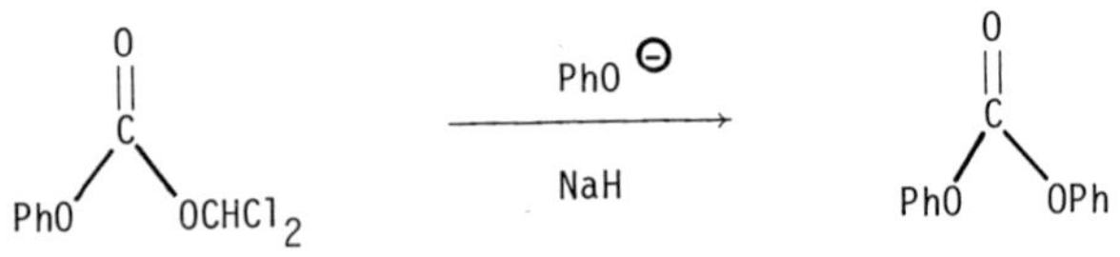

JOC (1976) <u>41</u> 2039

CH$_2$(COOMe)$_2$

+

Bz-Br

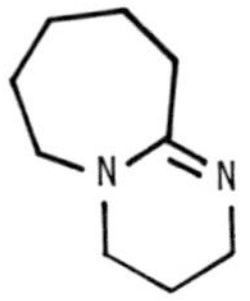

Bz-CH(COOMe)$_2$ 84%

Liebigs Ann Chem (1976) 348

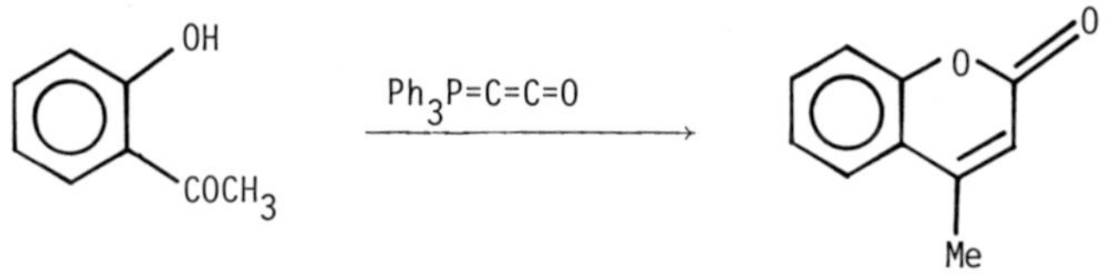

Angew Int Ed (1976) 15 115

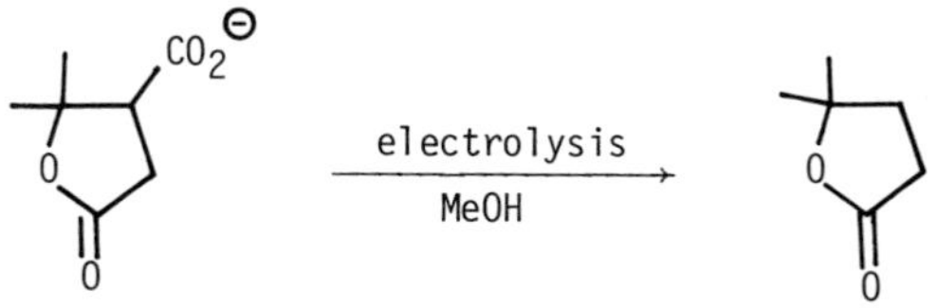

JOC (1974) 39 2486

$$\underset{\substack{\| \\ O}}{EtO-C-H} \quad + \quad NaOMe \quad + \quad O_2 \quad \xrightarrow[\text{cat.}]{\text{Se}} \quad \underset{\substack{\| \\ O}}{EtO-C-OMe}$$

Tetr Lett (1974) 804

PhCHO

+

$$\xrightarrow{\hspace{2cm}}$$

$$\underset{\substack{\| \\ N_2}}{\overset{OSiMe_3}{\underset{|}{PhCHC-COOEt}}}$$

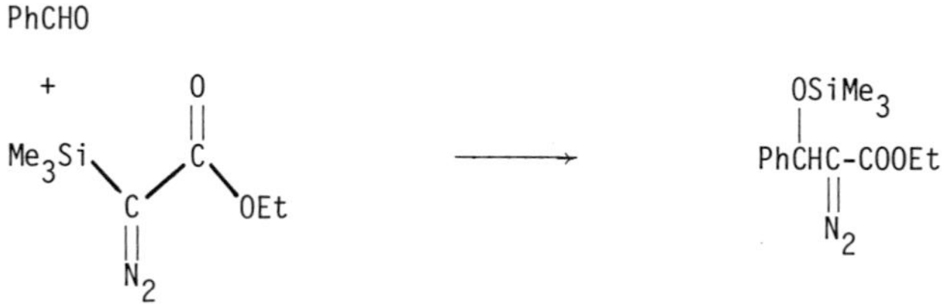

86%

JOC (1976) 41 3335

Conjugate reductions of unsaturated esters are listed in Section 74
(Alkyls from Olefins)

Section 114   Esters from Ethers

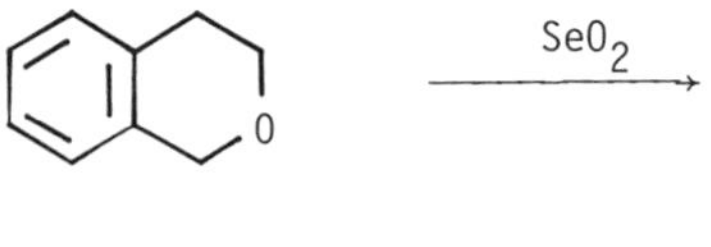

Org React (1976) 24, 261

$R_1R_2\overset{Li}{\underset{}{C}}CO_2Li$ +

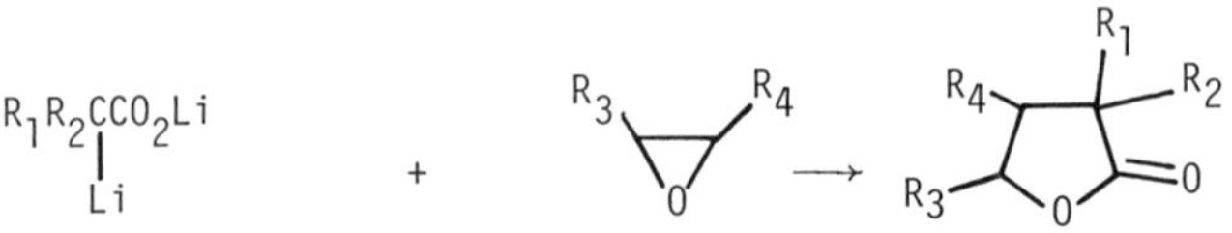

~ 80%

Aust J Chem (1974) 27 2205

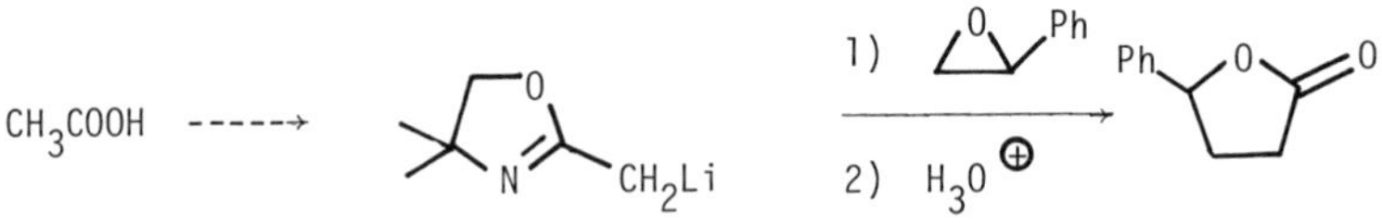

CH_3COOH ----->

89%

JOC (1974) 39 2783

R-OR' $\xrightarrow{\text{FeCl}_3\text{-Ac}_2\text{O}}$ ROAc + R'OAc

(allyl ethers not rearranged or isomerized)

JOC (1974) $\underline{39}$ 3728

Section 115   Esters from Halides

ArX $\xrightarrow[\text{MeOH, Et}_2\text{NH}]{\text{Pd}^{++}, \text{CO}}$ ArCOOMe

Bull Chem Soc Japan (1975) $\underline{48}$ 2075
JOC (1975) $\underline{40}$ 532

PhI + CO + EtOH + Et$_3$N + Pd(0) cat $\longrightarrow$ PhCOOEt   90%

MeNH$_2$ $\longrightarrow$ PhCONHMe

JOC (1974) $\underline{39}$ 3318
JOC (1974) $\underline{39}$ 3327

$\xrightarrow[\text{AgOAc}]{\text{AcOH}}$ 40-60%

JOC (1974) $\underline{39}$ 1761

RBr + Hg(NO$_3$)$_2$ $\longrightarrow$ RONO$_2$ 60-100%

28 cases

Tetrahedron (1974) $\underline{30}$ 2467

RBr + Hg(OAc)$_2$ + AcOH $\longrightarrow$ ROAc   88-98%

15 cases

Tetrahedron (1974) $\underline{30}$ 2467

RX + Hg(O$_2$CR')$_2$ $\xrightarrow{\text{B(O$_2$CR')$_3$}}$ RO$_2$CR'

RX + NaO$_2$CR' $\xrightarrow{\text{HMPA}}$ RO$_2$CR'   Comparison of these methods.

RX + AgO$_2$CR' $\xrightarrow{\text{THF}}$ RO$_2$CR'

JOC (1974) $\underline{39}$ 3721

JOC (1975) $\underline{40}$ 1186

HO_2CCH_2COOEt $\xrightarrow[\text{2) } PhCH_2Cl]{\text{1) LiICA, THF}}$ $PhCH_2CH_2COOEt$ 75%

JOC (1975) <u>40</u> 2556

Related methods: Carboxylic Acids from Halides (Section 25)

Section 116   <u>Esters from Hydrides</u>

This section contains examples of the reaction RH → RCOOR' or R'COOR
(R=alkyl, allyl, aryl, etc.) and ArH → Ar-X-COOR (X=alkyl chain)

Tetr Lett (1975) 3973

OMe
OMe
bis-(2,2'-bipyridine)-silver(II)-peroxydisulfate
HOAc
OMe
OAc
OMe

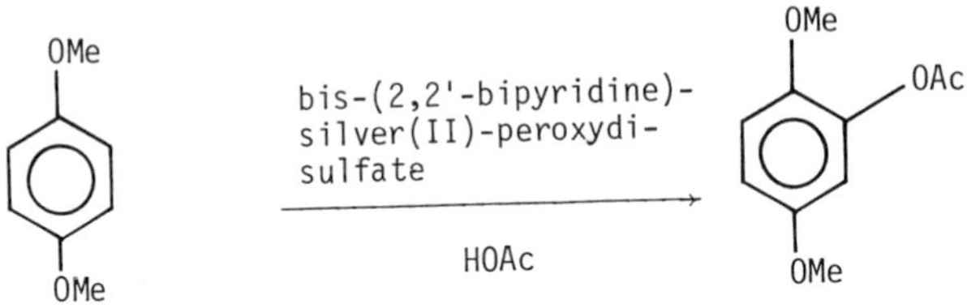

183% (based on Ag)

Acta Chem Scand B (1975) 29 629

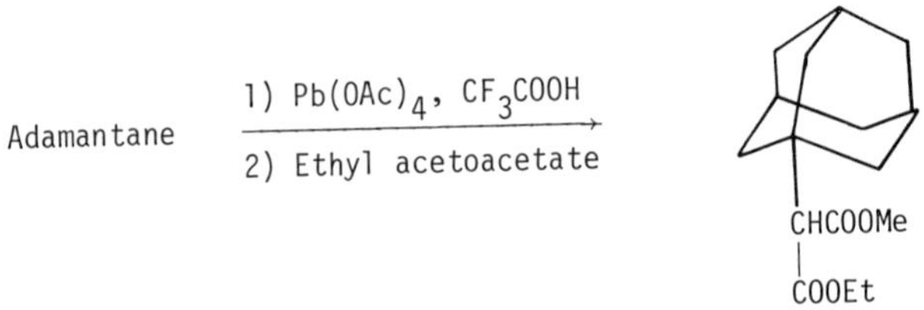

Adamantane
1) $Pb(OAc)_4$, CF_3COOH
2) Ethyl acetoacetate
CHCOOMe
COOEt

87%

Synthesis (1976) 32

Also via: Carboxylic acids, Section 26; Alcohols, Section 41

Section 117 Esters from Ketones

O
$K_2Cr_2O_7$
H_2SO_4/Et_2O
O
O

80%

Tetr Lett (1975) 1841

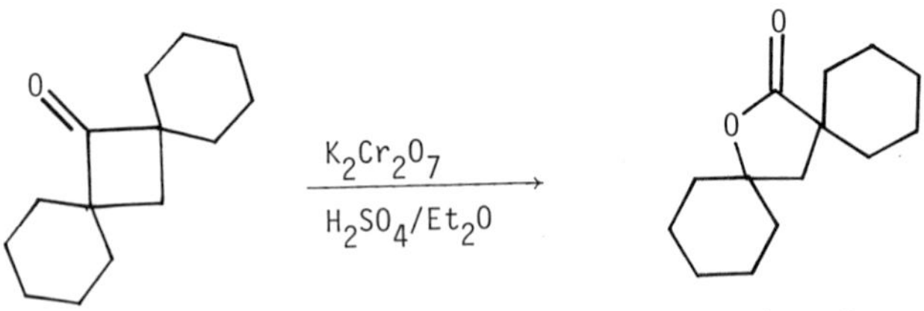

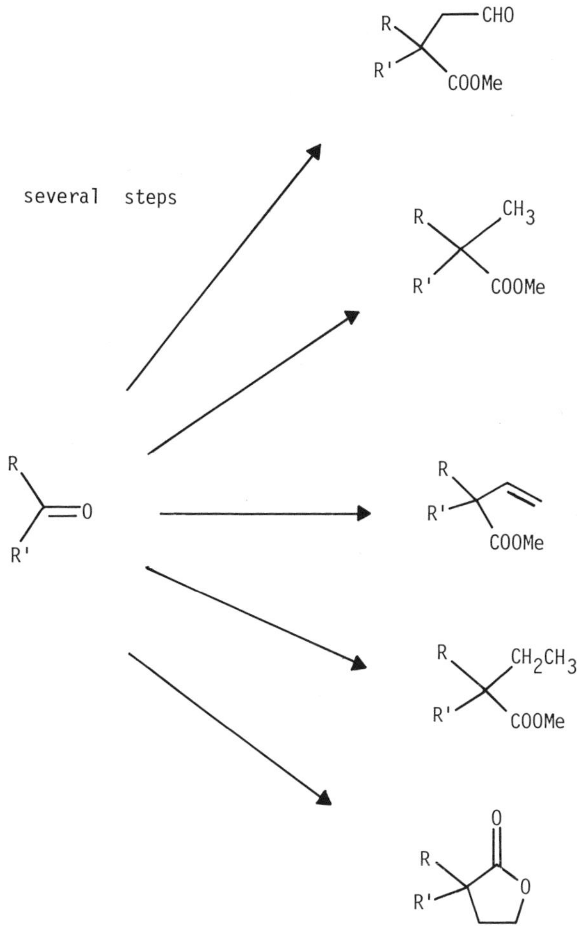

several steps

JACS (1975) 97 2218, 2224

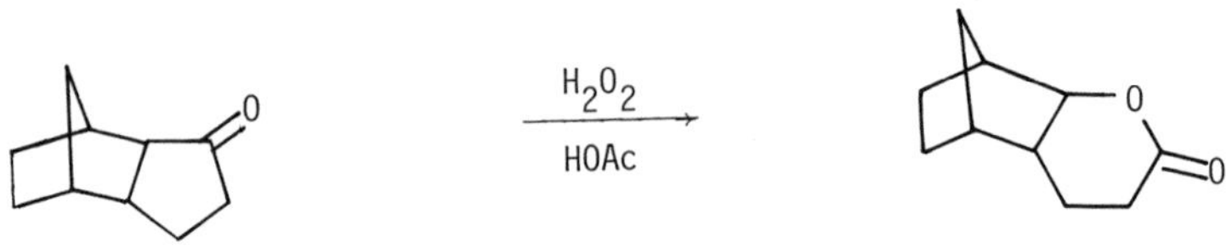

$$\xrightarrow[\text{HOAc}]{H_2O_2}$$

85%

Synthesis (1975) 404

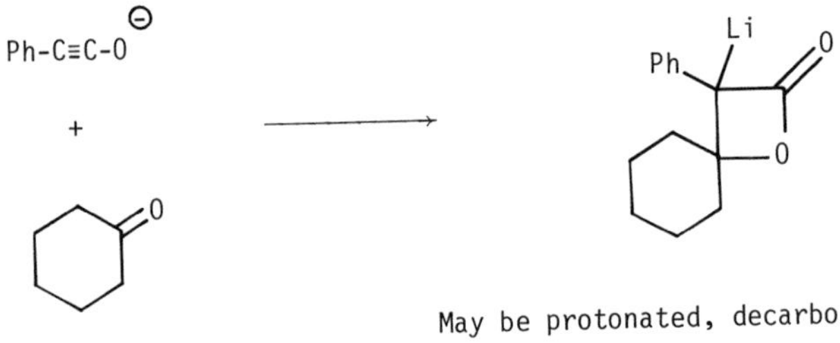

Ph-C≡C-O$^{\ominus}$

+

51%

May be protonated, decarboxylated,
or further alkylated.

Angew Int Ed (1975) 14 765

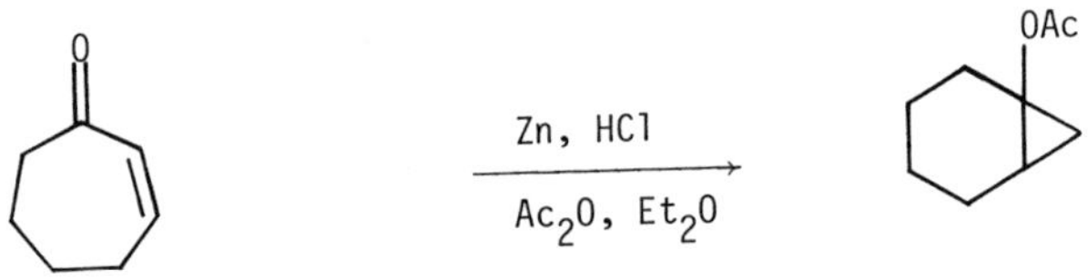

$$\xrightarrow[\text{Ac}_2O, \text{ Et}_2O]{\text{Zn, HCl}}$$

40%

Helv Chim Acta (1976) 59 962

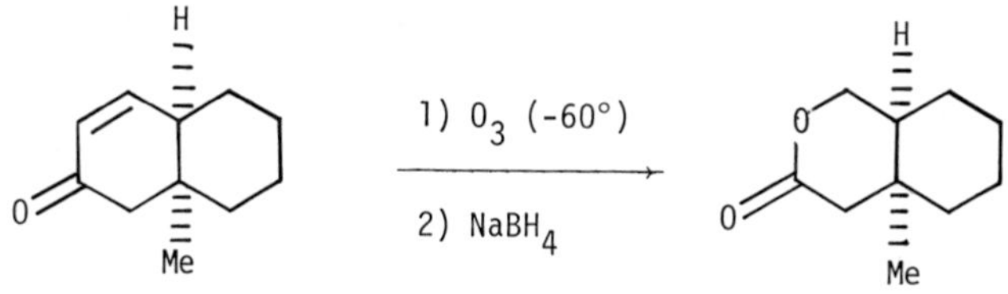

JOC (1975) 40 2970

Also via Carboxylic acids, Section 27

Section 118   Esters from Nitriles

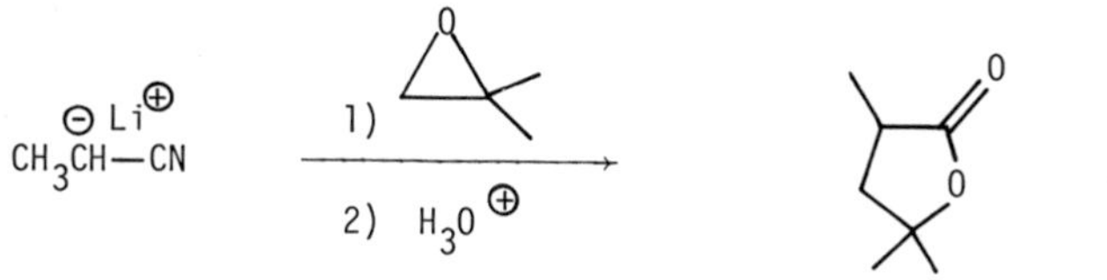

Synthesis (1976) 238

Section 119   Esters from Olefins

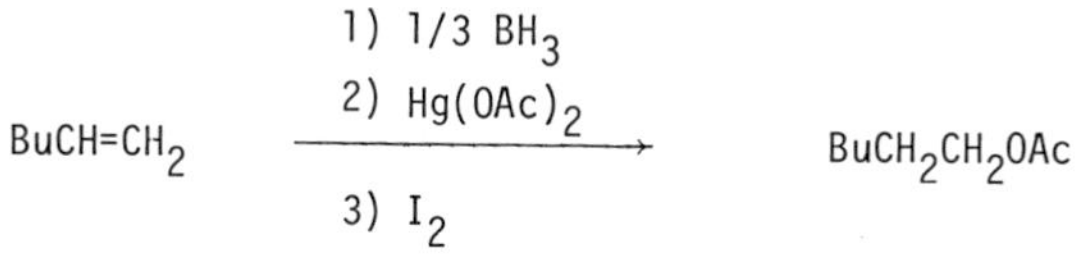

BuCH=CH$_2$ 1) 1/3 BH$_3$
               2) Hg(OAc)$_2$ ⟶   BuCH$_2$CH$_2$OAc   86%
               3) I$_2$

JOC (1974) 39 834

Me   n-C$_5$H$_{11}$
$\diagup\!\diagdown$

$\xrightarrow[\text{Rh(O-CO-\underline{t}-Bu)}_2]{\text{N}_2\text{CHCOO-}\underline{t}\text{-Bu}}$

$$\underset{\text{Me}\qquad\qquad\underline{n}\text{-C}_5\text{H}_{11}}{\overset{\overset{\text{O}}{\underset{\|}{\text{C-O-}\underline{t}\text{-Bu}}}}{\underset{\text{H}\qquad\qquad\text{H}}{\triangle}}}$$ 89%

Synthesis (1976) 600

$\xrightarrow[\text{CH}_3\text{OH}]{\text{O}_3}$ COOMe / COOMe   83%

Angew Int Ed (1975) 14 716

$$\underset{\text{R}}{\text{CH}_2\text{=CHCH}_2\overset{|}{\text{CH}}\text{COOH}}$$

$\xrightarrow[\text{2) H}_2\text{O}_2]{\text{1) disiamyl borane}}$ 54-84%

R = H, 1°, 2°, 3° alkyl

Chem Pharm Bull (1976) 24 538

$$\underset{\text{R}^2}{\overset{\text{R}^1}{\diagup}}\!\!=\!\!\underset{\text{R}^4}{\overset{\text{R}^3}{\diagdown}} \quad + \quad \underset{\text{R}^6}{\overset{\text{R}^5}{\diagup}}\!\!\text{CH-}\overset{\overset{\text{O}}{\|}}{\text{C}}\text{-OH}$$

$\xrightarrow{\text{Mn(OAc)}_3}$ 30-60%

JACS (1974) 96 7977
JOC (1974) 39 3456

Also via Alcohols, Section 44

Section 120   Esters from Miscellaneous Compounds

No additional examples

Section 120A   Protection of Esters

No additional examples

Chapter 9 PREPARATION OF ETHERS AND EPOXIDES

Section 121   <u>Ethers and Epoxides from Acetylenes</u>

No examples

Section 122   <u>Ethers and Epoxides from Carboxylic Acids</u>

No additional examples

Section 123   <u>Ethers and Epoxides from Alcohols and Phenols</u>

$$ROH \ + \ CH_3I \ + \ NaH \ \longrightarrow \ ROCH_3 \qquad \text{high yields}$$

Synthesis (1974) 434

Br—⟨benzene ring⟩—OH

+

EtOH

$$\xrightarrow[\text{Ph}_3\text{P}]{\begin{array}{c}\text{N-CO}_2\text{Et}\\ \|\\ \text{N-CO}_2\text{Et}\end{array}}$$

OEt—⟨benzene ring⟩—Br   90%

JCS Perkin I (1975) 461

$\text{RBr} + \text{HgClO}_4 + \text{R'OH} \longrightarrow \text{ROR'}$ 74-98%

23 cases

Tetrahedron (1974) $\underline{30}$ 2467

$\text{ArOH} + \text{RX} \xrightarrow[\text{cat.}]{\text{phase transfer}} \text{ArOR}$

Tetrahedron (1974) $\underline{30}$ 1379

Polyphenol + Me_2SO_4 + NaOAc $\xrightarrow[\text{dioxane}]{\text{acetone-}}$ methylation of most acidic OH group

Indian J Chem (1974) $\underline{12}$ 893

$\text{PhOH} \xrightarrow[\begin{array}{c}\text{N-COOEt}\\ \|\\ \text{N-COOEt}\end{array}]{\text{CH}_3\text{OH, Ph}_3\text{P}} \text{PhOCH}_3$ 98%

Chem and Ind (1975) 281

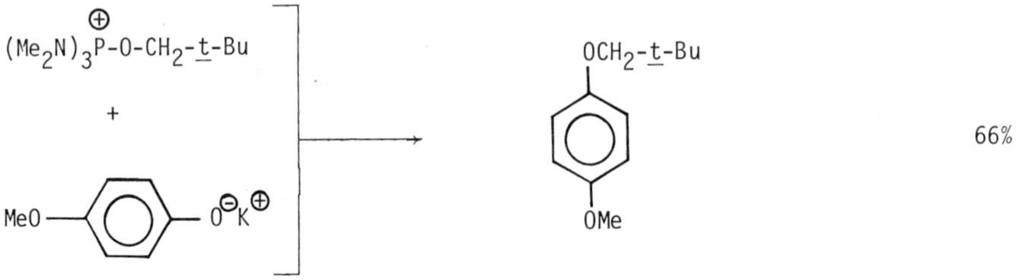

$(Me_2N)_3\overset{\oplus}{P}-O-CH_2-\underline{t}-Bu$

+

MeO—⟨ ⟩—$O^{\ominus}K^{\oplus}$

OCH$_2$-$\underline{t}$-Bu

⟨ ⟩

OMe

66%

Angew Int Ed (1975) <u>14</u> 370

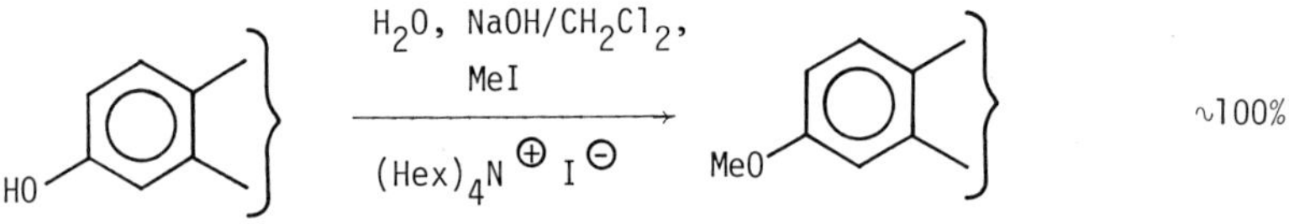

H_2O, $NaOH/CH_2Cl_2$,
MeI

$(Hex)_4\overset{\oplus}{N} I^{\ominus}$

∿100%

Steroids (1976) <u>28</u> 481

Related methods: Protection of Alcohols and Phenols (Section 45A)

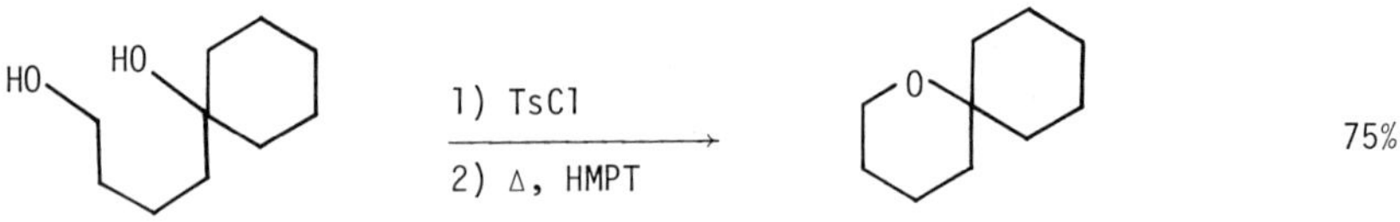

1) TsCl

2) Δ, HMPT

75%

Tetr Lett (1975) 2731

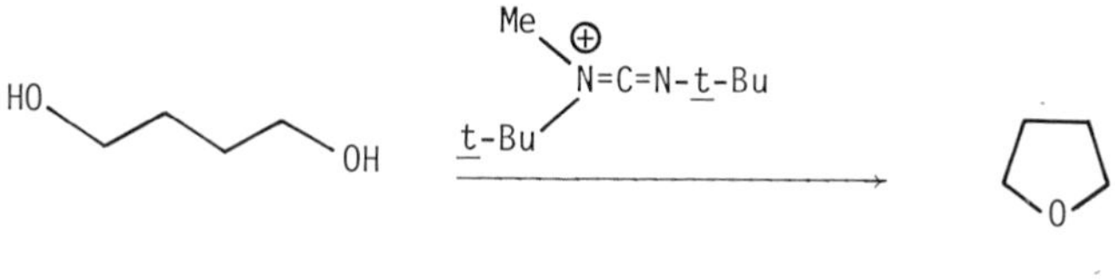

JACS (1975) 97 465

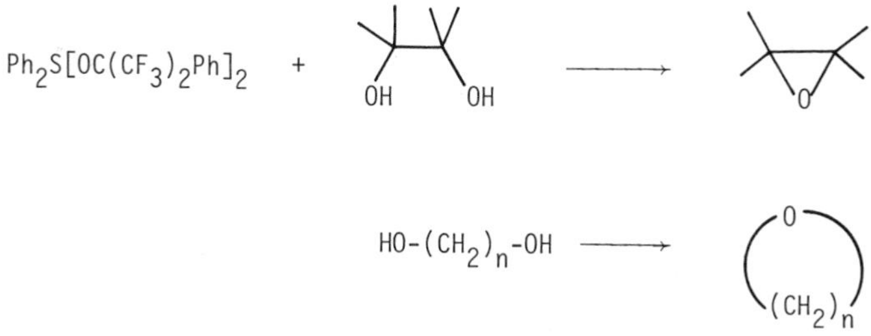

moderate to high yields

JACS (1974) 96 4604

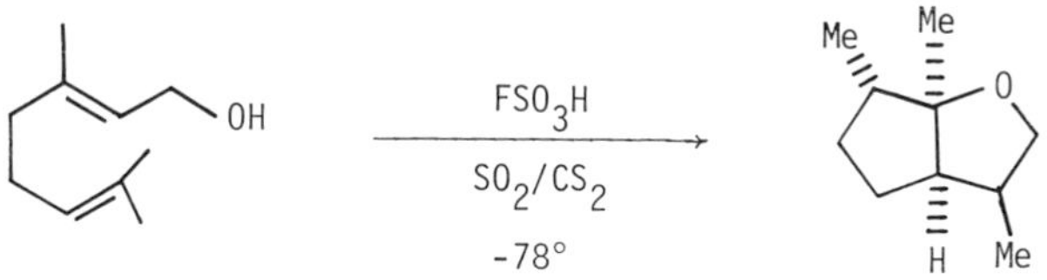

JCS Perkin I (1974) 1637

$$\xrightarrow{\text{NaBiO}_3}$$

molecular weight $\sim 10^4$

JOC (1975) 40 1515

Section 124   Ethers and Epoxides from Aldehydes

$$RCHO + R'OH \longrightarrow R-\overset{\overset{\displaystyle OR'}{|}}{\underset{\underset{\displaystyle H}{|}}{C}}-OR' \xrightarrow[\text{THF}]{\text{B}_2\text{H}_6} RCH_2OR' \qquad 60\text{-}90\%$$

Can J Chem (1974) 52   888

$$RCHO + Et_3SiH \xrightarrow[\text{CHCl}_3]{\text{AcOH}} (RCH_2)_2O$$

JOC (1974) 39 2740

PhCHO

+ $\xrightarrow[\text{TEBA}]{\text{50\% aq. NaOH}}$

$TsCH_2Cl$

60%

JOC (1975) 40 266

PhCHO $\xrightarrow[\substack{(-)\text{-N,N-dimethylephedrinium} \\ \text{bromide}}]{\substack{\oplus \ominus \\ Me_3SI, NaOH}}$

77%

(R)-67% ee

JACS (1975) 97 1626

Section 125   Ethers and Epoxides from Alkyls, Methylenes and Aryls

No examples of the preparation of ethers and epoxides by replacement of alkyl, methylene and aryl groups occur in the literature. For the conversion of RH → ROR' (R,R'=alkyl) see Section 131 (Ethers from Hydrides)

Section 126   Ethers and Epoxides from Amides

No additional examples

Section 127   Ethers and Epoxides from Amines

No additional examples

Section 128   Ethers and Epoxides from Esters

No additional examples

Section 129   Ethers and Epoxides from Ethers and Epoxides

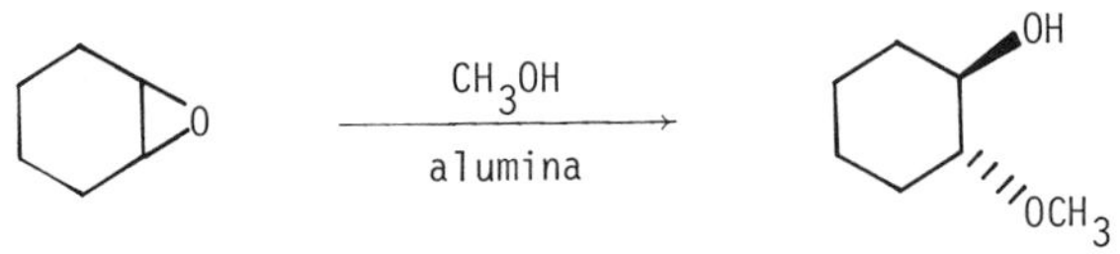

Tetr Lett (1975) 3597

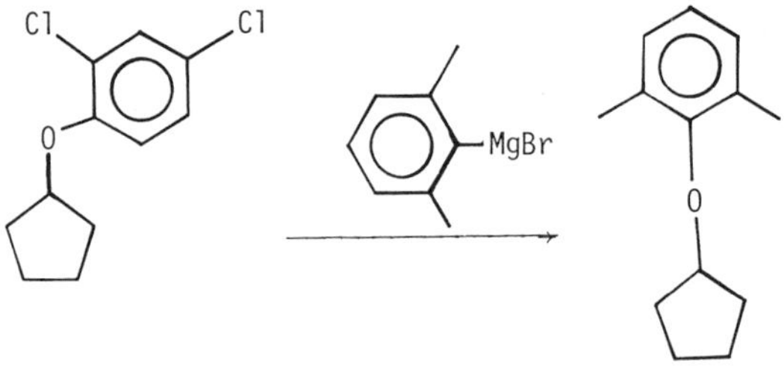

66%

Chem Lett (1975) 1051

79%

Section 130   Ethers from Halides

BzCl

\+ $\xrightarrow[\text{Bu}_4\text{N}\oplus\text{HSO}_4\ominus]{\text{50\% NaOH}}$ Bu-O-Bz   92%

Bu-OH

Tetr Lett (1975) 3251

MeOCH$_2$CH$_2$OH $\xrightarrow[\text{2) BzBr, MeCN}]{\text{1) TlOEt}}$ MeOCH$_2$CH$_2$OBz   81%

Angew Int Ed (1975) <u>14</u> 762

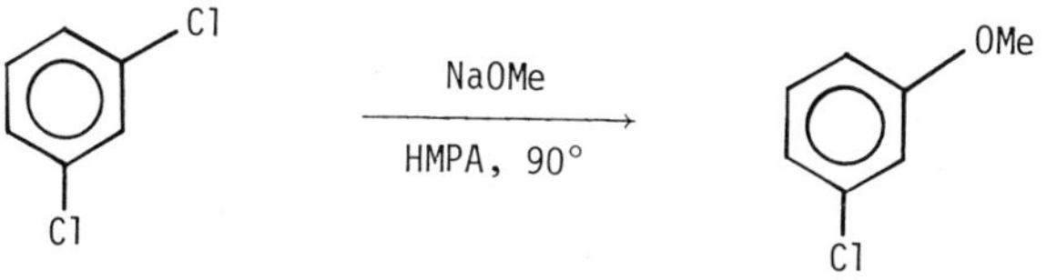

$\xrightarrow[\text{HMPA, 90°}]{\text{NaOMe}}$

87%

JOC (1976) <u>41</u> 732

$\xrightarrow{\text{BuLi}}$

25°

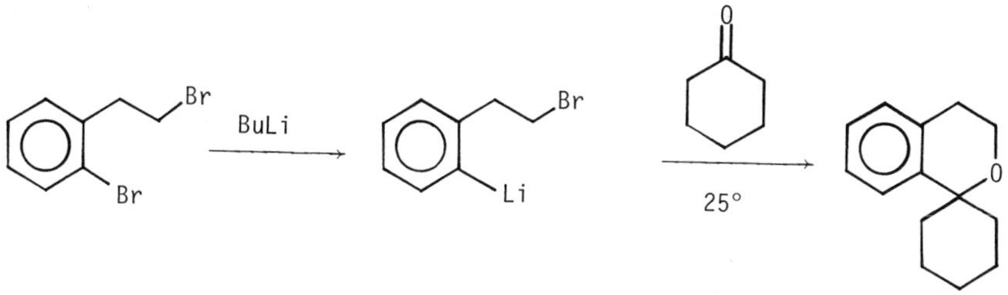

75%

JOC (1976) <u>41</u> 1184

Section 131   Ethers from Hydrides

No additional examples

Section 132   Ethers and Epoxides from Ketones

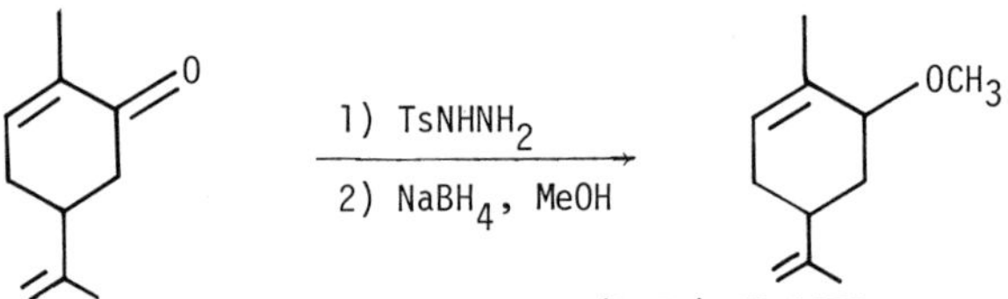

1) TsNHNH$_2$

2) NaBH$_4$, MeOH

JOC (1976) 41 1755

77%

2) AgBF$_4$, MeI

3) KO-$\underline{t}$-Bu, DMSO

+

PhCHO

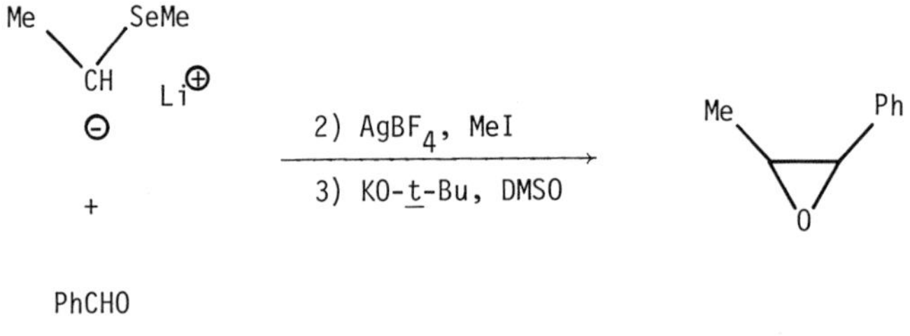

78%

Angew Int Ed (1975) 14 350, 700

NTs
‖
Ph-S-CH$_2$CH$_3$ ───── 1) NaH, DMSO / 2) PhCOCH$_3$ ─────→

Ph, Me — epoxide — Me   54%

Synthesis (1976) 35

Ph$_2$CO
+
TsCH$_2$Cl ───── 50% aq. NaOH / TEBA ─────→

Ph, Ph — epoxide — Ts   90%

JOC (1975) 40 266

R^{1}CHBrC-S-t-Bu   NaH
        ‖
        O

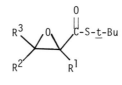

$$\underset{R^2}{\overset{O}{\|}}\underset{R^3}{C}$$

R^3, R^2 — epoxide — R^1, C-S-t-Bu
                              ‖
                              O

JOC (1975) 40 3173

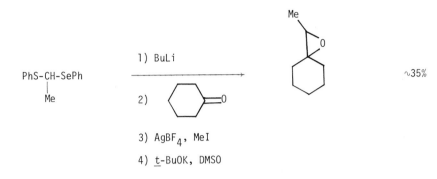

Tetr Lett (1975) 1617

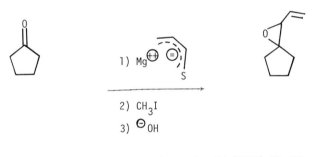

Angew Int Ed (1976) 15 437

Section 133   Ethers and Epoxides from Nitriles

No additional examples

Section 134   Ethers and Epoxides from Olefins

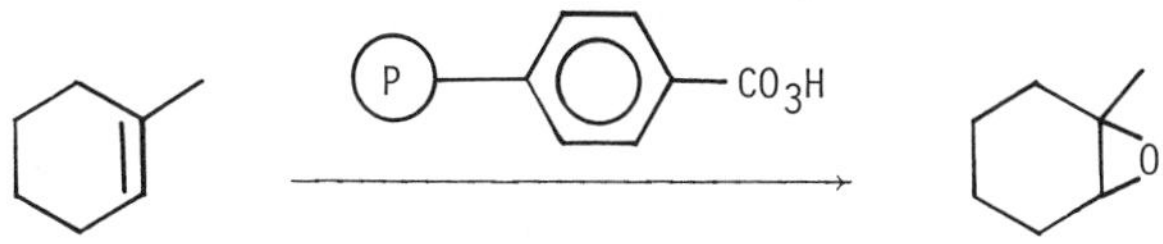

65%

JCS Perkin I (1976) 605

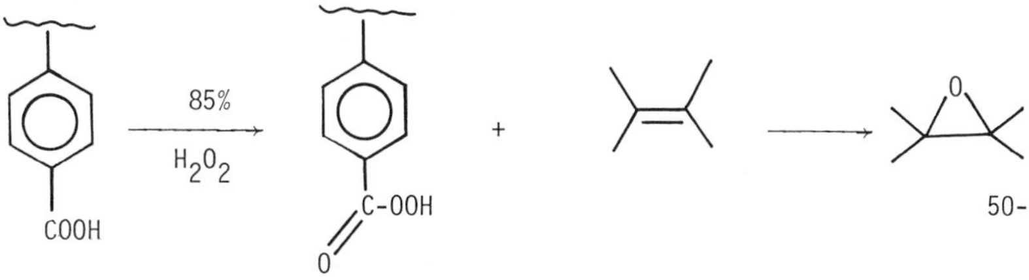

50-90%

Use of polymer supported peracids to epoxidize olefins.

JCS Chem Comm (1974) 1009

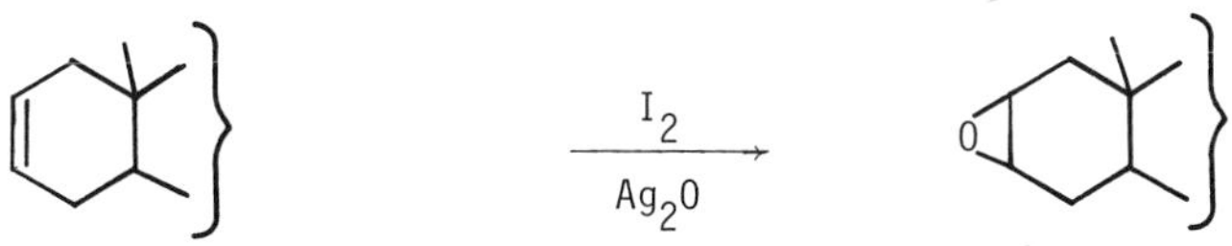

~80%

Tetr Lett (1976) 207

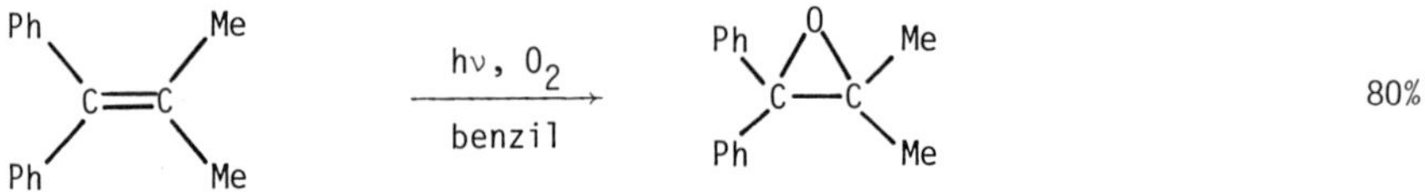

80%

JACS (1976) 98 4193

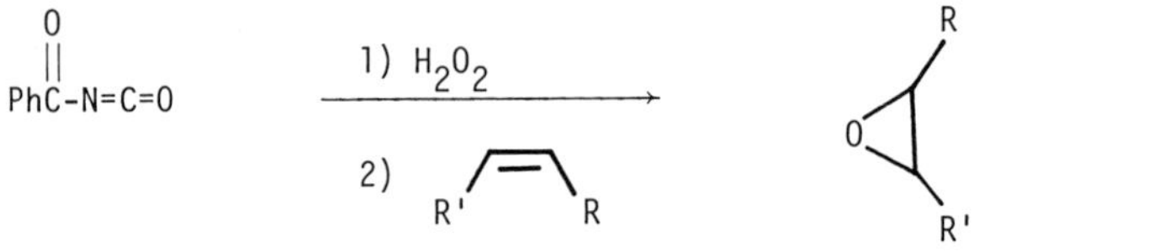

JCS Chem Comm (1974) 711

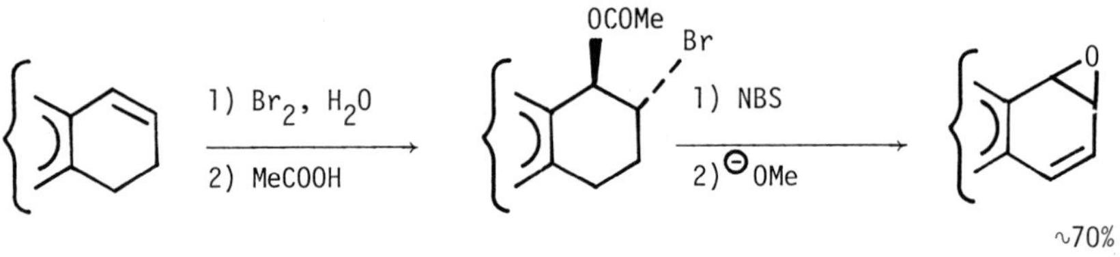

~70%

JACS (1975) 97 3185

55%

Tetr Lett (1976) 3779

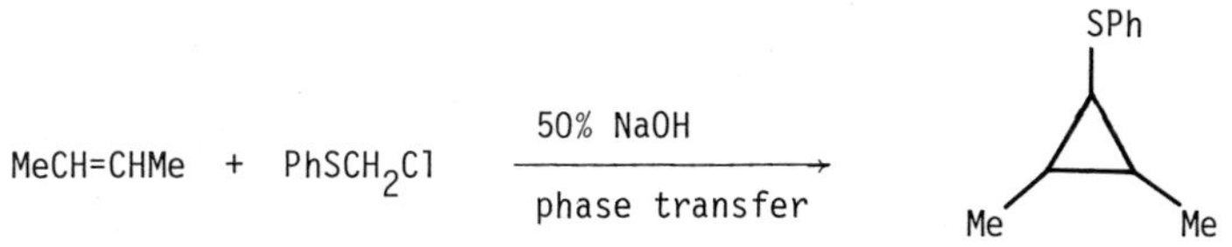

MeCH=CHMe   +   PhSCH$_2$Cl $\xrightarrow[\text{phase transfer}]{\text{50\% NaOH}}$

SPh

Me   Me

79%

Tetr Lett (1975) 4247

Review: "Metal Ketenides in the Catalytic Epoxidation of Olefins"

Chem and Ind (1975) 154

Section 135   Epoxides from Miscellaneous Compounds

No additional examples

Chapter 10 PREPARATION OF HALIDES AND SULFONATES

Section 136 Halides and Sulfonates from Acetylenes

$CH_3\text{-}C{\equiv}C\text{-}CH_2CH_3 \xrightarrow{MoCl_5}$ 38%

JACS (1975) 97 1599

Section 137 Halides from Acid Halides

$PhCOCl + SF_4 \xrightarrow{95°} PhCF_3$ 95%

Org React (1974) 21 1

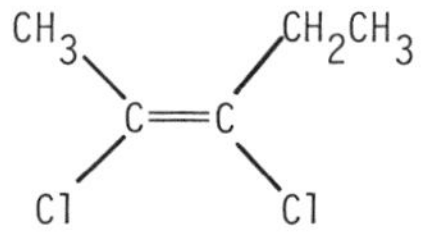

Section 138   Halides and Sulfonates from Alcohols

$$geranyl-OH \xrightarrow[\quad CCl_4 \quad]{Ph_3P} geranyl-Cl \qquad \sim 80\%$$

$$\underline{or} \xrightarrow[\quad LiCl, HMPA/Et_2O \quad]{MeLi, TsCl}$$

Org Synth (1974) 54 63, 68

$$PhCH_2OH \xrightarrow[\quad CCl_4 \quad]{\text{(P)}-\text{phenyl}-PPh_2} PhCH_2Cl \qquad 88\%$$

JCS Chem Comm (1975) 622

$$\underline{n}-BuOH \xrightarrow[\quad ZnCl_2 \quad]{PCl_3, DMF} \underline{n}-BuCl \qquad 87\%$$

Synthesis (1976) 398

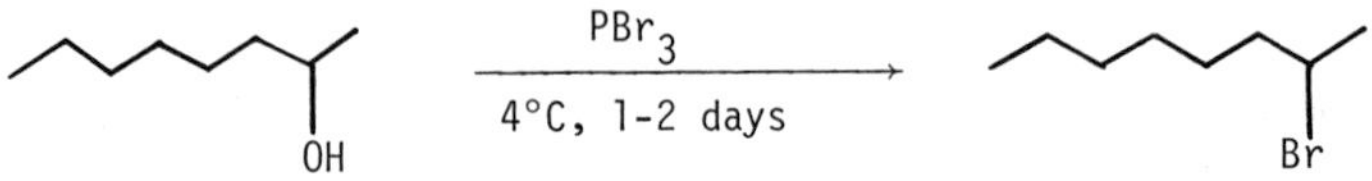

67%

97% retention of configuration

JOC (1976) 41 1071

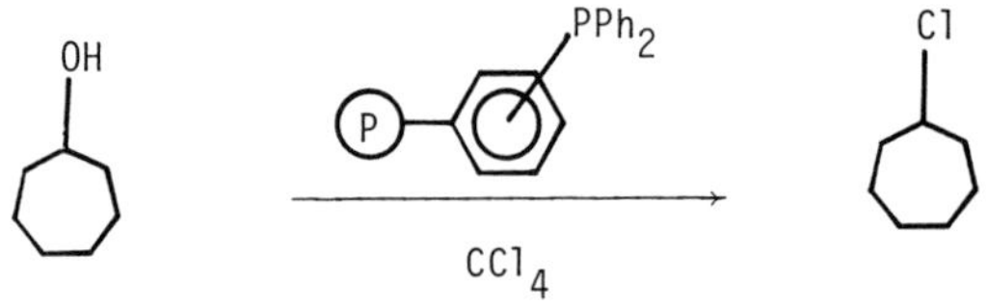

92%

JOC (1975) 40 1669

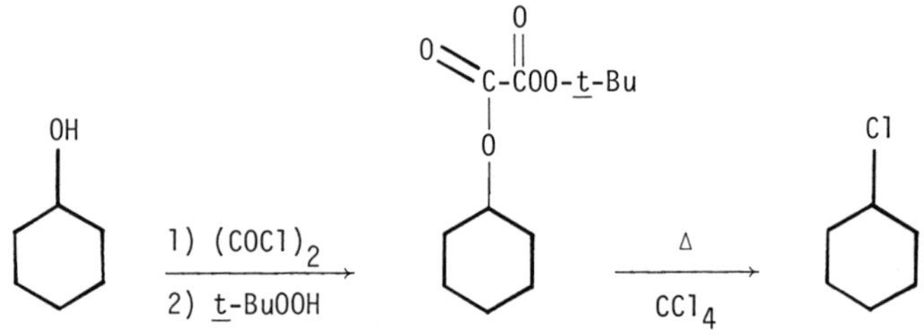

48%

JACS (1975) 97 2281

R-OH + PCl₅ $\xrightarrow[\text{CaCO}_3]{\text{CHCl}_3}$ R-Cl

R = 3° alkyl   retention of configuration

Aust J Chem (1976) 29 133

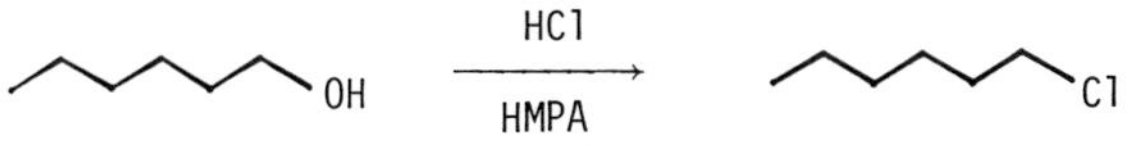

Can J Chem (1975) 53 3620

n-Hexyl-OH   +   aq. HCl $\xrightarrow[\text{cat.}]{\text{phase transfer}}$ n-Hexyl Cl   >90%

Synthesis (1974) 37

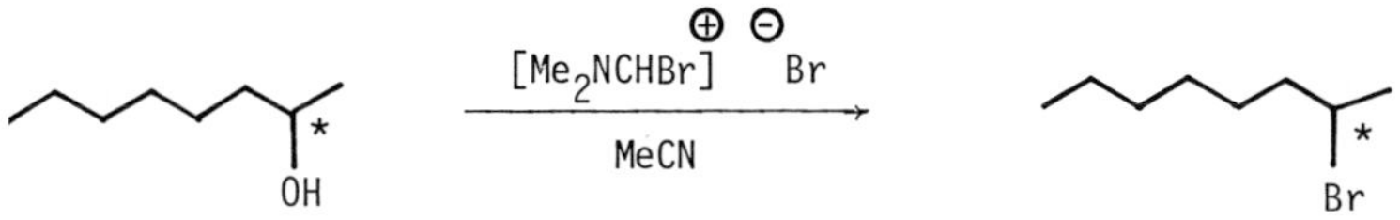

∿80% yield

∿75% ee (retention)

JCS Perkin I (1976) 754

BzOH $\xrightarrow[\text{CH}_2\text{Cl}_2]{\text{Et}_2\text{NSF}_3}$ BzF   100%

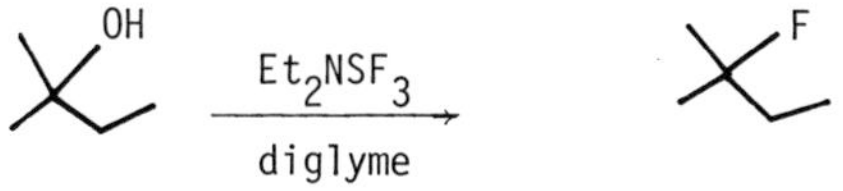

JOC (1975) 40 574

$C_6F_5CH_2OH$ + SF_4 $\xrightarrow{85°}$ $C_6F_5CH_2F$ 80%

Org React (1974) 21 1

ROH + SeF_4 $\longrightarrow$ RF   40-90%

JACS (1974) 96 925

JACS (1975) 97 2566   98%

Section 139   Halides and Sulfonates from Aldehydes

PhCHO + SF_4 $\xrightarrow{160°}$ $PhCHF_2$ 81%

Org React (1974) 21 1

Section 140   Halides and Sulfonates from Alkyls

No additional examples

For the conversion RH→RHal see Section 146 (Halides from Hydrides)

Section 141　　Halides and Sulfonates from Amides

No additional examples

Section 142　　Halides from Amines

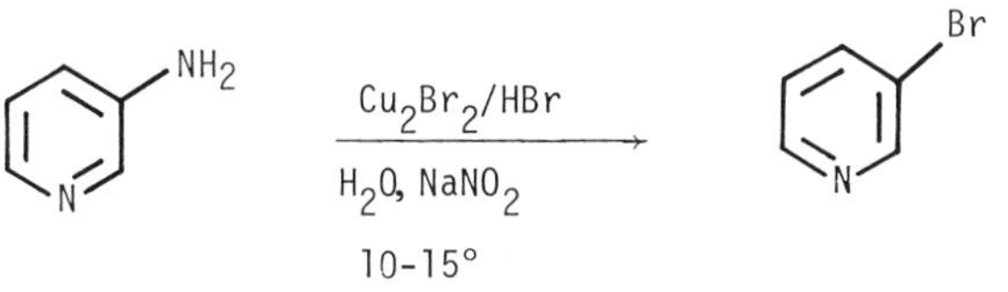

Synthesis (1974) 292

Section 143　　Halides and Sulfonates from Esters

No additional examples

Section 144　　Halides from Ethers

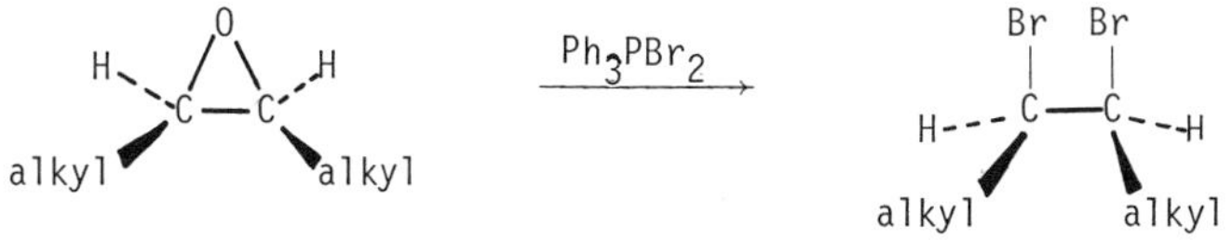

JOC (1976) 41 3279

Section 145   Halides from Halides and Sulfonates

$$RX \quad + \quad KF \quad \xrightarrow[\text{120-160°}]{\text{phase-transfer catalyst}} \quad RF \quad \text{good yield}$$

Synthesis (1974) 428

$$\underline{n}\text{-}C_{16}H_{33}Br \quad \xrightarrow{\text{resin-NMe}_3 \overset{\oplus}{} F \overset{\ominus}{}} \quad \underline{n}\text{-}C_{16}H_{33}F \qquad 77\%$$

Synthesis (1976) 472

1) Li, Ph$_2$
2) FClO$_3$

Helv Chim Acta (1976) 59 1027

$$\begin{array}{c} \text{CH}_3 \\ | \\ \text{Hex-CHOMs} \\ (R) \end{array} \quad \xrightarrow[\text{phase-transfer catalyst}]{\text{KCl}} \quad \begin{array}{c} \text{CH}_3 \\ | \\ \text{Hex-CHCl} \\ (S) \end{array} \qquad 74\%$$

89% ee

Synthesis (1975) 430

R-Cl $\xrightarrow[\text{NaBr, EtBr}]{}$ R-Br   ∿80%

R = 1° alkyl

Bull Chem Soc Japan (1976) <u>49</u> 1989

$\underline{n}\text{-C}_{16}\text{H}_{33}\text{OTHP}$ $\xrightarrow{\text{PPh}_3\text{Br}_2}$ $\underline{n}\text{-C}_{16}\text{H}_{33}\text{Br}$ 87%

Synth Comm (1976) <u>6</u> 21

$\xrightarrow{\text{MgBr}_2}$ 83%

Angew Int Ed (1975) <u>14</u> 824

$\underline{t}$-BuCl   +   FeCl$_3$ +   I$^{\ominus}$ $\longrightarrow$ $\underline{t}$-BuI   95%

Tetr Lett (1974) 2691

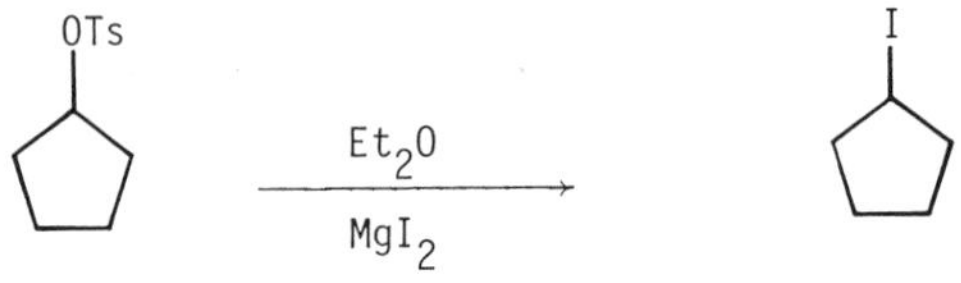

80%

Bull Soc Chim France (1976) 169

$Me_2C(Cl)Et$ $\xrightarrow[ZnCl_2]{NaI,\ CS_2}$ $Me_2C(I)Et$ 96%

JCS Perkin I (1976) 416

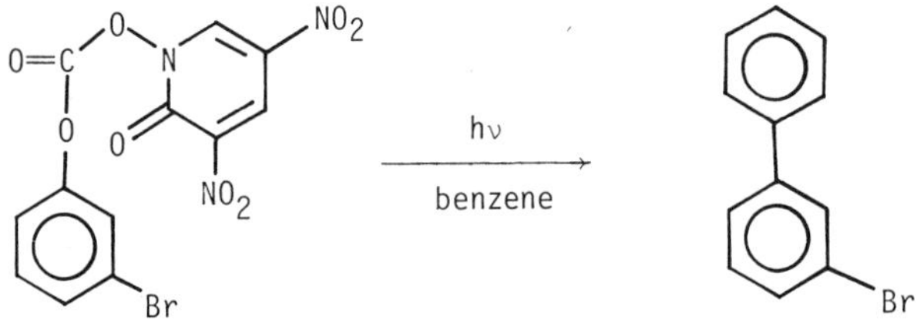

56%

JOC (1976) 41 24

$PhSO_2CH_2CH_2OH$ $\xrightarrow[2)\ Hexyl\ Br]{1)\ BuLi}$ $PhSO_2CH(Hex)CH_2OH$ ~100%

Bull Soc Chim France (1976) 519

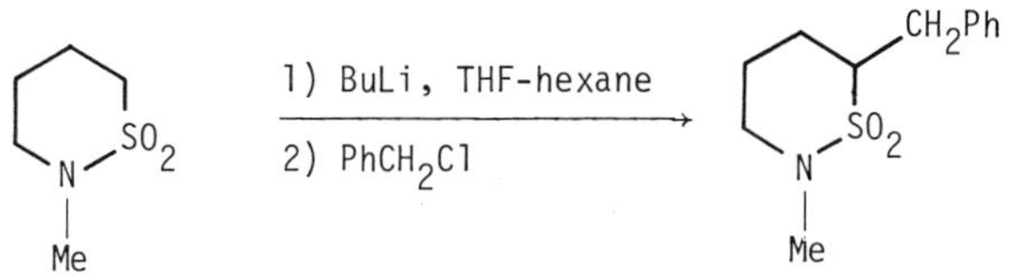

1) BuLi, THF-hexane

2) PhCH$_2$Cl

65%

JOC (1975) 40 1342

HCBr$_3$ $\xrightarrow{\begin{array}{c}\text{1) LDA, THF-Et}_2\text{O-110}^\circ \\ \text{2) BuI}\end{array}}$ BuCBr$_3$ 90%

Bull Soc Chim France (1975) 1797

Section 146   Halides from Hydrides

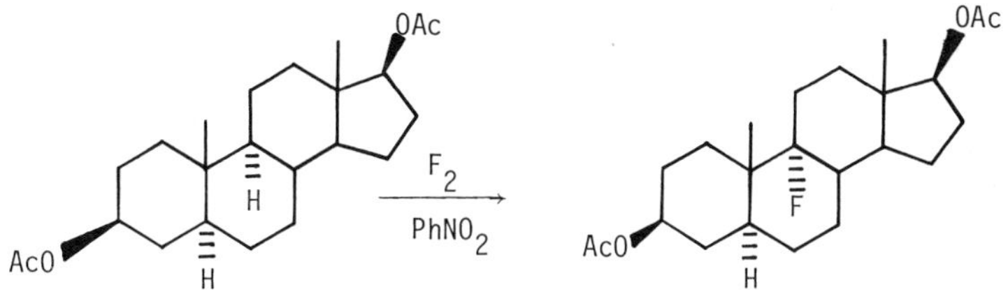

$\xrightarrow{\begin{array}{c}\text{F}_2 \\ \text{PhNO}_2\end{array}}$

50%

JACS (1976) 98 3036

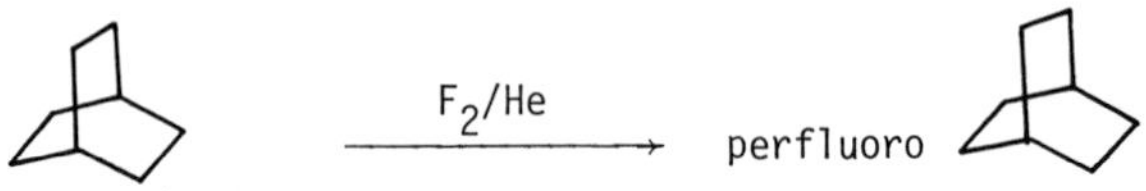

F$_2$/He → perfluoro   3.5%

JACS (1975) 97 513

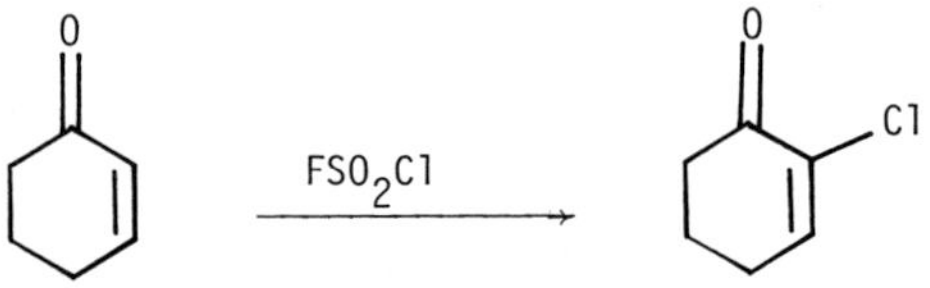

FSO$_2$Cl   70%

Synthesis (1976) 33

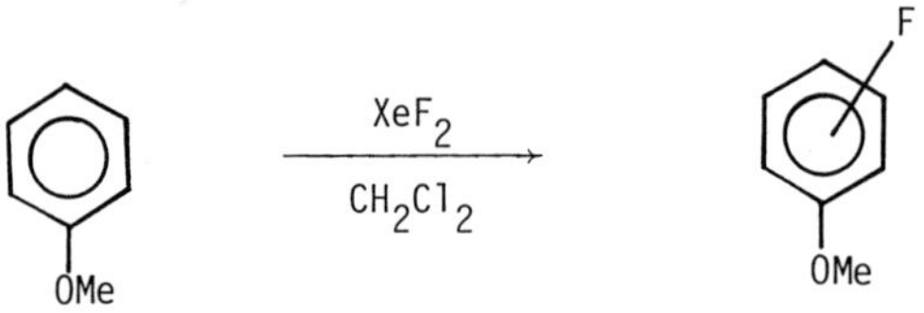

XeF$_2$ / CH$_2$Cl$_2$ 71%

10:1:8
o:m:p

JOC (1975) 40 807

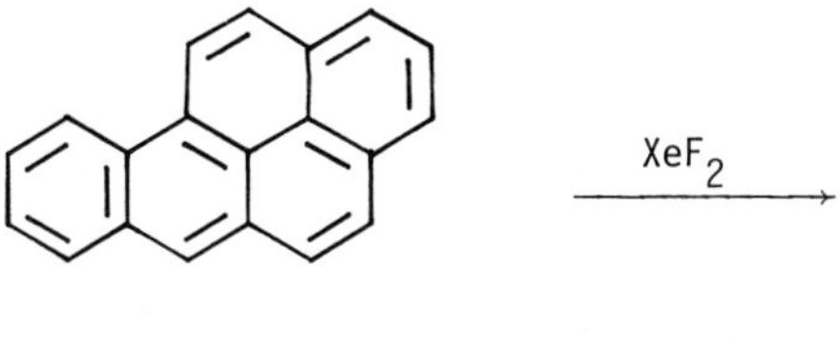

$$XeF_2$$

26%

Experientia (1976) 15 417

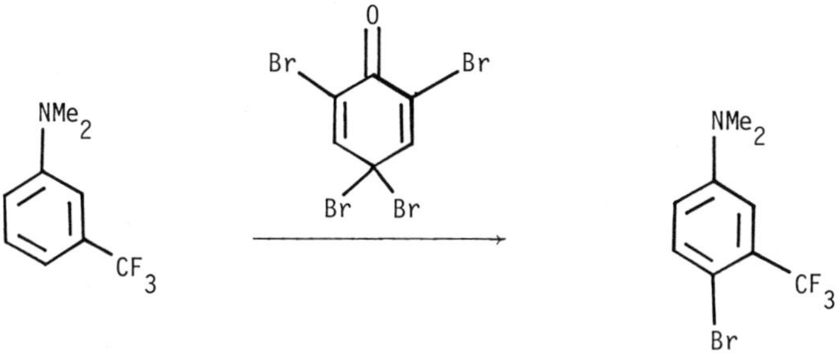

~85%

Org Synth (1976) 55 20

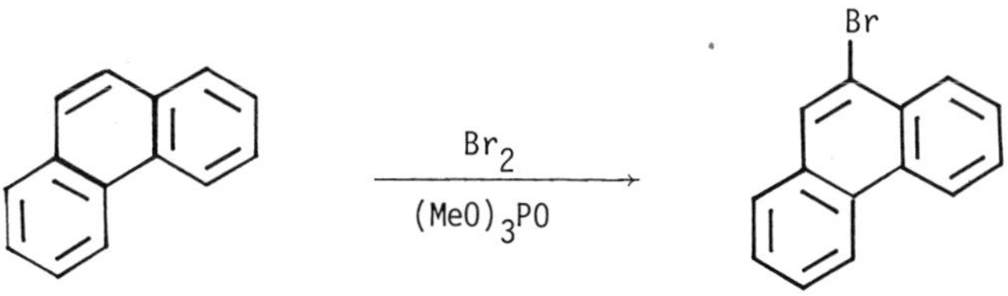

$$\frac{Br_2}{(MeO)_3PO}$$

70%

Synthesis (1976) 621

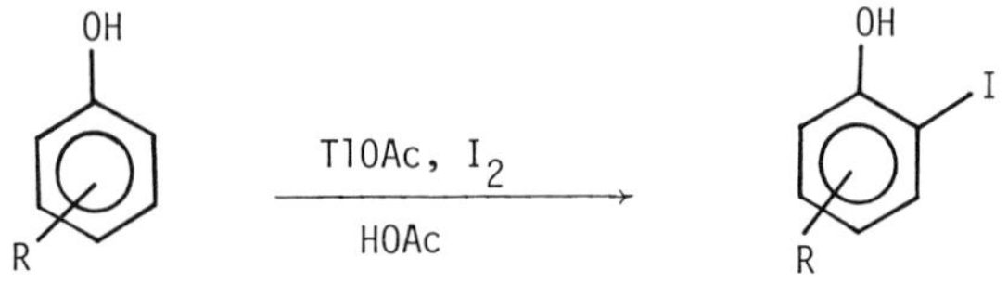

R = H, alkyl, aryl

JCS Perkin I (1976) 1161

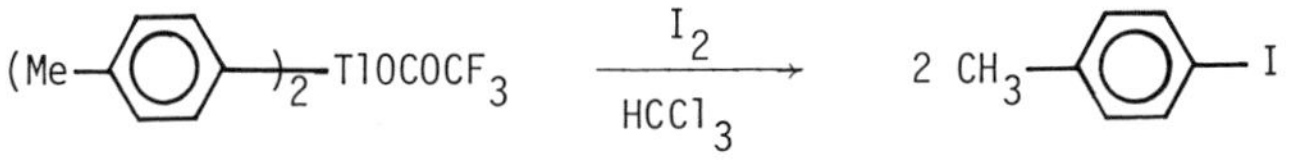

JOC (1975) 2351

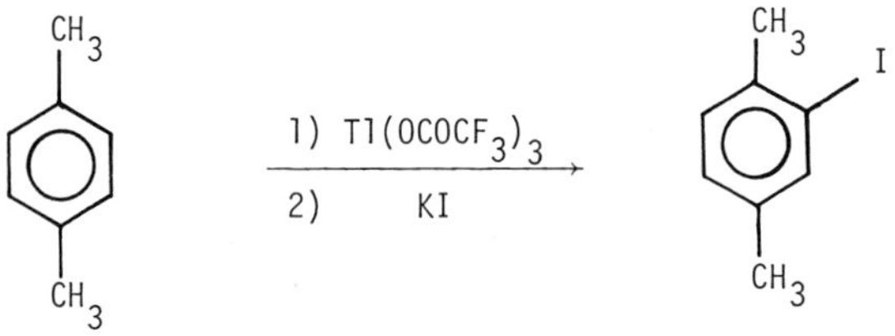

Org Synth (1976) 55 70

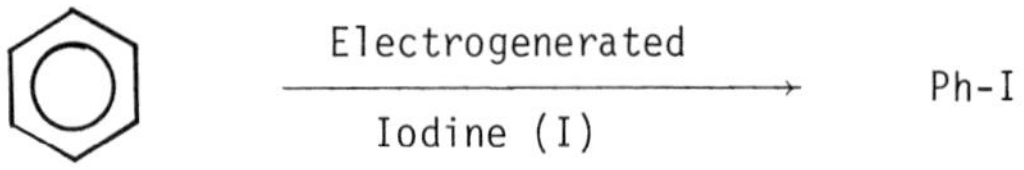

JACS (1976) 98 1515

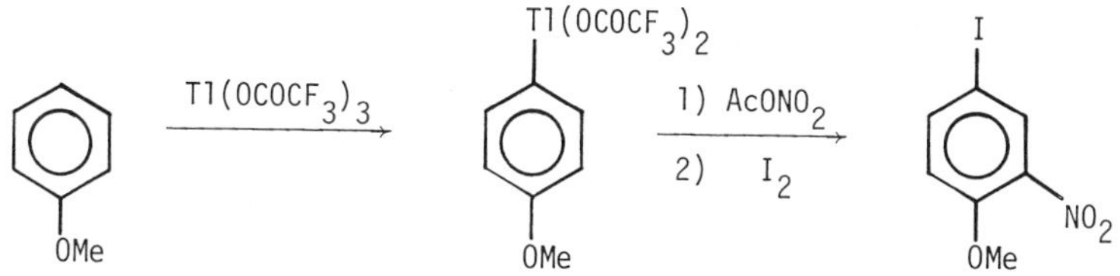

JOC (1975) $\underline{40}$ 3441

Section 147   Halides from Ketones

JACS (1974) $\underline{96}$ 925

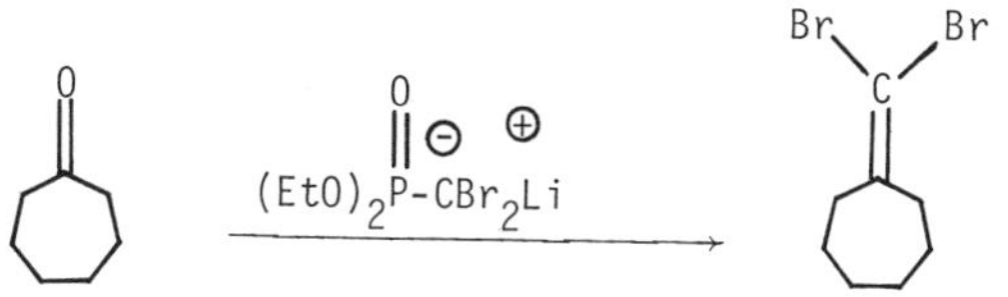

67%

Synthesis (1976) 197

Section 148   Halides and Sulfonates from Nitriles

No examples

Section 149   Halides from Olefins

For allylic halogenation see Section 146 (Halides from Hydrides)
For halocyclopropanations see Section 74 (Alkyls from Olefins)

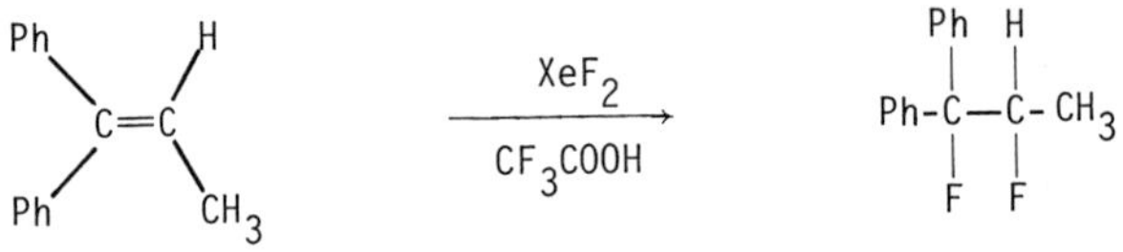

JOC (1976) 41 4002

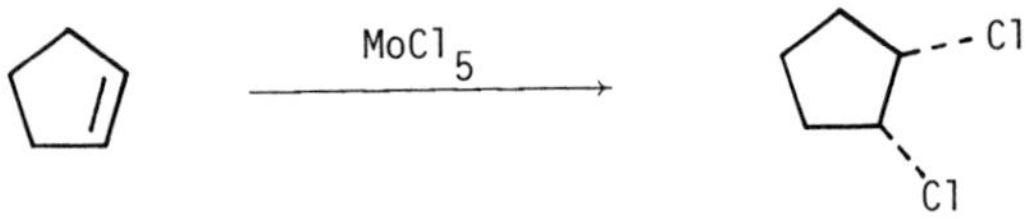

JACS (1975) 97 1599

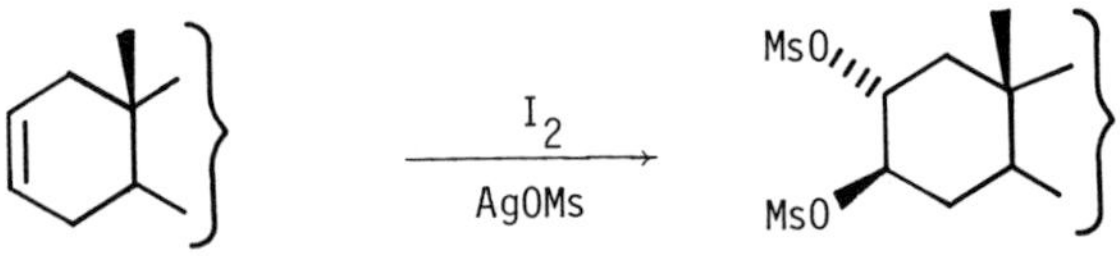

Steroids (1975) 25 619

$$\xrightarrow[\text{2) Br}_2]{\text{1) LiAlH}_4,\ \text{ZrCl}_4}$$

92%

J Organometal Chem (1976) 122 C25

CpZr(H)Cl + $\longrightarrow$ $\xrightarrow{\text{Br}_2}$ 96%

JACS (1974) 96 8115

$$\xrightarrow[\text{NaOCH}_3]{\text{I}_2,\ \text{CH}_3\text{OH}}$$

$CH_3CHCH_2CH_3$
|
I

R config., 84% ee

49%

JACS (1976) 98 1290

$H_2C=CH(CH_2)_8COOMe$ $\xrightarrow[\text{2) I}_2,\ \text{NaOCH}_3]{\text{1) B}_2\text{H}_6,\ \text{THF}}$ $I-CH_2(CH_2)_9COOMe$ 68%

Synthesis (1976) 114

Section 150   Halides from Miscellaneous Compounds

$$\underset{\underset{H}{|}}{\overset{\overset{n\text{-}C_9H_{19}}{|}}{Me\text{-}C\text{-}SeR}} \xrightarrow{\quad Br_2 \quad} \underset{\underset{H}{|}}{\overset{\overset{n\text{-}C_9H_{19}}{|}}{Me\text{-}C\text{-}Br}} \qquad\qquad 84\%$$

Tetr Lett (1976) 2647

$$\underset{\underset{SH\ \ NH_2}{|\ \ \ \ |}}{Me_2C\text{---}CHCOOH} \xrightarrow[\text{liquid HF}]{\quad CF_3OF \quad} \underset{\underset{F\ \ \ NH_2}{|\ \ \ |}}{Me_2C\text{---}CHCOOH} \qquad\qquad 94\%$$

JOC (1976) 41 3107

Review: "Advances in the Synthesis and Investigation of Organo-
          fluorine Compounds"

                               Russ Chem Rev (1975) 44 339

Review: "Some Recent Developments in Organic Chlorine Chemistry"

                               Chem and Ind (1975) 249

Chapter 11   PREPARATION

OF

HYDRIDES

This chapter lists hydrogenolysis and related reactions by which functiona
groups are replaced by hydrogen, e.g. $RCH_2X \rightarrow RCH_2$-H or R-H

Section 151   Hydrides from Acetylenes

No examples of the reaction $RC\equiv CR \rightarrow RH$ occur in the literature

Section 152   Hydrides from Carboxylic Acids

No additional examples

Section 153   Hydrides from Alcohols and Phenols

This section lists examples of the hydrogenolysis of alcohols and phenols,
$ROH \rightarrow RH$

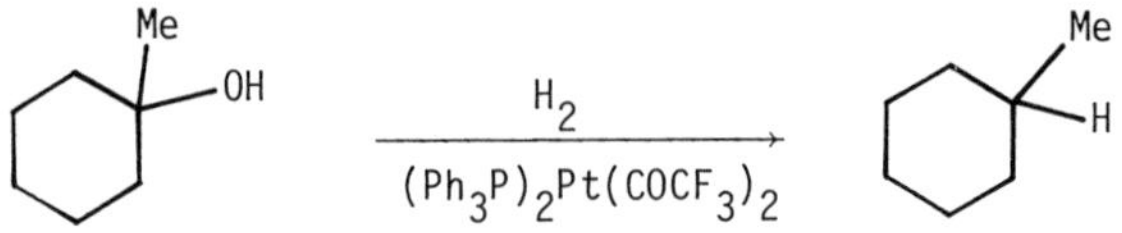

83%

Doklady Chem (1974) <u>219</u> 888

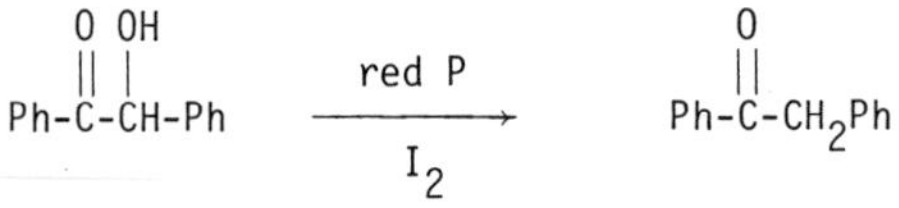

90%

Synthesis (1975) 161

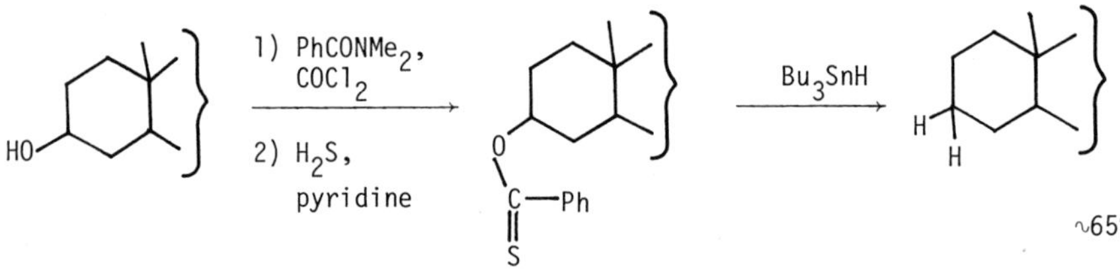

∿65%

JCS Perkin I (1975) 1574

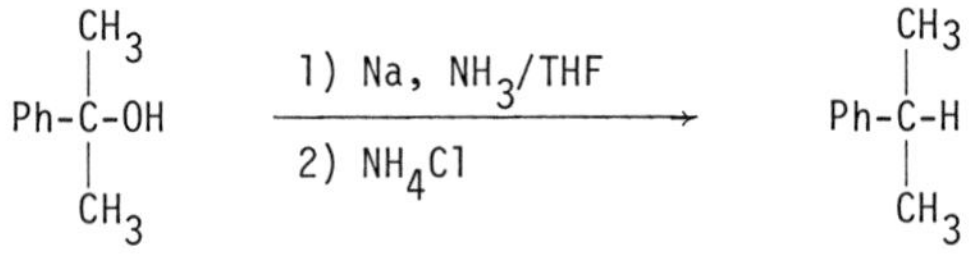

95%

JOC (1975) <u>40</u> 3151

Ph —|—|— OH $\xrightarrow[\text{1-NapPhMeSiH}]{\text{BF}_3}$ Ph —|—|— H   86%

Tetr Lett (1976) 2955

Ph —CH=CH— CH$_2$OH $\xrightarrow{\text{Ph}_3\text{P}\cdot\text{I}_2}$ Ph —CH=CH— CH$_3$ 45%

Chem Ber (1976) 109 1586

HO ⟨⟩ $\xrightarrow[\text{TiCl}_4]{\text{LiAlH}_4}$ ⟨⟩   65%

Chem Pharm Bull (1976) 24 825

n-C$_8$H$_{17}$OH + RNCNR $\xrightleftharpoons{\text{CuCl}}$ n-C$_8$H$_{17}$-O-C$\begin{smallmatrix}\nearrow \text{NR}\\ \searrow \text{NHR}\end{smallmatrix}$ $\xrightarrow[\text{Pd/C}]{\text{H}_2}$ octane   90%

Chem Ber (1974) 107 1353
Chem Ber (1974) 107 907

PhOH ⟶ PhOSO$_3$ $^{\ominus}$ NHR$_2$Ph $^{\oplus}$

1) KOH
⟶ PhH   94%
2) RaNi

JCS Perkin I (1975) 169

1) 5-chloro-1-
   phenyltetrazole
⟶
2) H$_2$, Pd/C

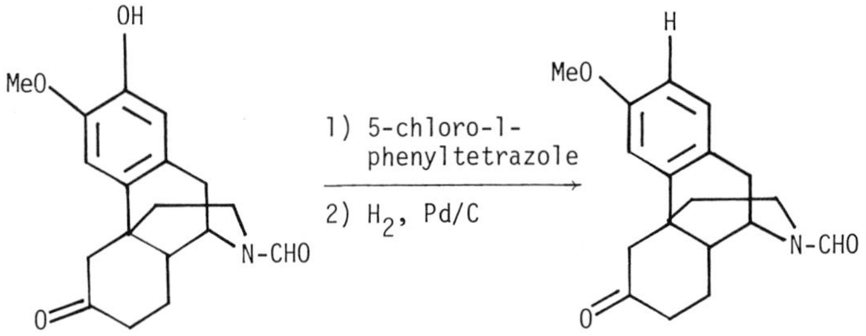

∿50%

Rec Trav Chim (1976) 95 43

Also via Halides and Sulfonates, Section 160

Section 154   Hydrides from Aldehydes

No additional examples

For the conversion RCHO → RMe etc. see Section 64 (Alkyls from Aldehydes)

Section 155   Hydrides from Alkyls, Methylenes and Aryls

No additional examples

Section 156   Hydrides from Amides

No additional examples

Section 157   Hydrides from Amines

This section lists examples of the conversion $RNH_2 \rightarrow RH$

$ArCH_2NH_2$

\+

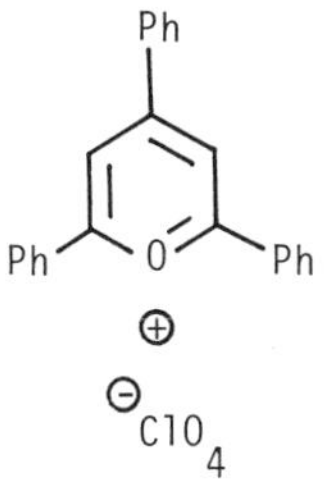

$\begin{array}{c} \text{2) NaBH}_4 \\ \hline \text{3) 200°} \end{array}$ ⟶ $ArCH_3$ 50%

Tetr Lett (1976) 2689

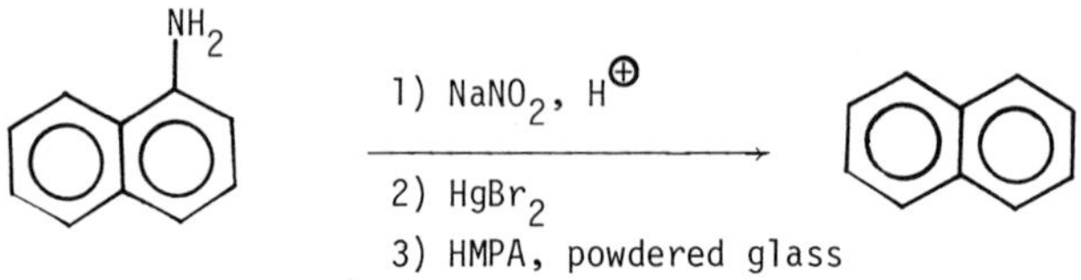

JOC (1974) <u>39</u> 1317

$$C_{10}H_{21}NH_2 \longrightarrow C_{10}H_{21}NTs_2 \xrightarrow[\text{HMPA}]{\text{NaBH}_4} C_{10}H_{22} \qquad 91\%$$

JOC (1975) <u>40</u> 2018

Section 158   Hydrides from Esters

This section lists examples of the reactions RCOOR'→RH and RCOOR'→R'H.

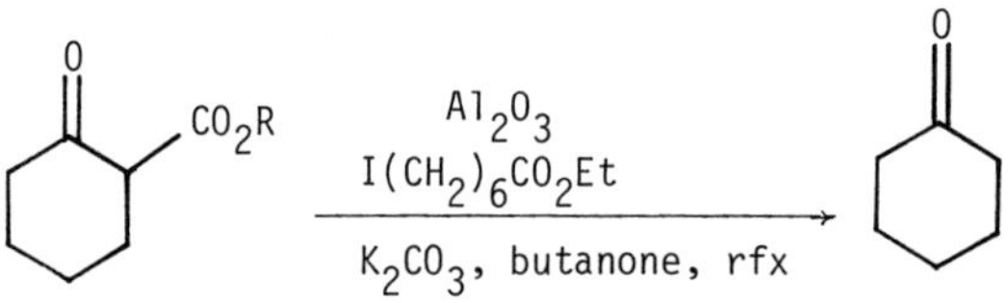

R = H, Me, Et

Tetr Lett (1976) 2707

$$CH_3(CH_2)_6\overset{O}{\overset{||}{C}}HCCH_3 \quad \xrightarrow[\underline{o}\text{-xylene}]{} \quad CH_3(CH_2)_6CH_2\overset{O}{\overset{||}{C}}CH_3 \qquad 97\%$$
$$\underset{COOEt}{}$$

same conditions   98%

Synth Comm (1975) <u>5</u> 341

hν
HMPA, H$_2$O   >60%

JCS Chem Comm (1975) 439

Section 159   <u>Hydrides from Ethers</u>

No additional examples

Section 160   Hydrides from Halides and Sulfonates

This section lists the reduction of halides and sulfonates RX → RH

$$\xrightarrow[\text{H}_2\text{O/THF}]{2\text{VCl}_2}$$

96%

Synthesis (1976) 807

$R-CCl_2-COO$

1) CH_3MgBr

2) $H_3O^{\oplus}$

$R-\overset{\text{Cl}}{\underset{\text{H}}{\text{C}}}-COO$ ∿80%

$R'X$

$R-\overset{\text{Cl}}{\underset{\text{R'}}{\text{C}}}-COO$ ∿80%

R = alkyl

R' = alkyl, acyl

Synthesis (1975) 533

$Ph(CH_2)_3Br$ $\xrightarrow[\text{NaBH}_4, \ h\nu]{\text{Bu}_3\text{SnCl (0.1 equiv)}}$ $PhCH_2CH_2CH_3$ 88%

JOC (1975) 40 2554

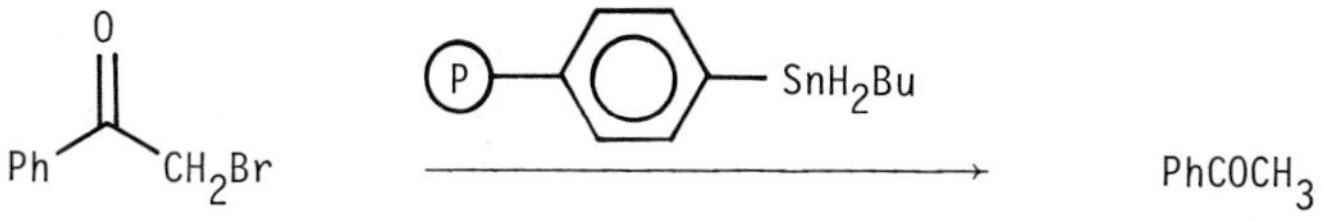

JOC (1975) 40 1966

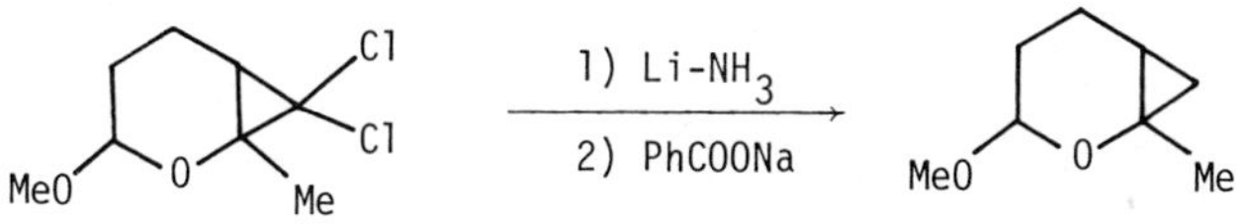

JOC (1975) 40 2238

JOC (1976) 41 1393

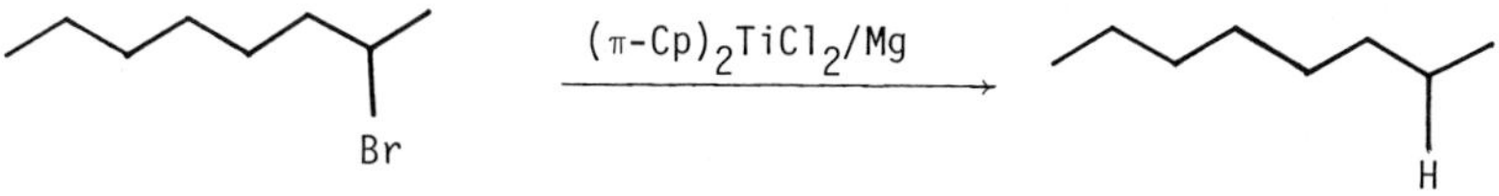

JOC (1975) 40 3159

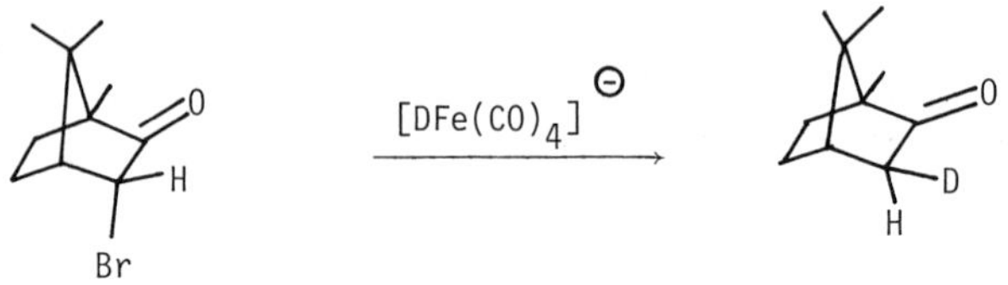

60%

Tetr Lett (1975) 2257

PhCH$_2$Br + K (s-Bu)$_3$BH + CuI $\longrightarrow$ PhCH$_3$ <90%

JCS Chem Comm (1974) 762

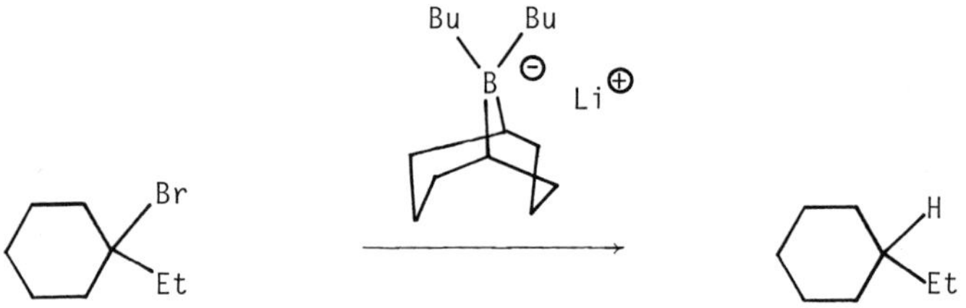

81%

JACS (1975) 97 2558

RC(Br)$_2$R' + Zn(Cu) $\xrightarrow[\text{D}_2\text{O}]{\text{THF}}$ R-C(D)(Br)-R' + R-C(D)(D)-R'

60%   20%

JOC (1974) 39 2300

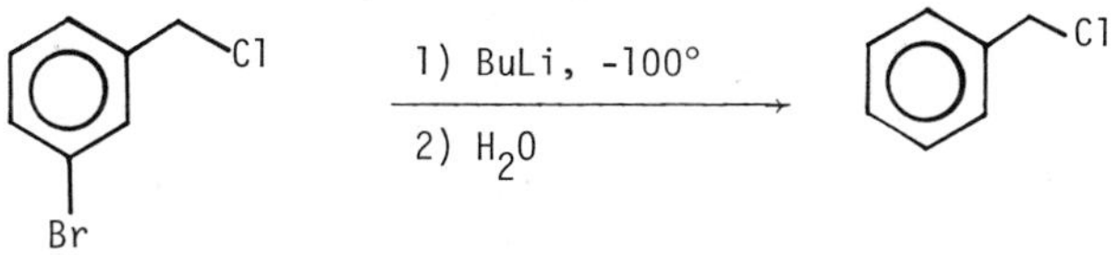

1) BuLi, -100°

2) H_2O

88%

JOC (1976) 41 1184, 1187

$PhCCl=CH_2$ →[t-BuLi / Et_2O] $PhCH=CH-t-Bu$ 71%

Tetr Lett (1975) 2935

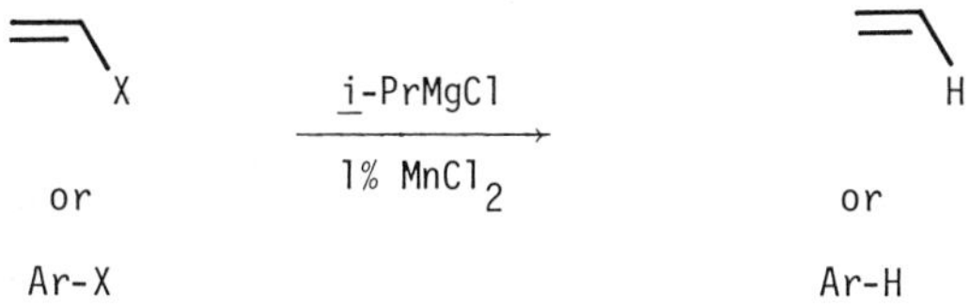

or

Ar-X

→[i-PrMgCl / 1% $MnCl_2$]

or

Ar-H

J Organometal Chem (1976) 113 107

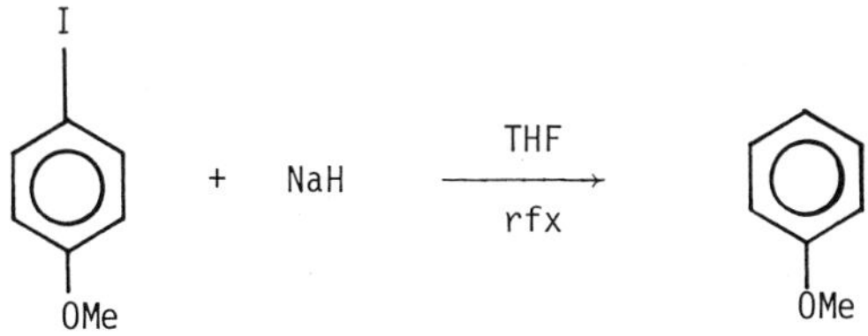

+ NaH →[THF / rfx]

87%

JOC (1974) 39 1425

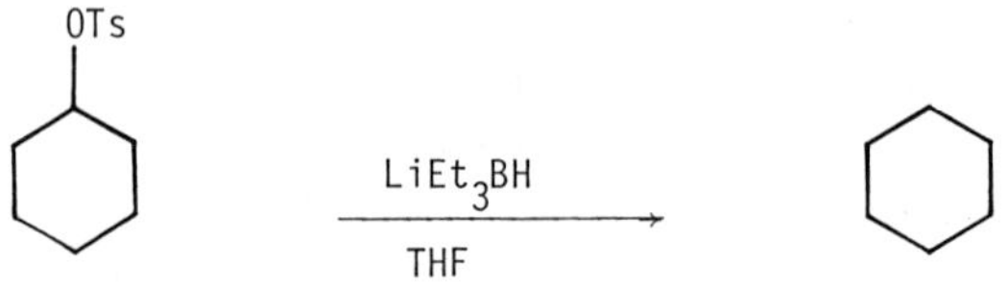

OTs
$LiEt_3BH$
THF

80%

JOC (1976) 41 3064

CO_2Me
SO_2Ph
Na(Hg), MeOH
Na_2HPO_4
CO_2Me

88%

Tetr Lett (1976) 3477

H
OTs
C
$(CH_2)_{11}$
NaI/Zn
H
H
C
$(CH_2)_{11}$

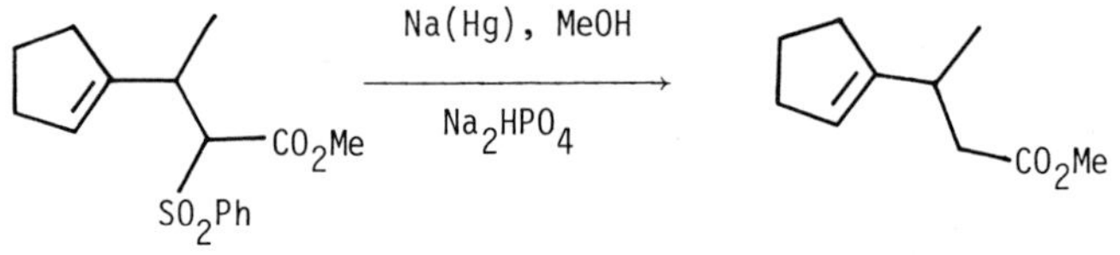

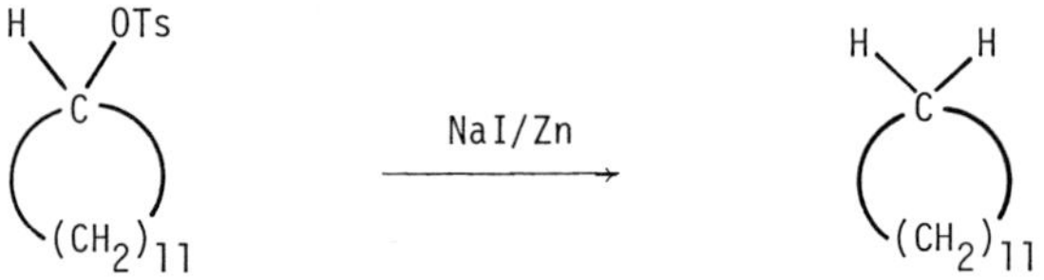

95%

Tetr Lett (1976) 3325

Section 161 Hydrides from Hydrides

No additional examples

Section 162   Hydrides from Ketones

No additional examples

For the conversion $R_2CO \rightarrow R_2CH_2$ or R_2CHR' see Section 72 (Alkyls and Methylenes from Ketones)

Section 163   Hydrides from Nitriles

No additional examples

Section 164   Hydrides from Olefins

No additional examples

Section 165   Hydrides from Miscellaneous Compounds

No additional examples

Chapter 12   PREPARATION

OF

KETONES

$$C \equiv CCH_3$$

1) [bicyclic boron reagent]

2) H_2O_2, $^{\ominus}OH$

$$\underset{O}{\overset{\parallel}{CH_2CCH_3}}$$

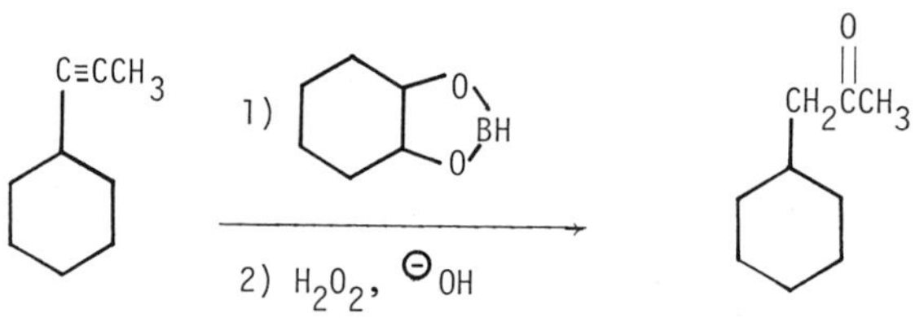

JACS (1975) <u>97</u> 5249

R_3B

1) $LiC \equiv CH$

2) HCl

3) $NaOH$, H_2O_2

RCOMe

70-90%

JACS (1975) <u>97</u> 5017
Tetr Lett (1975) 3327 (more functional-
                        ized alkynes)

HC≡C-CH$_2$
|
CH$_2$OSO$_2$CF$_3$

$\xrightarrow[\text{2)}\quad\text{H}_2\text{O}]{\text{1) TFA, NaOCOCF}_3}$

~35%

Org Synth (1974) 54 84

PhCH$_2$OC≡CH $\xrightarrow{\Delta}$

75%

Tetr Lett (1975) 3275

Ph-C≡C-$\overset{\overset{\displaystyle O}{\|}}{C}$-CH$_3$ $\xrightarrow[\text{NaCN}]{\text{CH}_3\text{OH}}$ Ph-$\overset{\overset{\displaystyle OMe}{|}}{\underset{\underset{\displaystyle OMe}{|}}{C}}$-CH$_2$COCH$_3$

86%

JOC (1976) 41 3765

Section 167   Ketones from Carboxylic Acids and Acid Halides

1) 2.5 LDA

2) O$_2$

3) H$_3$O$^\oplus$

60%

Tetr Lett (1975) 4611

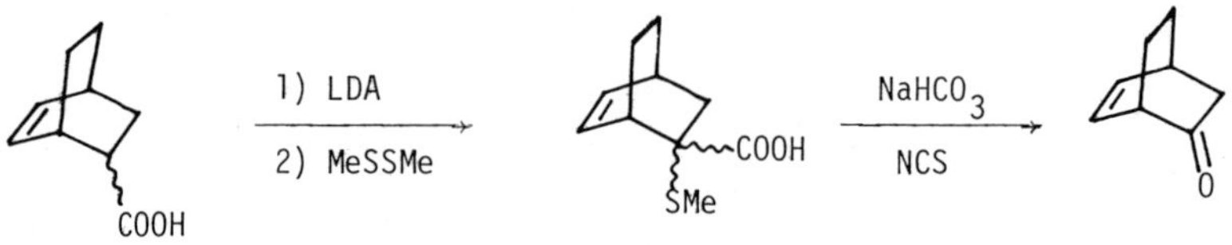

$$\xrightarrow[\text{2) MeSSMe}]{\text{1) LDA}}$$

$$\xrightarrow[\text{NCS}]{\text{NaHCO}_3}$$

44%

JACS (1975) 97 3528

$CH_2=CHCMe_2COOH$

MeLi

Et$_2$O

MeLi

80% THF/20% Et$_2$O

$H_2C=CHCMe_2COMe$ 55%

$Me_2C=CHCH_2COMe$ 50%

Tetr Lett (1975) 3179

$EtCO_2Li$ + PhLi $\xrightarrow[\text{rfx 24 hrs.}]{\text{Et}_2\text{O}}$ EtCOPh   82%

JOC (1975) 40 1770

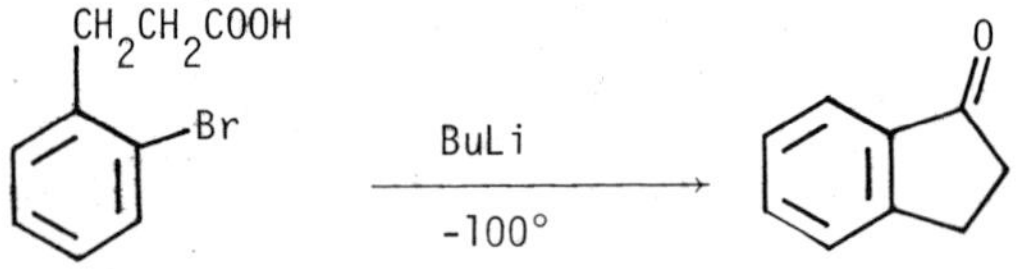

$\xrightarrow[-100°]{\text{BuLi}}$

76%

JOC (1975) 40 2394

$$\underline{n}\text{-}C_4H_9CO(CH_2)_4COCl \xrightarrow{\text{(}\underline{t}\text{-BuOCuBu-}\underline{t}\text{)Li}} \underline{n}\text{-}C_4H_9CO(CH_2)_4CO\text{-}\underline{t}\text{-Bu} \quad 73\%$$

Org React (1975) $\underline{22}$ 253

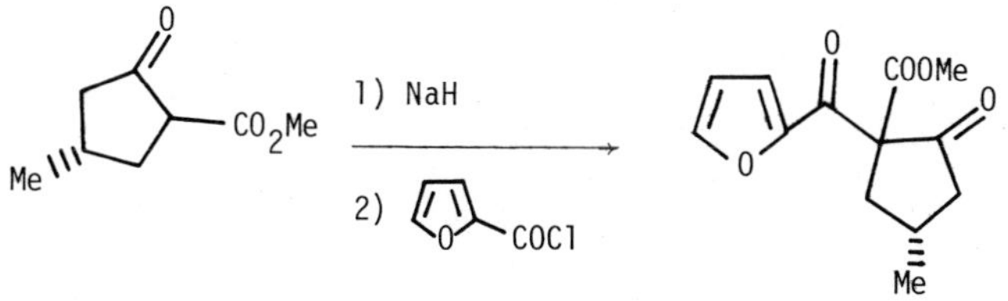

JOC (1974) $\underline{39}$ 3241

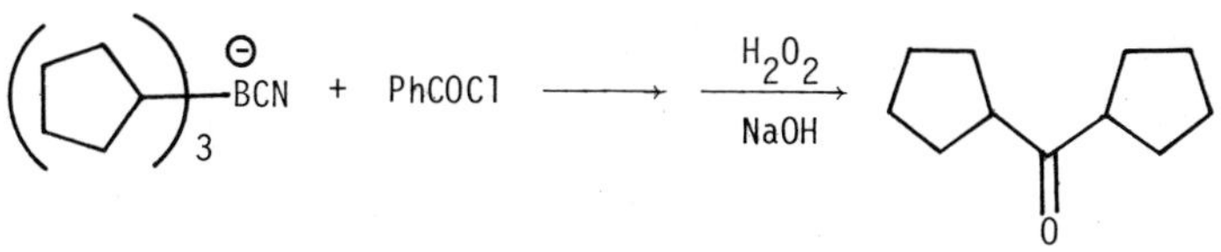

JCS Perkin I (1975) 129, 142

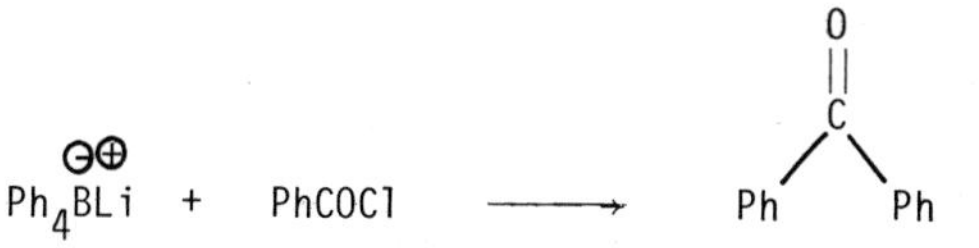

JCS Chem Comm (1975) 138

LiBBzBu$_3$ +   PhCOCl $\xrightarrow{\text{THF/Hexane}}$ [structure: Bz–CO–Ph]   88%

JOC (1975) 40 1676

[structure: 1,3-dioxane–CH$_2$CH$_2$MgBr]

+

CH$_3$(CH$_2$)$_5$COCl $\xrightarrow{\text{2) H}_2\text{0, oxalic acid}}$ CH$_3$(CH$_2$)$_5$CCH$_2$CH$_2$CHO   82%

JOC (1976) 41 560

n-Hept-MgBr $\xrightarrow[\text{2) }\underline{t}\text{-BuCOCl}]{\text{1) MnI}_2}$ [structure: n-Hept–C(O)–t-Bu]   80%

Tetr Lett (1976) 3155

Ph-C-Cl $\xrightarrow[\text{THF}]{\text{PhS[Me}_3\text{C]CuLi}}$ Me$_3$C-C-Ph   85%

Org Synth (1976) 55 122

CH$_2$=CH$_2$ +   PhCOCl $\xrightarrow{\text{L}_3\text{RhH(CO)}}$ PhCOEt   50-80%

JACS (1974) 96 4721

BuLi + RhCl(CO)(PPh$_3$)$_2$ + PhCOCl $\longrightarrow$ PhCOBu   90%

JACS (1975) $\underline{97}$ 5448

$$(MeCO)_2C=SMe_2 \quad \xrightarrow[\text{2) PhCOCl}]{\text{1) BuLi, THF}} \quad PhCOCH_2\overset{\overset{\displaystyle SMe_2}{||}}{C}COMe \qquad 55\%$$

JCS Chem Comm (1975) 289

PhCOOH

1) $\underline{i}$-PrMgX
2) hydrolysis

$\underline{i}$-PrCOPh
quant.

Gazz Chim Ital (1975) $\underline{105}$ 907

$$\begin{array}{c} O \\ || \\ Ph-C-Br \end{array}$$

+

Et$_2$Hg

$\xrightarrow[\text{HMPA}]{\text{Pd(PPh}_3)_4}$

$$\begin{array}{c} O \\ || \\ Ph-C-Et \end{array} \qquad 86\%$$

Chem Lett (1975) 951

PhCOCl + PhH $\xrightarrow{\text{CrCl}_3\cdot\text{5 DMSO}}$ PhCOPh   80%

J Gen Chem USSR (1974) $\underline{44}$ 2316

Section 168 Ketones from Alcohols and Phenols

H
OH
C
Ph
Ph
CrO_3
HMPT
O
Ph
Ph

100%

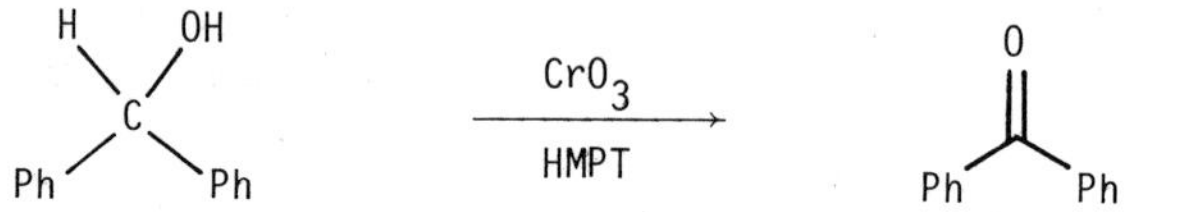

Synthesis (1976) 394

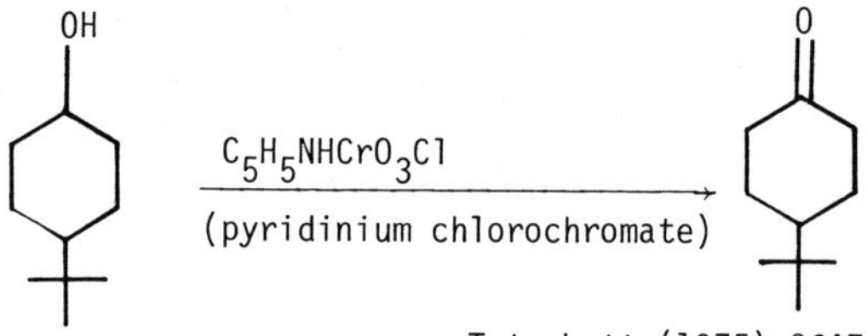

OH
$C_5H_5NHCrO_3Cl$
(pyridinium chlorochromate)
O

97%

Tetr Lett (1975) 2647

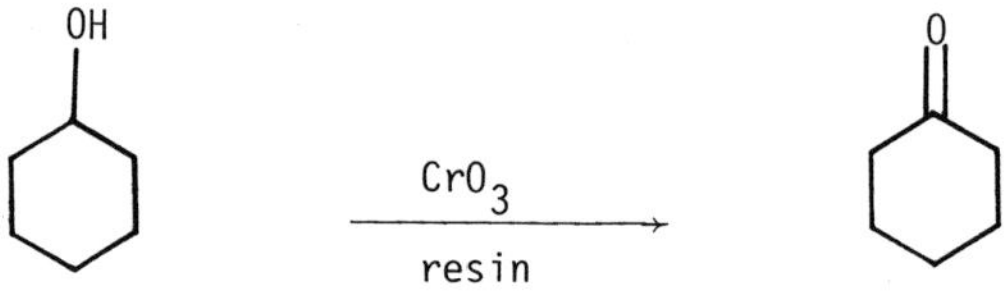

OH
CrO_3
resin
O

77%

JACS (1976) 98 6737

$$\underset{\text{OH}}{\underset{|}{\text{RCHR'}}} + Na_2Cr_2O_7 \xrightarrow[\text{conc. } H_2SO_4]{\text{DMSO}} \overset{\text{O}}{\overset{||}{\text{RCR'}}}$$

80-90%

JOC (1974) 39 3304

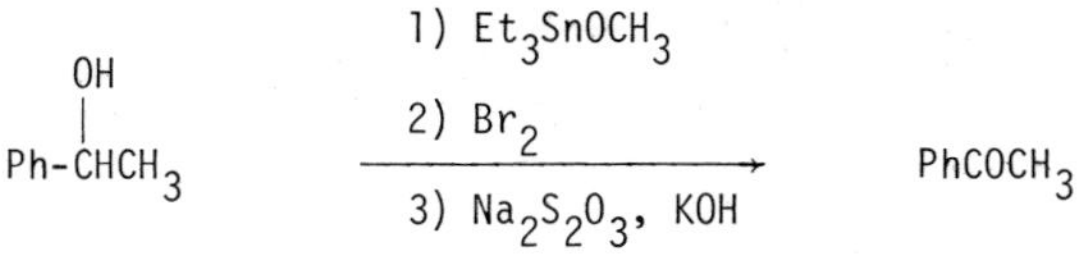
OH
Ph-CHCH3
1) Et3SnOCH3
2) Br2
3) Na2S2O3, KOH
PhCOCH3

78%

Chem Lett (1975) 145

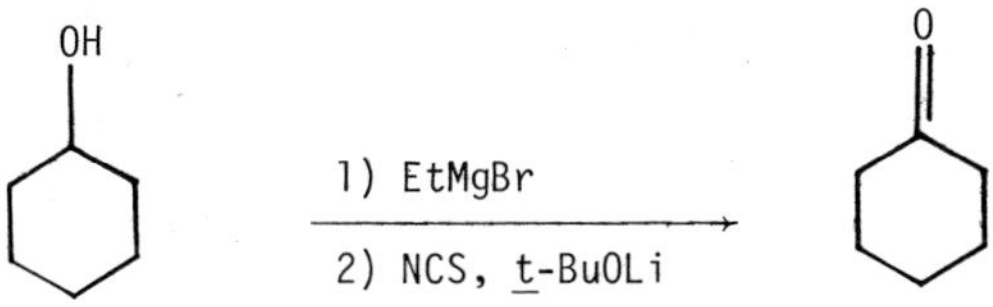
OH
1) EtMgBr
2) NCS, t-BuOLi
O

80%

Chem Lett (1975) 691

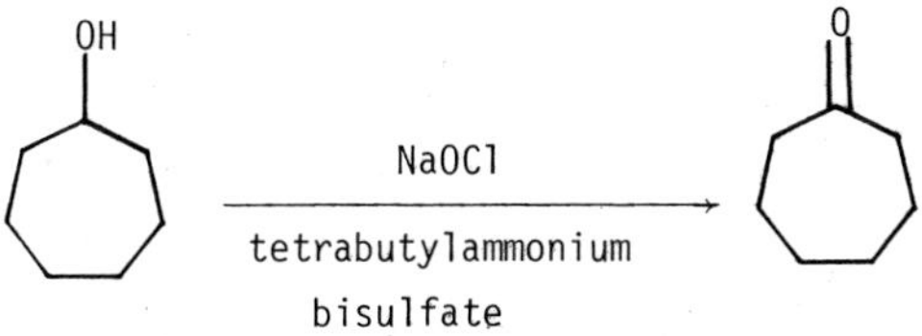
OH
NaOCl
tetrabutylammonium
bisulfate
O

89%

Tetr Lett (1976) 1641

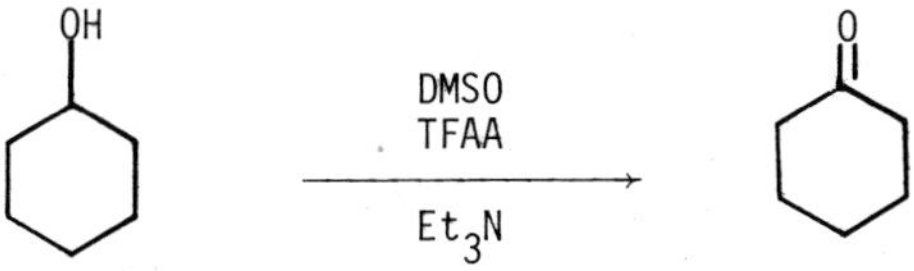
OH
DMSO
TFAA
Et3N
O

65%

JOC (1976) 41 957

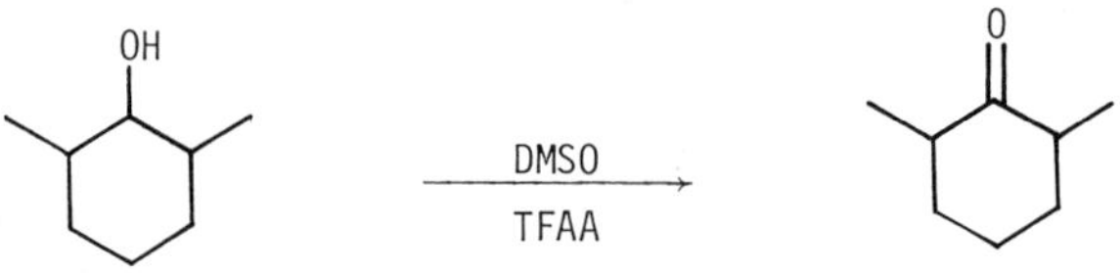

JOC (1976) 41 3329   89%

Synthesis (1975) 445   65%

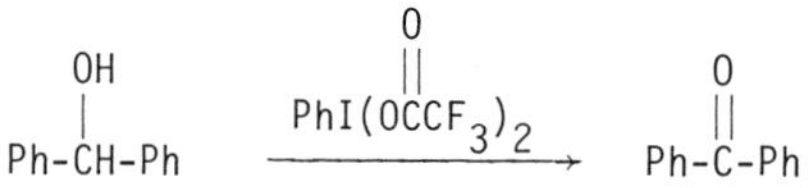

1) COCl$_2$, (epoxide)

2) Et$_3$N, DMSO

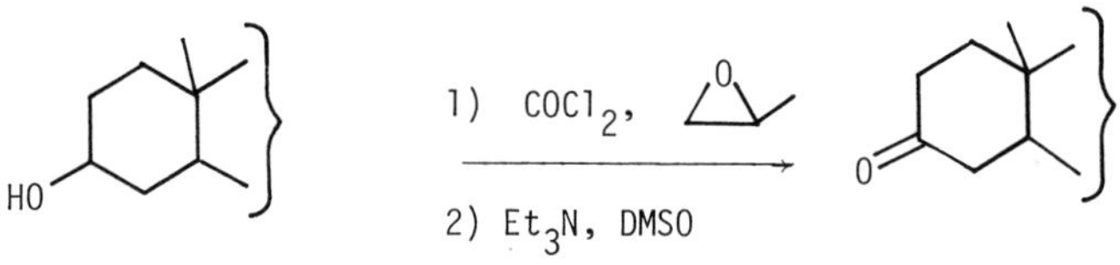

JCS Perkin I (1975) 1614   ~70%

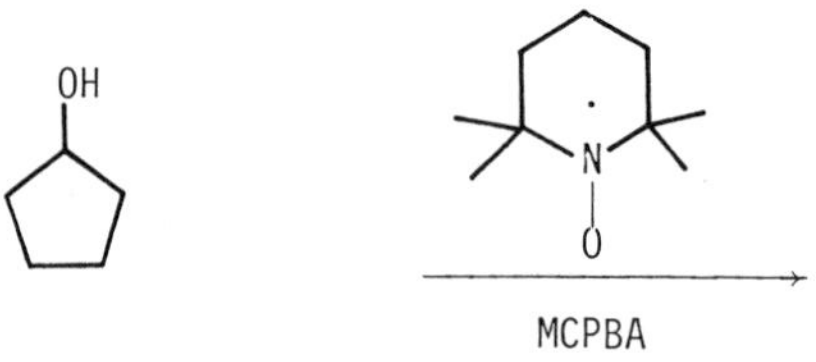

MCPBA

JOC (1975) 40 1860   77%

Tetr Lett (1975) 4115

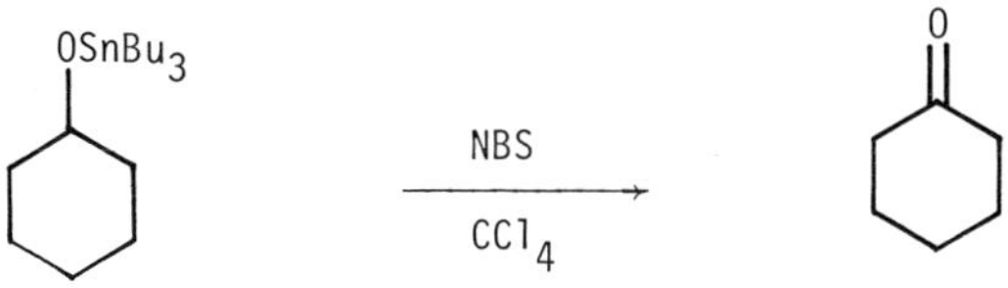

98%

JACS (1976) 98 1629

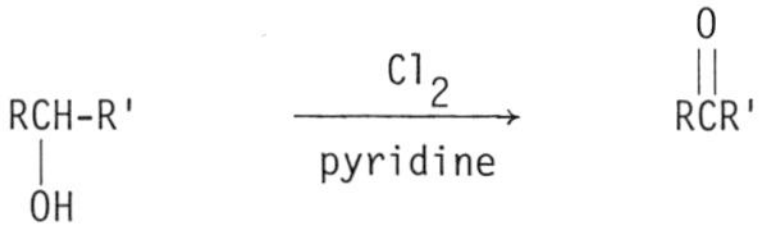

85%

$2°>1°$

Tetr Lett (1974) 3059

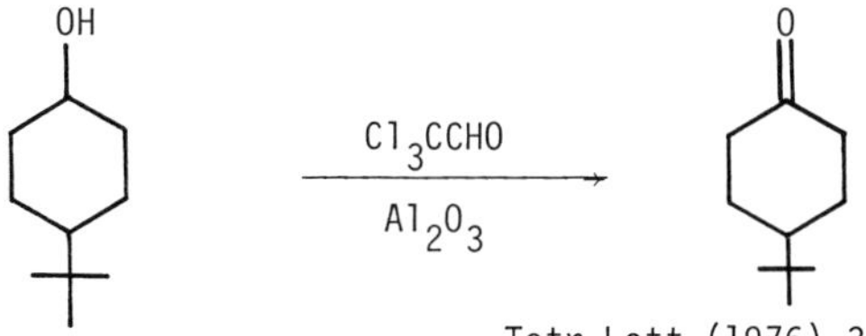

Tetr Lett (1976) 3499

71%

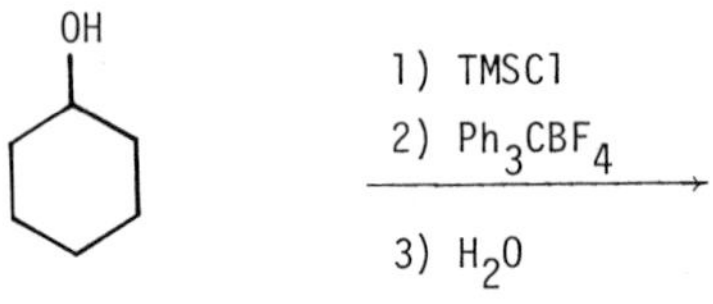

JOC (1976) <u>41</u> 1479

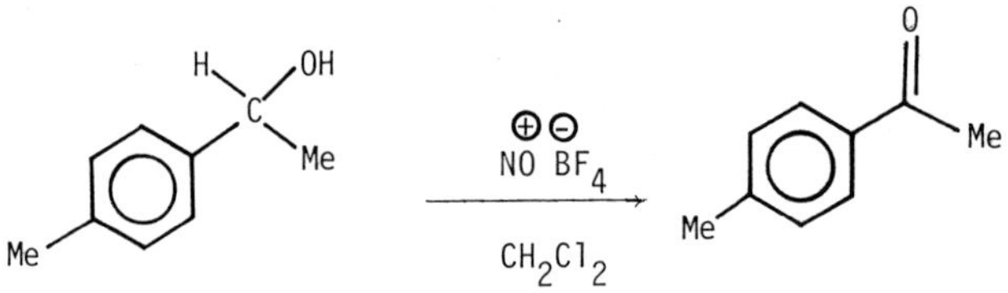

92%

Synthesis (1976) 609

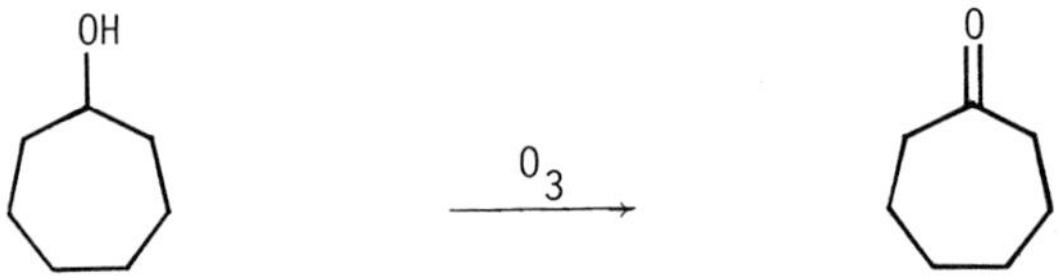

78%

JOC (1976) <u>41</u> 889

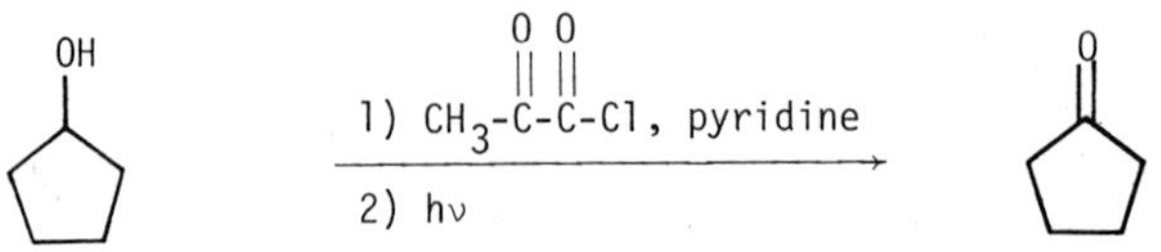

74%

JOC (1976) <u>41</u> 3030

84%

$$\underset{\underset{Ph-CHOH}{|}}{CH_3} \xrightarrow[\substack{Bu_4NBF_4 \\ \text{electrolysis}}]{MeCN} PhCOCH_3 \qquad 82\%$$

$$\underset{\underset{\underline{t}\text{-Bu}}{|}}{\overset{Et}{\underset{|}{Ph\text{-}C\text{-OH}}}} \xrightarrow[\text{conditions}]{\text{same}} Ph\text{-}\overset{O}{\overset{||}{C}}\text{-Et} \qquad 82\%$$

JACS (1975) 97 4012

MnO$_2$ / CHCl$_3$

Indian J Chem (1976) 146 143

(Bu$_3$Sn)$_2$OBr$_2$ / CH$_2$Cl$_2$ 86%

Tetr Lett (1976) 4597

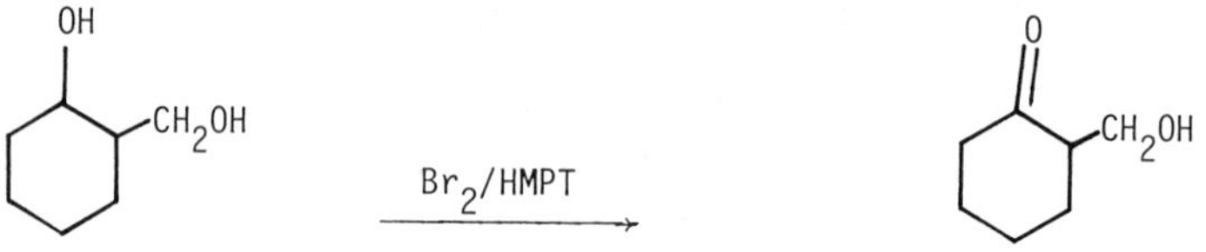

Synthesis (1976) 811

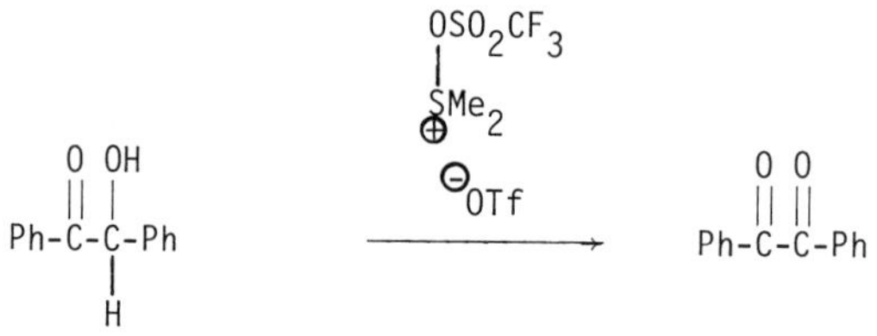

Tetr Lett (1975) 273

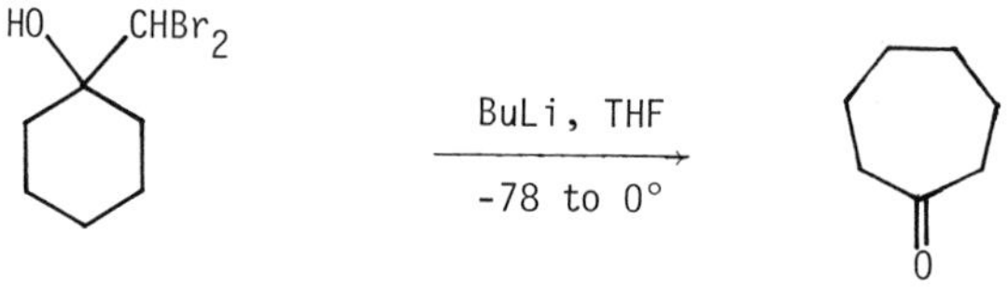

JACS (1974) 96 6510

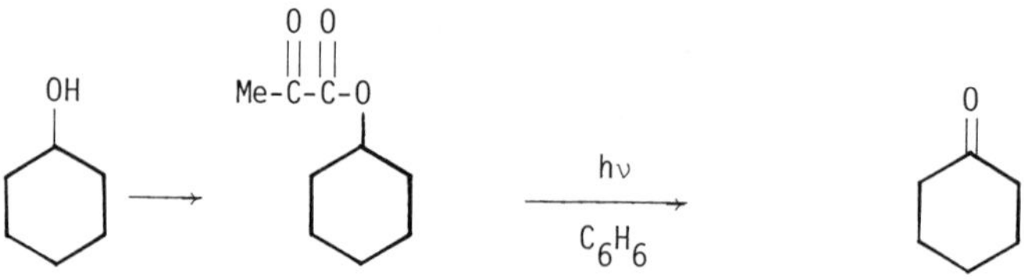

Synth Comm (1976) 6 281

$$R-\underset{\underset{H}{|}}{\overset{\overset{O}{|}}{C}}-R' \quad + \quad O_2 \quad \xrightarrow[\substack{Rose \\ Bengal}]{h\nu} \quad R-\overset{\overset{O}{||}}{C}-R' \qquad \text{high yield}$$

JACS (1974) 96 585

$$R^1CH=CH-\underset{\underset{OH}{|}}{CH}-R^2 \quad \xrightarrow[110°]{RuHCl(PPh_3)_3} \quad R^1CH_2CH_2\overset{\overset{}{C}}{\underset{||}{}O}-R^2 \qquad 80\text{-}90\%$$

Tetr Lett (1974) 4133

$$Et_2\underset{\underset{OH}{|}}{C}-CH(SPh)_2 \quad \xrightarrow[LDA]{CuTf, PhH} \quad EtCOCH-SPh \underset{\underset{Et}{|}}{} \qquad 88\%$$

JACS (1975) 97 4749

Related methods: Aldehydes from Alcohols and Phenols (Section 48)

Section 169   Ketones from Aldehydes

1) BuLi
2) CH_3CHICH_3
3) $H_3O^{\oplus}$

~95%

JOC (1975) 40 231

ArCHO $\xrightarrow{\begin{array}{l}\text{1) KCN, Et}_2\text{NH} \\ \text{2) Base} \\ \text{3) Ar'CH}_2\text{X}\end{array}}$ $\underset{\underset{\overset{|}{\text{CH}_2\text{Ar'}}}{\overset{\overset{|}{\text{NEt}_2}}{\text{Ar-C-CN}}}}{}$ $\xrightarrow{\text{H}_3\text{O} \oplus}$ $\underset{}{\overset{\overset{O}{\|}}{\text{Ar-C-CH}_2\text{Ar'}}}$ ~50%

Tetrahedron (1975) <u>31</u> 1219

Ph-CHO

+

Me$_3$SiCN

$\longrightarrow$ $\underset{\overset{|}{\text{CN}}}{\overset{\overset{|}{\text{OTMS}}}{\text{Ph-CH}}}$ $\xrightarrow{\begin{array}{l}\text{1) LDA} \\ \text{2)} \text{-Br} \\ \text{3) HCl} \\ \text{4) NaOH}\end{array}}$ 48%

Synthesis (1975) 180

$\xrightarrow[\text{PhNO}_2, \; \Delta]{\text{CuBr}_2}$

~50%

JCS Perkin I (1976) 2241

$\xrightarrow{\begin{array}{l}\text{1) Me}_3\text{SiCN, ZnI}_2 \\ \text{2) LDA, PhCH}_2\text{CH}_2\text{Br} \\ \oplus \\ \text{3) H}_3\text{O}\end{array}}$

54%

Synthesis (1976) 416

$$Me_2CO + PhCHO \xrightarrow{KHFe(CO)_4} MeCOCH_2CH_2Ph \qquad 70\%$$

JCS Perkin I (1975) 1273

Section 170 Ketones from Alkyls and Methylenes

No additional examples

Section 171 Ketones from Amides

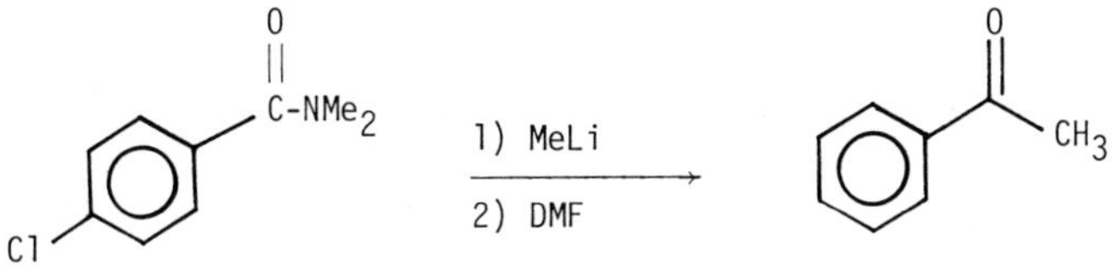

O
C-NMe2
Cl
1) MeLi
2) DMF
O
CH3

JOC (1976) 41 3651

N(SO2PhNO2)2
DMSO
NaHCO3
O

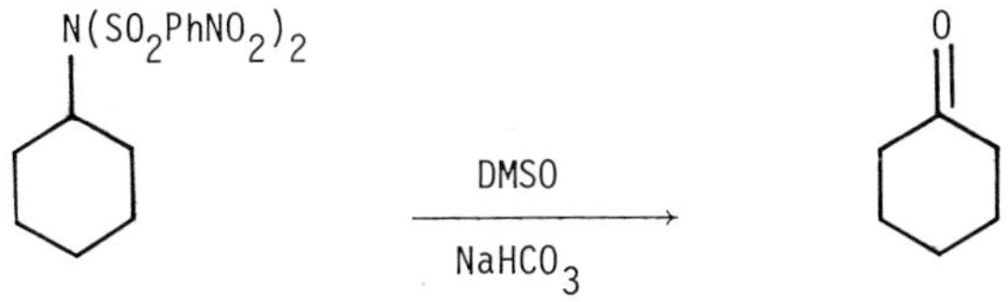

67%

Tetr Lett (1975) 3107

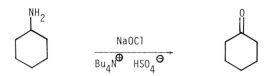

JOC (1976) <u>41</u> 3651

Section 172   Ketones from Amines

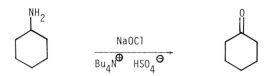

Tetr Lett (1976) 1641

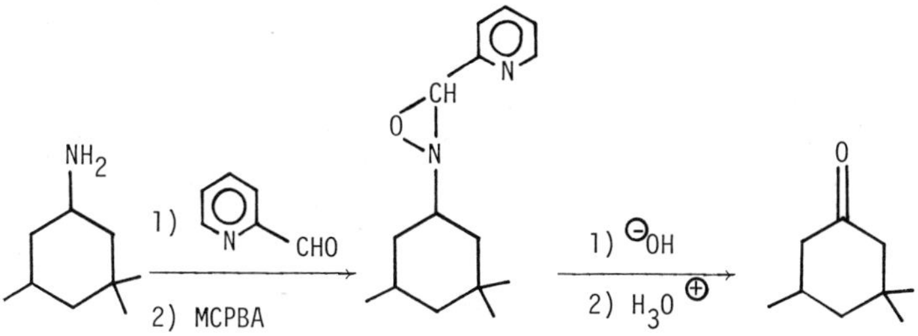

NH2
1)
N
CHO
2) MCPBA
CH
O
N
1) ⊖OH
2) H3O⊕
O

~40%

JACS (1975) 97 6900

H
NH2
1) acetone
2) MCPBA
3) H3O⊕
O

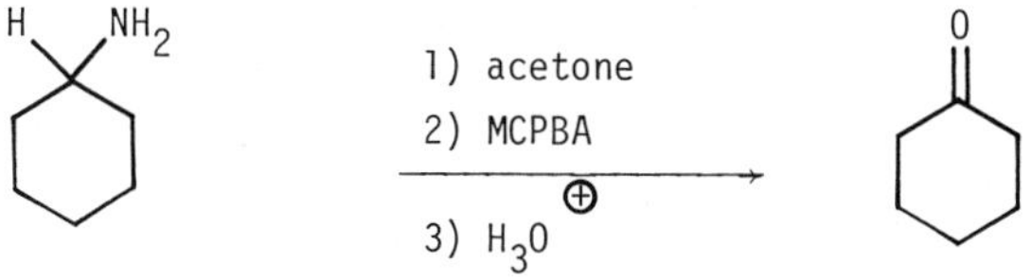

~30%

Aust J Chem (1975) 28 2547

NH2
CH
Ph Ph
O
Ph-Se-Cl
NH
Ph Ph

97%

Synth Comm (1976) 6 285

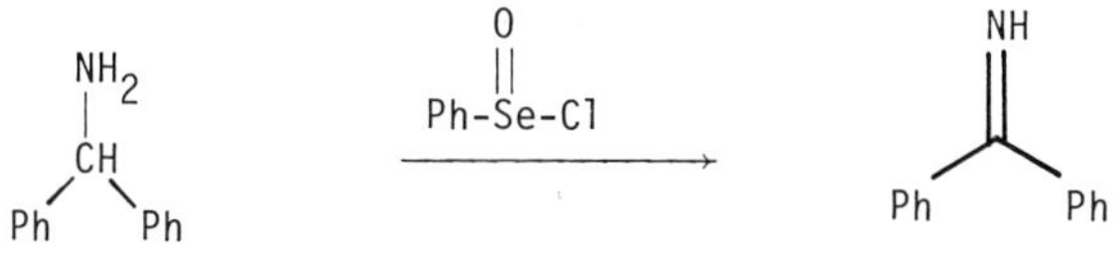

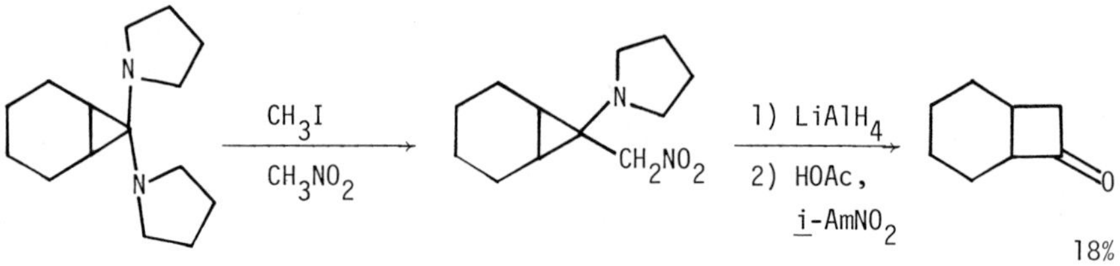

JOC (1976) <u>41</u> 153

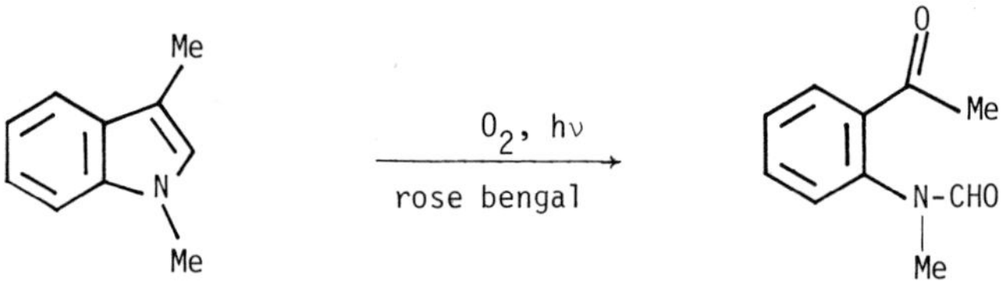

90%

Synthesis (1976) 255

Section 173   Ketones from Esters

$$\underset{RCCH_2C-OR}{\overset{O\ \ \ O}{||\ \ ||}} \quad + \quad DABCO \quad \xrightarrow[\text{rfx}]{\text{xylene}} \quad RCOCH_3 \qquad \qquad \sim80\%$$

JOC (1974) <u>39</u> 2647

$$MeCH_2COOEt \quad \xrightarrow[\text{THF}]{\text{KH}} \quad MeCH_2COCH(Me)COOEt \qquad \qquad 95\%$$

Synthesis (1975) 326

$$BuCO_2Et \xrightarrow[300°]{ThO_2} BuCOBu \qquad\qquad 80\%$$

Bull Soc Chim France (1974) 1455

$$\underset{RCH_2\overset{\displaystyle O}{\overset{\|}{C}}-OR'}{} \xrightarrow{\text{several steps}} \underset{R-\overset{\displaystyle O}{\overset{\|}{C}}-CH\overset{R''}{\underset{R''}{\diagdown}}}{}$$

JACS (1975) <u>97</u> 439

Review: "Reaction of Organomagnesium Compounds with Thiocarbonyl Compounds"

Bull Soc Chim France (1975) 1439

Section 174   <u>Ketones from Ethers and Epoxides</u>

86%

JACS (1976) <u>98</u> 6717

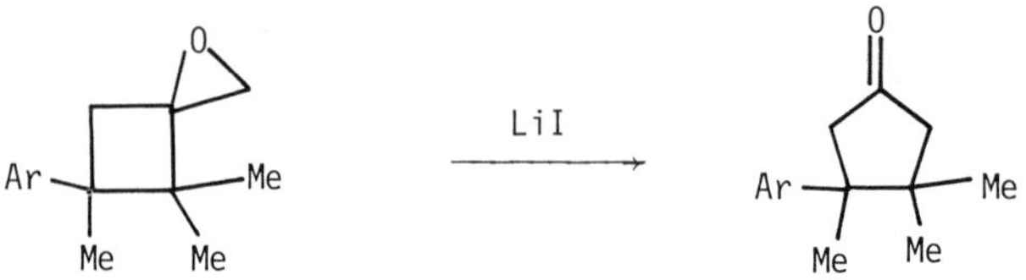

78%

Comptes Rendus C (1975) 280 791

Section 175   Ketones from Halides

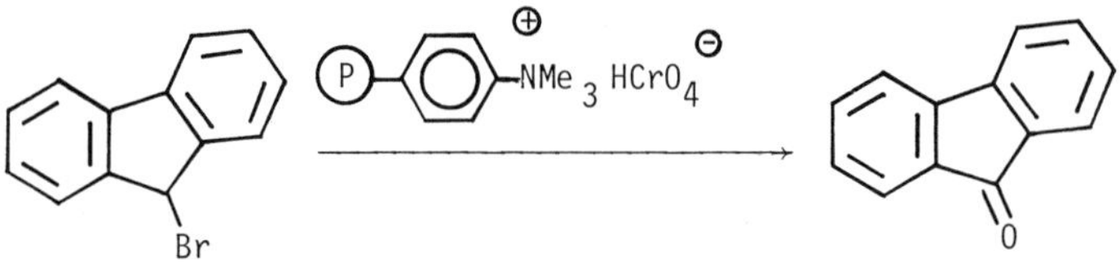

97%

Tetr Lett (1976) 3985

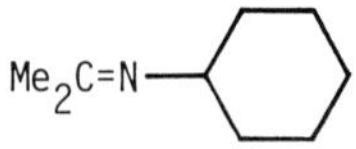

$Me_2C=N-$⬡   1) LDA, THF $\underline{n}-C_8H_{17}CH_2COCH_3$ 80%
                  2) $\underline{n}-C_8H_{17}Cl$
                  3) $H_3O^{\oplus}$

Liebigs Ann Chem (1975) 719

PhCHCOOH   +   Zn   +   MeCN   ⟶ $MeCOCH_2Ph$ <80%
|
Br

J Organometal Chem (1974) 81 139

$$\text{MeO}-\underset{2}{\overbrace{\Big(\!(\text{Ni}\overset{Br}{\diagdown}\!\Big)}} \quad + \quad \text{PhI} \quad \xrightarrow[\text{2) H} \oplus]{\text{1) DMF}} \quad \text{PhCH}_2\text{COCH}_3 \qquad\qquad 91\%$$

JACS (1974) <u>96</u> 3250

$$\text{RMgX} \quad + \quad \text{1 eq. HMPA} \quad \xrightarrow[\text{30 atm}]{\text{CO}} \quad
\begin{array}{l}
\xrightarrow{R = 3°} \quad \underset{\underset{\text{OH}}{|}\ \ \underset{\text{O}}{\|}}{\text{RCH-C-R}} \\[2em]
\xrightarrow{R = 1°} \quad R_2\text{CHCOR} \\[2em]
\xrightarrow{R = 2°} \quad \text{Mixtures}
\end{array}
\qquad 30\text{-}50\%$$

Tetr Lett (1974) 3377

$$\underline{n}\text{-C}_5\text{H}_{11}\text{Br} \quad + \quad \text{Na}_2\text{Fe(CO)}_4 \quad + \quad \text{CH}_2\text{CH}_2 \quad \xrightarrow[\text{2) HOAc}]{\text{1) THF}} \quad \underline{n}\text{-C}_5\text{H}_{11}\text{COEt} \qquad 95\%$$

JACS (1975) <u>97</u> 6863

$$\text{ArX} \quad + \quad \text{KCH}_2\text{COMe} \quad \xrightarrow[\text{NH}_3(1)]{\text{h}\nu} \quad \text{ArCH}_2\text{COMe} \qquad (\text{S}_{\text{RN}}1 \text{ mechanism})$$

Chem Pharm Bull (1975) <u>23</u> 2621

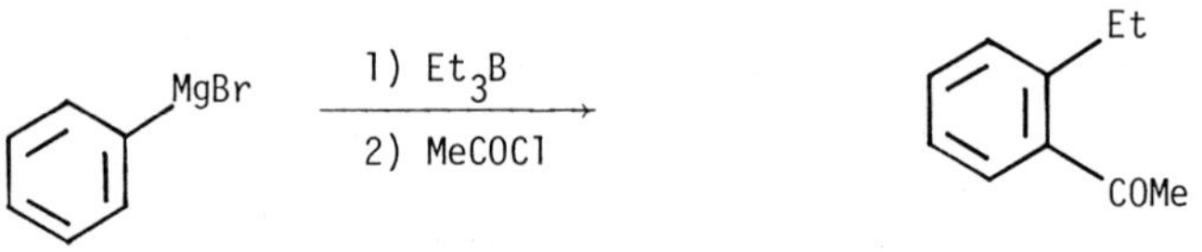

MgBr
1) Et_3B
2) MeCOCl
Et
COMe

69%

Tetr Lett (1975) 4239

Review: "Disodium Tetracarbonylferrate - A Transition Metal Analog of a Grignard Reagent"

Accts Chem Res (1975) 8 342

Cl
O
Cl
+ Li CH
SMe
S-Me
O
2) $H_3O^{\oplus}$
O
O

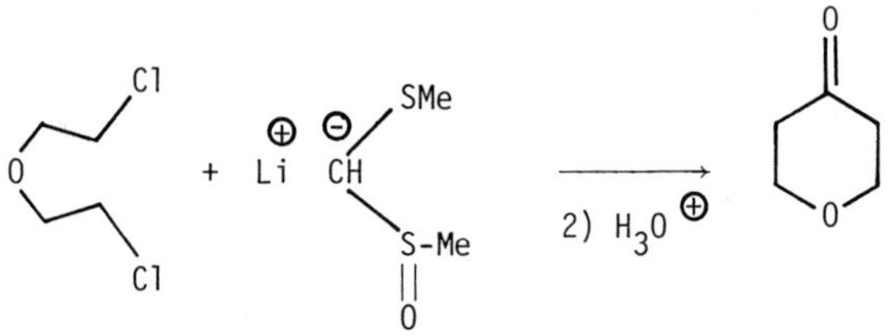

66%

Tetr Lett (1975) 2767

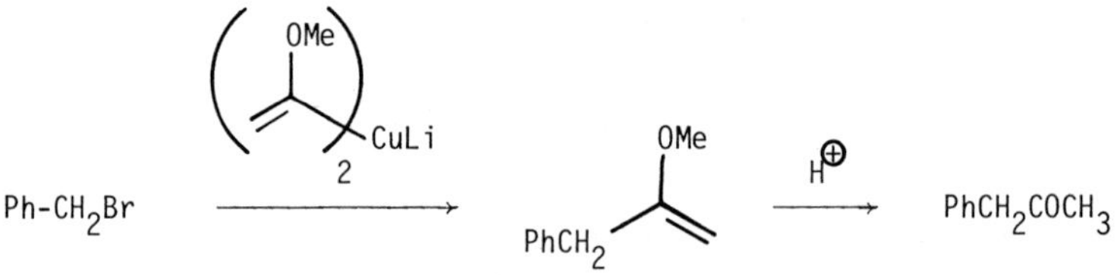

OMe
CuLi
2
$Ph-CH_2Br$
OMe
$PhCH_2$
$H^{\oplus}$
$PhCH_2COCH_3$

72%

JACS (1975) 97 3822

$MeSCH_2SOMe$ + $Br(CH_2)_3Br$ $\xrightarrow[\text{2) } H_3O^{\oplus}]{\text{1) KH}}$ 76%

Tetr Lett (1974) 3653

30-40%

Tetr Lett (1974) 3295
Angew Int Ed (1974) 13 277

$\xrightarrow[\substack{\text{2) } Ph_2CO \\ \text{3) } H^{\oplus}}]{\text{1) BuLi, THF, -90°}}$

90%

Chem Ber (1975) 108 2368

$\underline{n}-C_6H_{13}I$ + $\overset{\overset{\text{O}}{\underset{||}{}}}{MeSCHSMe}^{(-)}$ $\xrightarrow{}$ $\xrightarrow{HCl}$ $C_6H_{13}COC_6H_{13}$ 62%

Synthesis (1974) 117

$$CH_2=\underset{\underset{OMe}{|}}{C}-CH_2-S-\overset{\overset{S}{||}}{C}-NMe_2 \quad \xrightarrow[\substack{2)~PhCH_2Br \\ \\ 3)~hydrolysis}]{1)~LDA} \quad CH_3-\overset{\overset{O}{||}}{C}CH_2CH_2Ph \qquad 98\%$$

Tetr Lett (1975) 4027

$$NC-CH_2\underset{\underset{S}{||}}{S}CNMe_2 \quad \xrightarrow[\substack{2)~NaOH,~Amyl~Br}]{1)~NaOH,~MeI} \quad \underset{Am}{\overset{Me}{\diagdown}}\underset{\diagup}{\overset{|}{C}}\overset{\overset{\overset{S}{||}}{SCNMe_2}}{}{}_{\diagdown CN} \quad \xrightarrow{NBS} \quad \underset{Am}{\overset{Me}{\diagdown}}C=O \quad 68\%$$

Tetr Lett (1976) 2967

$$(MeCO)_2C=SMe_2 \quad \xrightarrow[\substack{2)~PhCH_2Br}]{1)~BuLi} \quad PhCH_2CH_2CO\overset{\overset{SMe_2}{||}}{C}COMe \qquad 73\%$$

JCS Chem Comm (1975) 289

Related methods: Ketones from Ketones (Section 177), Aldehydes from Halides
                  (Section 55)

Section 176   Ketones from Hydrides

This section lists examples of the replacement of hydrogen by ketonic
groups, RH → RCOR'. For the oxidation of methylenes $R_2CH_2 \rightarrow R_2CO$ see
Section 170 (Ketones from Alkyls and Methylenes)

$$H_2N-\overset{\overset{\displaystyle O}{\|}}{C}-NHCOCH_3$$

polyphosphoric acid

51%

Chem and Ind (1976) 1069

Review: "Friedel Crafts Acylation"

Chem and Ind (1974) 727

Section 177   Ketones from Ketones

This section contains alkylations of ketones and protected ketones, ketone transpositions and annelations, ring expansions and ring openings, and dimerizations. Conjugate reductions and reductive alkylations of enones are listed in Section 74 (Alkyls from Olefins).

t-BuOK

-70°

MeI

88%

Tetrahedron (1974) 30 3263

PhCH$_2$COMe

RX, NaOH

phase transfer
catalyst

PhCHRCOMe   high yields

Tetr Lett (1975) 3757
JCS Chem Comm (1975) 393

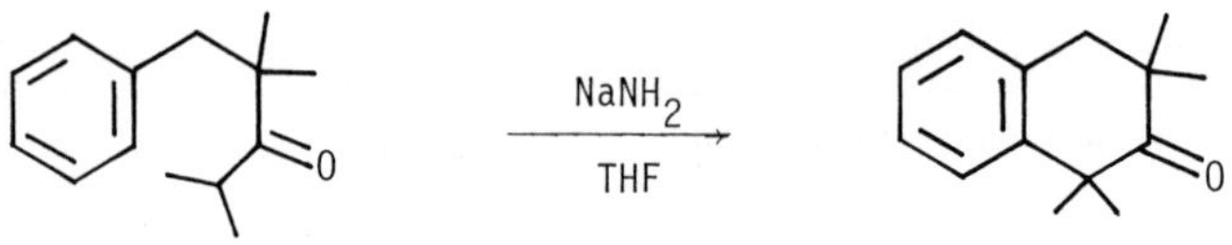

NaNH$_2$ / THF

90%

Synthesis (1974) 201

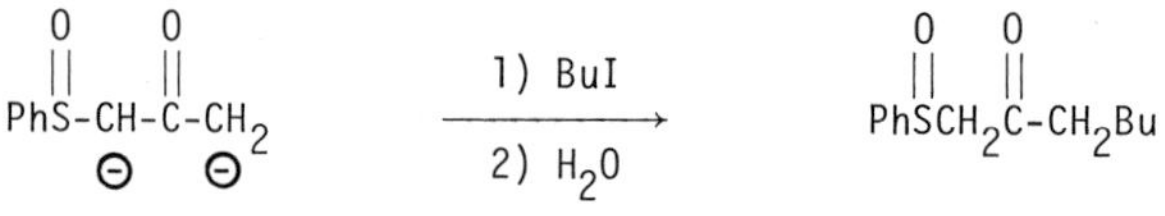

PhS-CH-C-CH$_2$ 1) BuI   2) H$_2$O   PhSCH$_2$C-CH$_2$Bu

78%

JOC (1974) $\underline{39}$ 732

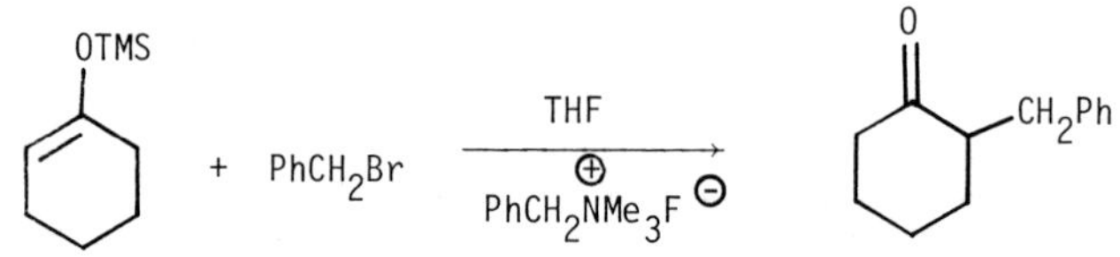

OTMS   + PhCH$_2$Br

THF / PhCH$_2$NMe$_3$F$^{\ominus}$ ⊕

63%

JACS (1975) $\underline{97}$ 3257

Ag$_2$O

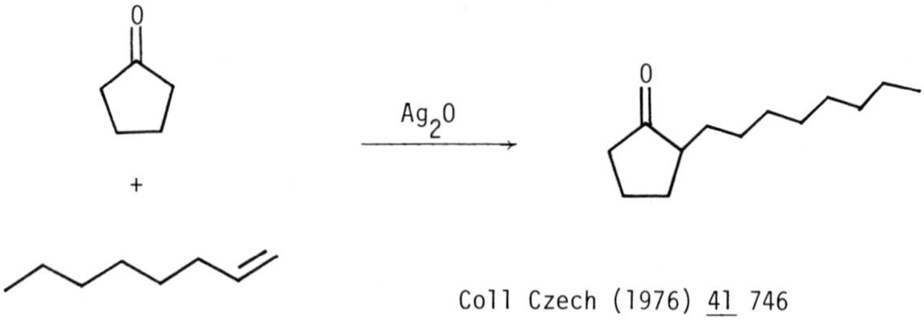

∿69%

Coll Czech (1976) $\underline{41}$ 746

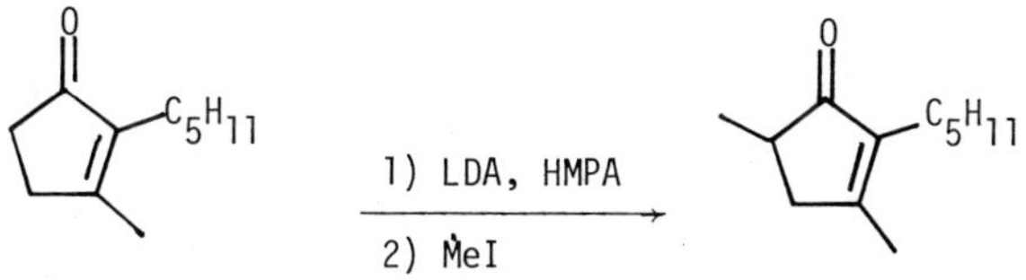

O
C5H11
1) LDA, HMPA
2) MeI
O
C5H11

74%

Synth Comm (1975) 5 435

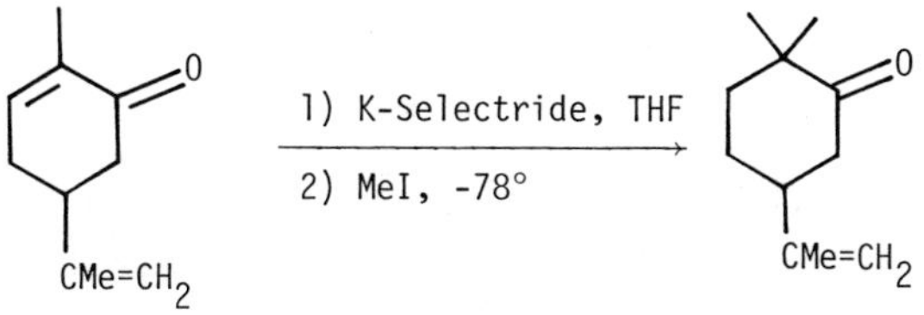

O
CMe=CH2
1) K-Selectride, THF
2) MeI, -78°
O
CMe=CH2

98%

JOC (1975) 40 146

O
OTMS
O
O
O

JACS (1975) 97 1619

O
SPh
1) LDA, THF
2) MeI
O
Me
SPh

93%

Tetr Lett (1974) 3955

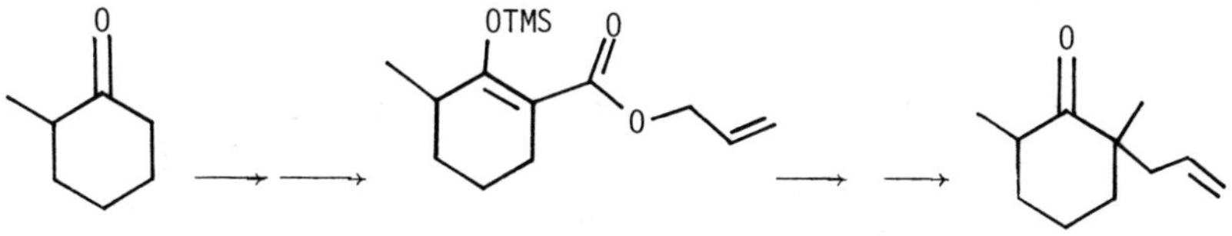

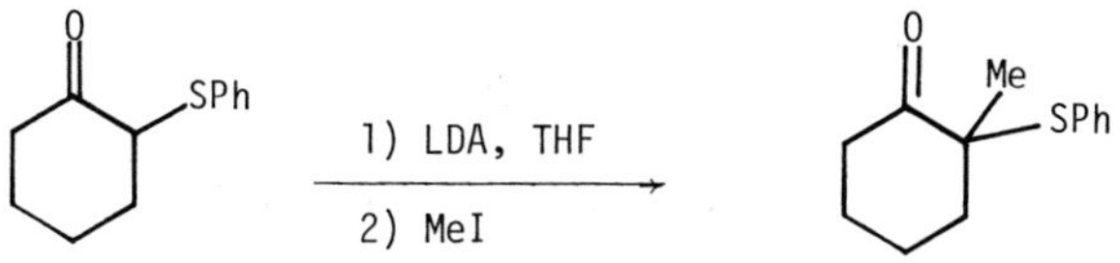

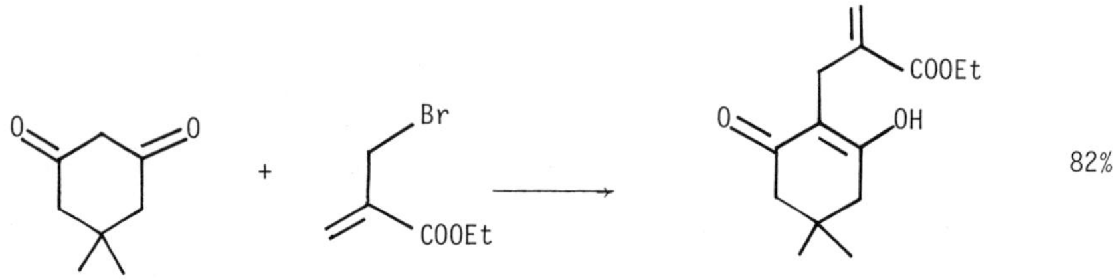

82%

Rec Trav Chim (1974) <u>93</u> 153

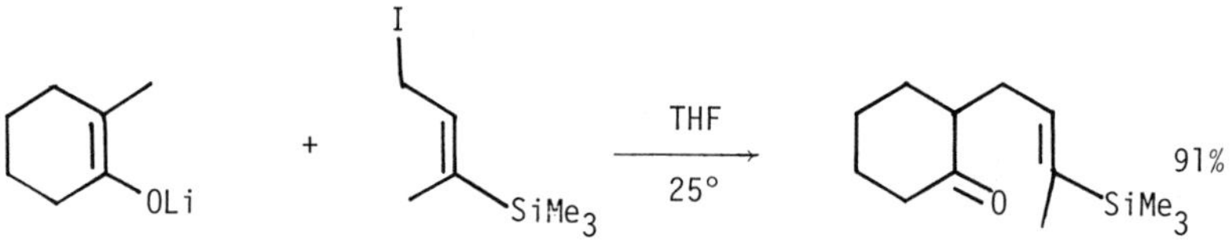

91%

JACS (1974) <u>96</u> 3684

90%

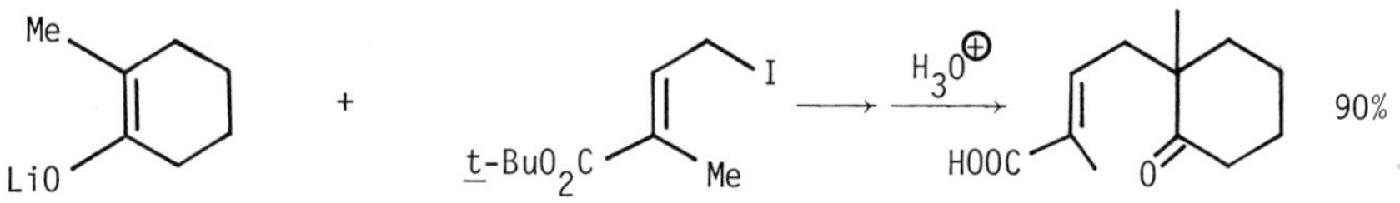

JACS (1974) <u>96</u> 6524

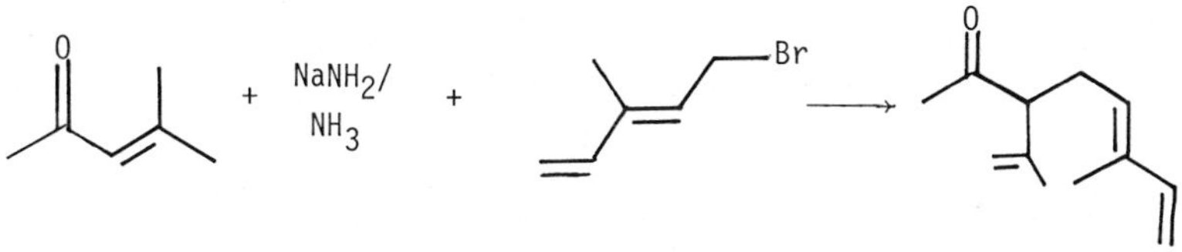

JACS (1974) 96 7573

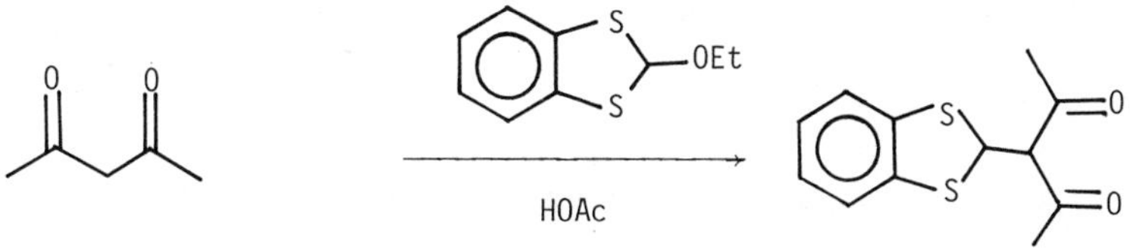

JCS Perkin I (1976) 540

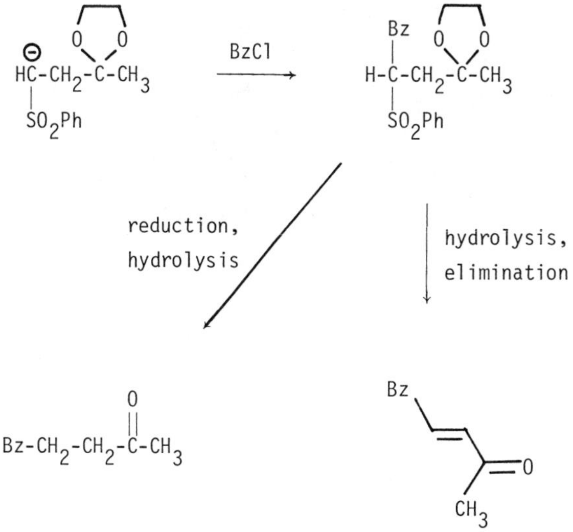

Bull Soc Chim France (1975) 1363

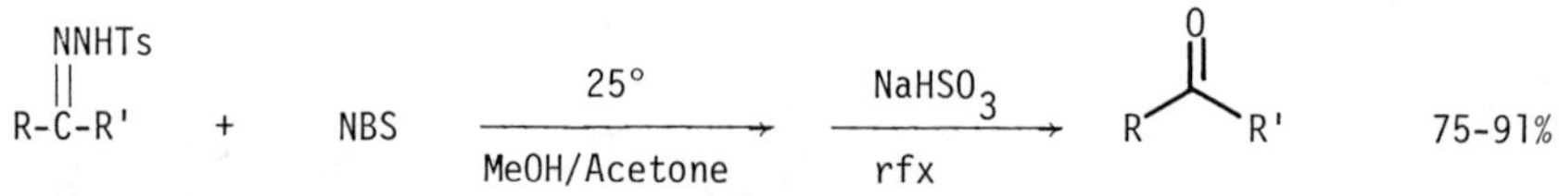

NNHTs
R-C-R'
+
NBS
25°
MeOH/Acetone
NaHSO3
rfx
O
R
R'
75-91%

JOC (1974) 39 3504

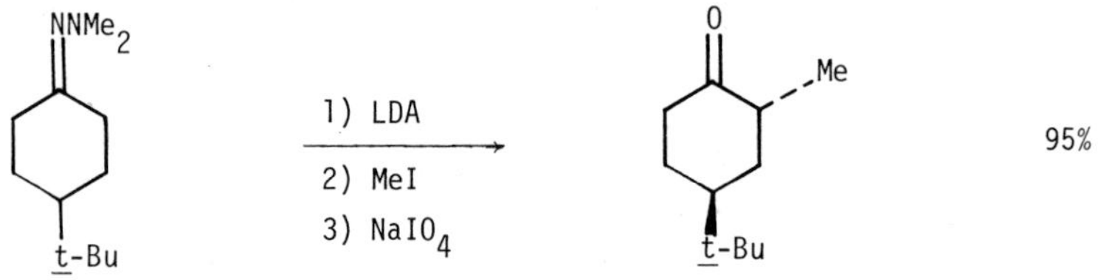

NNMe2
t-Bu
1) LDA
2) MeI
3) NaIO4
O
Me
t-Bu
95%

Tetr Lett (1976) 3

N
N
OMe
1) LDA
2) n-PrI
3) O3
O
n-Pr
*
(R) 87% ee
73%

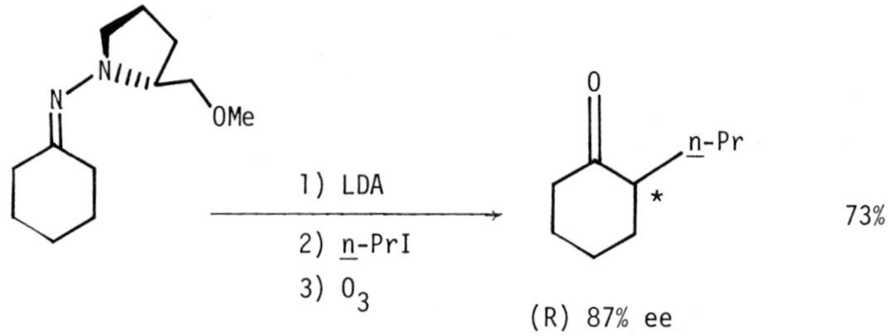

Angew Int Ed (1976) 15 549

1) BuLi (2 moles)

2) ⟋⟍ Br

3) $H_3O^{\oplus}$

65%

JOC (1976) 41 439

$$\underset{\text{NOH}}{R-\overset{\|}{C}-R'} \quad + \quad (Ph_3P)_2 \; PdO_2 \quad \xrightarrow[\text{benzene}]{} \quad \underset{\text{O}}{R-\overset{\|}{C}-R'} \qquad 74\text{-}98\%$$

Tetr Lett (1974) 797

For the preparation of enamines from ketones see Section 356 (Amine - Olefin)

A 1,2-alkylative carbonyl transposition:

60-75%

also:

$$\underset{\text{O}}{RCH_2\overset{\|}{C}R'} \quad \longrightarrow \quad \underset{\text{O}}{R\overset{\|}{C}-\overset{R''}{\underset{R'}{C}H}}$$

JACS (1975) 97 439

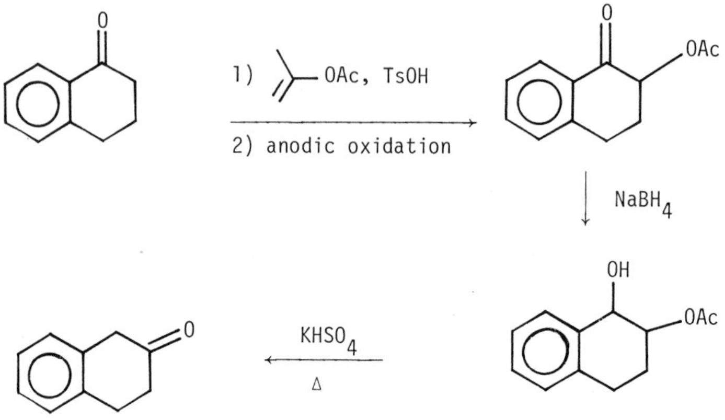

∿30% overall

Chem Lett (1976) 1319

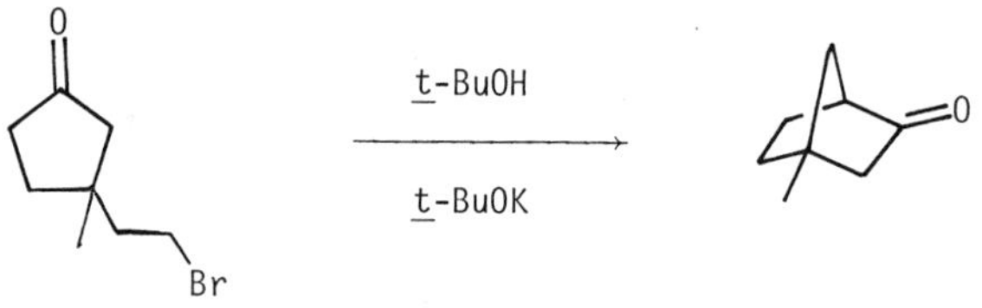

65%

JOC (1974) 39 573

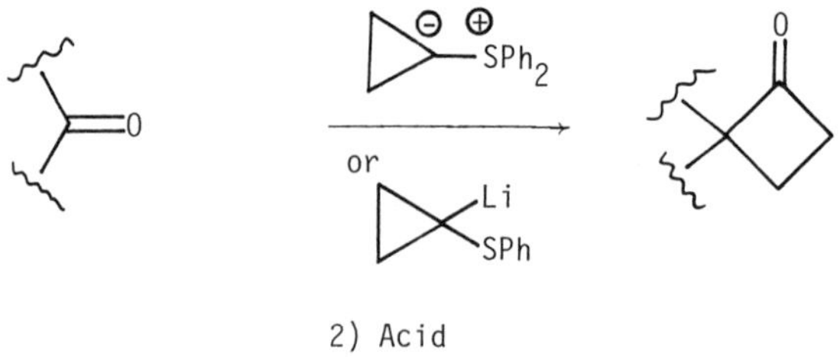

2) Acid

JACS (1975) 97 2218, 2224

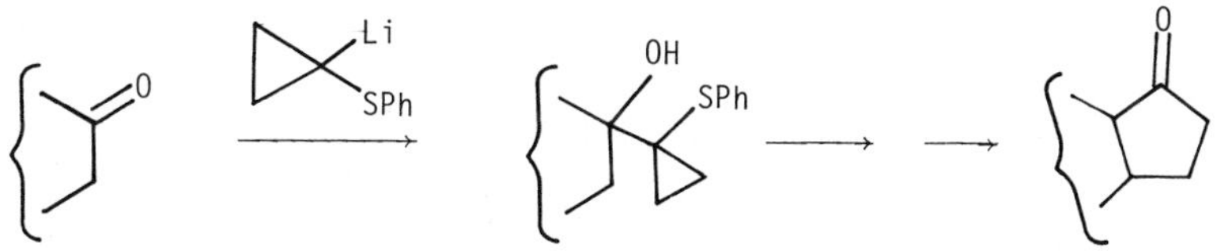

JACS (1976) <u>98</u> 248

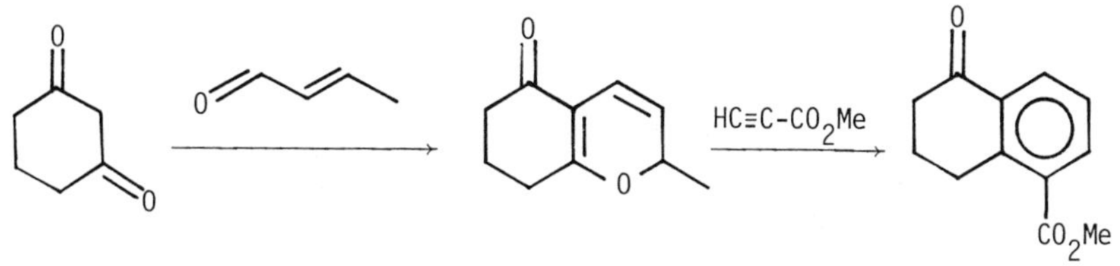

70% (second step)

JOC (1976) <u>41</u> 2918

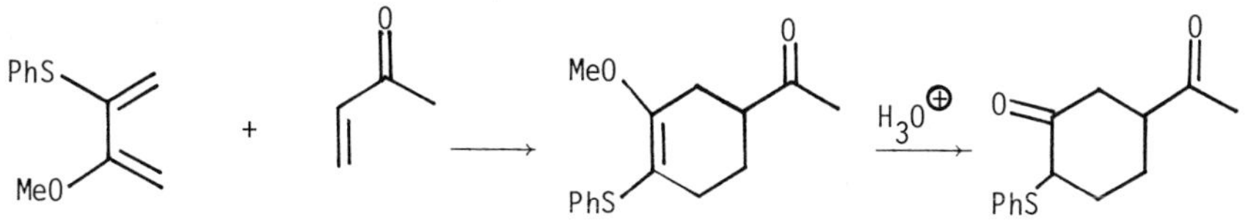

75%

Can be alkylated further, dehydrosulfenylated.

JACS (1976) <u>98</u> 5017

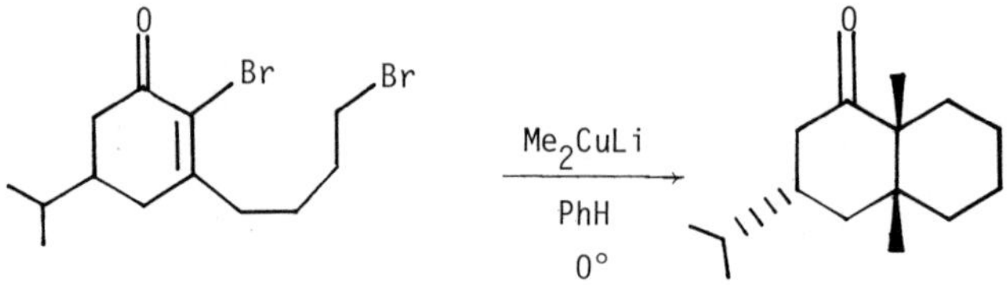

25%

Tetr Lett (1974) 2591

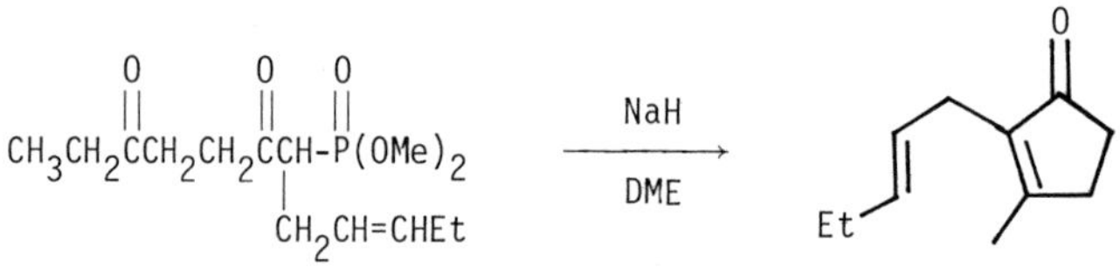

68%

Synth Comm (1975) 5 1

Review: "The Thermal Cyclization of Unsaturated Carbonyl Compounds"

Synthesis (1975) 1

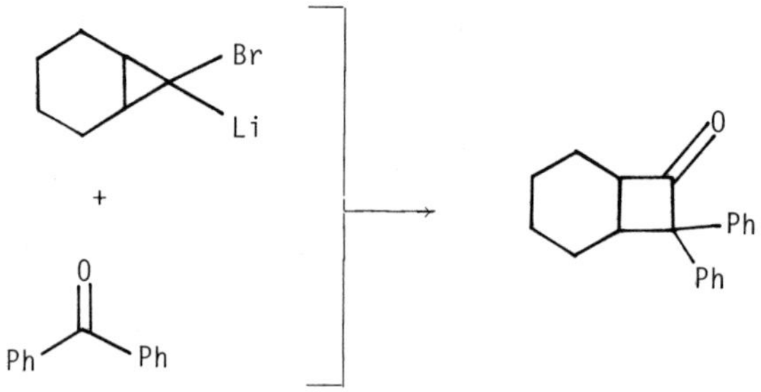

85%

Chem Ber (1975) 108 2368

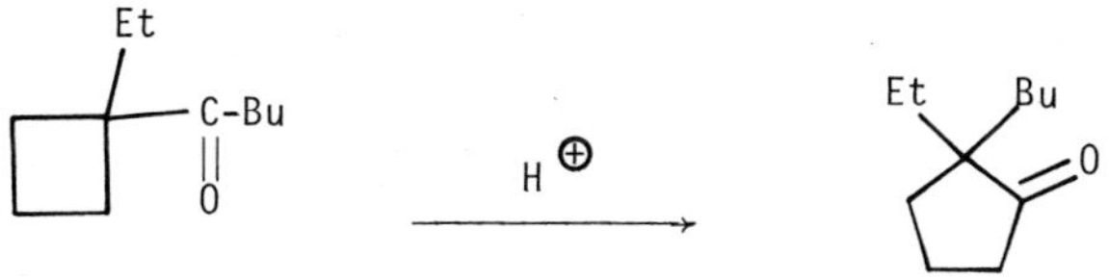

91%

Comptes Rendus C (1975) 280 309

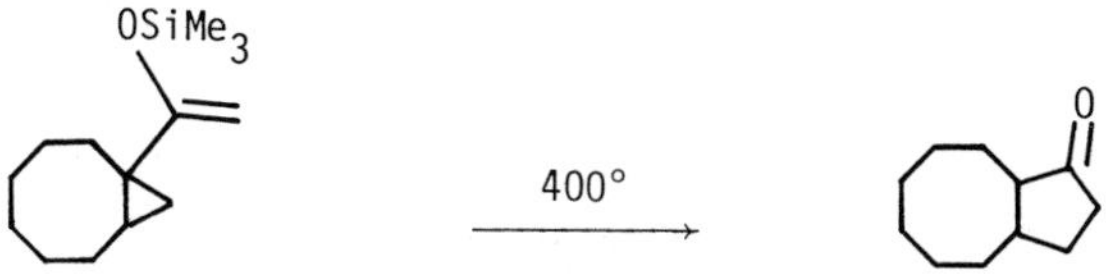

99%

Widely varying yields with other ring sizes.

JOC (1975) 40 858

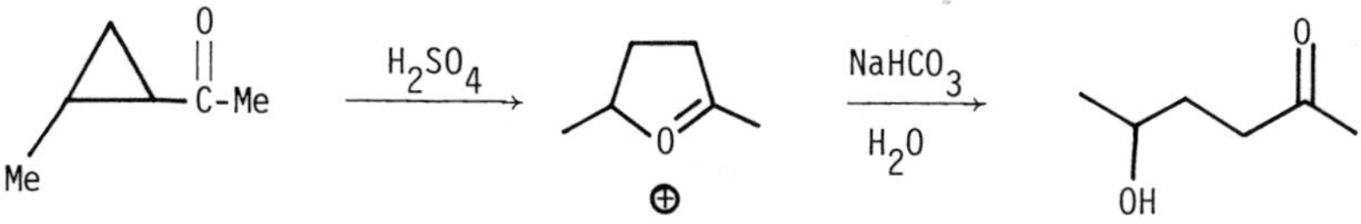

45%

Tetr Lett (1975) 1531

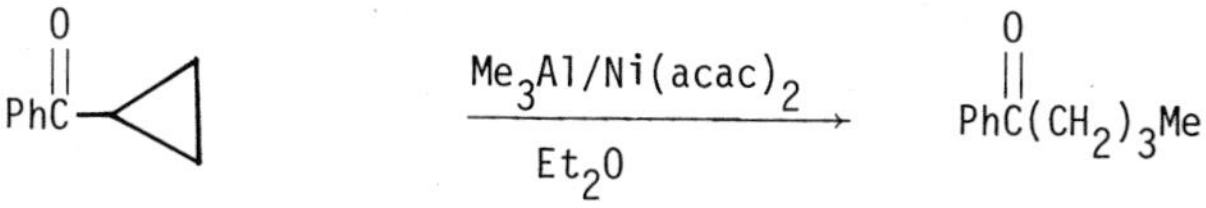

55%

Aust J Chem (1975) 28 821

$$CH_3\overset{O}{\overset{||}{C}}-\underset{\underset{Na}{|}}{CH}-\overset{O}{\overset{||}{C}}-CH_2Li \xrightarrow[\text{2) }H_3O^{\oplus}]{\text{1) }CH_2Br_2} CH_2(CH_2\overset{O}{\overset{||}{C}}CH_2\overset{O}{\overset{||}{C}}CH_3)_2 \qquad \sim9\%$$

JOC (1976) <u>41</u> 2772

$$2 \text{ (cyclopentene OTMS)} \xrightarrow[\text{DMSO}]{Ag_2O} \text{(biscyclopentanone)} \qquad 61\%$$

JACS (1975) <u>97</u> 649

$$2 \text{ PhCOCH}_3 \xrightarrow[\text{2) } CuCl_2,\text{ DMF}]{\text{1) LDA, THF}} Ph-\overset{O}{\overset{||}{C}}-CH_2CH_2-\overset{O}{\overset{||}{C}}-Ph \qquad 83\%$$

JACS (1975) <u>97</u> 2912

Review: "Inversion of Carbonyl Reactivity"

Chem and Ind (1974) 687

Review: "The Use of Kinetically Generated Unstable Enolate Ions in the Regiospecific Formation of Carbon-Carbon Bonds"

Pure and Appl Chem (1975) <u>43</u> 553

Ketones may also be alkylated and homologated via olefinic ketones (Section 374)

Related methods: Aldehydes from Aldehydes (Section 49)

Section 178   Ketones from Nitriles

$$\text{Ph, Me, CN, H} \xrightarrow[\text{3) Ag}_2\text{O}]{\substack{\text{1) Me}_3\text{SiCl} \\ \text{2) Br}_2}} \text{Ph-CO-Me}$$

~70%

JOC (1974) 39 2799
Tetr Lett (1974) 3029

82%

JOC (1975) 40 267

$$\text{(CH}_3)_2\text{CH-CHBrCOOH} + \text{PhCN} \xrightarrow[\text{2) H}_3\text{O}^{\oplus}]{\text{1) Zn}} \text{(CH}_3)_2\text{CH-CH}_2\text{COPh}$$

74%

J Organometal Chem (1974) 81 139

$$\underline{t}\text{-BuCN} + \underline{i}\text{-BuBr} \xrightarrow[\text{EtLi}]{\text{LDA}} \xrightarrow{\text{H}_3\text{O}^{\oplus}} \underline{t}\text{-BuCOCHMe}$$
$$\qquad\qquad\qquad\qquad\qquad\qquad\qquad\qquad\underset{\underline{i}\text{-Bu}}{|}$$

73%

Comptes Rendus (1975) 280 889

I-(CH$_2$)$_3$CMe$_2$CN $\xrightarrow[\text{Et}_2\text{O}]{\text{Mg}}$ 79%

J Organometal Chem (1975) <u>87</u> 25

PhCN $\xrightarrow[\text{Ni(acac)}_2]{\text{Me}_3\text{Al}}$ PhCOMe   80%

Aust J Chem (1974) <u>27</u> 2577

Section 179   <u>Ketones from Olefins</u>

RCH=CHR $\xrightarrow[\text{2) } ^{\ominus}\text{OH}]{\text{1) HgSO}_4\text{, H}_2\text{O}}$ $\overset{\text{O}}{\overset{||}{\text{RCCH}_2\text{R}}}$ 30-70%

JOC (1974) <u>39</u> 3445

$\xrightarrow[\text{3) CrO}_3]{\begin{array}{l}\text{1) Hg(NO}_3)_2\\\text{2) NaBH}_4\end{array}}$ 80%

JOC (1974) <u>39</u> 2674

$$\xrightarrow[\text{H}^{\oplus},\ \text{Hg(OAc)}_2]{\text{Na}_2\text{Cr}_2\text{O}_7}$$

82%

JOC (1975) 40 3577

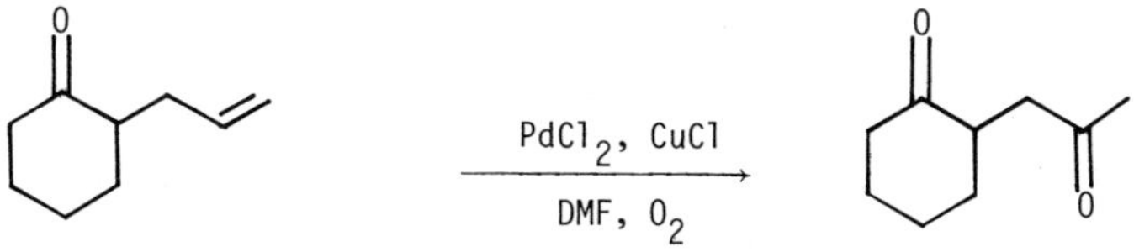

$$\xrightarrow[\text{DMF, O}_2]{\text{PdCl}_2,\ \text{CuCl}}$$

68%

Tetr Lett (1976) 2975

$ArSO_2N_3$ +

$$\xrightarrow{\quad\quad} \xrightarrow{\text{H}_3\text{O}^{\oplus}}$$

JCS Perkin I (1974) 2169

$Cp_2Zr(H)Cl$ +

$$\xrightarrow{\text{CH}_3\text{COCl}}$$

80%

JACS (1974) 96 8115

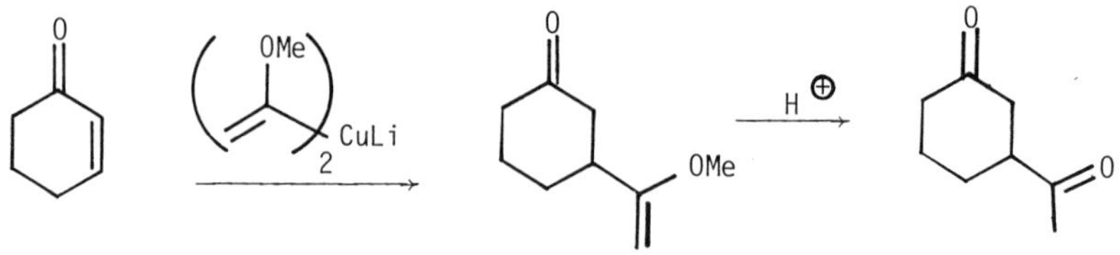

JACS (1975) 97 3822

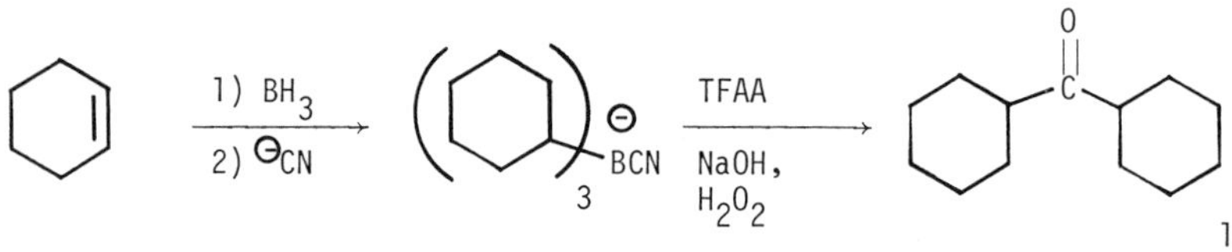

JCS Perkin I (1975) 129

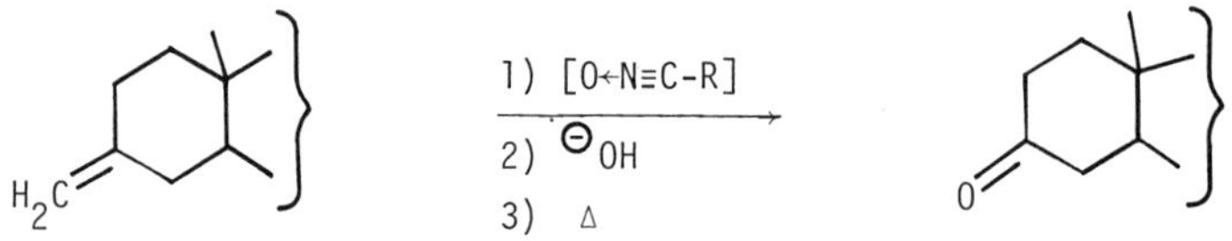

JCS Chem Comm (1976) 209, 210

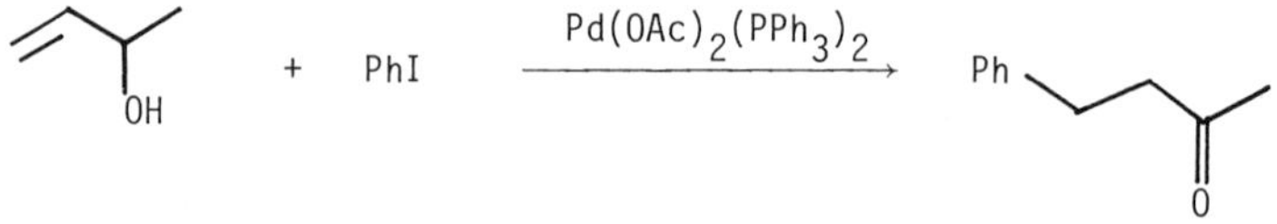

JOC (1976) 41 265
JOC (1976) 41 273

See also section 134 (Ethers and Epoxides from Olefins) and Section 174
(Ketones from Ethers and Epoxides)

Section 180   Ketones from Miscellaneous Compounds

Conjugate reductions and reductive alkylations of enones are listed
in Section 74 (Alkyls from Olefins).

1) [Ni(dppp)$_2$Cl$_2$]
2) H$_3$O$^\oplus$

$$PhCH_2\overset{O}{\overset{\|}{C}}-CH_3$$

75%

Chem Lett (1976) 1239

NiEt$_2$(bpy)

CO, -78°
THF

$$Et\overset{O}{\overset{\|}{\underset{}{C}}}Et$$

100%

Chem Lett (1976) 1217

NaOCH$_3$
EtOH

52%

Helv Chim Acta (1976) 59 522

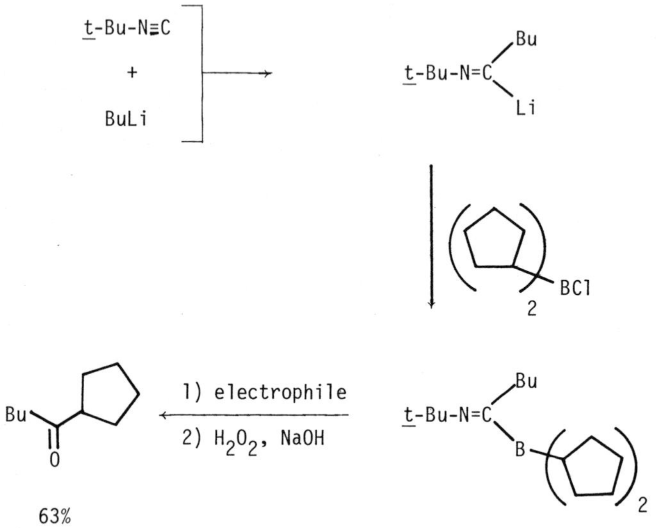

$\underline{t}$-Bu-N≡C

+

BuLi

$\underline{t}$-Bu-N=C〈Bu / Li

$\underline{t}$-Bu-N=C〈Bu / B(⬠)$_2$

1) electrophile

2) H$_2$O$_2$, NaOH

Bu-C(=O)-⬠

63%

Bull Chem Soc Japan (1975) $\underline{48}$ 3682

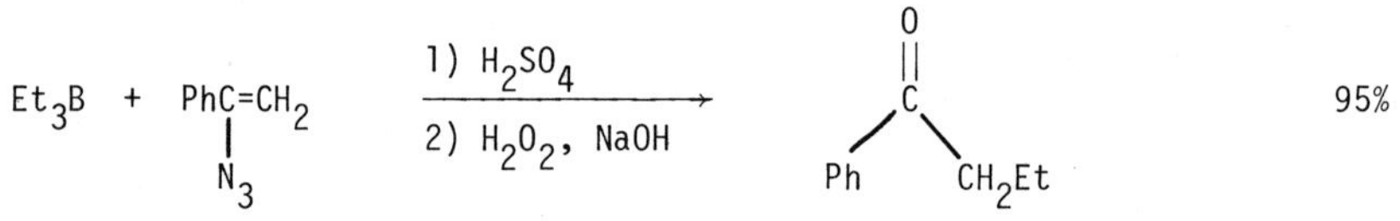

Et$_3$B + PhC=CH$_2$ / N$_3$

1) H$_2$SO$_4$

2) H$_2$O$_2$, NaOH

Ph-C(=O)-CH$_2$Et   95%

Tetr Lett (1975) 2195

PhCH$_2$CHCH$_3$ / NO$_2$

VCl$_2$, HCl

DMF/H$_2$O

PhCH$_2$CCH$_3$ (C=O)   65%

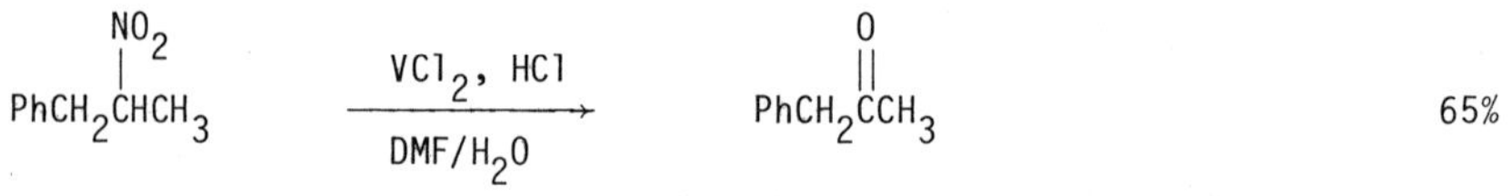

Tetr Lett (1976) 2533

RCHR'   1) MeO $\ominus$
|
NO$_2$ 2) O$_3$ 65-88%

             3) Me$_2$S

JOC (1974) 39 259

Section 180A   Protection of Ketones

This section covers ketals, thioketals, and hydrazones. See also Section 363
(Ether-Ether) for ketals and Section 367 (Ether-Olefin) for enol ethers.
Some of the methods in Section 60A (Protection of Aldehydes) are also appli-
cable to ketones.

74%

TsOH   Ph

JCS Chem Comm (1975) 432

                                                              OEt
                                                              OEt
              EtOH, HC(OEt)$_3$ 78%
              montmorillonite

Bull Soc Chim France (1975) 2558

              SMe
              SMe
H$_2$N   C-CH$_3$ SO$_2$Cl$_2$, SiO$_2$ H$_2$N   C-CH$_3$ 100%
                          H$_2$O, CH$_2$Cl$_2$ O

Synthesis (1976) 678

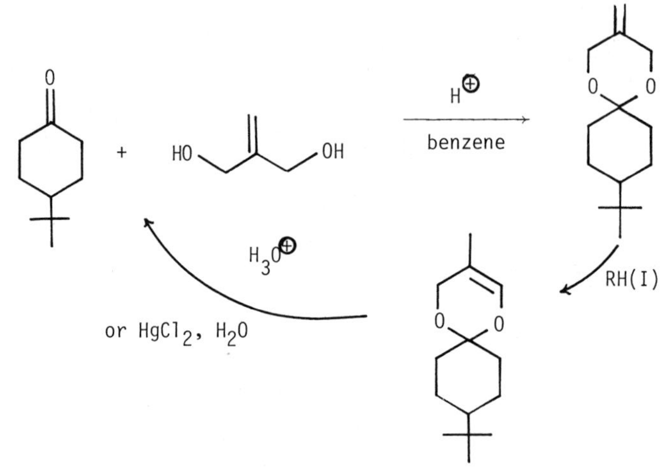

Tetr Lett (1975) 3775

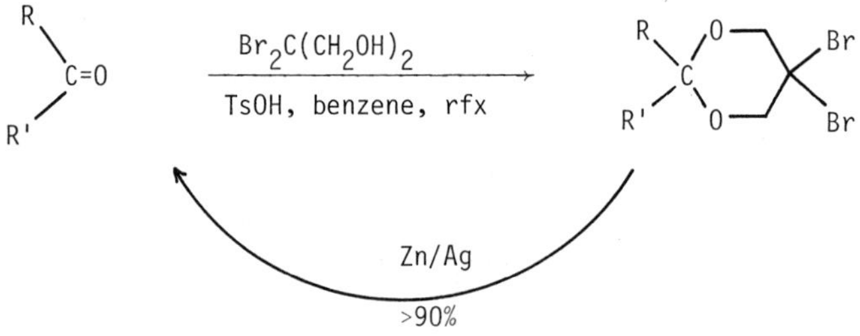

Tetr Lett (1976) 4577

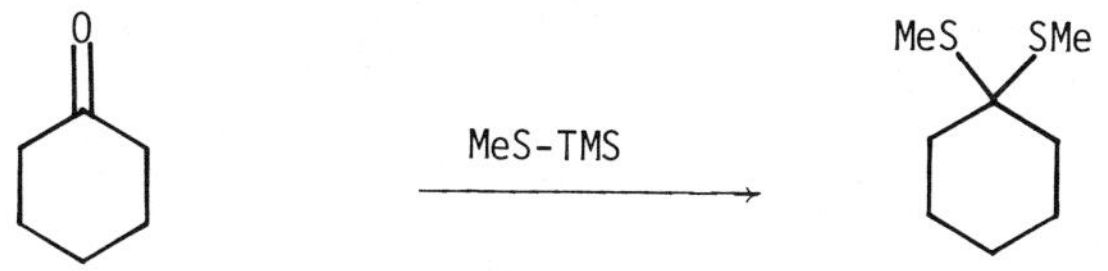

MeS-TMS

No acid catalyst required.

JACS (1975) 97 3229

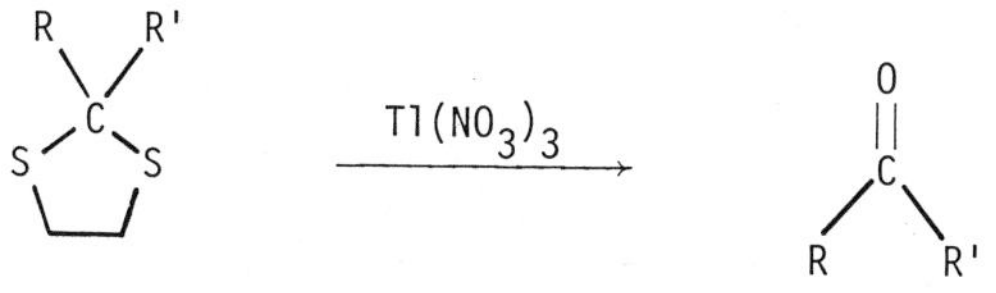

90%

Tl(NO$_3$)$_3$

R,R' = H, alkyl, aryl, cyclic

73-99%

Chem Pharm Bull (1976) 24 1115

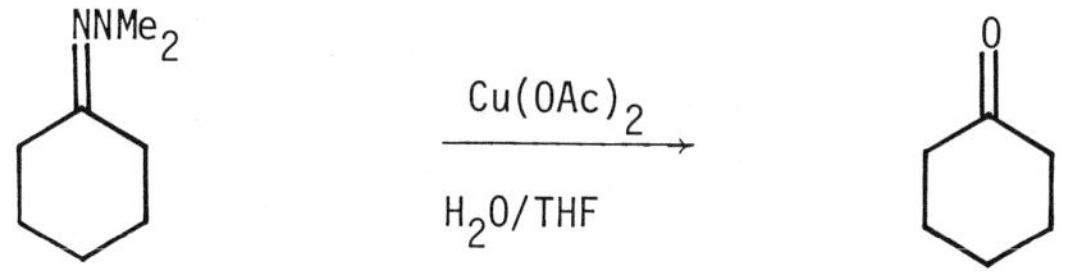

$\xrightarrow{\text{Cu(OAc)}_2}{\text{H}_2\text{O/THF}}$

Tetr Lett (1976) 3667

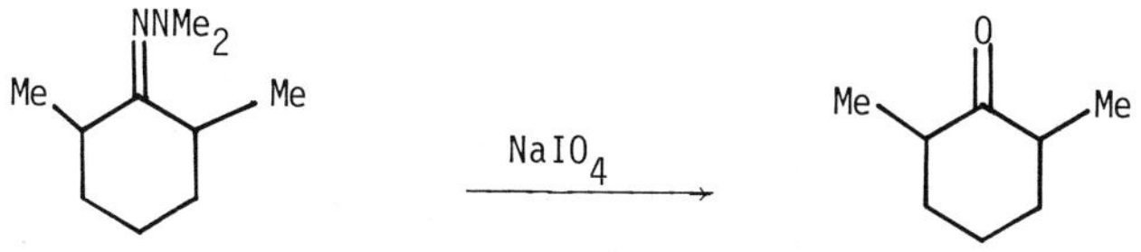

96%

NaIO$_4$

Tetr Lett (1976) 3

100%

NNHTs

$$\xrightarrow[\text{or } WF_6]{\text{MoOCl}_3}$$

74%

Synthesis (1976) 808, 809

NNHTs

$$\xrightarrow{\text{Na}_2\text{O}_2}$$

72%

Synthesis (1976) 611

$$\xrightarrow[\text{NaOAc/HOAc}]{\text{TiCl}_3}$$

~90%

Chem and Ind (1975) 87

NNHTs

$$\xrightarrow[\text{BF}_3\cdot\text{Et}_2\text{O}]{\text{acetone/H}_2\text{O}}$$

83%

Synthesis (1976) 456

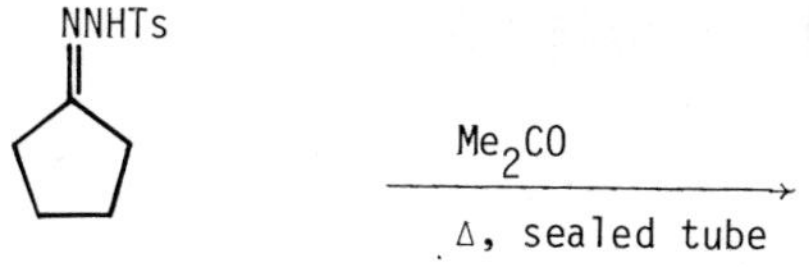

JOC (1975) <u>40</u> 3302

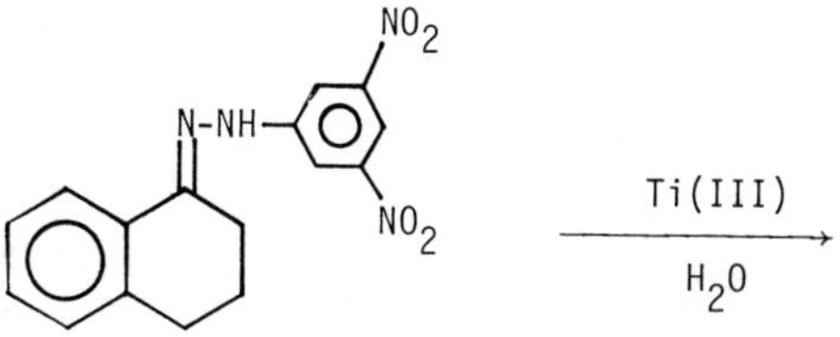

75%

JOC (1975) <u>40</u> 1502

98%

Chapter 13   PREPARATION

OF

NITRILES

Section 181   Nitriles from Acetylenes

No additional examples

Section 182   Nitriles from Carboxylic Acids and Acid Halides

$$BuCH_2COOH \xrightarrow[\text{NaNO}_2]{(CF_3CO)_2O} BuC \equiv N$$

60%

JOC USSR (1975) 11 653

Section 183   Nitriles from Alcohols

HO—⟨○⟩—CH$_2$OH $\xrightarrow[\text{DMF, 110°C}]{\text{NaCN}}$ HO—⟨○⟩—CH$_2$CN

67%

JOC (1976) 41 2502

Section 184   Nitriles from Aldehydes

Ph⤴⤵=NOH (H) $\xrightarrow{\text{Me-C}\equiv\overset{+}{\text{N}}\text{Et } ^-\text{BF}_4}$ Ph⤴⤵CN

56%

Synthesis (1975) 401
296

$$\text{(furan)}-CH=NOH \xrightarrow[\text{Et}_3N]{\text{PhOSOCl}} R-C\equiv N \qquad 92\%$$

Synthesis (1975) 502

$$\xrightarrow[\text{or}]{\overset{+}{Me_2}N=CH-SMe \quad I^-} \qquad 71\%$$

(2,4-dinitrofluorobenzene)

Synth Comm (1975) 5 299

$$\xrightarrow{\overset{+}{Me_2}SCl \quad Cl^-} \qquad 89\%$$

Synth Comm (1975) 5 423

$$C_6H_{13}CHO + H_2NO-\text{(2,4,6-trinitrophenyl)} \xrightarrow[2) \; ^-OH]{1) \; H^+} C_6H_{13}CN \qquad 91\%$$

JOC (1975) 40 126

$$\xrightarrow{(CF_3SO_2)_2O} Ph-C\equiv N \qquad 90\%$$

Tetr Lett (1976) 603

$$\xrightarrow[2) \; \Delta]{1) \; TsOH} Cl-\text{(phenyl)}-CN \qquad 81\%$$

Z Chem (1976) 16 17

89%

Helv Chim Acta (1976) <u>59</u> 2786

$$RCHO + NH_2OSO_3H \xrightarrow[65°]{H_2O} RCN$$

60-90%

Tetr Lett (1974) 3187

$$MeCH=NOH + Me_2N=CCl_2^+ \ Cl^- \xrightarrow[\substack{CHCl_3 \\ 2-4 \ hr}]{rfx} RC≡N$$

82-97%

Synthesis (1974) 563

$$Ph-CHO \longrightarrow PhCH=NTs \xrightarrow[HMPA]{NaCN} PhC≡N$$

77%

Tetr Lett (1976) 1781

$$RCHO \longrightarrow RCH=NOH \xrightarrow[\substack{CuCl_2 \\ Et_3N}]{DCC} RCN$$

55-92%
(44 cases)

Chem Ber (1974) <u>107</u> 1221

$$RCH=NOH + R'C(OEt)_3 \xrightarrow{H^+} RCN$$

>90%

JOC (1974) <u>39</u> 3424

$$Et_2CH-CH=NNMe_2 \xrightarrow[\text{2) } H_2O]{\text{1) Li, } Et_2NH, \text{ HMPT}} Et_2CH-CN \qquad 71\%$$

Synthesis (1976) 237 and 238

$$Ar-CHO + \text{[pyridinone structure, Ph, Ph substituents]} \xrightarrow{\text{2) } \Delta} Ar-CN \qquad \sim 80\text{-}90\%$$

Tetr Lett (1976) 2691

Section 185　　**Nitriles from Alkyls, Methylenes and Aryls**

No additional examples

Section 186　　**Nitriles from Amides**

$$Polymer-PCl_2 \xrightarrow{RCONH_2} RCN$$

JACS (1974) **96** 6469

$$PhCONH_2 \xrightarrow{Ph_3P(OTf)_2} PhCN \qquad 70\%$$

Tetr Lett (1975) 277

Section 187　　**Nitriles from Amines**

$$PhCH_2NH_2 \xrightarrow{PhSeOCl} PhCN \qquad 85\%$$

Synth Comm (1976) **6** 285

Section 188   Nitriles from Esters

+ Ph$_3$P=CHCN $\longrightarrow$ —CH$_2$CN   89%

Aust J Chem (1975) <u>28</u> 1097; 2499

Section 189   Nitriles from Ethers and Epoxides

No additional examples

Section 190   Nitriles from Halides

68%

Synthesis (1975) 605

Polystyrene

$\xrightarrow[\text{aq NaCN}]{\text{n-C}_8\text{H}_{17}\text{Br}}$ n-C$_8$H$_{17}$CN   92%

JACS (1975) <u>97</u> 5956

Br(CH$_2$)$_4$Br + Et$_4$N$^+$CN $\xrightarrow{\text{CH}_2\text{Cl}_2}$ NC-(CH$_2$)$_4$CN   90%

Synthesis (1975) 605

CH$_3$(CH$_2$)$_5$Br $\xrightarrow[\substack{\text{18-crown-6}\\\text{CH}_3\text{CN}}]{\text{KCN}}$ CH$_3$(CH$_2$)$_5$CN   100%

JOC (1974) <u>39</u> 3416

$$\text{n-BuBr} \xrightarrow[\text{Bu}_3\text{N}]{\text{NaCN}} \text{n-BuCN} \qquad\qquad 100\%$$

Synth Comm (1976) <u>6</u> 193

$$\text{PhI + KCN} \xrightarrow{\text{Pd(0) cat}} \text{PhCN} \qquad\qquad 93\%$$

Chem Lett (1975) 277

Bull Chem Soc Japan (1975) <u>48</u> 3298
(Pd II catalyst)

Section 191   <u>Nitriles from Hydrides</u>

50%

JACS (1976) <u>98</u> 271

Section 192   <u>Nitriles from Ketones</u>

No additional examples

Section 193   <u>Nitriles from Nitriles</u>

Reductive alkylations of of unsaturated nitriles are found in Section 74
(Alkyls from Olefins).

96%

Synthesis (1975) 180

PhCH$_2$CN $\xrightarrow[\text{2) MeI}]{\text{1) MeLi, THF}}$ PhC(Me)$_2$CN   95%

Tetrahedron (1975) $\underline{31}$ 153

CH$_2$=CHCH$_2$CN $\xrightarrow[\text{2) } \diagup\!\!\diagup\text{Br}]{\text{1) i-Pr}_2\text{NLi}}$ CH$_2$=CHCH-CN   98%

Tetr Lett (1975) 4647

$\underset{NC}{\overset{Me}{>}}$C=CH$_2$ + H-C-CN $\xrightarrow[\text{H}_2\text{O, NaOH}]{\text{TEBA}}$ [cyclopropane structure: Me, Me on ring carbons, NC, Cn]   75%

Synthesis (1976) 387

[p-nitro benzonitrile] + Me$_2$$\bar{\text{C}}$NO Li$^+$ $\longrightarrow$ [Me$_2$CNO$_2$ substituted benzonitrile]   82%

CN   JOC (1976) $\underline{41}$ 1560

[(i-Bu)$_3$BMe]Cu + CH$_2$=CHCN $\xrightarrow[\text{2) H}_2\text{O}]{}$ [branched chain with CN]   88%

Tetr Lett (1976) 255

PhCH$_2$CN $\xrightarrow[\text{phase-transfer cat.}]{\text{EtBr, NaOH}}$ PhCHCN
                                                                       |
                                                                       Et

Org Synth (1976) $\underline{55}$ 91

Section 194   Nitriles from Olefins

$$Ph_2C=CHNO_2 \xrightarrow[\text{MeCN}]{Et_4NF} NCCH_2CPh_2CH_2NO_2$$

45%

Synthesis (1975) 162

$$(n\text{-octyl})_3B \xrightarrow[\substack{\text{CH}_3\text{CN} \\ \text{electrochemical} \\ \text{rdxn}}]{Et_4N^+I^-} (n\text{-octyl})CH_2CN$$

~50%

Chem Lett (1975) 523

Section 195   Nitriles from Miscellaneous Compounds

$$HOOC\text{-}CH_2\text{-}CH_2\text{-}\underset{\underset{COOH}{\overset{N\text{-}OH}{\|}}}{C} \xrightarrow{PhNCO} HOOC\text{-}CH_2CH_2CN$$

93%

Synthesis (1976) 418

$$\xrightarrow{NaOBr} Ph\text{-}NC$$

92%

Angew Int Ed (1976) 15 113

~100%

JCS Perkin I (1976) 1901

$$\underset{\text{Et-C-NH}_2}{\overset{\overset{\displaystyle NH}{\|}}{}} \quad \xrightarrow[\text{phase-transfer cat}]{\text{CHCl}_3,\ \text{NaOH}} \quad \text{Et-C}\equiv\text{N} \qquad\qquad 84\%$$

Z Chem (1975) $\underline{15}$ 301

Chapter 14   PREPARATION
OF
OLEFINS

Olefins from Acetylenes

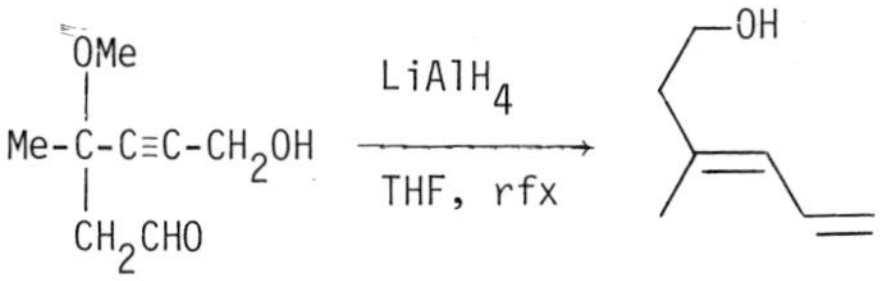

Me-C-C≡C-CH₂OH with OMe and CH₂CHO substituents → LiAlH₄, THF, rfx → product with -OH   93%

Acta Chem Scand B (1975) 29 609

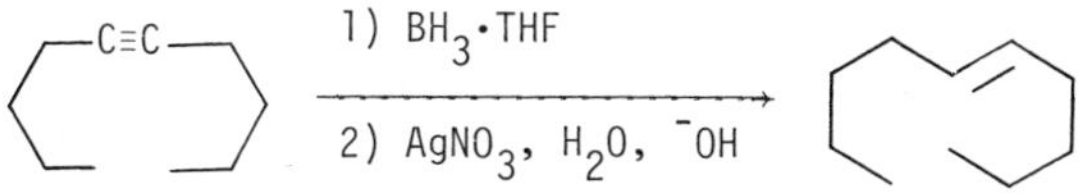

─C≡C─ (cyclic) → 1) BH₃·THF 2) AgNO₃, H₂O, ⁻OH → cyclic olefin   72%

Tetr Lett (1976) 4871

Ph-C≡CH → D₂, [RhOCOPh(cod)PPh₃] → Ph and H on one carbon, D and D: $Ph/D \, C=C \, H/D$

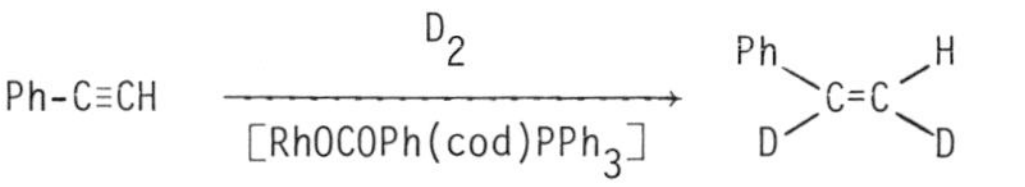

>95% cis isomer

No yields, few details

JCS Chem Comm (1975) 647

305

85%

JCS Chem Comm (1976) 596

$CH_3-C{\equiv}C-CH_2CH_3$ $\xrightarrow[\text{H}_2]{\text{PdCl}_2, \text{ DMF}}$

~100%

Chem Ber (1976) 109 531

$C_4H_9-C{\equiv}CH$ $\xrightarrow{\begin{array}{l}1)\ \underline{i}\text{-Bu}_2\text{AlH}\\2)\ \underline{n}\text{-BuLi}\\3)\ \text{BzBr}\end{array}}$

46%

Tetr Lett (1976) 1927

$C{\equiv}CH$ + Et_3B $\xrightarrow{\begin{array}{l}1)\ \text{BuLi}\\2)\ \text{Bu}_3\text{SnCl}\end{array}}$

71%

Tetr Lett (1976) 805

$C{\equiv}C$ $\xrightarrow{\text{LiAlH}_4/\text{TiCl}_4}$

73%

Tetr Lett (1976) 15

$[Rh(NBD)(PPhMe_2)_3]^+$

Me$_2$C-C≡CMe   →   97%
|
OH   acetone, H$_2$

H, H
\C=C/
Me$_2$C, Me
OH

JACS (1976) <u>98</u> 2143

—C≡CH

1) DIBAL

2) CH$_3$Li, ⟋⟍Br

              68%

JOC (1976) <u>41</u> 2214

$(⬡)_3$B + BuC≡CLi

1) HCl

2) I$_2$/MeOH

$(⬡)_2$C=CHBu   76%

Synthesis (1975) 376

$\overset{-}{\underset{Li^+}{H-BBu_2C≡CBu}}$

1) Me$_2$SO$_4$

2) H$^+$

H, Bu
\C=C/
Bu/ \Me

         74%

(83% E isomer)

Tetr Lett (1975) 1633

C≡CCH$_3$
⬡ (cyclohexyl)

1) ⬡benzodioxaborole BH

2) CH$_3$COOH, 100°

H, H
\C=C/
CH$_3$/ \⬡

         (value)

JACS (1975) <u>97</u> 5249

MeC≡CSMe + PhMgX

CuI
────
THF

Me, SMe
\C=C/
Ph/

         95%

Rec Trav Chim (1974) <u>93</u> 24

Ph-C≡C-CH$_3$ $\xrightarrow[\text{2 BuLi}]{\text{CuI}}$ 100%

JOC (1976) $\underline{41}$ 4089

PhC≡C-H $\xrightarrow[\text{2) H}_2\text{0}]{\text{1) [PrCuBr]MgBr}}$ 95%

Rev Trav Chim (1976) $\underline{95}$ 299, 304

Ph-C≡CH $\longrightarrow$ $\xrightarrow[\text{HMPA}]{\text{PdCl}_2, \text{ LiCl}}$ 99%

JOC (1976) $\underline{41}$ 2241

Section 197   Olefins from Carboxylic Acids and Acid Halides

Ph$_2$CCH$_2$COOH $\xrightarrow[\text{CHCl}_3]{\text{DMF acetal}}$ Ph$_2$C=CH$_2$ 88%
|
OH

Tetr Lett (1975) 1545

+ $\xrightarrow[\text{CH}_2\text{Cl}_2]{\text{AlCl}_3}$

Synthesis (1976) 737

1) LDA

2)

3) SOCl$_2$/Py
4) 140°

63%

JOC (1974) 39 1650

Et$_3$BC≡CPh + EtCOOH ⟶ EtCH=CHPh   75%

Tetr Lett (1974) 2961

$$\xrightarrow[\text{Cu(OAc)}_2\text{, benzene}]{\text{Pb(OAc)}_4}$$

80%

Tetr Lett (1976) 2079

Section 198   Olefins from Alcohols

CH$_2$=C-CH(OH)CH$_2$OH
    |
 (CH$_2$)$_2$CH=CMe$_2$

$$\xrightarrow[\text{2) Zn}]{\text{1) PBr}_3\text{, CuBr}}$$

CH$_2$=C-CH=CH$_2$
    |
 (CH$_2$)$_2$CH=CMe$_2$

58%

JACS (1975) 97 3252

$(PhCH_2)_2\overset{\displaystyle|}{\underset{\displaystyle OH}{C}}-CH_2SPh$ $\xrightarrow[\text{R}_3\text{N, PhH, diox}]{(\text{TiCl}_4 + \text{LiAlH}_4)}$ $(PhCH_2)_2C=CH$ 96%

Chem Lett (1975) 871

$PhSCMe_2CHOHR$ $\xrightarrow[\text{dark}]{\text{p-TsOH, PhH}}$ $RCH(SPh)CMe=CH_2$ high yield

JCS Chem Comm (1975) 8211

$CH_3\underset{\displaystyle OH}{\overset{\displaystyle|}{C}}H-\underset{\displaystyle NO_2}{\overset{\displaystyle|}{C}}HCH_3$ $\xrightarrow[\text{Et}_3\text{N}]{\text{MeSO}_2\text{Cl}}$ $CH_3CH=\underset{\displaystyle NO_2}{\overset{\displaystyle|}{C}}CH_3$ 67%

JOC (1975) 40 2138

$\xrightarrow{\text{active Ti}^\circ}$ 85%

JOC (1976) 41 896

$\xrightarrow[\text{2) Na, xylene}]{\text{1) Me}_2\text{NPCl}_2}$ 85%

90% cis

Synth Comm (1975) 5 293

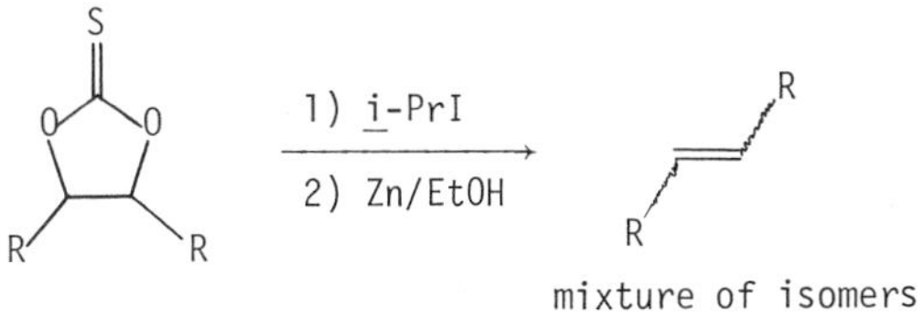

S
O
O
R
R
1) i-PrI
2) Zn/EtOH
R
R
mixture of isomers

40-80%

JOC (1974) 39 3641

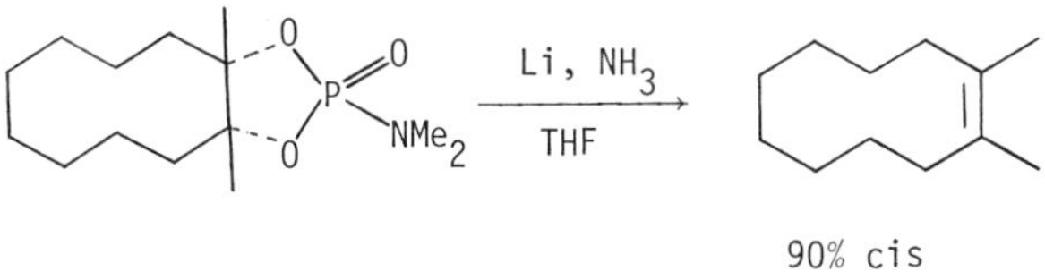

O
O
P
O
NMe2
Li, NH3
THF
90% cis

98%

Synth Comm (1975) 5 293

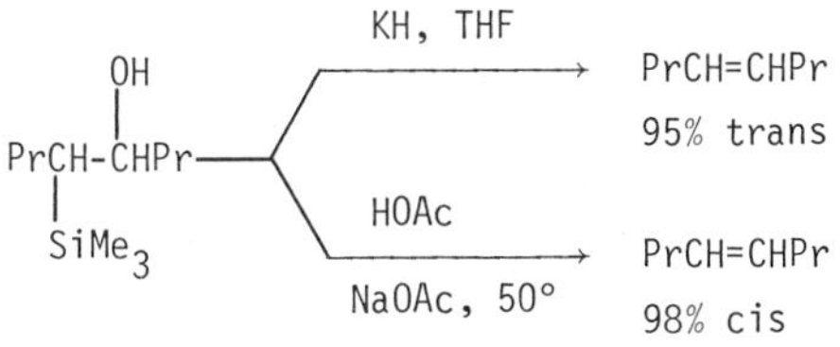

OH
PrCH-CHPr
SiMe3
KH, THF
PrCH=CHPr
95% trans
HOAc
NaOAc, 50°
PrCH=CHPr
98% cis

96%

85%

JACS (1975) 97 1464

HO
CH2SePh
MeSO2Cl
Et3N, CH2Cl
CH2

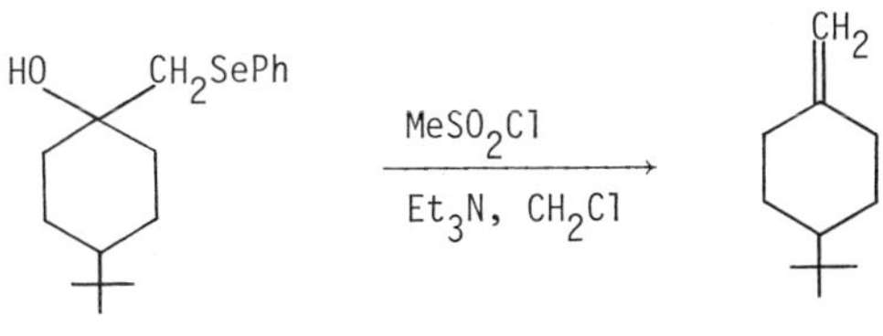

98%

JCS Chem Comm (1975) 790

96%

99%

Tetr Lett (1974) 1133

Section 199   Olefins from Aldehydes

$PhCHO + Ph_3\overset{+}{P}CH_2Ph \xrightarrow[\substack{CH_2Cl_2, \\ 18\text{-crown-}6}]{t\text{-BuOK}} PhCH=CHPh$ 96%

Synthesis (1975) 784

$(EtO)_2P(O)CH_2Ph \xrightarrow[\substack{2)\ CCl_4 \\ 3)\ \underline{i}\text{-PrCHO}}]{1)\ BuLi} i\text{-PrCH=}\underset{\underset{Cl}{|}}{C}\text{-Ph}$ 69%

Synthesis (1975) 658

$Ph_3\overset{+}{P}CH_3 + PhCHO \xrightarrow[NaOH/H_2O]{PhH} PhCH=CH_2$ 95%

Tetr Lett (1974) 2587

$$\text{>}\!\!-(CH_2)_4CH=PPh_3 + \underline{n}\text{-}C_{10}H_{21}CHO \xrightarrow{\text{HMPA}} \text{>}\!\!-(CH_2)_4CH=CH(CH_2)_9CH_3$$

94% (E)

Tetr Lett (1974) 207

$$PhSO_2CH_3 + PhCHO \xrightarrow[\text{CH}_2\text{Cl}_2, \text{ TEBA}]{\text{50\% aq NaOH}} PhSO_2CH=CHPh \qquad 86\%$$

Synthesis (1975) 453

$$O_2NCH_2CH_2COOEt \xrightarrow[\text{n-BuNH}_2, \text{ EtOH}]{\text{PhCHO}} \underset{H}{\overset{Ph}{>}}C=C\underset{NO_2}{\overset{CH_2COOEt}{<}} \qquad 45\%$$

J Prakt Chem (1975) 317 337

$$PhCH=CH_2 + PhCONHCHCOOMe \xrightarrow[\text{2) NaOH}]{\text{1) H}_2\text{SO}_4, \text{ PhH}} PhCH=CHCH(NHCOPh)COOH \qquad 42\%$$

OMe

Tetr Lett (1975) 3737

$$MeCOCH(OMe)SO_2Ph \xrightarrow[\text{2) CH}_2\text{O}]{\text{1) NaH}} CH_2=C\underset{SO_2Ph}{\overset{OMe}{<}} \qquad 75\%$$

Liebigs Ann Chem (1975) 1484

$$PhNHSO_2CH_2COOH + PhCHO \xrightarrow[\text{PhMe, rfx}]{\text{Py, NH}_4\text{OAc}} PhNHSO_2CH=CH\text{-}Ph \qquad 67\%$$

Synthesis (1975) 321

Hex-CHO $\xrightarrow{\begin{array}{c}1)\ Me_3Si\bar{C}HHex^+Li\\ \hline 2)\ H_3O^+\end{array}}$

Hex—CH=CH—Hex   45%

Angew Int Ed (1976) <u>15</u> 161

$\xrightarrow[\text{"proton sponge"}]{TiCl_4-LiAlH_4}$ 94%

retinal $\xrightarrow{\hspace{3cm}}$ β-carotene   90%

Chem Lett (1976) 1127

retinal + LiAlH$_4$ + TiCl$_3$ $\longrightarrow$ carotene   85%

(RCHO $\longrightarrow$ RCH=CHR)

JACS (1974) <u>96</u> 4708

PhCHO $\xrightarrow{H_2CCa_2I_2}$ $\underset{H}{\overset{Ph}{>}}C=CH_2$ 11%

Bull Chem Soc Japan (1976) <u>49</u> 1177

Ph—CH=CH—CHO $\xrightarrow[\text{2)}\ \underset{O}{\overset{O}{\bigcirc}}BH \atop NaOAc]{1)\ TsNHNH_2}$ Ph—CH$_2$—CH=CH$_2$ 53%

JOC (1976) <u>41</u> 574

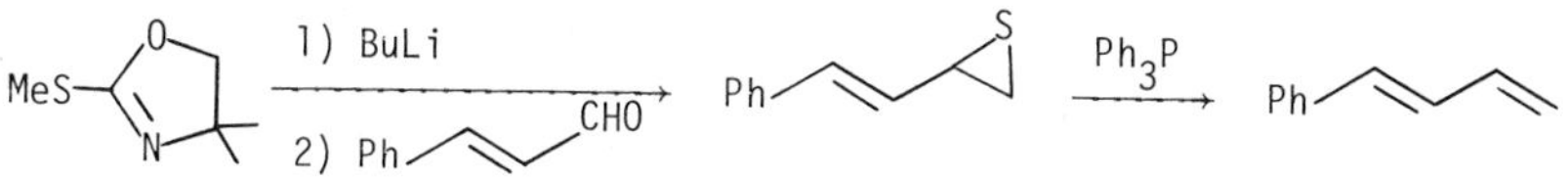

Tetr Lett (1975) 2861

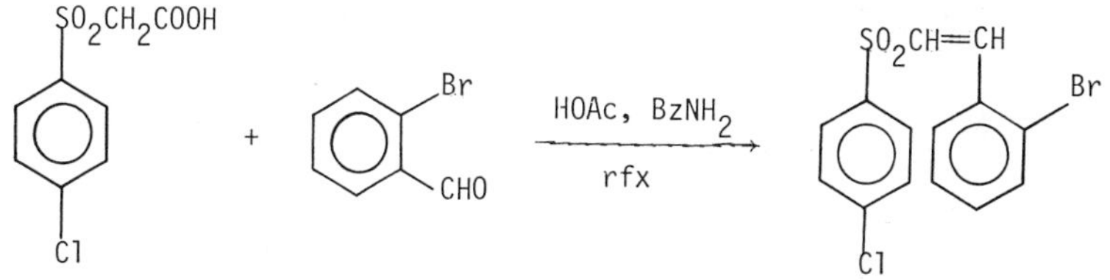

Bull Chem Soc Japan (1975) <u>48</u> 1091

Section 200   Olefins from Aryls

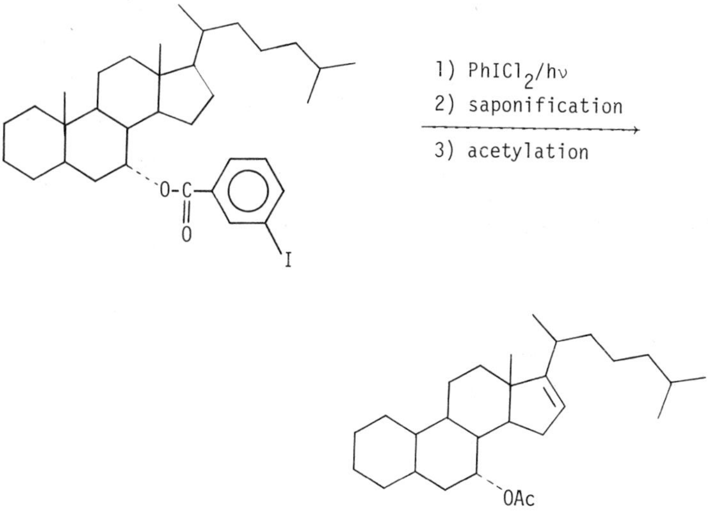

1) PhICl$_2$/hv
2) saponification
3) acetylation

54%

JACS (1975) <u>97</u> 6580

Related methods: Alkyls and Aryls from Alkyls and Aryls (Section 65)

Alkyls and Aryls from Olefins (Section 74)

Section 201   Olefins from Amides

No additional examples

Section 202   Olefins from Amines

$CH_2CH_2OCHPh_2$
|
NMe_2

1) MeI
2) anion-exchange resin
————————————————→
3) Δ

$H_2C=CHOCHPh_2$ ~80%

Org Synth (1976) 55 3

Section 203   Olefins from Esters

PhCOOEt $\xrightarrow[\text{DMSO}]{\text{xs } Ph_3P=CH_2}$ $PhCMe=CH_2$ 56%

Tetr Lett (1975) 1439

Collidine
————————→
Δ

58%

JCS Perkin I (1976) 884

Review: Olefins via Phosphonates   Chem Rev (1974) <u>74</u> 87

Section 204   <u>Olefins from Epoxides</u>

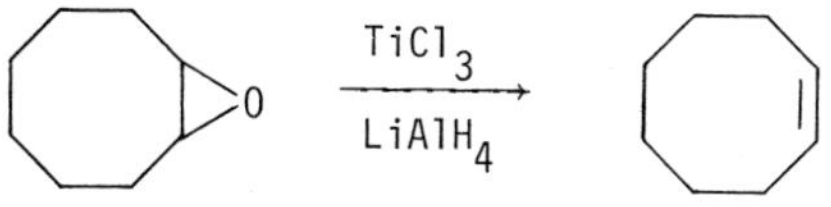

$$\xrightarrow[\text{LiAlH}_4]{\text{TiCl}_3}$$

53%

JOC (1975) <u>40</u> 2555

$$\xrightarrow{\text{LiCuPr}_2}$$

acid

base

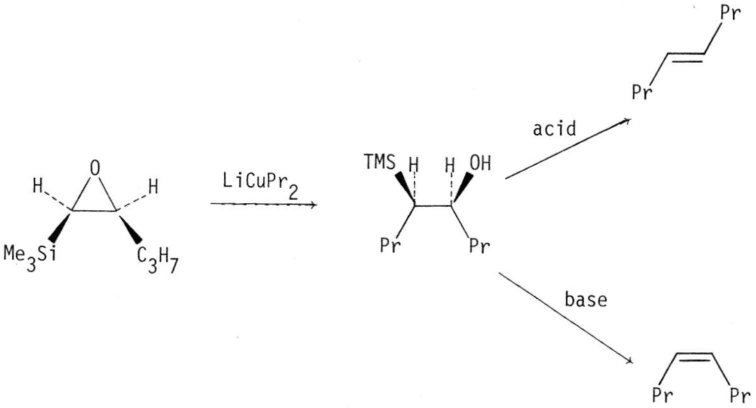

~90%

JOC (1975) <u>40</u> 2263

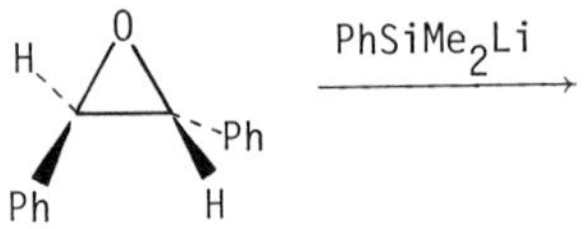

$$\xrightarrow{\text{PhSiMe}_2\text{Li}}$$

75%

Synthesis (1976) 199

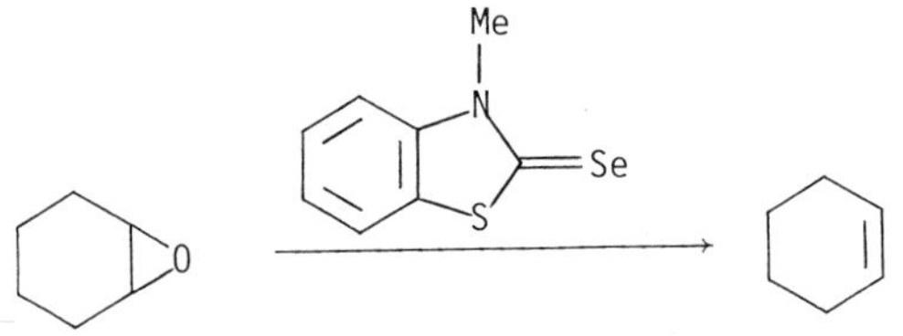

100%

Synthesis (1976) 200

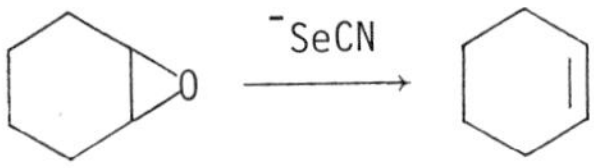

$$\xrightarrow{^-\text{SeCN}}$$

100%

JCS Perkin I (1975) 1216

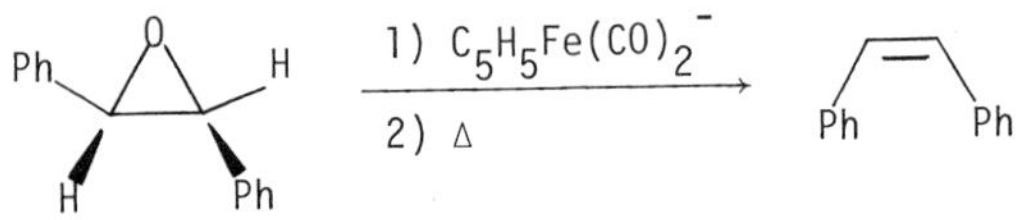

1) $C_5H_5Fe(CO)_2^-$

2) Δ

Tetr Lett (1975) 4009

$$\xrightarrow[\text{KOMe}]{\text{Me}_3\text{SiSiMe}_3}$$

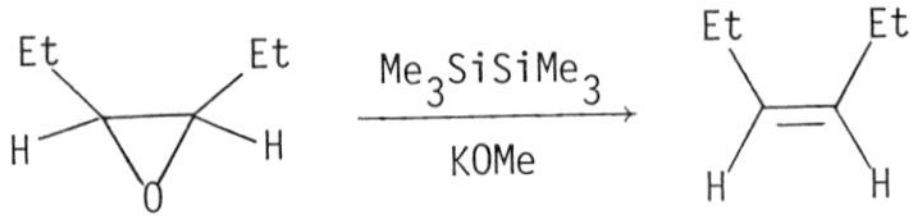

86%

JACS (1976) 98 1265

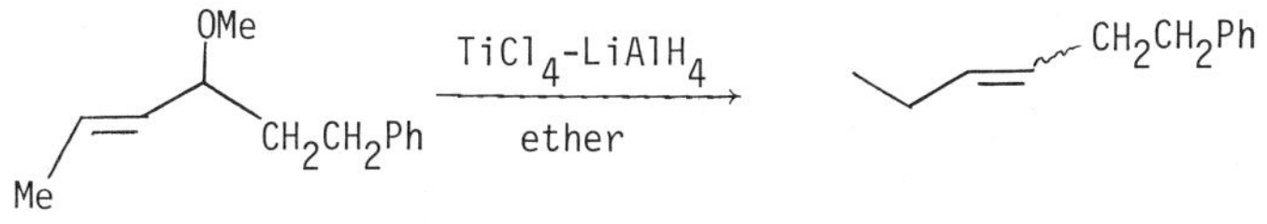

95%

Chem Lett (1976) 737

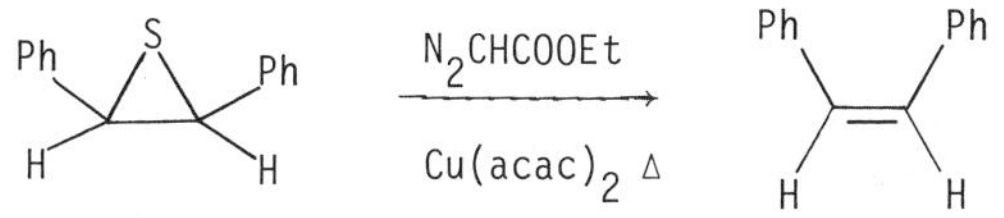

94%

JACS (1975) **97** 2553

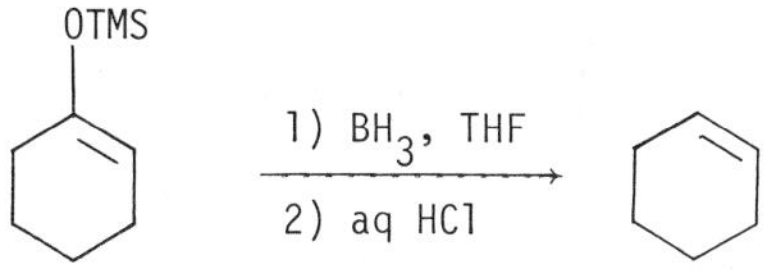

76%

Tetr Lett (1975) 4005

Section 205   Olefins from Halides and Sulfonates

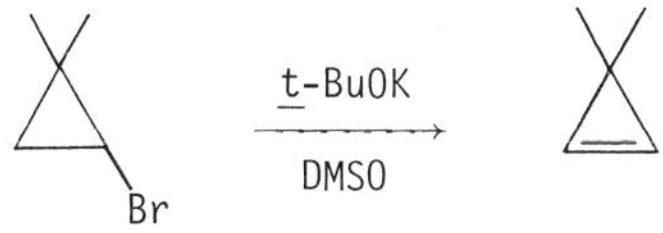

84%

Synthesis (1974) 190

RCH_2CH_2Br $\xrightarrow[\text{2) } H_2O_2, \text{ THF}]{\text{1) } o\text{-}NO_2C_6H_4Se^-}$ $RCH=CH_2$

JOC (1975) **40** 947

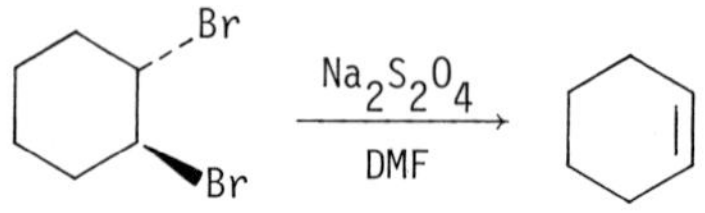
Br
Br
$Na_2S_2O_4$
DMF

60%

Acta Chem Scand B (1976) 30 366

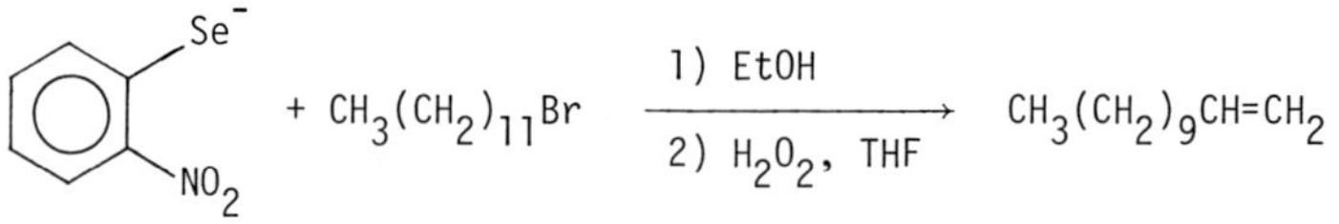
Se⁻
NO_2
+ $CH_3(CH_2)_{11}Br$
1) EtOH
2) H_2O_2, THF
$CH_3(CH_2)_9CH=CH_2$

~70%

JOC (1975) 40 947

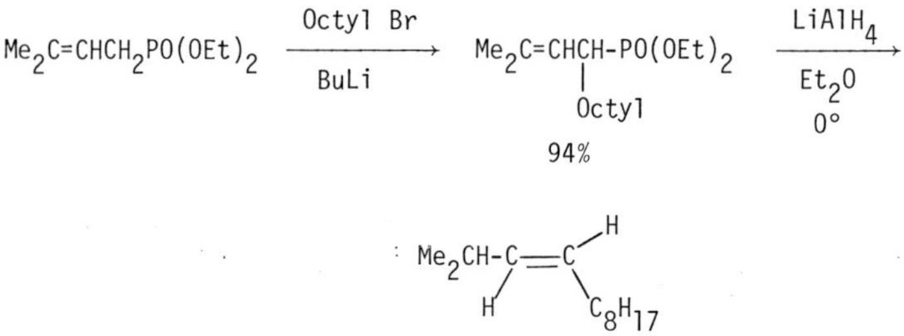
$Me_2C=CHCH_2PO(OEt)_2$
Octyl Br
BuLi
$Me_2C=CHCH-PO(OEt)_2$
Octyl
94%
$LiAlH_4$
Et_2O
0°
$Me_2CH-C=C$
H
H
C_8H_{17}

84%

Angew Int Ed (1974) 13 407

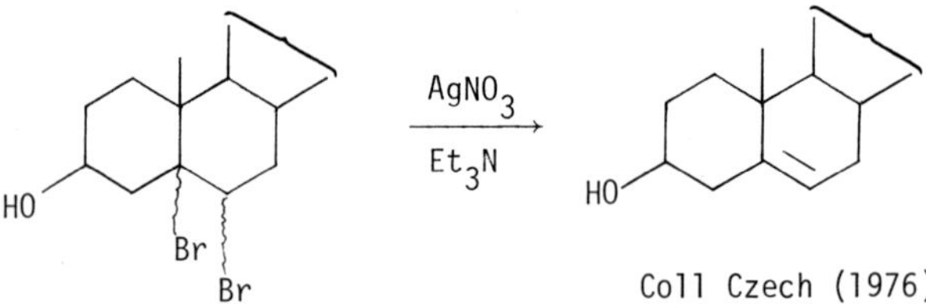
HO
Br
Br
$AgNO_3$
Et_3N
HO

95%

Coll Czech (1976) 41 2040

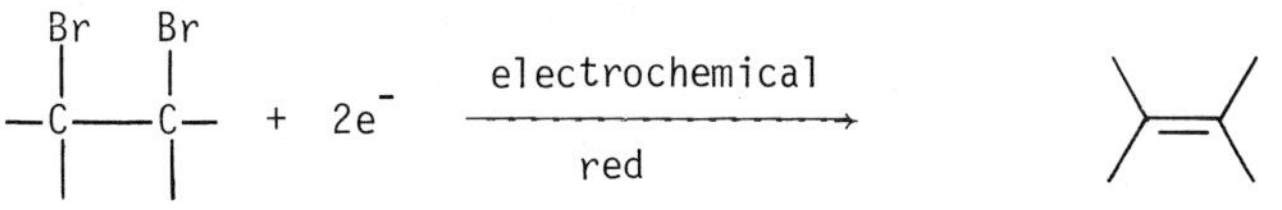

trans periplanar elimination
high yield

JOC (1974) 39 2408

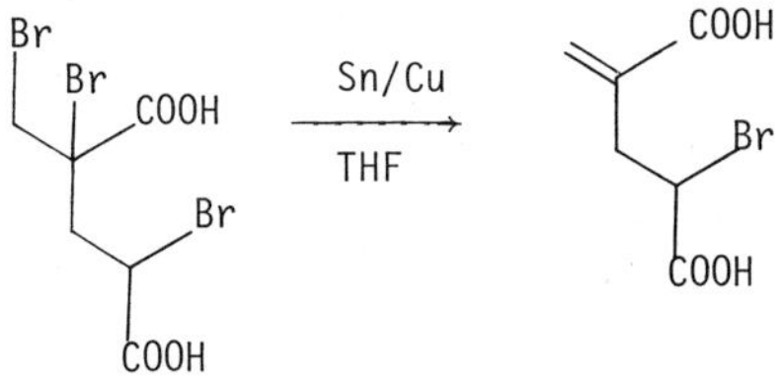

81%

JOC (1976) 41 4035

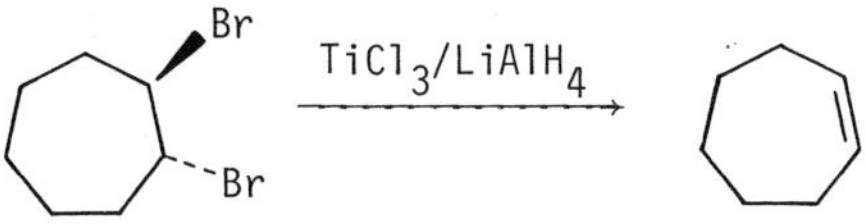

81%

Synthesis (1976) 607

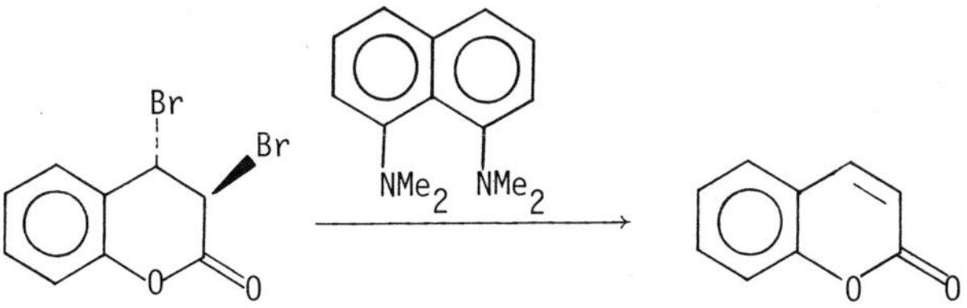

92%

Synth Comm (1975) 5 87

n-C$_8$H$_{17}$CHBrCH$_2$(OH) $\xrightarrow{\text{TiCl}_4\text{-LiAlH}_4}$ n-C$_8$H$_{17}$CH=CH$_2$ 74%

JOC (1975) 40 3797

$$\text{Synthesis (1975) 397}$$

$$\text{Synth Comm (1975) } \underline{5} \text{ 87}$$

$$Et_2CH\text{-}CHCCl_3 \xrightarrow[\substack{Et_2O/THF}]{\text{BuLi, -50°}} Et_2CH\text{-}CH\text{=}CCl_2 \qquad 75\%$$
$$\underset{\substack{| \\ OMe}}{}$$

$$\text{J. Organometal Chem (1975) } \underline{97} \text{ 355}$$

$$PhCHBrCHBrPh \; + \; R_4P^+Br^- \xrightarrow[\substack{NaI \\ H_2O/PhMe \quad 90\%}]{Na_2S_2O_3} PhCH\text{=}CHPh \qquad 86\%$$

$$\text{Synthesis (1975) 397}$$

$$\text{Tetr Lett (1976) 4439}$$

$$PhS(\rightarrow O)CH_2Ph + LiICA \xrightarrow[rfx]{RCH_2X} RCH{=}CHPh$$

JOC (1975) 40 2014

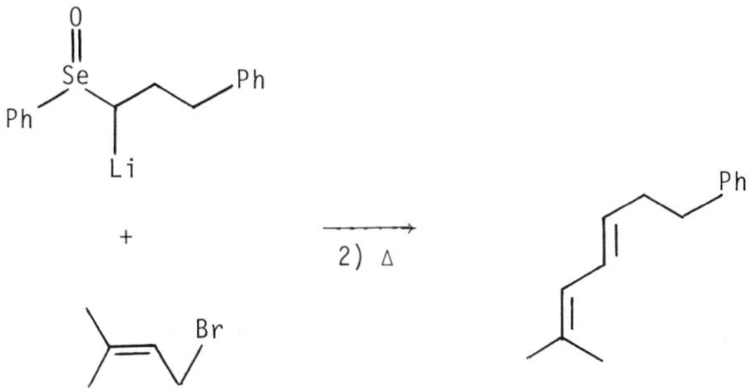

O
Se
Ph
Ph
Li
+
Br
2) Δ
Ph

64%

JACS (1975) 97 3250

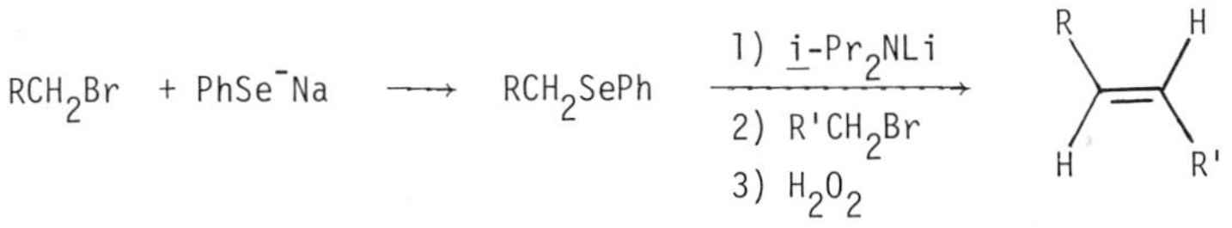

RCH2Br + PhSe⁻Na
RCH2SePh
1) i-Pr2NLi
2) R'CH2Br
3) H2O2
R
H
H
R'

60-80%

JCS Chem Comm (1974) 990

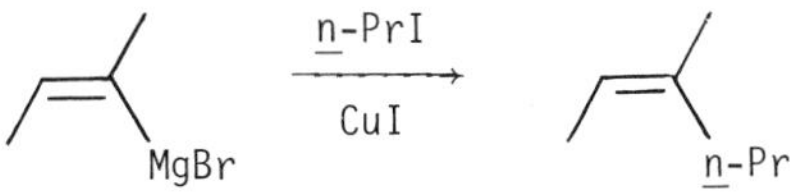

MgBr
n-PrI
CuI
n-Pr

97%

Tetr Lett (1976) 3225

$$\underset{\underset{Br}{\overset{+}{|}}}{PhCH_2AsPPh_3} \xrightarrow[\text{2) PhCHO}]{\text{1) NaH, THF}} PhCH=CH-\bigcirc-Br \qquad 50\%$$

J Organometal Chem (1975) <u>96</u> 237

$$Br-\bigcirc-CH_2Br \xrightarrow[\text{2) } CH_2O, H_2O, \ ^-OH]{\text{1) } Ph_2P} Br-\bigcirc\diagup \qquad 98\%$$

Synth Comm (1976) <u>6</u> 53

$$CH_3(CH_2)_7I \ + \ \underset{Me}{\diagup}\!\!\diagdown^{Li} \xrightarrow[\text{THF}]{25°} \underset{Me}{\diagup}\!\!\diagdown^{(CH_2)_7CH_3} \qquad 100\%$$

Tetr Lett (1974) 3809

Section 206   <u>Olefins from Hydrides</u>

No additional examples

Section 207   <u>Olefins from Ketones</u>

$$\xrightarrow[\text{Zn}]{\text{ClSiMe}_3} \qquad 37\text{-}72\%$$

JCS Perkin I (1975) 809

O
OSiMe3
1) B2H6, THF
2) H3O+

76%

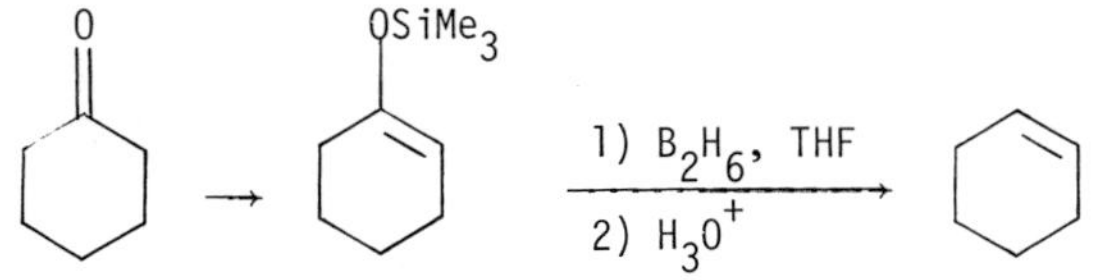

Tetr Lett (1975) 4005

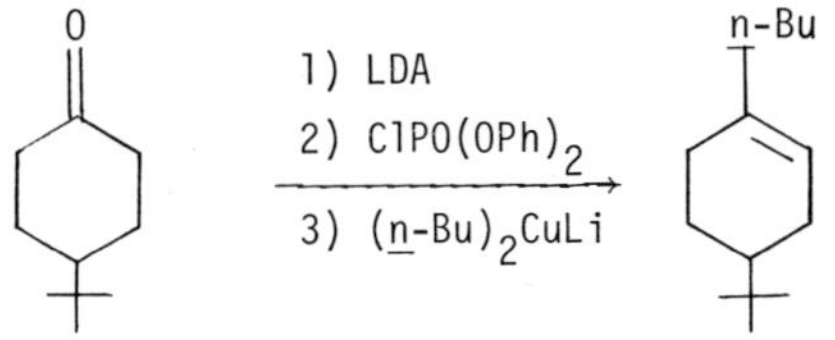

O
1) LDA
2) ClPO(OPh)2
3) (n-Bu)2CuLi
n-Bu

59%

Tetr Lett (1976) 4405

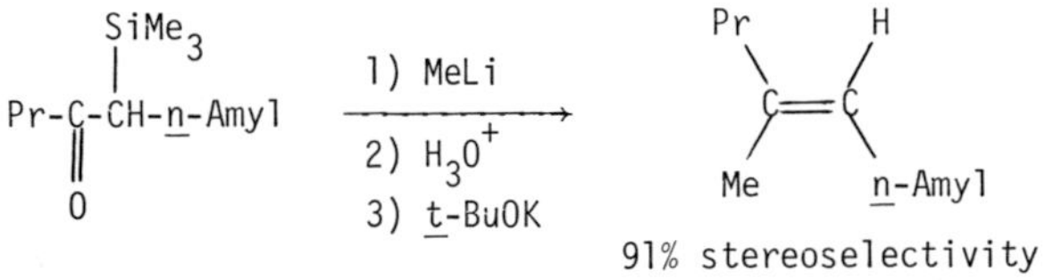

SiMe3
Pr-C-CH-n-Amyl
O
1) MeLi
2) H3O+
3) t-BuOK
Pr
H
Me
n-Amyl
91% stereoselectivity

59%

JOC (1976) 41 2940

PPh2O
1) n-BuLi
2)
O

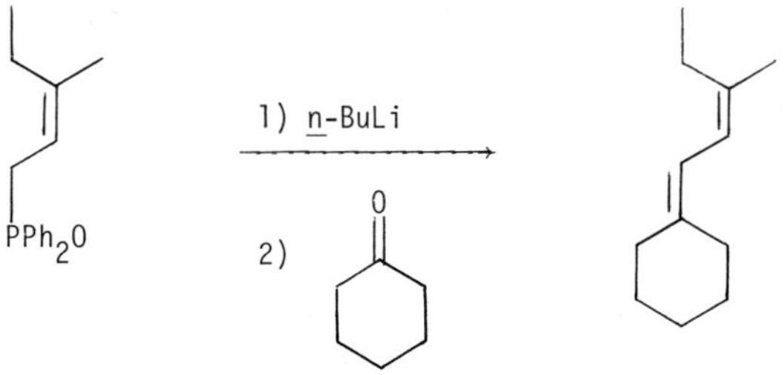

67%

JCS Perkin I (1976) 2386

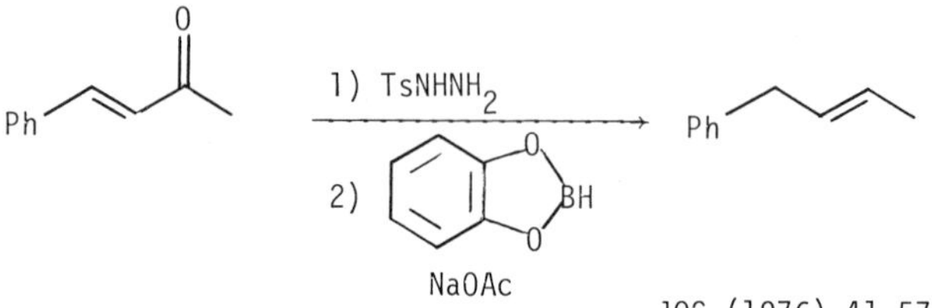

JOC (1976) <u>41</u> 574

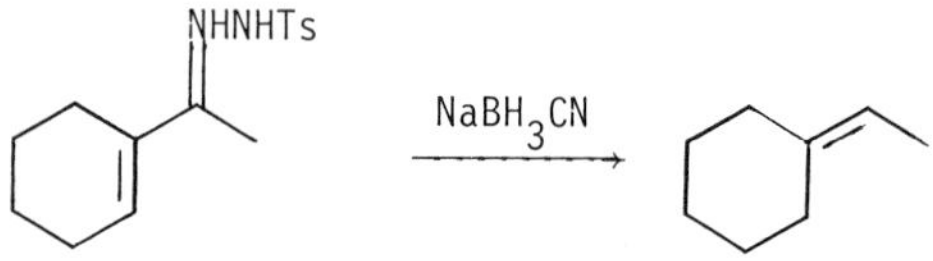

72%

JOC (1975) <u>40</u> 923

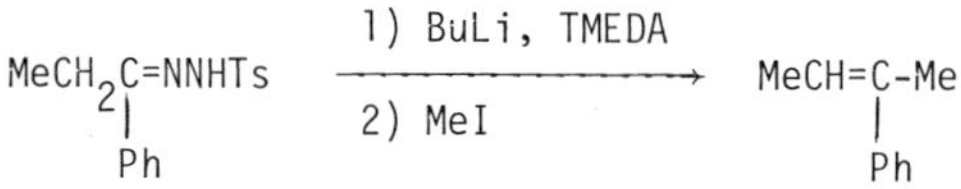

79%

$$\underset{\underset{Ph}{|}}{MeCH_2C=NNHTs} \quad \xrightarrow[\text{2) MeI}]{\text{1) BuLi, TMEDA}} \quad \underset{\underset{Ph}{|}}{MeCH=C-Me}$$

Tetr Lett (1975) 1811

95%

Org React (1976) <u>23</u> 405

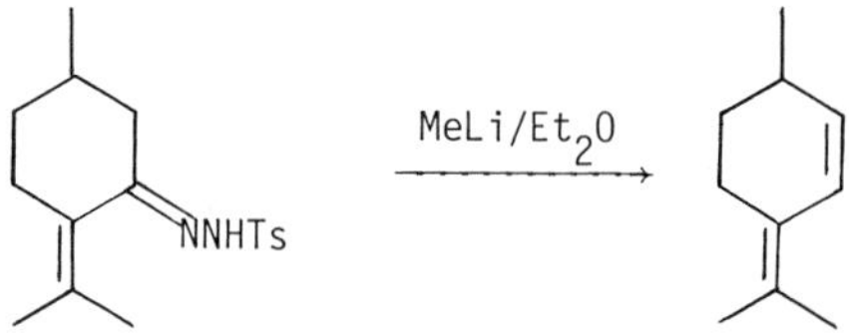

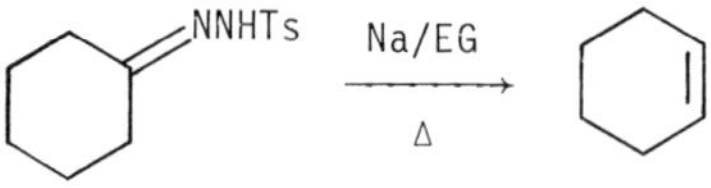

100%

Org React (1976) <u>23</u> 405

100%

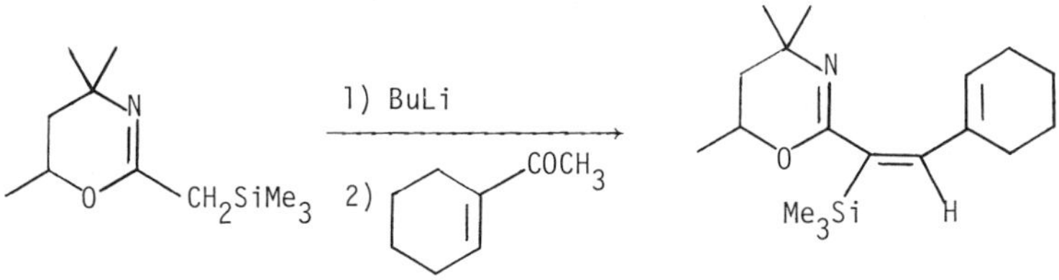

Tetr Lett (1976) 4041

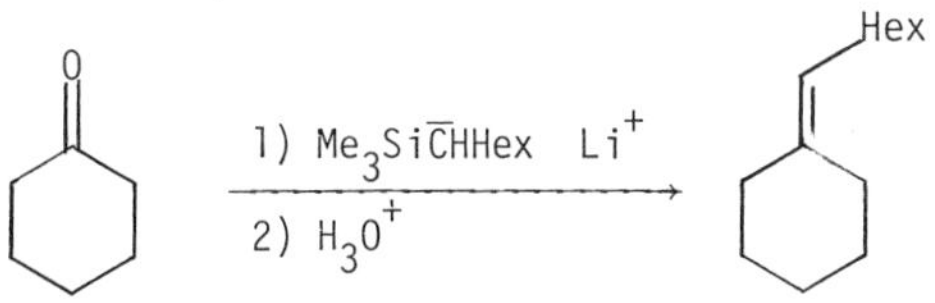

Angew Int Ed (1976) 15 161

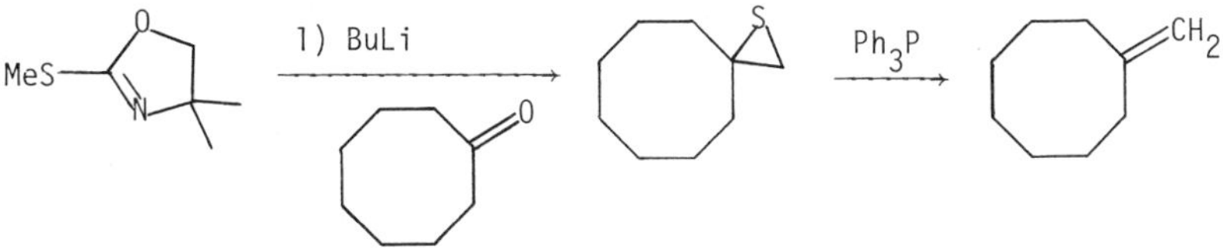

Tetr Lett (1975) 2861

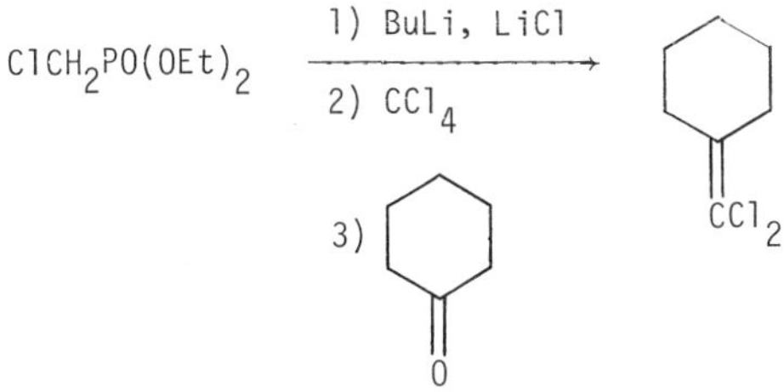

Synthesis (1975) 535

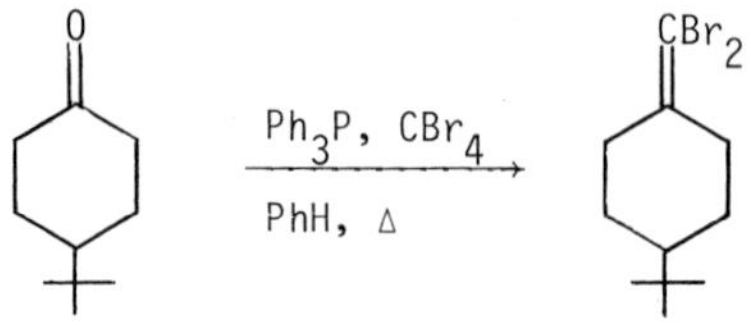

81%

Tetr Lett (1975) 1373

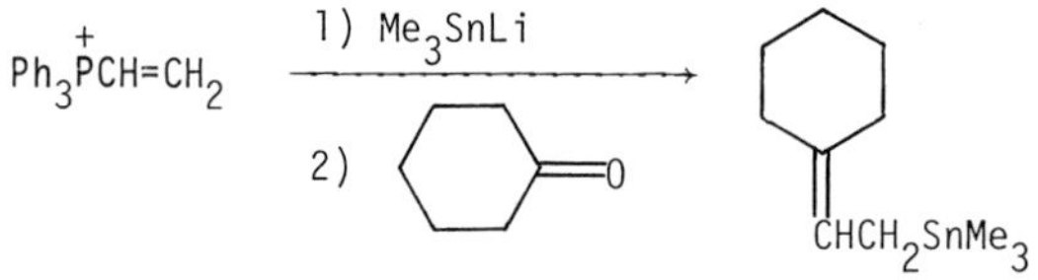

>90%

JCS Chem Comm (1975) 630

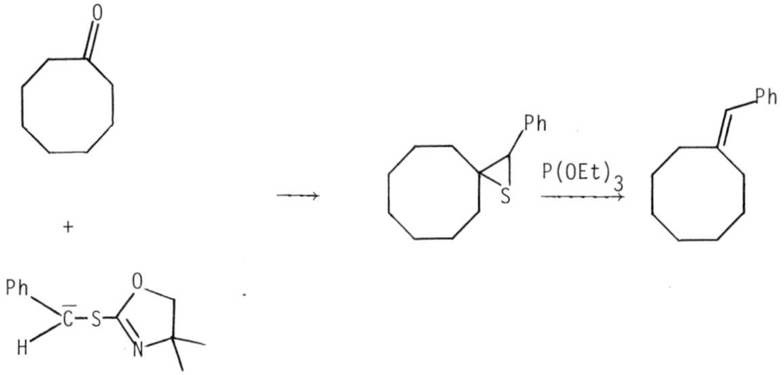

72%

JOC (1976) 41 1735

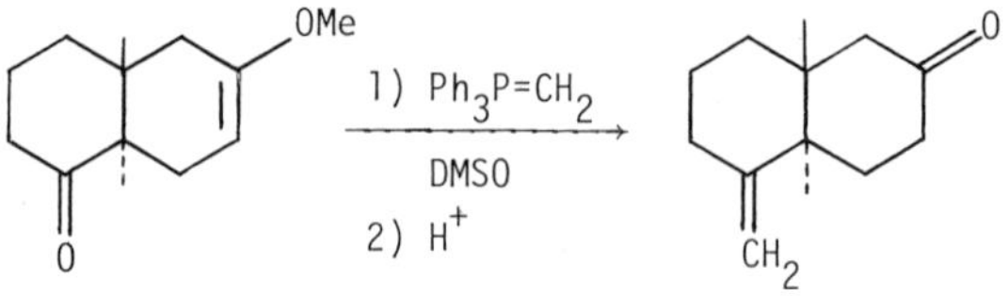

Tetrahedron (1974) 30 2961

$(EtO)_2P(O)CH_2SOMe$ $\xrightarrow{\begin{array}{c} \text{1) BuLi, THF, -78°} \\ \hline \text{2) Ph}_2\text{CO} \end{array}}$ $MeSOCH=CPh_2$ 84%

JOC (1975) 40 1979

$MeCOCH_2SiMe_3$ $\xrightarrow{\begin{array}{c} \text{1) LiCH}_2\text{CN} \\ \hline \text{2) H}^+ \end{array}}$ $\underset{\overset{|}{CH_2CN}}{MeC=CH_2}$ 75%

Synth Comm (1975) 5 15

Review: Bis-Wittig Reactions in the Synthesis of Nonbenzenoid Aromatic
         Ring Systems

Synthesis (1975) 765

$\xrightarrow[\text{"proton sponge"}]{\text{TiCl}_4\text{-LiAlH}_4}$

94%

Chem Lett (1976) 1127

Hex-CHO $\xrightarrow{\begin{array}{l} \text{1) MeSeH} \\ \text{2) BuLi, Hex-CHO} \\ \text{3) TsOH, pentane/rfx} \end{array}}$

∿65%

Tetr Lett (1976) 1385

$\xrightarrow[\Delta]{\text{TiCl}_3/\text{LiAlH}_4}$

∿50%

Tetr Lett (1976) 3265

2 (cyclohexanone) $\xrightarrow{\text{active Ti°}}$ (dicyclohexylidene)   85%

JOC (1976) <u>41</u> 896

$Ph_2CO \xrightarrow[\text{TiCl}_3]{\text{LiAlH}_4} Ph_2C=CPh_2$ 95%

JACS (1974) <u>96</u> 4708

Review: Organic Chemistry of Low Valent Titanium

Accts Chem Res (1974) <u>7</u> 281

Section 208   Olefins from Nitriles

No additional examples

Section 209   Olefins from Olefins

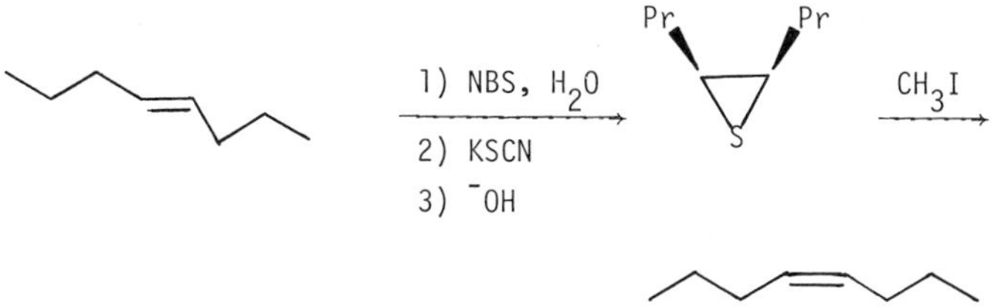

1) NBS, H$_2$O
2) KSCN
3) $^-$OH

CH$_3$I

~50%

Tetr Lett (1975) 2709

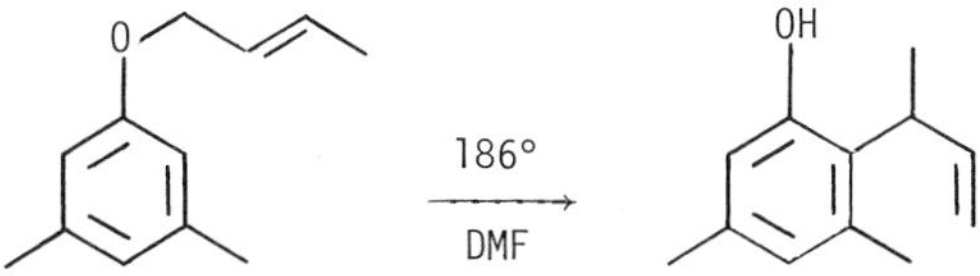

186°

DMF

91%

Org React (1975) 22 1

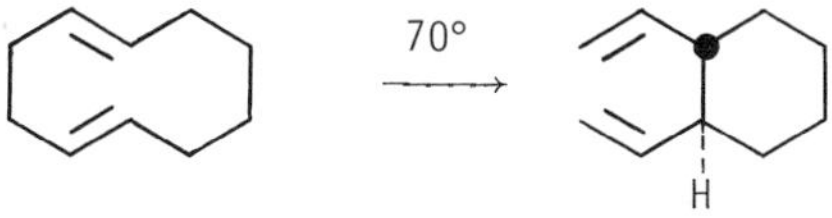

70°

100%

Org React (1975) 22 1

1) Ph$_2$PLi
2) AcOH-H$_2$O$_2$
3) NaH, DMF

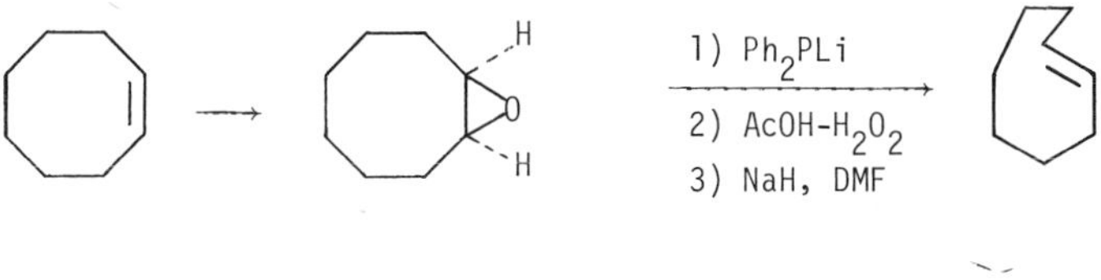

76%

JCS Chem Comm (1974) 142

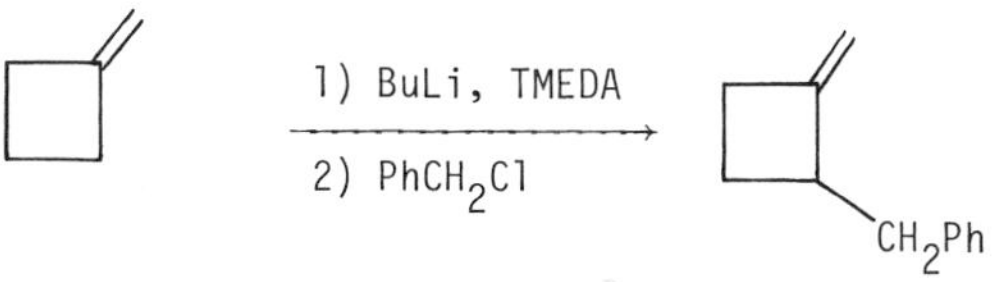

1) BuLi, TMEDA
2) PhCH$_2$Cl

CH$_2$Ph

71%

Tetr Lett (1975) 3047

$$\text{TMS-C=CH}_2 \quad \xrightarrow{\text{t-BuLi}} \quad \text{TMS-CH=CHBu} \qquad\qquad 85\%$$
$$\underset{|}{\overset{}{\text{Cl}}}$$

J Organometal Chem (1974) <u>81</u> C9

$$\text{CH}_2\text{=CH}_2 \ + \ \text{CH}_2\text{=C(Me)CH}_2\text{MgX} \quad \xrightarrow[\text{Et}_2\text{O}]{} \quad \text{MeCH}_2\text{CH}_2\text{C(Me)=CH}_2 \qquad 75\%$$

Liebigs Ann Chem (1975) 103; 119; 1176

$$\text{CH}_2\text{=CH}_2 \quad \xrightarrow[\text{2) HCl}]{\text{1) } \underline{t}\text{-Bu}_2\text{Zn, PhH}} \quad \text{CH}_2\text{=C}\underset{\text{Bu-}\underline{t}}{\overset{\text{Me}}{<}} \qquad 70\%$$

Liebigs Ann Chem (1975) 1162

Section 210   Olefins from Miscellaneous Compounds

22%

JOC (1976) <u>41</u> 3947

Hex Pr
 | | SOCl$_2$ Hex   Pr
H—C———C—Pr   —————————→   \   /
 | | Et$_3$N   C=C   65%
MeSe OH   /   \
                                   H   Pr

Tetr Lett (1976) 3743

CH(CH$_3$)SPh

          1) CF$_3$SO$_3$CH$_2$CO$_2$Et
        —————————————————————→   57%
          2) 50°, DMF

Tetr Lett (1976) 3487

              KOAc, 95% EtOH
PhCH$_2$NO$_2$ ————————————————→   PhCH=CHPh   81%
                    Δ

JOC (1975) 40 187

Review: Methods for the Preparation of Bridgehead Olefins

Angew Int Ed (1975) 14 528

Section 210A   Protection of Olefins

The protection of isolated double bonds is considered in this section.

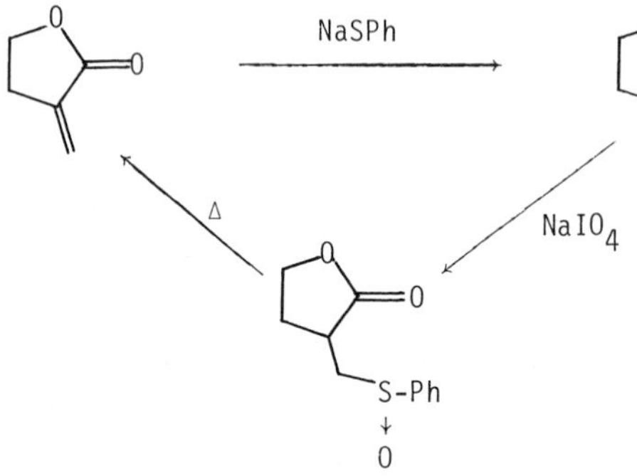

Individual yields >95%

JOC (1975) 40 1181

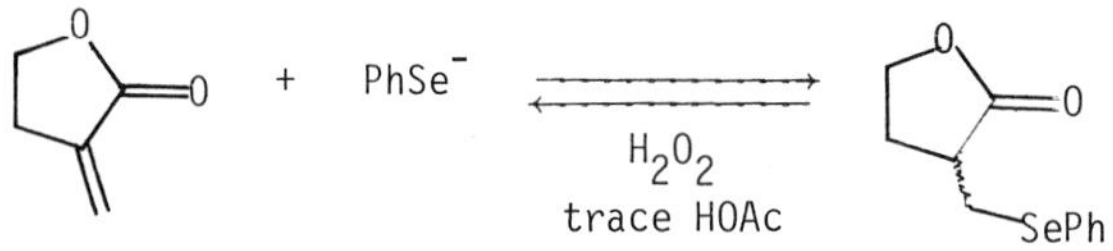

Tetr Lett (1974) 1869

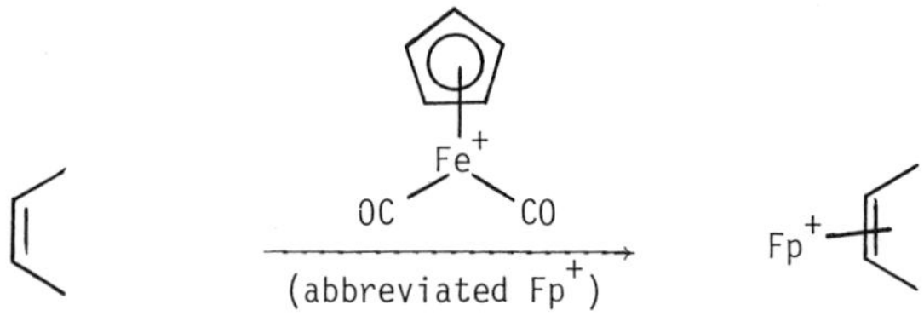

Stable to halogens, Hg(OAc)$_2$, cat. H$_2$, etc.

JACS (1975) 97 3254

∿60%

∿80%

Tetrahedron (1976) 32 765

Protects the diene system so the side chain double bond can be hydroxylated by OsO_4.   Removed with $FeCl_3$.

J Organometal Chem (1975) 102 507

Sections 211 to 299 are reserved for future additions (e.g., the preparation of nitro compounds).

Chapter 15   PREPARATION

OF

DIFUNCTIONAL COMPOUNDS

Section 300   <u>Acetylene - Acetylene</u>

Thexyl$_2$ BX $\xrightarrow{\begin{array}{c}\text{1) BuC}\equiv\text{CLi, THF}\\\text{2) I}_2\text{, -78°}\end{array}}$ BuC≡C-C≡CBu   90%

JCS Chem Comm (1975) 857

3   (CH$_2$CH$_2$ with two C≡CH groups) $\xrightarrow[\text{pyr, 55°}]{\text{Cu(OAc)}_2}$ ~30%

Org Synth (1974) <u>54</u> 1

(cyclohexyl)$_2$BSMe $\xrightarrow{\begin{array}{c}\text{1) Hex-C}\equiv\text{C-Li}\\\text{2) Bu-C}\equiv\text{C-Li}\end{array}}$ Hex-C≡C-C≡C-Bu   61%

Tetr Lett (1976) 4385

Section 301   <u>Acetylene - Carboxylic Acid</u>

No additional examples

Section 302   <u>Acetylene - Alcohol</u>

$HC \equiv CH$ $\xrightarrow{\begin{array}{l}\text{1) BuLi, THF, } -78° \\ \text{2) Me}_2\text{CO}\end{array}}$ $Me_2C-C \equiv CH$ 94%

                                                 OH

JOC (1975) <u>40</u> 2250

$MeC \equiv CH$ $\xrightarrow{\begin{array}{l}\text{1) BuLi, TMEDA} \\ \text{2) BuBr, hexane} \\ \text{3) CH}_2\text{O}\end{array}}$ $BuCH_2C \equiv CCH_2OH$

JCS Chem Comm (1975) 817

JCS Chem Comm (1975) 287

Section 303   Acetylene - Aldehyde

60%
(last step)

Org Synth (1976) <u>55</u> 52

Section 304   Acetylene - Amide

$$Ph-C{\equiv}C-MgBr \xrightarrow[\text{2) } H_2O]{\text{1) } Me_2SiNCO} Ph-C{\equiv}C-\overset{\displaystyle O}{\overset{\|}{C}}-NH_2$$

48%

Tetr Lett (1975) 981

Section 305   Acetylene - Amine

No additional examples

Section 306   Acetylene - Ester

$$HC{\equiv}CCH_2OH \longrightarrow HC{\equiv}CCH_2OTHP \xrightarrow[\substack{\text{2) } Ac_2O \\ \text{3) } H^+}]{\text{1) } NaNH_2} AcOC{\equiv}CCH_2OH$$

~40%

Synthesis (1974) 357

$CH_3C\equiv CCOOH$

1) [2,2,6,6-tetramethylpiperidinyl-NLi]

2) [structure: (CH3)C=CH-CH2-Br]

3) MeI

$\longrightarrow$ [structure: C=C-COOMe]   ~50%

JOC (1975) <u>40</u> 269

$$Pr-\overset{\overset{\text{O}}{\|}}{C}-CH_2-COOEt$$

1) H_2NNH_2

2) $Tl(NO_3)_3$, MeOH

$\longrightarrow$ $Pr-C\equiv CCOOCH_3$ ~55%

Org Synth (1976) <u>55</u> 73

$TMSC\equiv CCH_2N=CHPh$

1) BuLi, THF, -70°

2) [structure: CH2=CH-C(=O)-OMe]

$\longrightarrow$ $TMSC\equiv C\underset{\underset{\text{N=CHPh}}{|}}{C}H(CH_2)_2COOMe$ 53%

Tetr Lett (1975) 3337

Section 307   <u>Acetylene - Ether, Epoxide</u>

$HC\equiv COEt$

1) $LiNH_2$

2) [epoxide]

$\longrightarrow$ $HCH(OH)CH_2C\equiv COEt$
                      90%

1) TsCl, KOH

2) KOH [epoxide]

$\longrightarrow$ $CH_2=CHC\equiv COEt$
                      74%

Rec Trav Chim (1974) <u>93</u> 92

Section 308   Acetylene - Halide

$(Me_2N)_2\overset{\overset{\displaystyle O}{\|}}{P}CH_2C\equiv CH$ + n-BuI $\xrightarrow[\text{2) PBr}_3]{\text{1) NaH, THF}}$ BuC≡CCH$_2$Br   ~70%

Synthesis (1974) 730

—CH=CCl$_2$ $\xrightarrow[\text{Et}_2\text{O, THF}]{\text{Et}_2\text{NLi}}$ —C≡C-Cl   80%

Synthesis (1975) 458

HOCH$_2$-C≡C-CH$_2$OH + 2 Ph$_3$PBr$_2$ $\xrightarrow{\text{MeCN}}$ BrCH$_2$-C≡C-CH$_2$Br   92%

Synthesis (1975) 255

Section 309   Acetylene - Ketone

PhCOCN + PhC≡CLi $\xrightarrow[\text{-70°}]{\text{Et}_2\text{O}}$ PhCOC≡CPh   61%

Bull Soc Chim Fr (1975) 779

TMSC≡CH $\xrightarrow[\text{2) PhCOCl}]{\text{1) CuI, t-BuOLi}}$ TMSC≡CCOPh   66%

JOC (1975) 40 131

Ph-C≡C-Cu + Cl-$\overset{\overset{\text{O}}{\|}}{\text{C}}$-CH$_2CH_2$-$\overset{\overset{\text{O}}{\|}}{\text{C}}$-OMe $\longrightarrow$ Ph-C≡C$\overset{\overset{\text{O}}{\|}}{\diagup}\diagdown$COOMe   52%

Comptes Rendus C (1976) 282 277

H-C≡C-$\overset{\overset{\text{OH}}{|}}{\text{CH}}$-Ph $\xrightarrow{\text{nickel peroxide}}$ H-C≡C-$\overset{\overset{\text{O}}{\|}}{\text{C}}$-Ph   98%

JOC USSR (1974) 10 2081

Section 310   Acetylene - Nitrile

No additional examples

Section 311   Acetylene - Olefin

+ RC≡CMgBr $\longrightarrow$ $\xrightarrow{\text{SOCl}_2}$ ≡ - R   70-80%

Ber (1974) 107 2985

$\underset{\text{Et}}{\overset{\underline{\text{n-Bu}}}{}}$C=C$\underset{\text{Cu}\cdot\text{MgX}_2}{\overset{\text{H}}{}}$ + $\underline{\text{n-Bu}}$C≡CH $\xrightarrow[\text{TMEDA}]{\text{Et}_2\text{O/THF}}$ $\underset{\text{Et}}{\overset{\underline{\text{n-Bu}}}{}}$C=C$\underset{\text{C≡C-Bu}}{\overset{\text{H}}{}}$ 78%

Tetr Lett (1975) 1465

$$\underset{Et}{\overset{n-Bu}{>}}C=C\underset{I}{\overset{H}{<}} + \underline{n}\text{-BuC}\equiv\text{CCu} \xrightarrow[80-100°]{Py} \underset{Et}{\overset{Bu}{>}}C=C\underset{C\equiv C\text{-Bu}}{\overset{H}{<}}$$ 85%

J Organometal Chem (1975) 93 415

$$\underline{n}\text{-C}_9\text{H}_{19}\text{-C}\equiv\text{CH} + \text{Et-C}\equiv\text{C-CH}_2\text{ZnBr} \xrightarrow{\text{2) H}_3\text{O}^+}$$ 45%

Comptes Rendus C (1976) 282 277

Section 312   Carboxylic Acid - Carboxylic Acid

$$\text{PhCH(COOH)}_2 \xrightarrow[\text{2) MeI}]{\text{1) 3 BuLi, THF, -50°}} \text{PhC(Me)(COOH)}_2$$ 88%

Tetr Lett (1975) 707

$$\text{BrCH}_2\text{CH}_2\text{Br} + \text{CH}_2\text{(COOEt)}_2 \xrightarrow[\text{TEBA}]{\text{50\% NaOH}}$$ 75%

J Org Chem (1975) 40 2969

$$\xrightarrow{\text{MeOMgOCO}_2\text{Me}}$$ 93%

JOC (1974) 39 1676

JOC (1975) $\underline{40}$ 1488

Section 313   Carboxylic Acid - Alcohol

Synthesis (1976) 825

JACS (1975) $\underline{97}$ 596

$Ph_2C=C(OTMS)_2$ $\xrightarrow[\text{2) } H_3O^+]{\text{1) MCPBA}}$ $Ph_2C-COOH$ | OH   81%

JOC (1975) <u>40</u> 3783

$\xrightarrow[\text{3) } H_3O^+]{\begin{array}{l}\text{1) BuLi}\\\text{2) RCHO}\end{array}}$

R   *
 \   /
  CH–CH$_2$–COOH
  |
  OH

<77% yield

<20% optical
yield

Tetr Lett (1974) 1333

Me_3SiCH_2COOH $\xrightarrow[\text{2)}]{\text{1) 2 i-Pr}_2\text{NLi}}$ $CH_2(OH)CH_2CH(SiMe_3)COOH$ 94%

JCS Chem Comm (1975) 537

$\xrightarrow[\text{2) } R_3SiCl]{\text{1) LiN(i-Pr)}_2}$ $\xrightarrow[\text{2) } NaBH_4]{\text{1) } O_3}$

94%

$HO\diagdown\diagup\diagdown\diagup\diagdown\diagup CH(CH_3)COOH$

Tetr Lett (1974) 2027

Section 314   Carboxylic Acid - Aldehyde

No additional examples

Section 315   Carboxylic Acid - Amide

$$PhCH=C \underset{NHCOCH_3}{\overset{CO_2H}{<}} \quad \xrightarrow[H_2]{\substack{Rh-DIOP \\ on\ polymer}} \quad PhCH_2CH \underset{NHCOCH_3}{\overset{CO_2H}{<}} \qquad \sim100\%$$

81% ee

JACS (1976) 98 5400

$$\xrightarrow[H_2O,\ rfx]{Zn,\ KOH} \quad Ph-CH_2-\underset{NHCOPh}{\overset{|}{CHCOOH}} \qquad 74\%$$

Tetr Lett (1975) 4051

$$HO-\underset{HN-COPh}{\overset{|}{CH}}-CO_2H \quad \xrightarrow[H_2SO_4]{PhH} \quad Ph-\underset{HNCOPh}{\overset{|}{CH}}-CO_2H \qquad 91\%$$

JCS Chem Comm (1975) 349

$$HO-\underset{NHCOPh}{\overset{|}{CHCOOH}} + Me_2CHSH \quad \xrightarrow[HOAc]{H_2SO_4} \quad Me_2CHS-\underset{NHCOPh}{\overset{|}{CHCOOH}} \qquad \sim80\%$$

Tetrahedron (1975) 31 863

Section 316   Carboxylic Acid - Amine

Review: "Advances in the Synthesis and Manufacture of α-Aminoacids"

Russ Chem Rev (1974) 43 745

>90% ee
up to 60% yield

JCS Chem Comm (1975) 988

$$\underset{HOOC-\overset{R}{\underset{|}{CH}}-NH_2 \cdot HCl}{} + H_3C-C\overset{OTMS}{\underset{NTMS}{}} \quad \xrightarrow[2)\ H_2O]{1)\ THF} \quad H_2N-\overset{R}{\underset{|}{CH}}-COOH \qquad \sim 95\%$$

Synthesis (1975) 113

$$Me_2NCH_2COOH \quad \xrightarrow[\substack{naphthalide \\ 2)\ PhCH_2Cl}]{1)\ Li^+} \quad \underset{NMe_2}{PhCH_2-\overset{}{\underset{|}{CHCOOH}}} \qquad 60\%$$

Compte Rendus (1974) 287 1383

Me_2CHCH_2COOH $\xrightarrow{\begin{array}{l}1)\ 2\ LDA \\ 2)\ MeONH_2 \\ 3)\ H_3O^+\end{array}}$ $Me_2CHCHCOOH$ 33%
$\quad\quad\quad\quad\quad\quad\quad\quad\quad\quad\quad\quad\quad\quad\quad\quad\quad\quad\quad | $
$\quad\quad\quad\quad\quad\quad\quad\quad\quad\quad\quad\quad\quad\quad\quad\quad\quad\quad\quad NH_2$

Chem Pharm Bull (1975) 23 167

$Ph-CH_2\overset{O}{\overset{||}{C}}SC_8H_{17}$ $\xrightarrow[\begin{array}{c}Na_2S_2O_4,\ NaHCO_3\end{array}]{\begin{array}{c}CO_2 \\ \text{Schrauzer's complex}\end{array}}$ $\left[PhCH_2\overset{O}{\overset{||}{C}}COOH\right]$ $\xrightarrow[Na_2S_2O_4]{NH_3}$

73%

$\quad\quad\quad\quad\quad\quad\quad\quad\quad\quad\quad\quad\quad\quad NH_2$
$\quad\quad\quad\quad\quad\quad\quad\quad\quad\quad\quad\quad\quad\quad |$
$\quad\quad\quad\quad\quad\quad\quad\quad\quad\quad PhCH_2CHCOOH$

Tetr Lett (1976) 4343

$CH_3-\overset{O}{\overset{||}{C}}-COOMe$ + L-Phe-O-t-Bu $\xrightarrow{\begin{array}{l}1)\ -H_2O \\ 2)\ H_2,\ Pd/C \\ 2)\ t\text{-}BuOCl\end{array}}$ $CH_3-\overset{*}{C}H-COOH$ 58%
$\quad\quad\quad\quad\quad\quad\quad\quad\quad\quad\quad\quad\quad\quad\quad\quad\quad\quad\quad\quad\quad | $
$\quad\quad\quad\quad\quad\quad\quad\quad\quad\quad\quad\quad\quad\quad\quad\quad\quad\quad\quad\quad\quad NH_2$

70% ee

(L predominates)

Tetr Lett (1976) 997; 1001

$\underset{\overset{|}{CO_2Et}}{\overset{Ph}{\diagdown}CH=N\diagdown}$ $\xrightarrow{\begin{array}{l}1)\ LDA \\ 2)\ \underline{n}\text{-octyl iodide} \\ 3)\ LDA \\ 4)\ \underline{i}\text{-PrI} \\ 5)\ H_3O^+\end{array}}$ $\underset{H_2N}{\overset{i\text{-}Pr}{\diagdown}}C\underset{\diagdown COOH}{\overset{n\text{-}Oct}{\diagup}}$

JOC (1976) 41 3491

Section 317   Carboxylic Acid - Ester

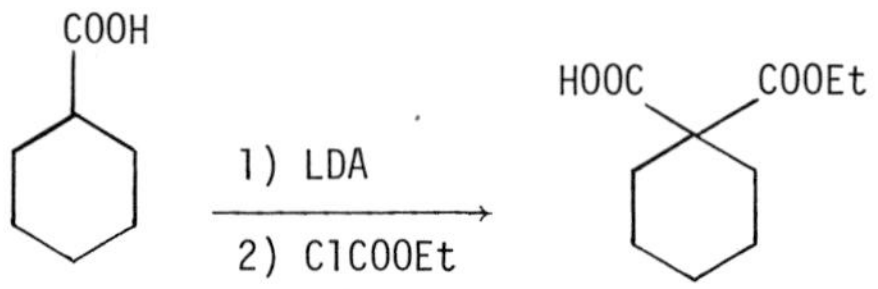

1) LDA

2) ClCOOEt

88%

Tetr Lett (1974) 2721

Section 318   Carboxylic Acid - Ether, Epoxide

No additional examples

Section 319   Carboxylic Acid - Halide

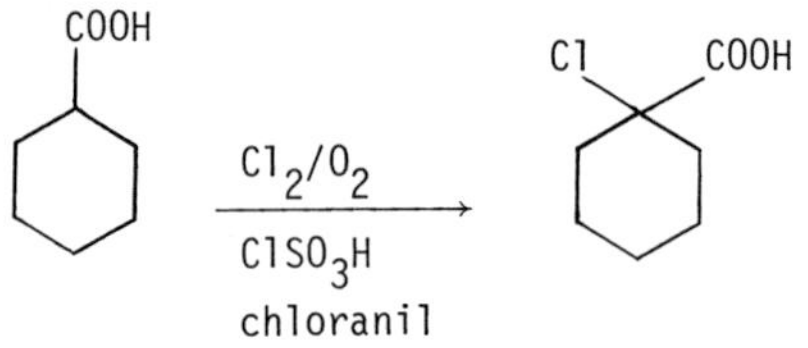

Cl_2/O_2

$ClSO_3H$

chloranil

73%

JOC (1975) 40 2960

$$RCH-COOH + NaNO_2 + (HF)_x F^- PyH^+ \longrightarrow RCHCOOH$$

(with NH$_2$ below the left carbon, F below the right carbon)

Synthesis (1974) 652

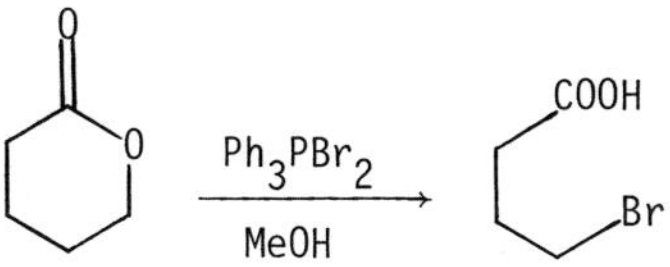

30%

JOC (1975) $\underline{40}$ 1640

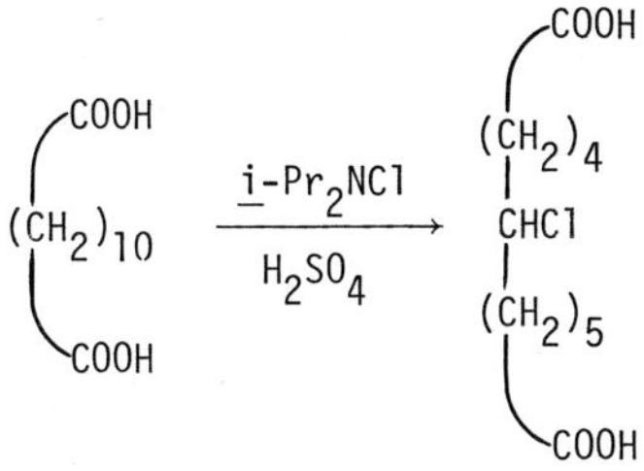

60%

Angew Int Ed (1976) $\underline{15}$ 306

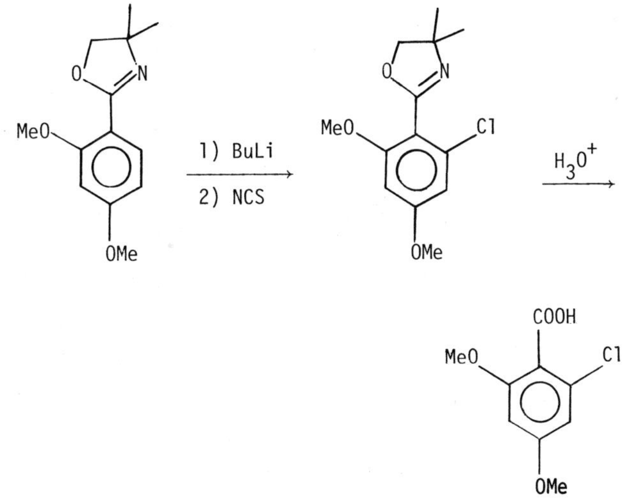

90%

JOC (1975) $\underline{40}$ 3158

$$CH_3(CH_2)_4\overset{\overset{\displaystyle O}{\|}}{C}-Cl \xrightarrow[CCl_4, \ H^+]{SOCl_2, \ NCS} CH_3(CH_2)_3\underset{\underset{\displaystyle Cl}{|}}{CH}-\overset{\overset{\displaystyle O}{\|}}{C}-Cl$$

87%

JOC (1975) <u>40</u> 3420

$$\underline{n}\text{-BuCH}_2\text{COOH} \xrightarrow[\text{2) }(CH_2CO)_2NBr, \ HBr]{\text{1) }SOCl_2} \underline{n}\text{-BuCH-COCl} \atop \underset{\displaystyle Br}{|}$$

~70%

Org Synth (1976) <u>55</u> 27

$$RCH_2COCl \xrightarrow[\text{2) }SOCl_2, \ HCl]{\text{1) }NCS} RCHClCOCl$$

~80%

Tetr Lett (1974) 3225

Also via: Haloesters (Section 359)

Section 320   <u>Carboxylic Acid - Ketone</u>

$$MeSCH=CHCMeCOOMe \atop \underset{\displaystyle CH_2C\equiv CH}{|} \xrightarrow[rfx]{HCl-HOAc}$$

93%

Tetr Lett (1975) 405

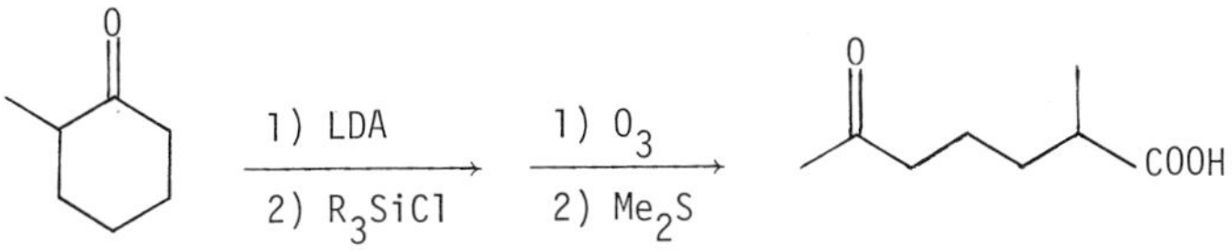

Tetr Lett (1974) 2027

$$MeCOCH_2COOMe + PhCONHCH(OH)COOH \xrightarrow{H_2SO_4} MeCOCHCOMeCH(NHCOPh)COOH$$ 71%

JCS Chem Comm (1975) 905

Section 321   Carboxylic Acid - Nitrile

$$PhCOCl + NaCN \xrightarrow[\text{cat., } CH_2Cl_2]{\text{phase transfer}} PhCOCN$$ 60%

Tetr Lett (1974) 2275

Also via: Cyanoesters (Section 361)

Section 322   Carboxylic Acid - Olefin

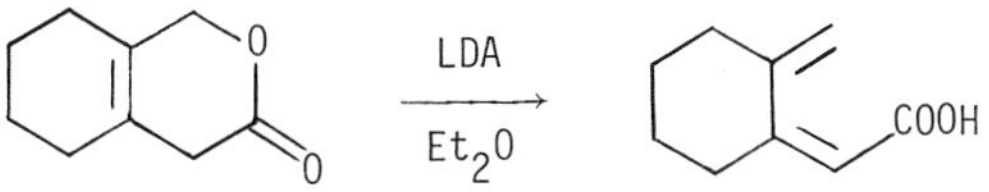

JCS Perkin I (1974) 2005

$PhCH_2COOH$ $\xrightarrow{\begin{array}{l}\text{1) } R_2NLi \\ \text{2) } PhCH_2CH_2Br \\ \text{3) } NaIO_4, \ 120°\end{array}}$ $PhC\text{-}COOH$ $\underset{CH_2}{\overset{\|}{}}$ 84%

JCS Chem Comm (1974) 135

Me_3SiCH_2COOH $\xrightarrow{\begin{array}{l}\text{1) 2 LDA} \\ \text{2) PhCHO}\end{array}}$ $PhCH\text{=}CHCOOH$ 88%

JCS Chem Comm (1975) 537

1) LDA (2 moles)

2) ⟋⟍Br

3) H_3O^+ 80%

JACS (1976) 98 4925

$Me_2C\text{=}CHCH_2SO_2CH_2COOMe$ $\xrightarrow[\text{KOH, } CCl_4]{\text{t-BuOH}}$ $Me_2C\text{=}CHCH\text{=}CHCOOH$ 80%

Synth Comm (1975) 5 315

$H_2C\text{=}CHCH\text{=}CH_2 + \underline{t}\text{-BuCu}, \ MgBr_2$ $\xrightarrow{CO_2}$ $\underline{t}\text{-BuCH}_2\underset{COOH}{\overset{}{CHCH}}\text{=}CH_2$ 95%

J Organometal Chem (1975) 92 C28

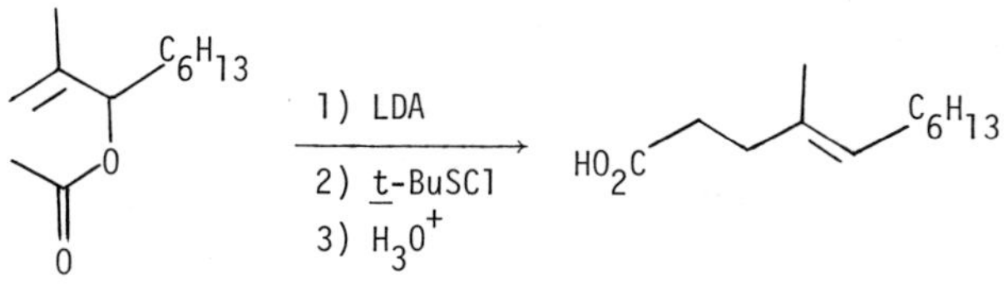

C_6H_{13}
O
O
1) LDA
2) t-BuSCl
3) H_3O^+
HO_2C
C_6H_{13}

80%

>98% E

JOC (1974) 39 3315

Also via: Hydroxy acids (Section 313)
Olefinic amides (Section 349)
Olefinic esters (Section 362
Olefinic nitriles (Section 376)

Section 323 Alcohol - Alcohol

1) $Tl(OAc)_3$, HOAc
2) KOH, MeOH
HO
HO

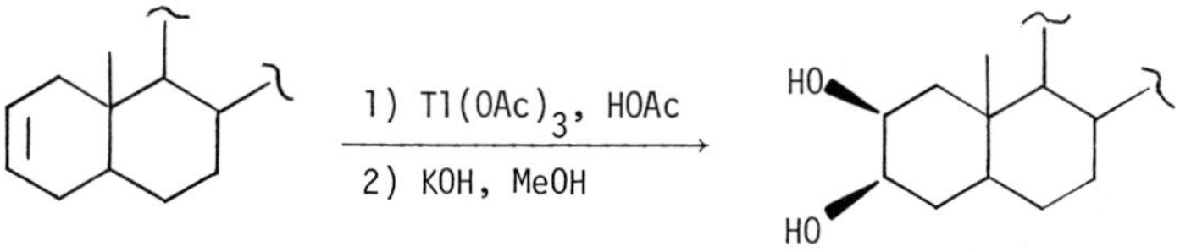

JCS Perkin I (1976) 1660

1) HOAc, I_2, KIO_3
2) KOAc
HO
HO

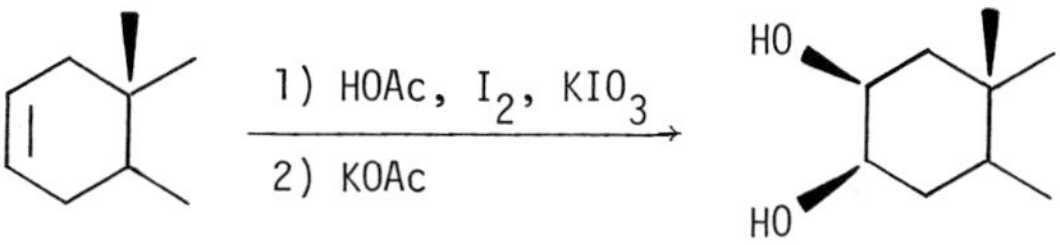

82%

Gazz Chim Ital (1975) 105 377

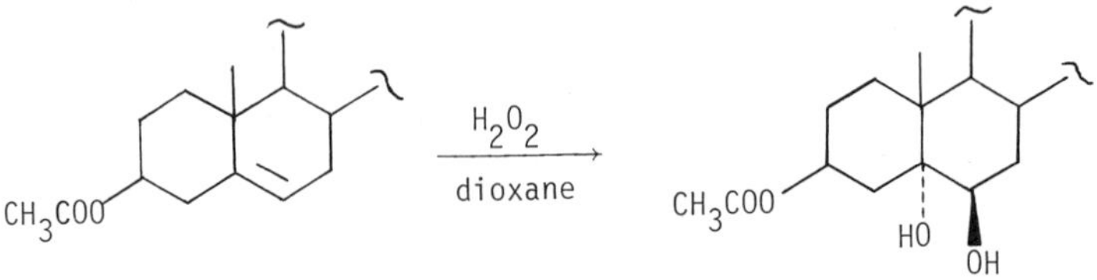

~60%

JOC USSR (1976) 12 1234

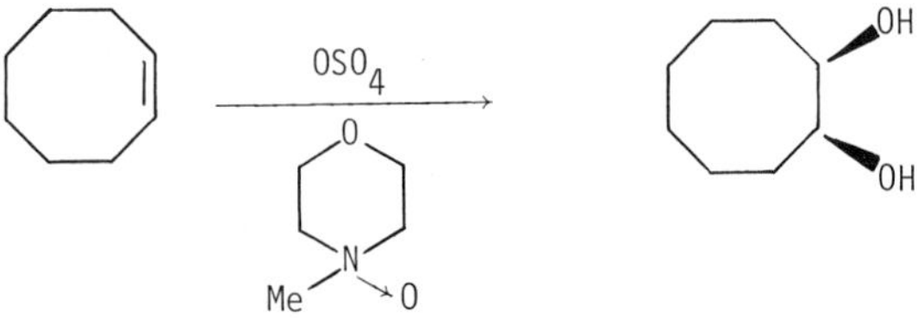

79%

Tetr Lett (1976) 1973

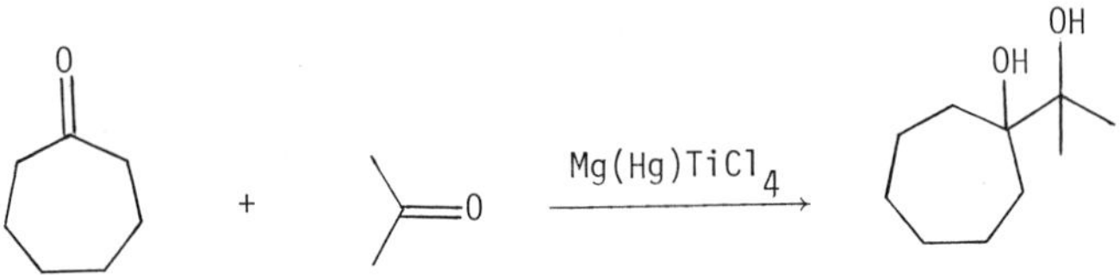

75%

JOC (1976) 41 260

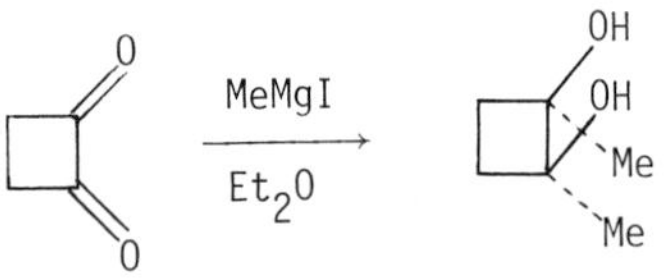

100%

Tetr Lett (1974) 30 1397

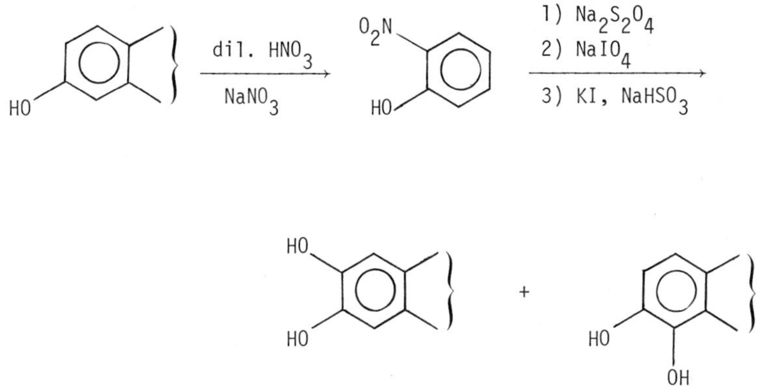

Steroids (1976) 28 733

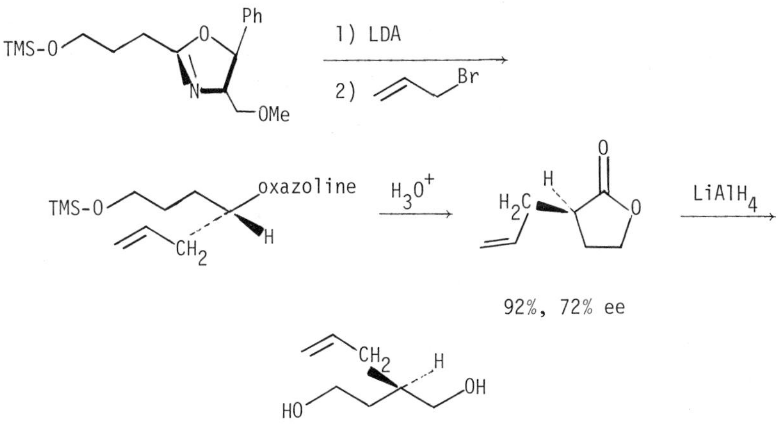

92%, 72% ee

JOC (1975) 40 1186

Also via: Hydroxyesters (Section 327)
           Diesters (Section 357)

Section 324   <u>Alcohol - Aldehyde</u>

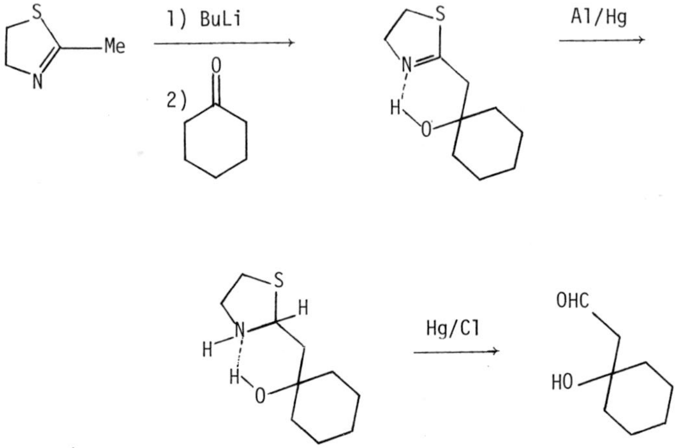

JOC (1975) <u>40</u> 2025

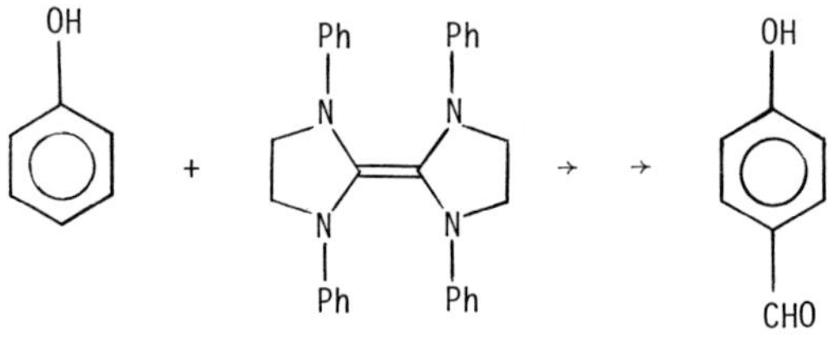

55%

Angew Int Ed (1976) <u>15</u> 169

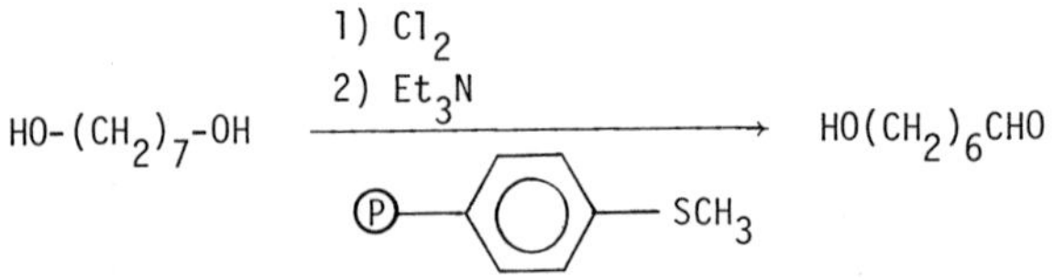

HO-(CH$_2$)$_7$-OH $\xrightarrow{\begin{array}{l}1)\ Cl_2\\2)\ Et_3N\end{array}}$ HO(CH$_2$)$_6$CHO   57%

JACS (1975) <u>97</u> 2232

Also via: Acetoxyladehydes (Section 336)

Section 325   Alcohol - Amide

$(\underline{i}\text{-Pr})_2\text{NCHO}$ $\xrightarrow[\text{2) Ph}_2\text{CO}]{\text{1) LDA}}$ $\text{Ph}_2\text{C-CON}(\underline{i}\text{-Pr})_2$ 92%
$\qquad\qquad\qquad\qquad\qquad\qquad\qquad\quad\;|$
$\qquad\qquad\qquad\qquad\qquad\qquad\qquad\;\;\text{OH}$

Can J Chem (1974) 52 185

$\xrightarrow[\substack{\text{2) O}_2 \\ \text{3) Na}_2\text{SO}_3}]{\text{1) LDA}}$ 84%

Tetr Lett (1975) 1731

$\xrightarrow[\text{2) H}_3\text{O}^+]{\text{1) Li-C-N(CH}_2\text{OMe)}_2}$ 44%

Angew Int Ed (1976) 15 293

$\text{Me}_2\text{CHCONEt}_2$ $\xrightarrow[\text{2) PhCHO}]{\text{1) LDA}}$ $\text{PhC(OH)HCMe}_2\text{CONMe}_2$ 85%

Comptes Rendus (1975) 281 893

$$\underset{\displaystyle \text{Ph-C-CH}_2\text{CONH}_2}{\overset{\displaystyle \overset{O}{\|}}{}} \quad \xrightarrow{\text{yeast}} \quad \underset{\displaystyle \text{Ph-CH-CH}_2\text{CONH}_2}{\overset{\displaystyle \overset{OH}{|}}{\underset{*}{}}}$$

26%

high optical purity

Also works with α-ketoamides

Aust J Chem (1976) <u>29</u> 2459

$$\xrightarrow[\text{2) Ph}_2\text{CO}]{\text{1) 2 BuLi, THF, 0°}}$$

88%

Can J Chem (1974) <u>52</u> 3206

$$\xrightarrow{\text{h}\nu}$$

~70%

JCS Chem Comm (1974) 743

$$\xrightarrow{\text{h}\nu}$$

90%

JCS Perkin I (1976) 2054

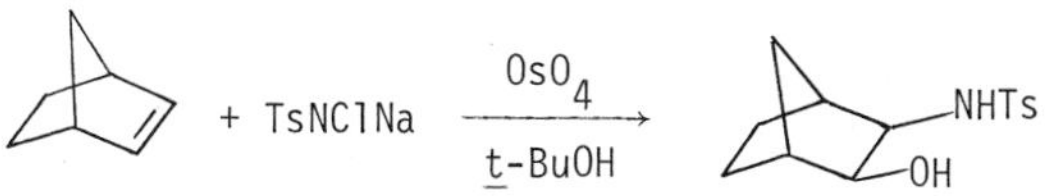

+ TsNClNa $\xrightarrow[\text{t-BuOH}]{\text{OsO}_4}$

64%

JOC (1976) <u>41</u> 177

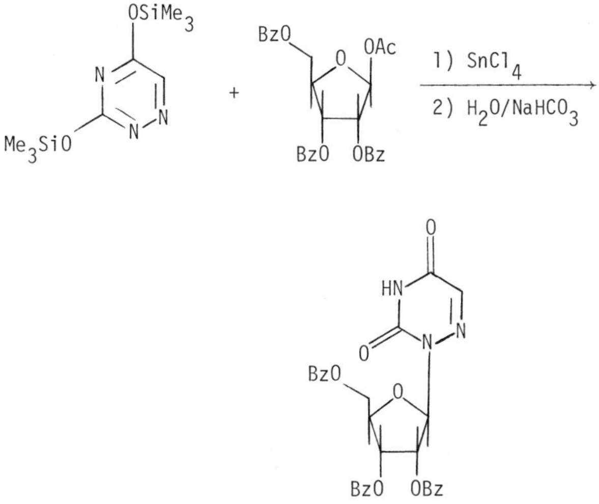

1) SnCl$_4$

2) H$_2$O/NaHCO$_3$

93% on 10 Kg scale

JOC (1974) <u>39</u> 3654
<u>39</u> 3660 (uracils)
<u>39</u> 3664 (disaccharides)
<u>39</u> 3668
<u>39</u> 3672 (5-azacytidines)

Glycosidation of Purines using SnCl$_4$/ClCH$_2$CH$_2$Cl

Tetr Lett (1974) 2141

Section 326   Alcohol - Amine

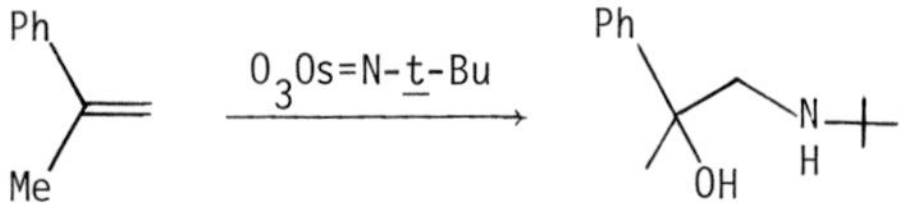

93%

JACS (1975) 97 2305

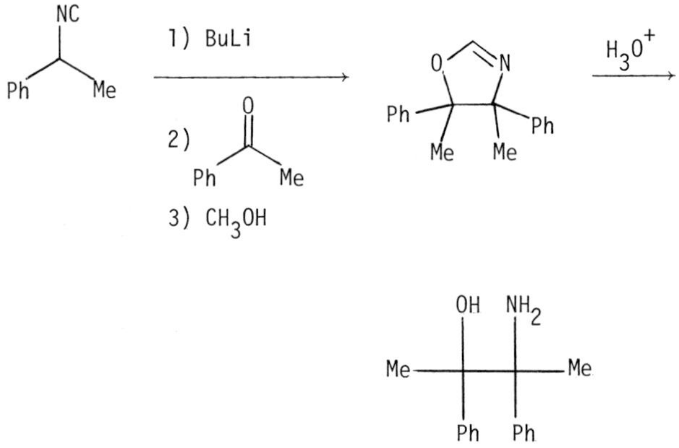

74%

Liebigs Ann Chem (1976) 183

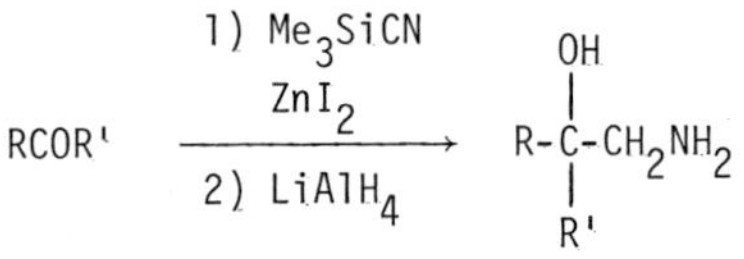

JOC (1974) 39 914

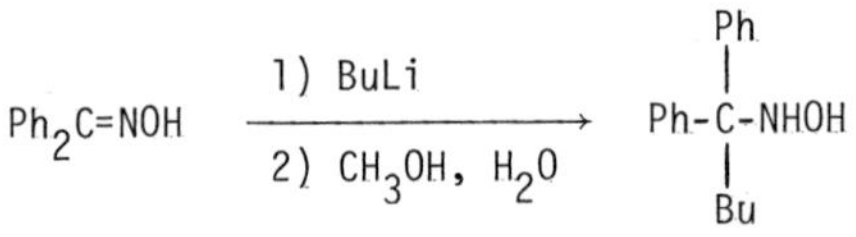

63%

Tetr Lett (1976) 233

Review: Methods of Synthesis of Pyrimidine Nucleosides

Russ Chem Rev (1973) 494

Review: The Silyl Method of Synthesis of Nucleosides and Nucleotides

Russ Chem Rev (1974) 140

Use of ArSO$_2$N⟨imidazole⟩ as condensing agent in polynucleotide synthesis.

JCS Chem Comm (1974) 325

Section 327   Alcohol - Ester

1) LDA

2) O$_2$

3) Na$_2$SO$_3$

76%

Tetr Lett (1975) 1731

$$RCH_2-\overset{O}{\overset{||}{C}}-OR' + LDA + MoO_5 \cdot Py \cdot HMPA \xrightarrow[\text{2) }H_2O]{\text{1) }-70°} RCH-\overset{O}{\overset{||}{C}}OEt$$
$$\underset{OH}{|}$$

50-80%

Works with lactones, ketones

JACS (1974) 96 5944

EtC=CHMe   +   MeCOCOOEt $\xrightarrow[\text{2) H}_2\text{O}]{\text{1) TiCl}_4}$ EtCOCH-C(OH)COOEt   87%
| | |
OSiMe$_3$ Me Me

Chem Lett (1975) 741

PhCHCOCOOEt + MeMgBr $\xrightarrow{\text{Et}_2\text{O}}$ PhCH-C-COOEt   90%
| | |
Cl   Cl Me

Comptes Rendus (1974) <u>279</u> 469

$\underset{\text{Ph-C-COOMe}}{\overset{\text{O}}{\overset{\|}{}}}$ $\xrightarrow[\text{R = (-)-menthyl}]{}$ Ph-CH-COOMe   82%
                                                          *

21% ee

JCS Chem Comm (1976) 101

$\underline{n}$-C$_6$H$_{13}$CHO + CH$_3$CHBrCOOEt + Zn $\longrightarrow$ $\underline{n}$-C$_6$H$_{13}$CHOH-CHCOOEt   73%
                                                                            |
                                                                           CH$_3$

Org React (1975) <u>22</u> 423

PhCHO $\xrightarrow[\text{Zn}]{\text{BrCH}_2\text{COOEt}}$ PhCH(OH)CH$_2$COOEt   97%

continuous flow apparatus

JOC (1974) <u>39</u> 269

PhCHO + BrCH$_2$CO$_2$Et $\xrightarrow{\text{"Rieke Zn"}}$ PhCH-CH$_2$CO$_2$Et   98%
$\qquad\qquad\qquad\qquad\qquad\qquad\qquad\qquad$ |
$\qquad\qquad\qquad\qquad\qquad\qquad\qquad\qquad$ OH

$\qquad\qquad\qquad\qquad\qquad\qquad\qquad$ Synthesis (1975) 452
$\qquad\qquad\qquad\qquad\qquad\qquad\qquad$ Use of activated In JOC (1975) 40 2253

+ Ti(OEt)$_4$ + H$_2$C=C=O $\longrightarrow$ $\begin{array}{c} \text{R'} \quad \text{O} \\ | \quad\quad || \\ \text{RC-CH C-OEt} \\ | \\ \text{OH} \end{array}$ 50-90%

$\qquad\qquad\qquad\qquad\qquad\qquad$ Helv Chim Acta (1974) 57 1703

PhCHO + MeCH=C$\begin{array}{c}\diagup\text{OTMS} \\ \diagdown\text{OMe}\end{array}$ $\xrightarrow[\text{2) H}_2\text{0}]{\text{1) TiCl}_4\text{, THF, -78°}}$ PhCH-CHCOOMe   88%
$\qquad\qquad\qquad\qquad\qquad\qquad\qquad\qquad\qquad\qquad\qquad$ | |
$\qquad\qquad\qquad\qquad\qquad\qquad\qquad\qquad\qquad\qquad\qquad$ OH Me

$\qquad\qquad\qquad\qquad\qquad\qquad$ Chem Lett (1975) 989

PhCH$_2$CH$_2$CHO + CH$_3$CO$_2$Et $\xrightarrow{\text{s-Bu(Me}_3\text{Si)}_2\text{COLi}}$ PhCH$_2$CH$_2$CHCH$_2$COOEt   77%
$\qquad\qquad\qquad\qquad\qquad\qquad\qquad\qquad\qquad\qquad\qquad\qquad$ |
$\qquad\qquad\qquad\qquad\qquad\qquad\qquad\qquad\qquad\qquad\qquad\qquad$ OH

$\qquad\qquad\qquad\qquad\qquad\qquad$ Tetr Lett (1976) 2253

AcOCH$_2$CH$_2$COMe $\begin{array}{c}\text{1) LiCH}_2\text{COOEt} \\ \xrightarrow{\hspace{2cm}} \\ \text{-78°, Et}_2\text{0} \\ \text{3) KOH, MeOH}\end{array}$ 94%

$\qquad\qquad\qquad\qquad\qquad\qquad$ Synthesis (1974) 719

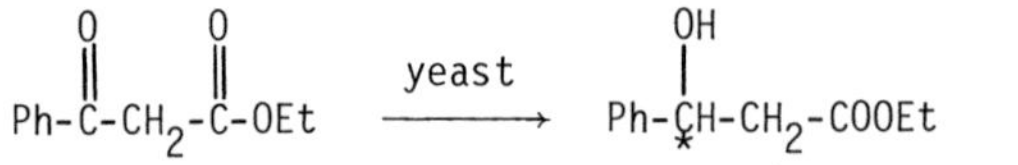

                                                                    70%

high optical yield

Also works with α-ketoesters

Aust J Chem (1976) <u>29</u> 2459

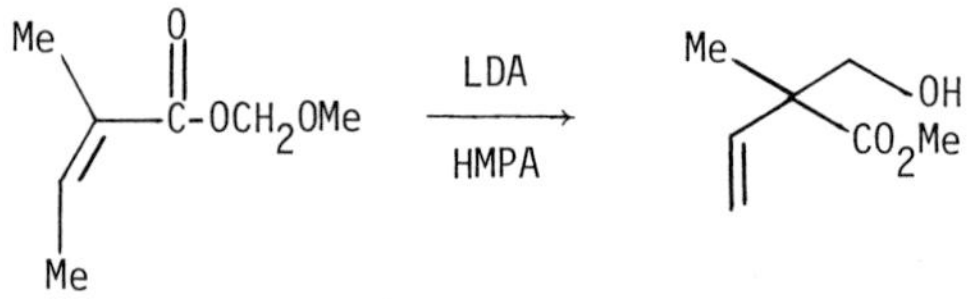

                                                                    61%

JOC (1976) <u>41</u> 585

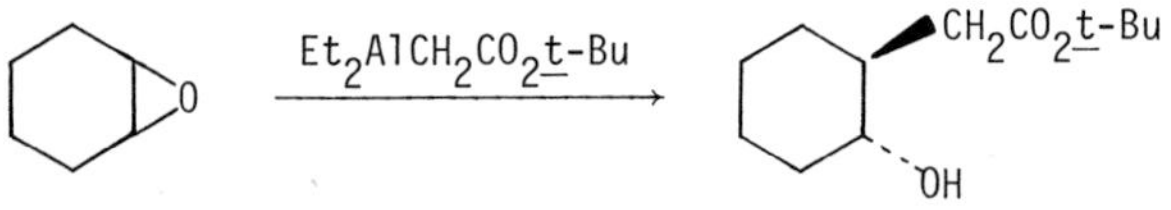

                                                                    34%

JOC (1976) <u>41</u> 1669

Also via: Hydroxyacids (Section 313)

Section 328   <u>Alcohol - Ether, Epoxide</u>

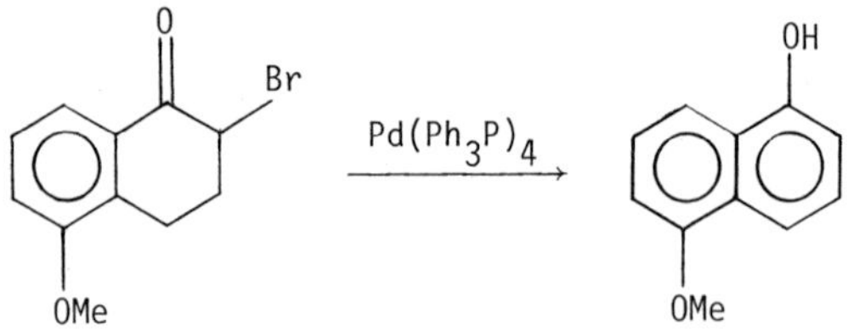

                                                                    94%

JOC (1975) <u>40</u> 2976

Section 329　　Alcohol - Halide

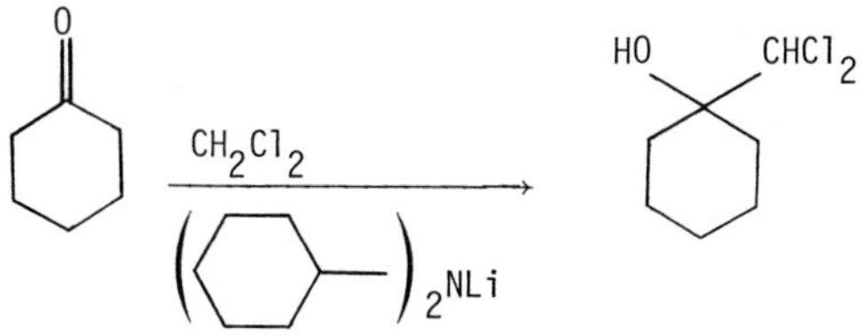

JACS (1974) 96 3010

$$R-CH=CH_2 \xrightarrow[\text{electrolysis}]{H_2O,\ Br^-} R-\overset{\overset{\displaystyle OH}{|}}{C}H-CH\ Br$$

R = 1° alkyl

~60%

J Gen Chem USSR (1975) 45 2072

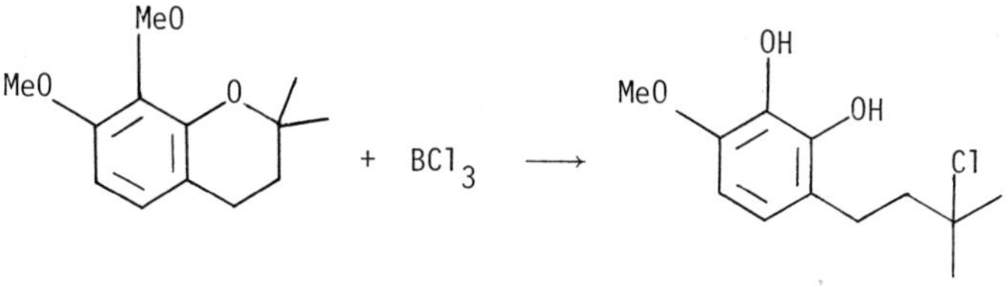

JCS Chem Comm (1974) 318

Section 330　　Alcohol - Ketone

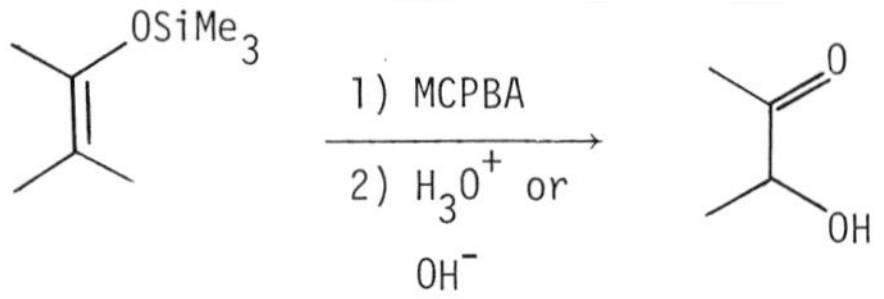

70%

Tetr Lett (1974) 4319

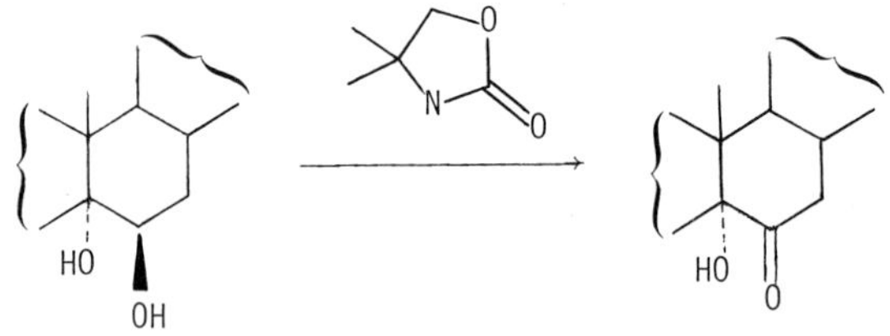

83%

Tetrahedron (1976) $\underline{32}$ 1097

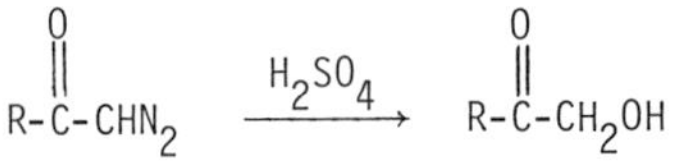

$$R-\overset{O}{\overset{||}{C}}-CHN_2 \xrightarrow{H_2SO_4} R-\overset{O}{\overset{||}{C}}-CH_2OH$$

~80%

Comptes Rendus $\underline{C}$ (1976) $\underline{282}$ 1125

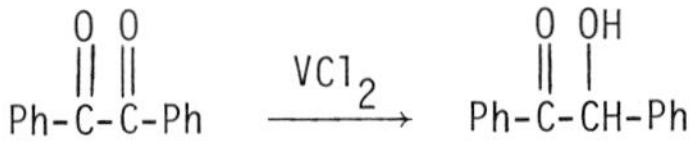

$$Ph-\overset{O}{\overset{||}{C}}-\overset{O}{\overset{||}{C}}-Ph \xrightarrow{VCl_2} Ph-\overset{O}{\overset{||}{C}}-\overset{OH}{\overset{|}{CH}}-Ph$$

100%

Synthesis (1976) 815

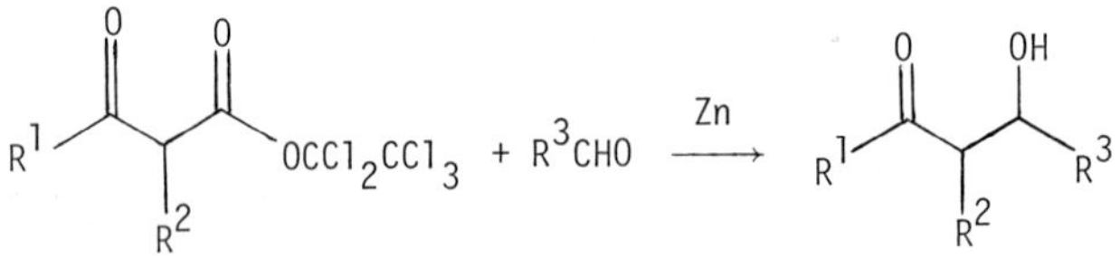

Chem Lett (1976) 95

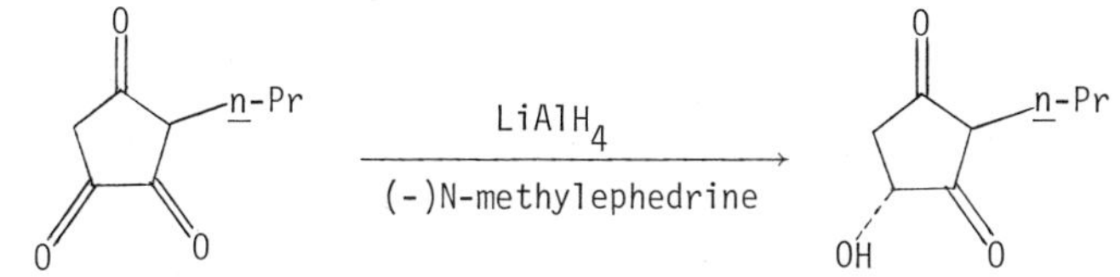

55%

58% ee

Tetr Lett (1976) 3165

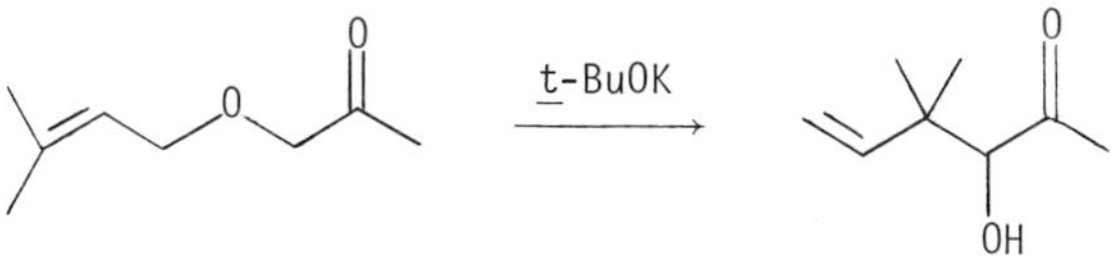

Helv Chim Acta (1974) 57 2084

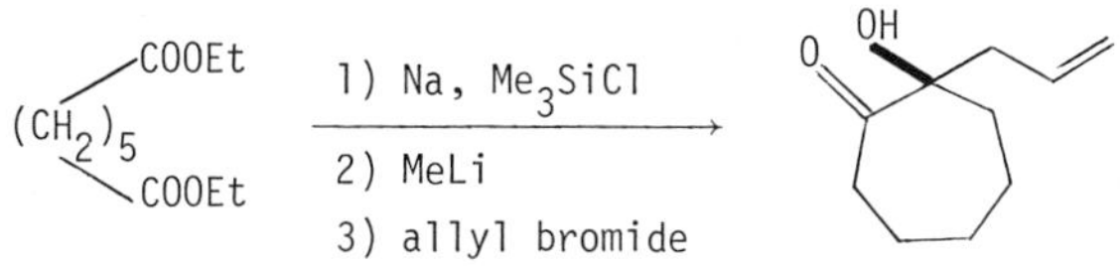

80%

Synthesis (1976) 341

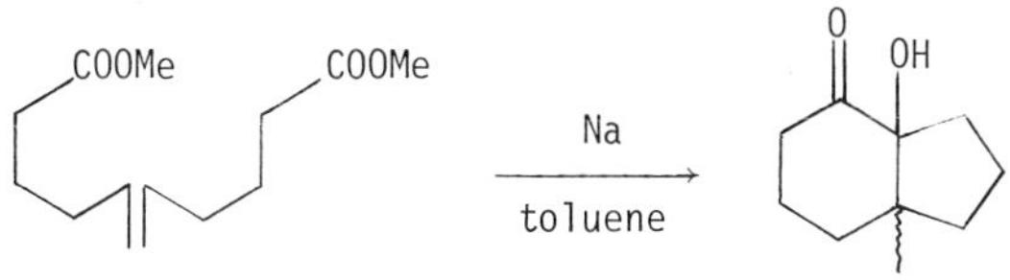

50%

JACS (1974) 96 5618

Ph   OAc
 \   /
  C=C   1) MeLi
 /   \   2) ZnCl$_2$ $\underset{\underset{Ph \quad OH}{|\quad\quad|}}{CH_3CO-C-C-Pr}$ ~55%
H   Me   3) CH$_3$CH$_2$CH$_2$CHO, NH$_4$Cl

$$\text{Org Synth (1974) }\underline{54}\text{ 49}$$

COOEt   OSiMe$_3$
  | Na/PhCH$_3$ |
  ·   ─────────   ═   ⟶
  | TMCS |
COOEt   OSiMe$_3$

                        90%

                    O
                    ‖
                         OH
                    77%

$$\text{Org React (1976) }\underline{23}\text{ 259}$$

      S   S
       \ /
        C   1) BuLi   O
       / \   ─────────   ‖
      R   H   2) R'COR"   R'   CH(OH)R'R"
        ↑   3) hydrolysis
     (RCHO)
                                 $$\text{JOC (1975) }\underline{40}\text{ 231 (full paper)}$$

   CN
    |
    ┌──OH   1) MeMgBr   ═O
    └──   ─────────   75%
                 2) H$^+$ OH

                        $$\text{Bull Soc Chim Fr (1975) 333}$$

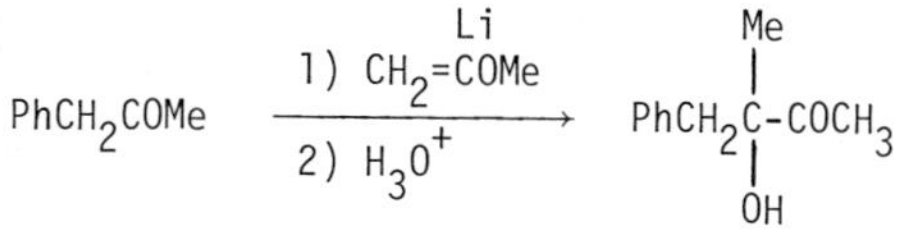

PhCH$_2$COMe $\xrightarrow[\text{2) H}_3\text{O}^+]{\text{1) CH}_2\text{=COMe}}^{\text{Li}}$ PhCH$_2$C-COCH$_3$ 72%

with structure showing Me and OH on central carbon

JACS (1974) **96** 7125

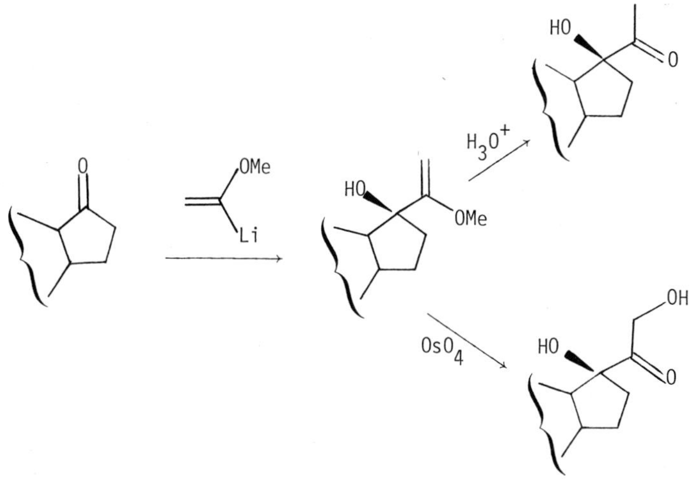

63%

54%

JOC (1976) **41** 2312

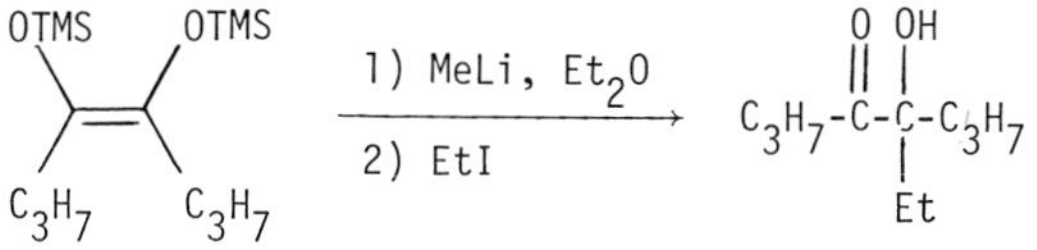

OTMS   OTMS structure with C$_3$H$_7$ C$_3$H$_7$ $\xrightarrow[\text{2) EtI}]{\text{1) MeLi, Et}_2\text{O}}$ C$_3$H$_7$-C-C-C$_3$H$_7$ with O, OH, Et   82%

Tetr Lett (1974) 3879
Tetr Lett (1974) 3883

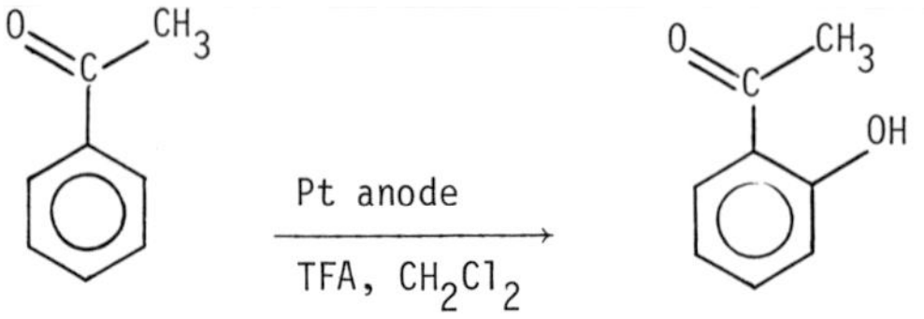

Pt anode
————————
TFA, CH$_2$Cl$_2$

85%

JCS Chem Comm (1975) 262

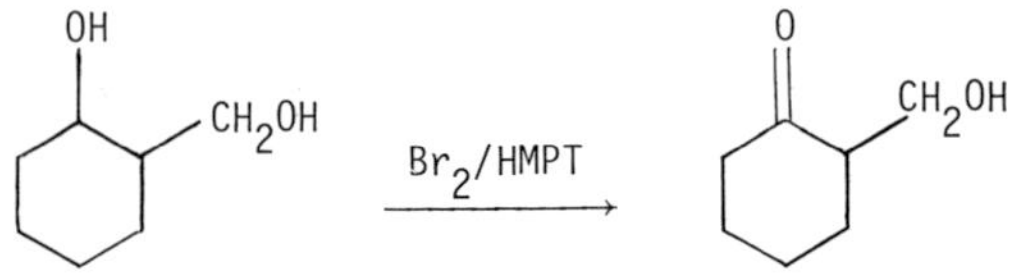

Br$_2$/HMPT
————————

93%

Synthesis (1976) 811

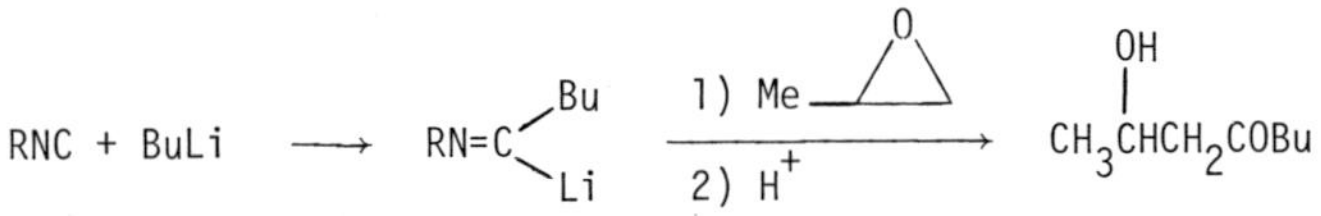

RNC + BuLi ⟶ RN=C$\overset{Bu}{\underset{Li}{}}$

1) Me—△(O)
————————————
2) H$^+$

CH$_3$CHCH$_2$COBu with OH

90%

JOC (1974) 39 600

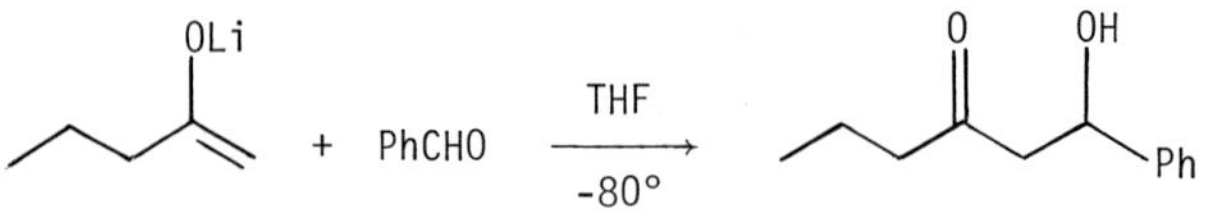

+ PhCHO

THF
————
-80°

80%

JOC (1974) 39 3459

1) MeLi
————————
2) CH$_2$O, -78°

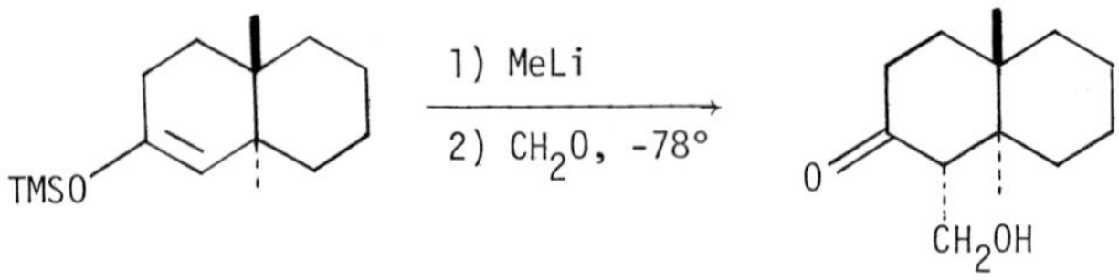

90%

JACS (1974) 96 7114

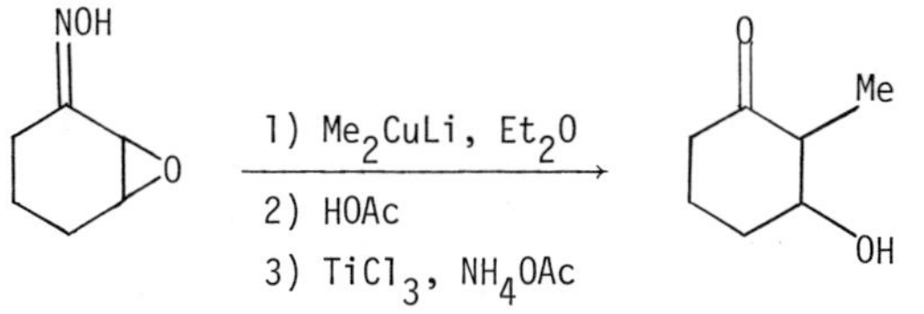

Tetr Lett (1975) 3117

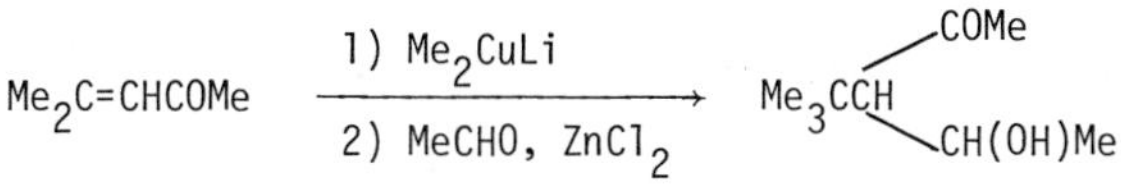

90%

$Me_2C=CHCOMe$

1) Me_2CuLi
2) $MeCHO$, $ZnCl_2$

Me_3CCH with COMe and CH(OH)Me

Tetr Lett (1975) 589

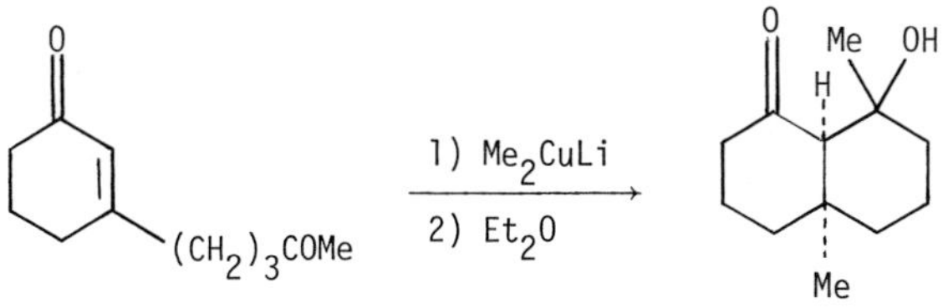

96%

1) Me_2CuLi
2) Et_2O

Helv Chim Acta (1975) 58 1808

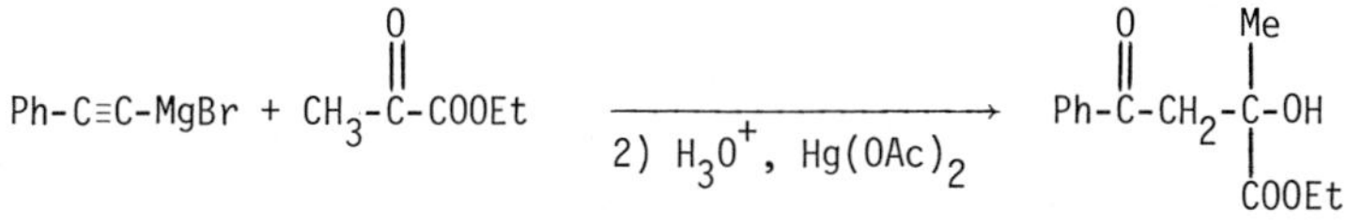

63%

$Ph-C\equiv C-MgBr + CH_3-C-COOEt$

2) H_3O^+, $Hg(OAc)_2$

$Ph-C-CH_2-C-OH$ with Me, COOEt

Synthesis (1976) 832

55%

+ PhCHO

$TiCl_4$

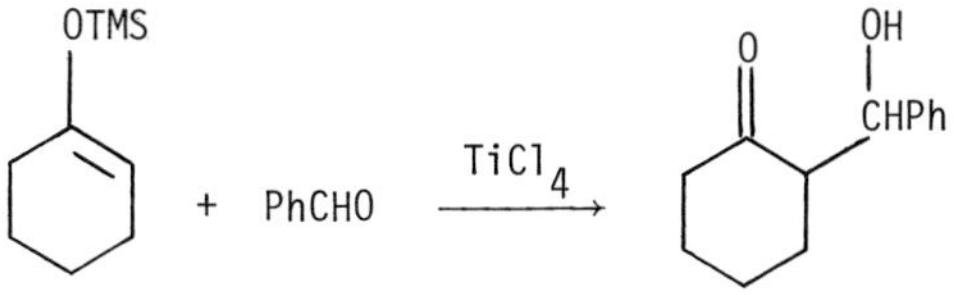

94%

JACS (1974) 96 7503

$$\text{PhCH}_2\text{CH}_2\overset{\text{O}}{\overset{\|}{\text{C}}}\text{CH}_3 + \text{Me}_2\text{CHCHO} \xrightarrow[]{\overset{\text{i-Bu}}{\overset{|}{(\text{Me}_3\text{Si})_2\overset{|}{\text{C}}\text{OLi}}}} \text{PhCH}_2\text{CH}_2\overset{\text{O}}{\overset{\|}{\text{C}}}\text{CH}_2\underset{\underset{\text{OH}}{|}}{\text{CHCHMe}}$$

71%

Tetr Lett (1976) 1817

$$\text{PhSOCH}_2\text{COCH}_3 \xrightarrow[\text{3) Ph}_2\text{CO}]{\begin{array}{l}\text{1) NaH}\\\text{2) BuLi}\end{array}} \text{Ph}_2\text{C(OH)CH}_2\text{COCH}_2\text{SOPh}$$

91%

Tetr Lett (1974) 107

$$\xrightarrow[\text{or Cl}_2/\text{DMSO}]{\text{Cl}_2/\text{PhSMe}}$$

80%

(no C-C bond cleavage)

Tetr Lett (1974) 287

+ LiCH(SCH₃)₂ $\longrightarrow$

$\xrightarrow{\text{H}_2\text{SO}_4}$

42%

Tetr Lett (1976) 759

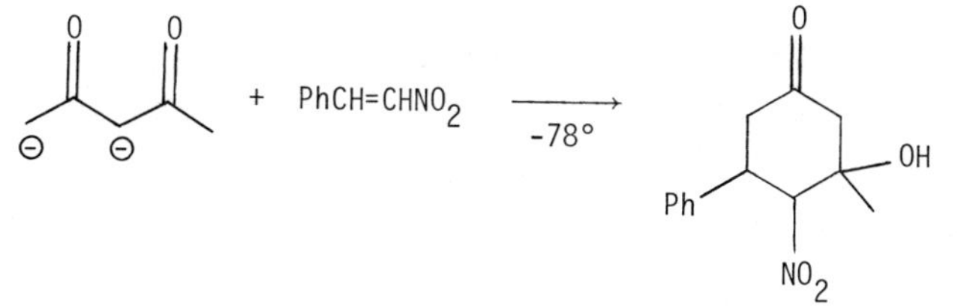

73%

Angew Int Ed (1974) 13 400

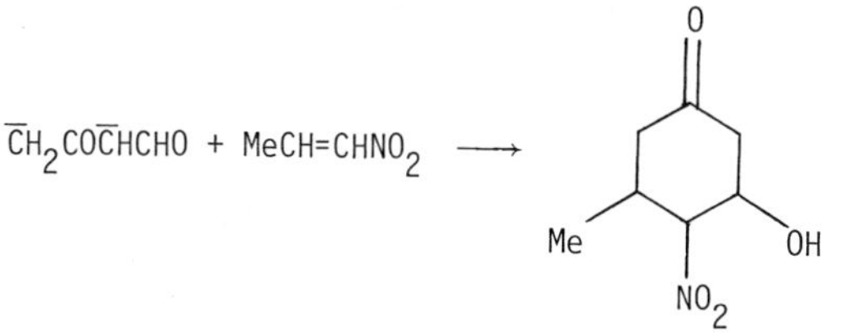

$\overline{C}H_2CO\overline{C}HCHO$ + MeCH=CHNO$_2$ ⟶

Ber (1975) 108 1961; 1924

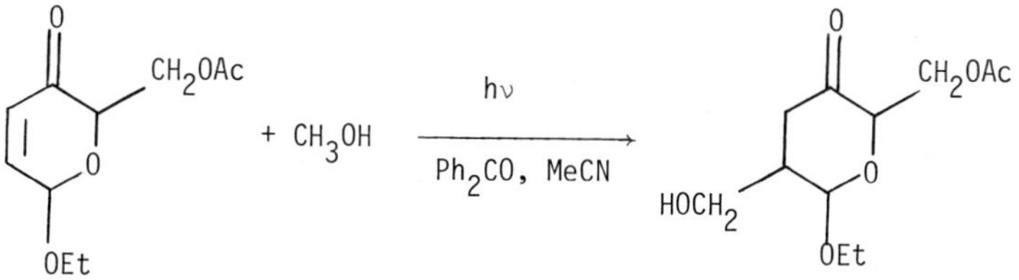

75%

Tetr Lett (1975) 297

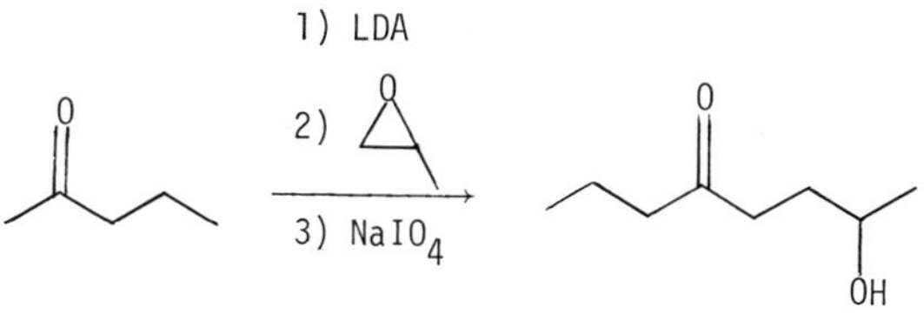

98%

Tetr Lett (1976) 11

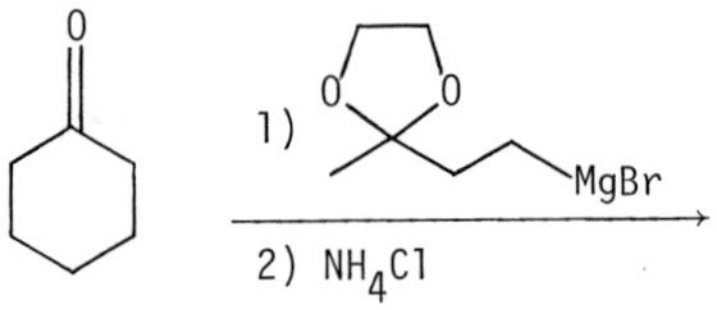

97%

Tetr Lett (1976) 3105

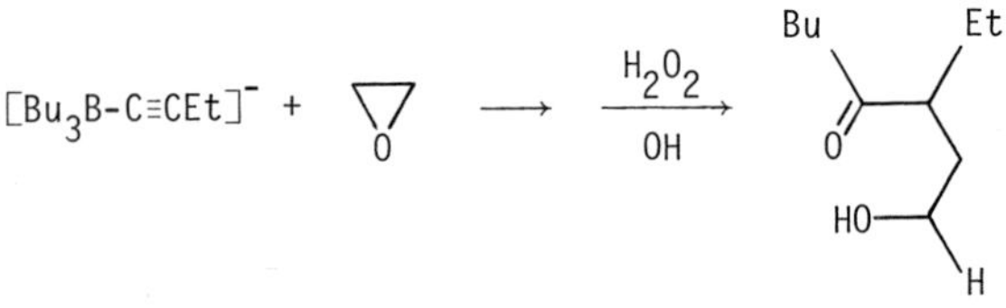

$[Bu_3B-C{\equiv}CEt]^-$ +

70%

Tetrahedron (1974) <u>30</u> 3037

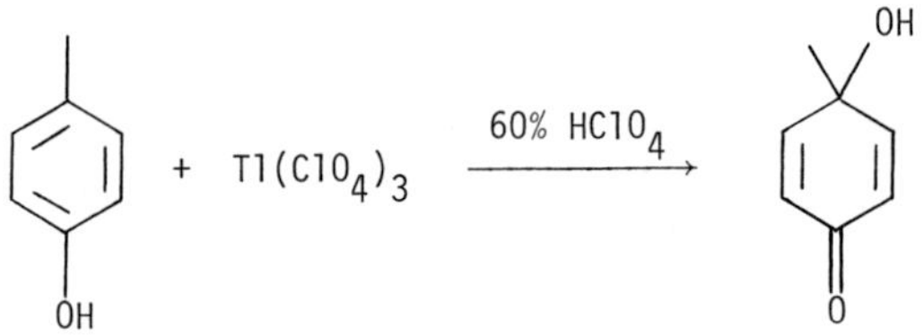

50%

Chem Comm (1974) 661

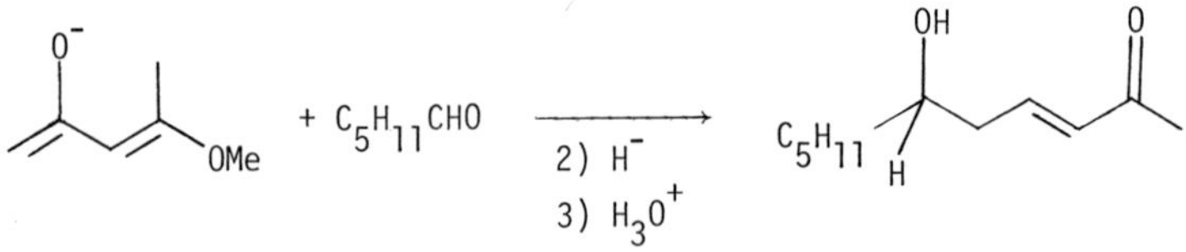

~50%

JACS (1976) <u>98</u> 2351

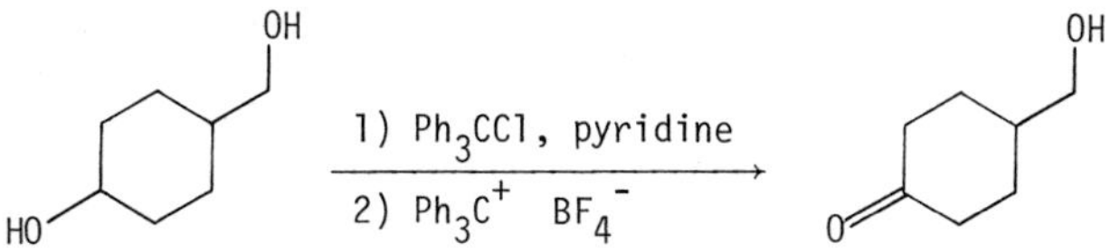

1) Ph$_3$CCl, pyridine

2) Ph$_3$C$^+$ BF$_4^-$

79%

JACS (1976) 98 7882

Also via: Acyloxyketones (Section 360)

Section 331   Alcohol - Nitrile

PhCH$_2$CH$_2$CH$_2$CN

1) LDA

2) MoO$_5$·Py·HMPA

PhCH$_2$CH$_2$CHCN
           |
           OH

57%

JOC (1976) 41 740

KNH$_2$

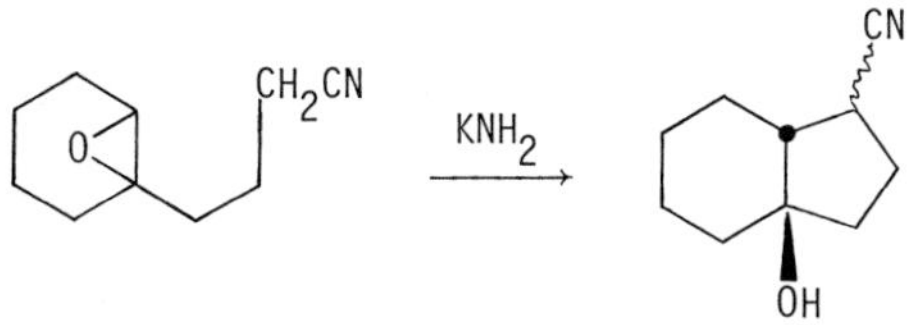

75%

JACS (1974) 96 5268

Also via: Cyanohydrin trimethylsilyl ethers (Section 366)
           Cyanohydrin esters (Section 361)

Section 332   Alcohol - Olefin

Allylic and benzylic hydroxylation (C=C-CH → C=C-C-OH, etc.) is listed in
Section 41 (Alcohols and Phenols from Hydrides)

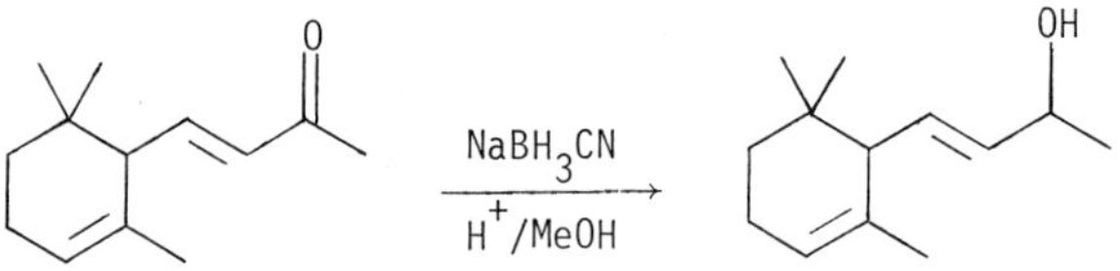

JOC (1975) 40 2530

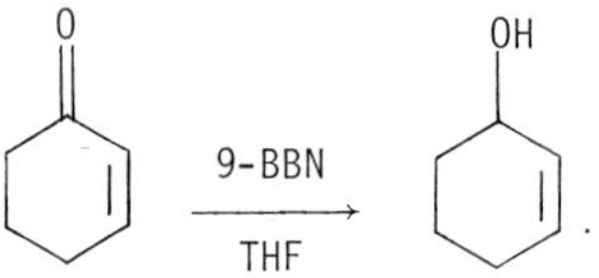

88%

JOC (1975) 40 1864

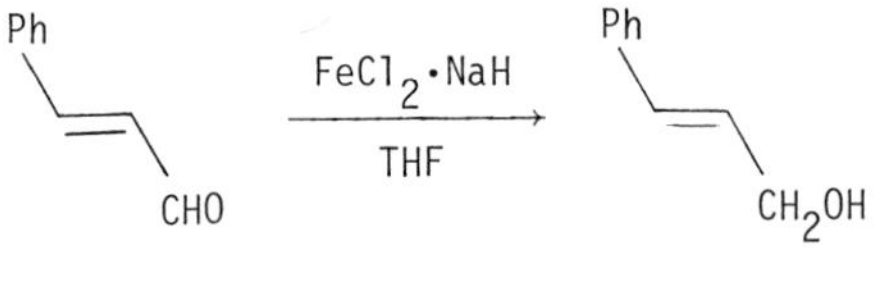

85%

Chem Lett (1976) 581

42%

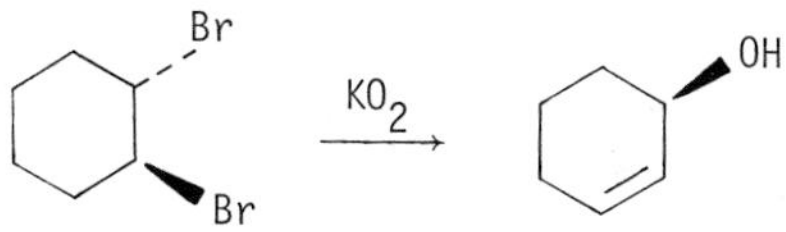

100%

Tetr Lett (1975) 3183

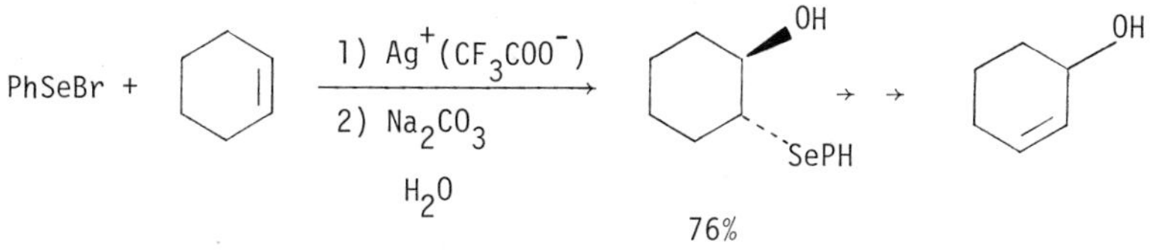

76%

JCS Chem Comm (1974) 100

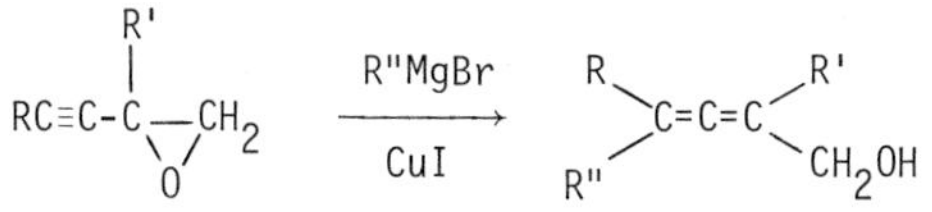

>90%

Rec Trav Chim (1974) $\underline{93}$ 47

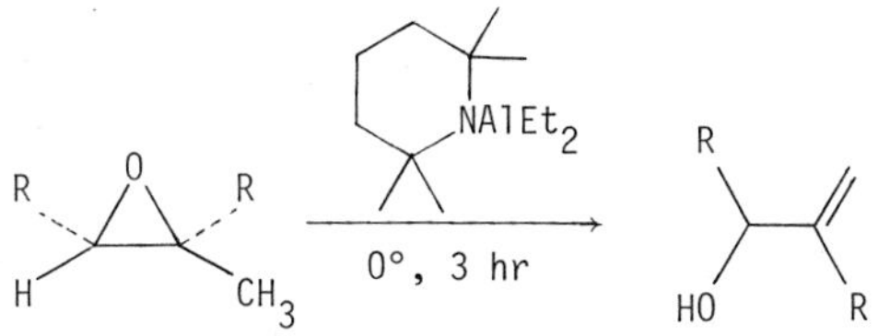

90%

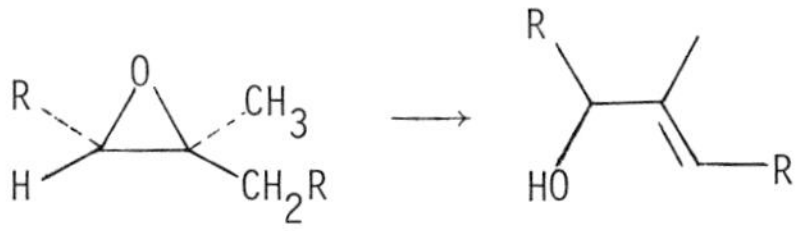

78%

JACS (1974) $\underline{96}$ 6513

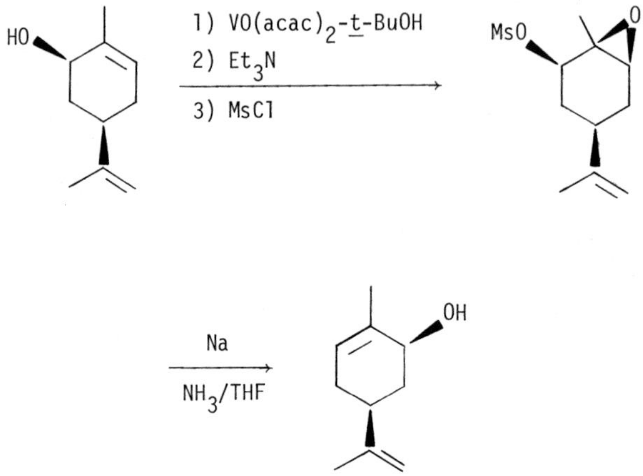
HO
1) VO(acac)$_2$-t-BuOH
2) Et_3N
3) MsCl
MsO
O
Na
NH_3/THF
OH

~30%

Tetr Lett (1976) 2621

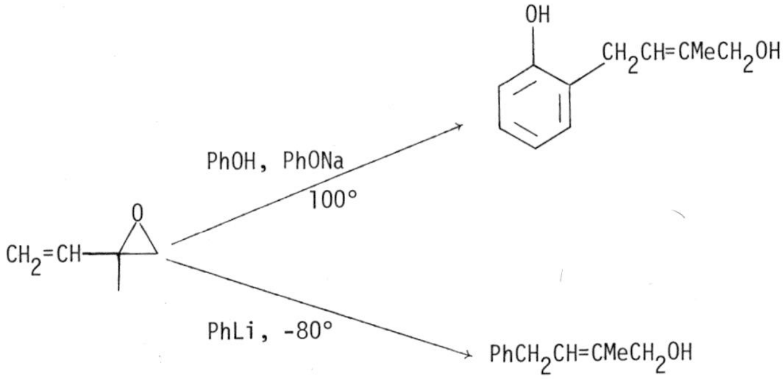
OH
$CH_2CH=CMeCH_2OH$
PhOH, PhONa
100°
O
$CH_2=CH-$
PhLi, -80°
$PhCH_2CH=CMeCH_2OH$

Tetr Lett (1975) 4419

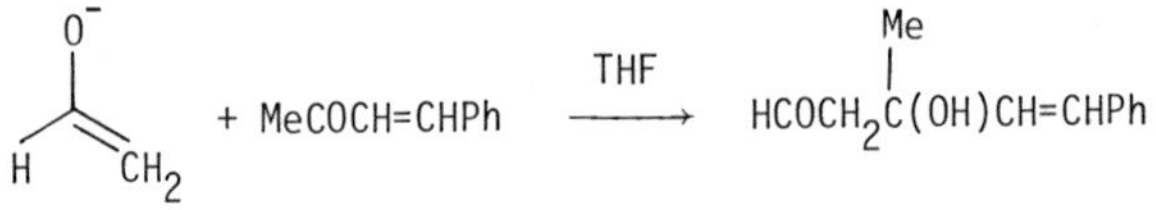
O^-
H
CH_2
+ MeCOCH=CHPh
THF
Me
$HCOCH_2C(OH)CH=CHPh$

98%

Comptes Rendus (1974) 278 533
279 961

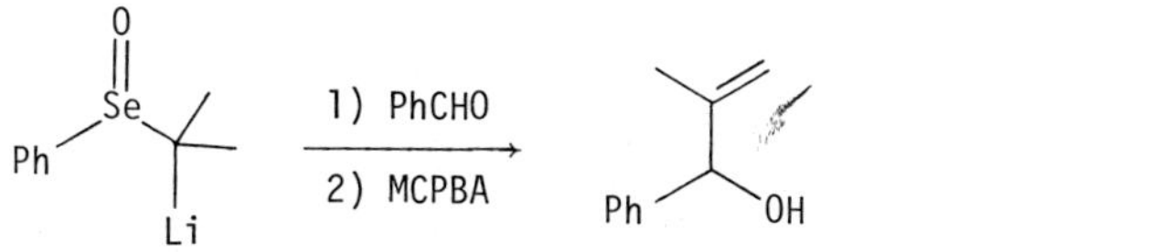

81%

JACS (1975) <u>97</u> 3250

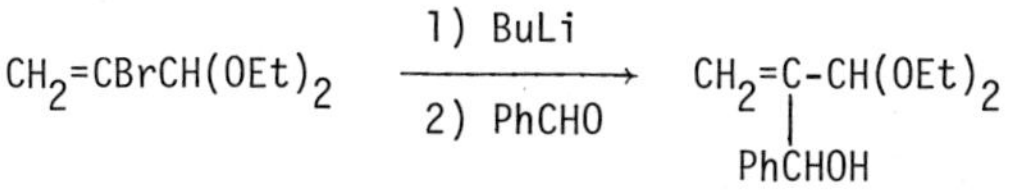

$CH_2=CBrCH(OEt)_2$　$\xrightarrow[\text{2) PhCHO}]{\text{1) BuLi}}$　$CH_2=\underset{\underset{PhCHOH}{|}}{C}-CH(OEt)_2$

82%

Tetr Lett (1974) 2751; 2755

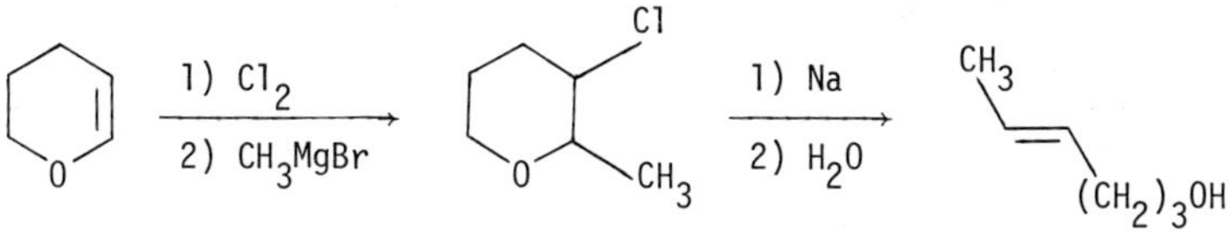

~60%

Org Synth (1976) <u>55</u> 62

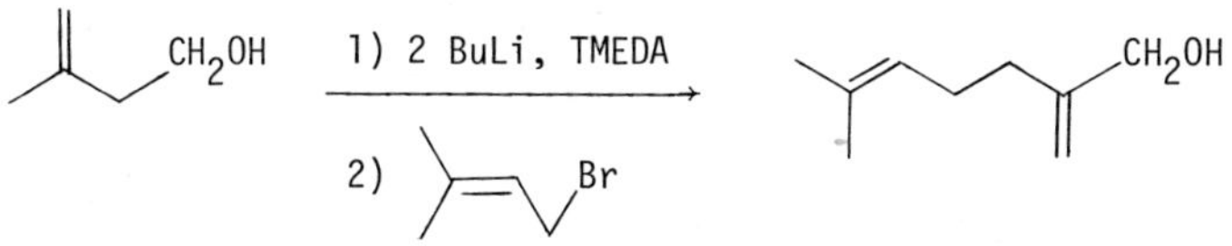

70%

Tetr Lett (1974) 2215

Ph⌐⌐⌐Br   +   [allyl SePh structure with Li, SePh] $\xrightarrow{\text{2) pyridine, } H_2O_2}$

[structure: Ph—CH_2CH_2—CH=CH—CH(OH)—CH_3]   80%

JOC (1975) <u>40</u> 2570

$\underset{\overset{|}{CH_3}}{PhSe(O)CHCH_3}$ $\xrightarrow[\begin{array}{c}2)\ PhCHO\\3)\ H^+,\ rfx\end{array}]{1)\ LDA,\ THF}$ $\underset{\overset{|}{Me}}{CH_2=C-CHPh}\overset{OH}{|}$ 81%

JACS (1975) <u>97</u> 3250

$HC\equiv CCH_2OH$ + $MeCH=CHCH_2ZnBr$ $\longrightarrow$ $H_2C=C\begin{array}{l}CH_2OH\\ \diagdown CH-CH=CH_2\\ \ \ |\\ \ \ Me\end{array}$ 93%

Comptes Rendus C (1975) <u>280</u> 1473

$Ph-C\equiv C-CH_2OH$ + $BuLi$ $\longrightarrow$ [alkene: Li, CH_2OLi, Ph, Bu] $\xrightarrow{H_2O}$ [alkene: CH_2OH, Ph, Bu]   90%

Acta Chem Scand <u>B</u> (1976) <u>30</u> 521

$$\underset{\underset{OTHP}{|}}{Me_2C-C\equiv CCH_2CH_2OH} \xrightarrow[Cu^+]{MeMgX} Me_2C=C=CMeCH_2OH \qquad 56\%$$

Tetr Lett (1975) 1509

$$PhC\equiv CCH_2OH \xrightarrow[CuI, Et_2O]{EtMgBr} \underset{H}{\overset{Ph}{\diagdown}}C=C\underset{CH_2OH}{\overset{Et}{\diagup}} \qquad 70\%$$

J Organometal Chem (1975) 91 C1

$$F_2C=CFCl \xrightarrow[\text{2) PhCOMe}]{\text{1) BuLi, THF, Et}_2\text{O, pentane, -135°}} \underset{\underset{Me}{|}}{\overset{\overset{OH}{|}}{PhC-CF=CF_2}} \qquad 83\%$$

Synthesis (1975) 122

$$CH_2=CBrCH_2OH \xrightarrow[\text{2) RCOR'}]{\text{1) }\underline{t}\text{-BuLi, Et}_2\text{O}} \underset{\underset{OH}{|}}{RR'CC(CH_2OH)=CH_2} \qquad 65\text{-}75\%$$

JOC (1975) 40 2975

$$Ph_2\bar{C}C\equiv C^-(Na)_2^+ \xrightarrow[\text{2) H}_3\text{O}^+]{\text{1) Ph}_2\text{CO}} \underset{\underset{OH}{|}}{Ph_2C=C=CHCPh_2} \qquad 70\%$$

JOC USSR (1975) 11 1691

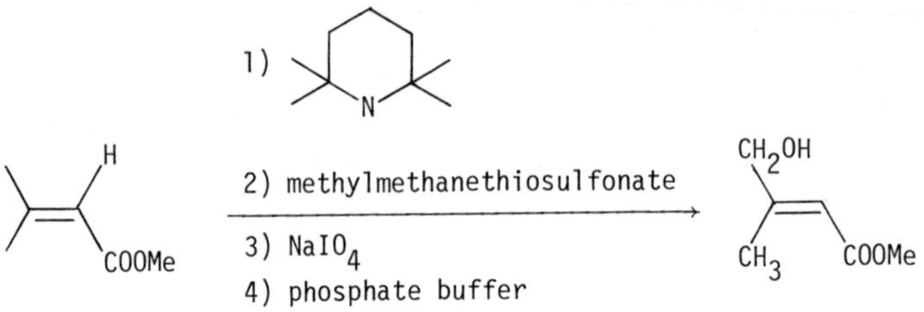

1)
N
H
COOMe
2) methylmethanethiosulfonate
3) NaIO4
4) phosphate buffer
CH2OH
CH3
COOMe

~30%

Tetr Lett (1976) 4215

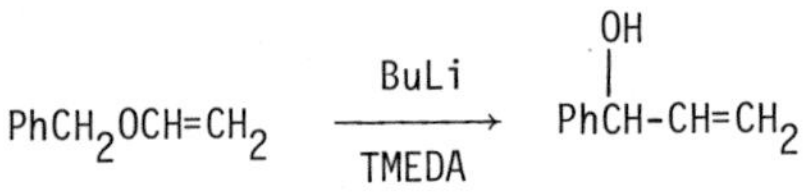

PhCH2OCH=CH2
BuLi
TMEDA
OH
PhCH-CH=CH2

90%

JACS (1974) 96 2576

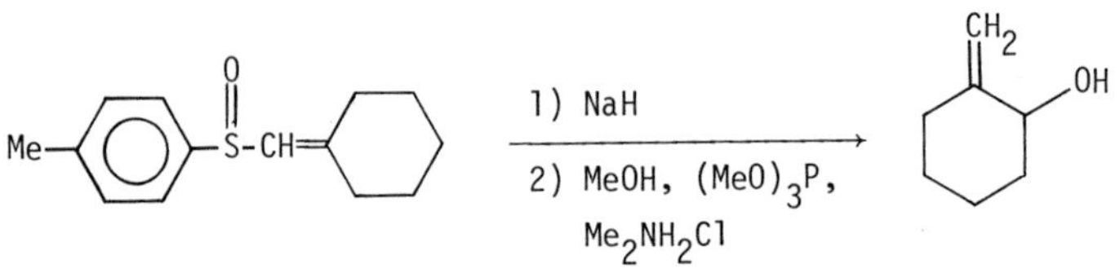

Me
O
S-CH=
1) NaH
2) MeOH, (MeO)3P,
Me2NH2Cl
CH2
OH

75%

Tetr Lett (1976) 2237

O
Et3N⁻ Li⁺
OH

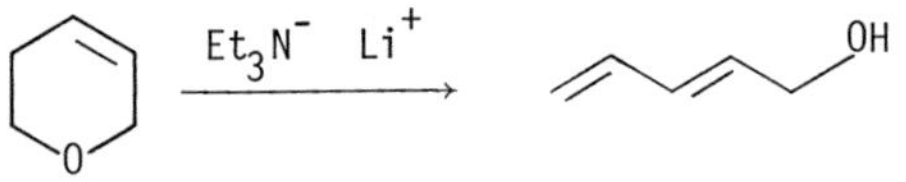

50%

Comptes Rendus C (1976) 282 391

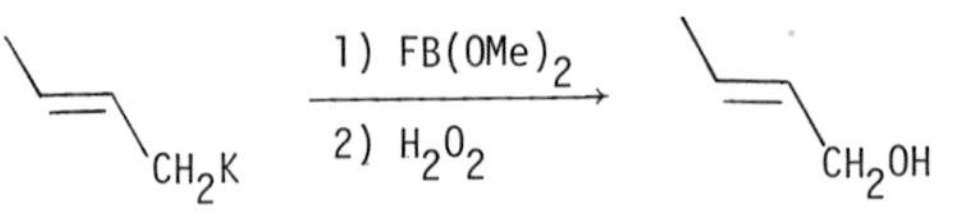

CH2K
1) FB(OMe)2
2) H2O2
CH2OH

58%

Helv Chim Acta (1975) 58 1094

Review: Preparation of α-Allenic Alcohols

Bull Soc Chim Fr (1975) 2369

$(CH_2=CH)_2CuLi$

Et_2O, -20°

72%

Tetr Lett (1974) 2439

$EtC \equiv CH$

1) $\underline{i}-Bu_2AlH$
2) MeLi
3) , Et_3N

84%

Synthesis (1975) 632

$MeC \equiv CH$

1) BuLi
2) Bu_3B
3)
4) HOAc

62%

Chem Lett (1975) 397

$H_2C=CHCH=CH_2$ + $\underline{t}$-BuCu, $MgBr_2$ $\xrightarrow{Me_2CO}$ $\underline{t}$-BuCH$_2$CHCH=CH$_2$

$Me_2\overset{|}{C}OH$

62%

J Organometal Chem (1975) <u>92</u> C28

100%

Me$_2$CO
TMEDA

Me$_2$CO
MBPE
[2.2.2]

60%

Tetr Lett (1974) 2665

PhCH$_2$CHO + CH$_2$=CH-CH$_2$SiMe$_3$ $\xrightarrow[\text{CH}_2\text{Cl}_2]{\text{AlCl}_3}$ PhCH$_2$-CH-CH$_2$CH=CH
                                                                        |
                                                                        OH

Tetr Lett (1976) 2449

$$\underset{\text{Ph-C-CH}_3}{\overset{\text{O}}{||}}$$ + $\overset{\text{Br}}{\diagup\!\!\!\diagdown}$ $\xrightarrow[\text{2) H}_3\text{O}^+]{\text{1) Zn}}$

97%

JOC (1976) 41 550

1) Me$_3$SiCH$_2$K, pentane

2) $\triangle$

83%

Synthesis (1975) 328

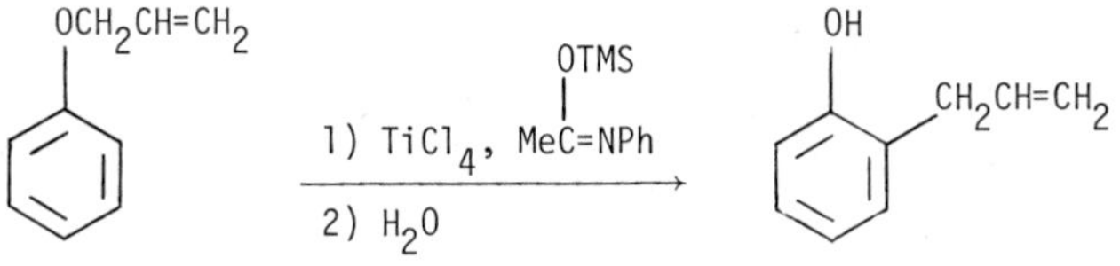

Chem Lett (1975) 1041

Also via: Acetylenes - Alcohols (Section 302)

Section 333   Aldehyde - Aldehyde

No additional examples

Section 334   Aldehyde - Amide

No additional examples

Section 335   Aldehyde - Amine

No additional examples

Section 336   Aldehyde - Ester

$$\underline{n}-C_3H_7CH_2CHO \xrightarrow[\substack{2) \ BrCH_2COOEt \\ 3) \ H_3O^+}]{1) \ (\underline{i}-Bu)_2NH} \underline{n}-C_3H_7\underset{\underset{CH_2COOEt}{|}}{CHCHO}$$

70%

Synth Comm (1974) <u>4</u> 147

$$\xrightarrow[\text{EtOH}]{\text{NOCl, SO}_2}$$

COOEt
CH=NOH·HCl

95%

JACS (1975) <u>97</u> 3848

Also via: Carboxyaldehydes (Section 314)

Section 337   Aldehyde - Ether, Epoxide

$$PhCH_2CH_2CHO + CH_2=CHOEt \xrightarrow[\text{2) EtOH}]{1) \ (\underline{i}-PrO)_4Ti, \ TiCl_4}$$

100%

$$PhCH_2CH_2\underset{\underset{O-\underline{i}-Pr}{|}}{CH}CH_2CH(OEt)_2$$

Chem Lett (1975) 569

Br2, -78°
NaOMe, MeOH
B(OH)2
CH-CH(OMe)2
Br

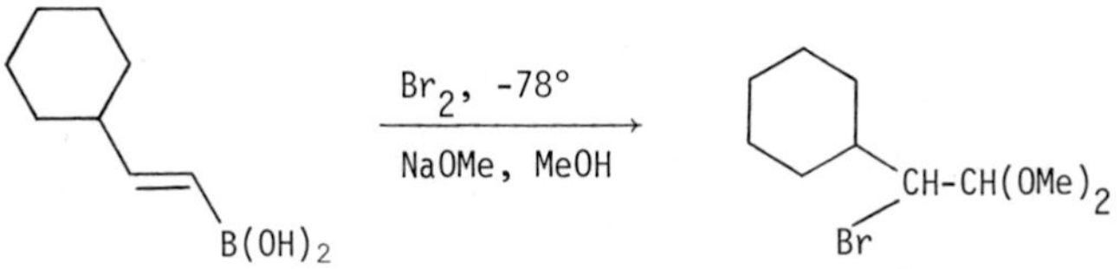

~70%

JOC (1975) 40 1189

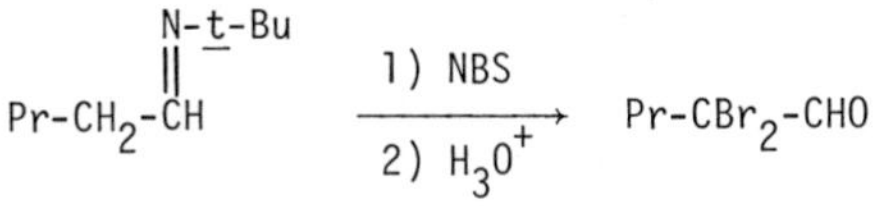

N-t-Bu
Pr-CH2-CH
1) NBS
2) H3O+
Pr-CBr2-CHO

58%

Synthesis (1975) 455

Section 339 Aldehyde - Ketone

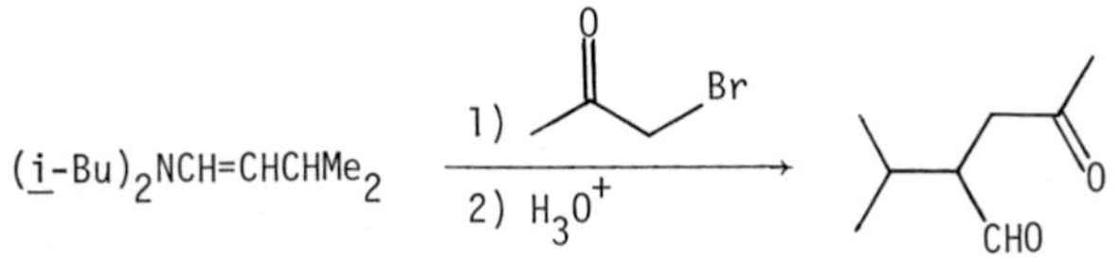

(i-Bu)2NCH=CHCHMe2
1)
Br
2) H3O+
O
CHO

Tetr Lett (1974) 3253

PhSO2CH2CH2CH
1) BuLi
2)
COOMe
3) Al-Hg
O

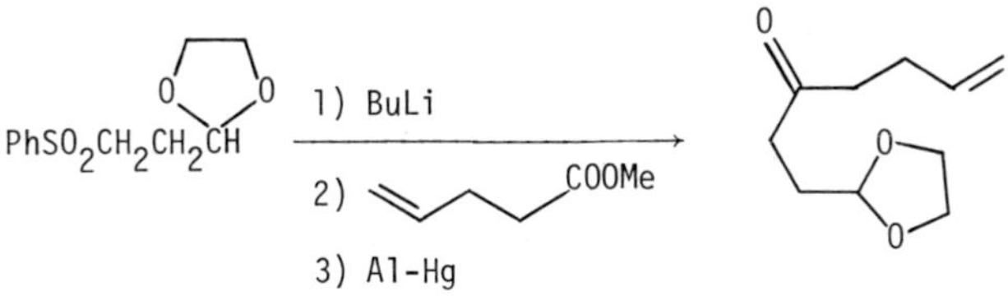

~70%

Tetr Lett (1975) 1397

Section 340   Aldehyde - Nitrile

KO⌇⌇CN + HS⌇⌇SH $\xrightarrow[\text{glyme}]{BF_3 \cdot Et_2O}$ [dithiolane]—CH$_2$CN   66%

Synth Comm (1974) **4** 331

[cyclohexanone with NOH and OH groups] $\xrightarrow[\text{pyridine/THF}]{Cl_2P=N-PCl_2 \text{ triazine}}$ $NC-(CH_2)_4CHO$ 85%

Synthesis (1975) 665

Section 341   Aldehyde - Olefin

For the oxidation of allylic alcohols to olefinic aldehydes see Section 48, Vol. 1 (Aldehydes from Alcohols)

Br—[furan]—CHO $\xrightarrow[\text{2) } H_3O^+]{\text{1) } Ph_3\overset{+}{P}\overset{-}{C}H\overset{O-}{C}H \text{ (dioxolane)}}$ Br—[furan]—CH=CH—CHO   58%

JCS Perkin I (1974) 37

BrCH$_2$—[dioxolane] $\xrightarrow{\begin{array}{l}\text{1) Ph}_3\text{P}\\\text{2) LiOMe}\\\text{3) C}_6\text{H}_{13}\text{CHO}\\\text{4) H}^+\end{array}}$ C$_6$H$_{13}$CH=CH-CHO   96%

JCS Perkin I (1974) 37

[pyridine-3-CHO] + Ph$_3$P=CHCHO $\longrightarrow$ [pyridine-3-CH=CHCHO]   56%

Angew Int Ed (1975) 14 1486

[cyclohexanone] $\xrightarrow{\begin{array}{l}\text{1) Me Si}\overset{\text{Me}}{\underset{\text{Li H}}{\text{C}}}\text{—C=N-\underline{t}-Bu}\\\text{2) oxalic acid, H}_2\text{O}\end{array}}$ [cyclohexylidene]=C(Me)-CHO   90%

Tetr Lett (1976) 7

[cycloheptanone-2-CHO] $\xrightarrow{\begin{array}{l}\text{1) PhSeNEt}_2\\\text{2) H}_2\text{O}_2\end{array}}$ [cycloheptenone-2-CHO]   67%

JOC (1975) 40 3313

PhSO$_2$C-CH$_2$-CH(dioxolane) , (CH$_2$)$_7$Me $\xrightarrow{\begin{array}{l}\text{1) aq HOAc}\\\text{2) base}\end{array}}$ Me(CH$_2$)$_7$CH=CH-CHO   85%

Tetr Lett (1975) 1007

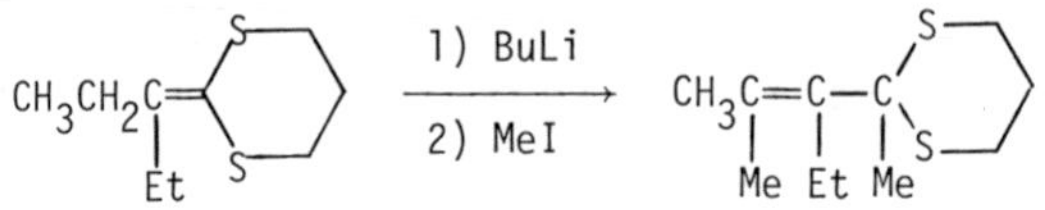

Tetr Lett (1974) 3171

$MeCH=CH-CH(OMe)_2$ + $CH_2=CHCH=CHOSiMe_3$ $\xrightarrow[\text{2) } H_2O]{\text{1) } TiCl_4,\ THF}$

86%

$MeCH=CHCH(OMe)CH_2CH=CHCHO$

Chem Lett (1975) 319

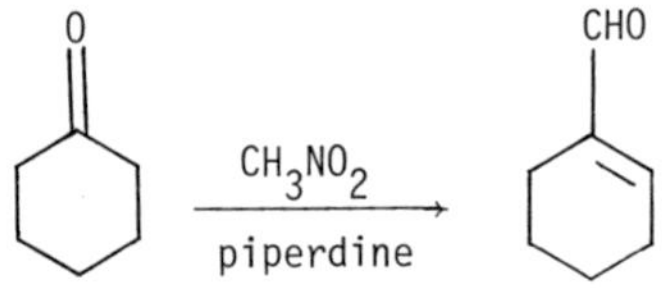

Synthesis (1974) 196

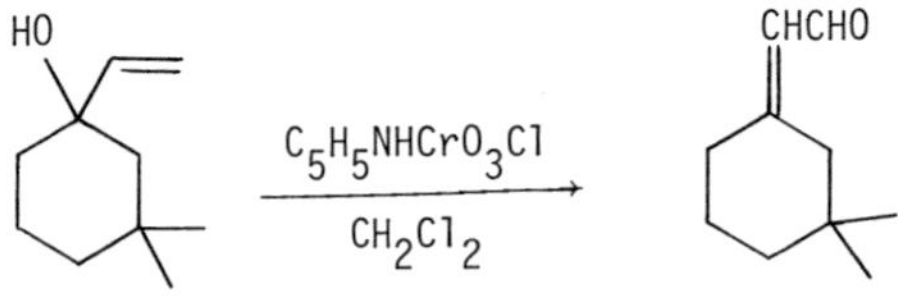

(mixture of E and Z)

Synth Comm (1976) 6 469

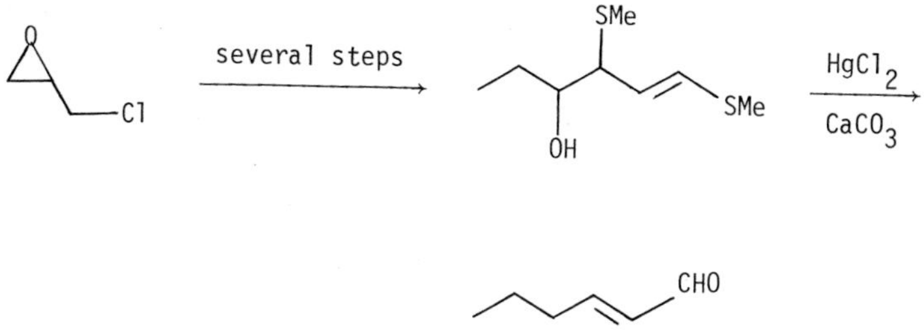

58%

Org Synth (1974) <u>54</u> 19

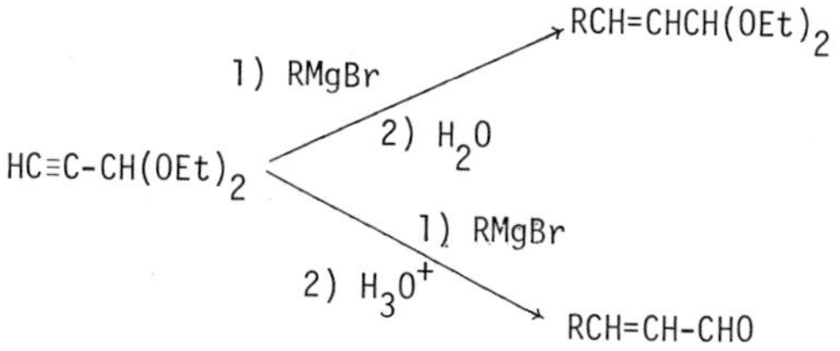

∼80%

R = Ph, <u>t</u>-Bu, <u>n</u>-Bu, cyclopentyl

Rec Trav Chim (1976) <u>95</u> 66

A 1,3-alkylative carbonyl transposition, e.g.:

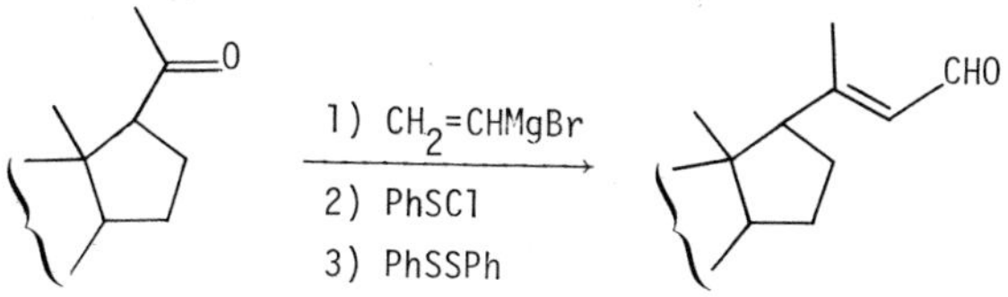

1) CH₂=CHMgBr

2) PhSCl

3) PhSSPh

44%

JACS (1975) <u>97</u> 4018

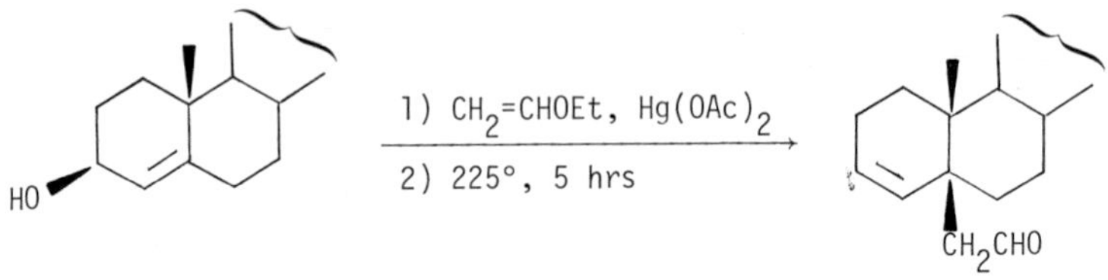

1) $CH_2=CHOEt$, $Hg(OAc)_2$

2) 225°, 5 hrs

CH_2CHO

50%

Org Synth (1974) 54 71

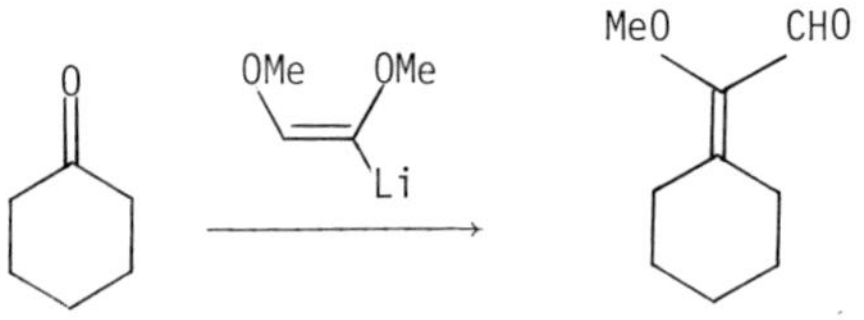

MeO   OMe

Li

MeO   CHO

68%

Synth Comm (1976) 6 119

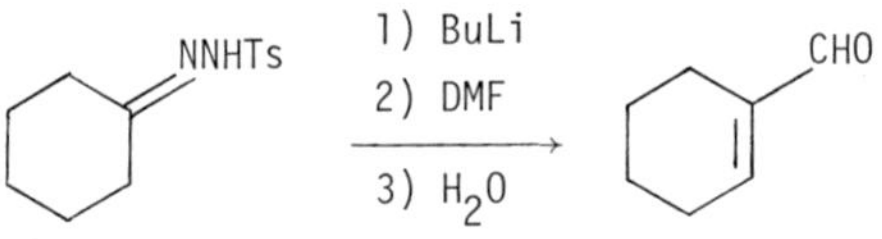

NNHTs

1) BuLi

2) DMF

3) H_2O

CHO

54%

Tetr Lett (1976) 2287

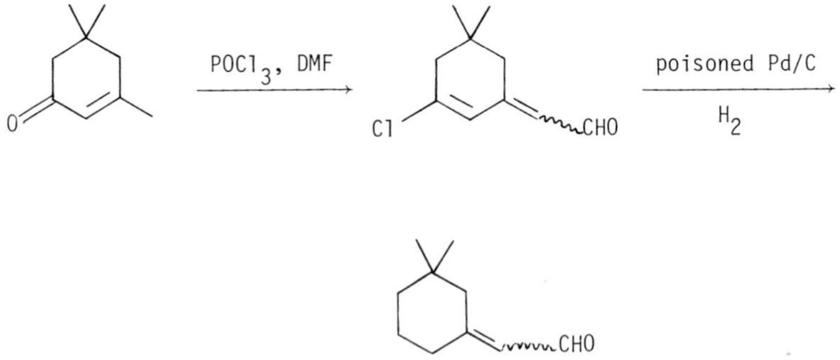

$POCl_3$, DMF

Cl   CHO

poisoned Pd/C

H_2

80%

CHO

Rec Trav Chim (1976) 95 308

~90%

Helv Chim Acta (1976) 59 1233

$\underline{t}$-BuOCH$_2$CH=CH$_2$ $\xrightarrow[\text{2) MeSSMe}]{\text{1) }\underline{\text{sec}}\text{-BuLi}}$ $\underline{t}$-BuOCHCH=CH$_2$ 72%
$\qquad\qquad\qquad\qquad\qquad\qquad\qquad\quad$ |
$\qquad\qquad\qquad\qquad\qquad\qquad\qquad\quad$ SMe

JACS (1974) 96 5560

H-C≡C-CH(OEt)$_2$ $\xrightarrow[\text{2) H}_2\text{O}]{\text{1) LiCu-}\underline{n}\text{-Bu}_2}$

94%

Tetr Lett (1976) 2313

PhCOCH$_3$ + HC-CH=N-NHPh $\xrightarrow[\text{2) hydrolysis}]{\text{1) EtOK, EtOH}}$ PhCOCH=CH-C-H ~80%

Chem Ber (1975) 108 88; 1756

PhCHO $\xrightarrow{\qquad\qquad}$

76%

Synth Comm (1976) 6 119

OHC-CHO $\xrightarrow[\text{DMF, 80°}]{\text{Ph}_3\text{P=CHCHO}}$ OHC-(CH=CH)$_2$CHO   77%

Chem Ber (1974) <u>107</u> 710

1) ▷—MgBr

2) DMSO, BF$_3$·Et$_2$O

Chem Lett (1976) 1297

PhCHCH$_2$CH=CHCHO $\xrightarrow[\text{CH}_2\text{Cl}_2]{\text{DBU}}$ PhCH=CH-CH=CHCHO   79%
|
OEt

Chem Lett (1975) 1167

+ PhCHO $\xrightarrow{h\nu}$ ... 35% ... $\xrightarrow{\text{TsOH}}$ ...   94%

JCS Chem Comm (1975) 206

OMe

+ $\xrightarrow[\substack{\text{2) MsCl} \\ \text{3) CH}_3\text{OH} \\ \text{4) H}_3\text{O}^+}]{}$ 95%

Li

Tetr Lett (1975) 3685

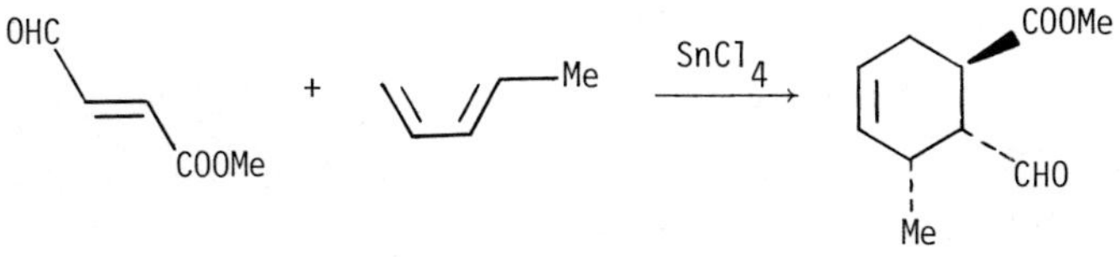

Can J Chem (1976) 54 3304

Also via: β-Hydroxyaldehydes (Section 324)

Section 342   Amide - Amide

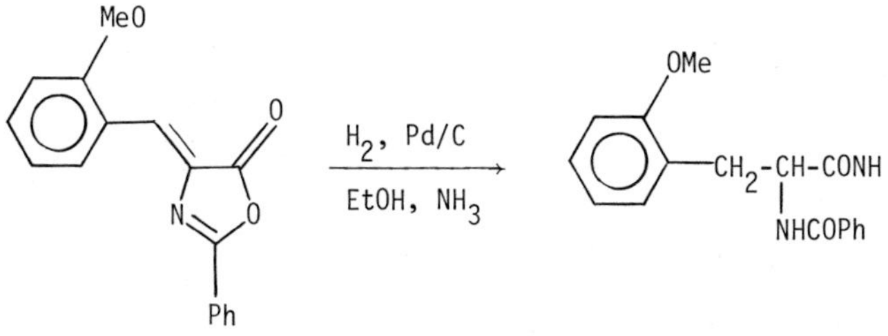

Synth Comm (1976) 6 227

Also via: Dicarboxylic acids (Section 312)
            Diamines (Section 350)

Section 343   Amide - Amine

No additional examples

Section 344   Amide - Ester

Can J Chem (1975) 52 3563

$Cl_3CCH=NCOOEt$ + MeMgX $\xrightarrow[Et_2O]{}$ $Cl_3CCHNHCOOEt$ 56%

                                                      |
                                                      Me

JCS Perkin I (1975) 2511

$CH_3CH_2(CH_2)n-COOR$ + e^- $\xrightarrow{CH_3CN}$ $CH_3CH-(CH_2)nCOOR$ 40-70%

                                                              $NHCOCH_3$

Electrolytic ω-1 oxidation

JOC (1974) 39 369

                                                                   53%

Chem Ber (1975) 108 500

R-CH-COOEt with NH$_2$ group   +   (epoxide) $\longrightarrow$ HO NHCHCOOEt (with R substituent)

$$\xrightarrow{\begin{array}{c}1)\ \underline{t}\text{-BuOCl}\\ \hline 2)\ Ag^+\end{array}}$$

47-70%

β-lactam with N-CHCOOEt and R substituent

R = alkyl, benzyl

JOC (1975) 40 1505

Ph-NCO   +   Me$_2$C-OH (with allyl group) $\xrightarrow{\text{Sn octanoate}_2}$ Ph-N-C-O-CMe$_2$ (with H, O, and allyl)

90%

Can J Chem (1976) 54 24

Section 345   Amide - Epoxide

No additional examples

Section 346   Amide - Halide

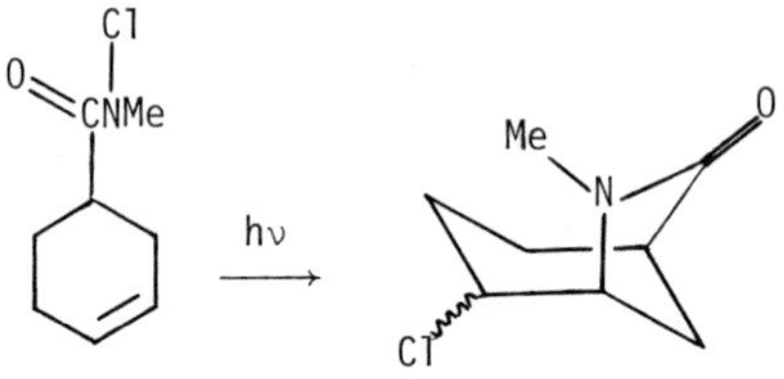

JOC (1975) <u>40</u> 1287

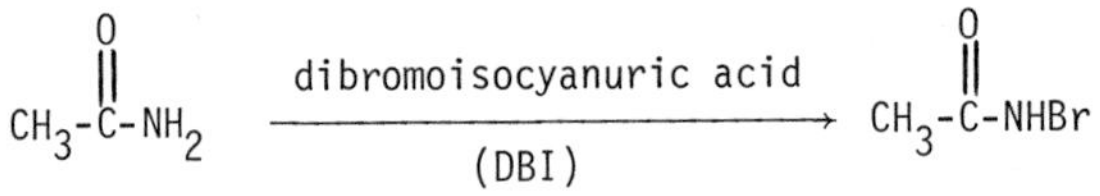

69%

$$CH_3-\overset{O}{\overset{\|}{C}}-NH_2 \xrightarrow[\text{(DBI)}]{\text{dibromoisocyanuric acid}} CH_3-\overset{O}{\overset{\|}{C}}-NHBr$$

Monatshefte (1975) <u>106</u> 611

83%

Section 347   Amide - Ketone

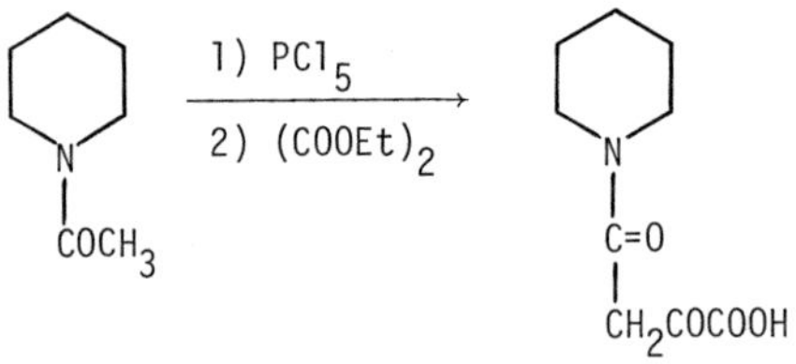

Experientia (1976) <u>32</u> 1491

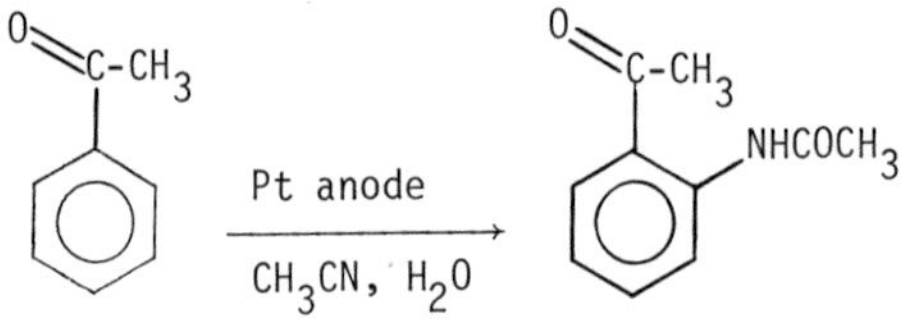

45%

JCS Chem Comm (1975) 262

70%

Also via: Ketoacids (Section 320)

Section 348   Amide - Nitrile

No additional examples

Section 349   Amide - Olefin

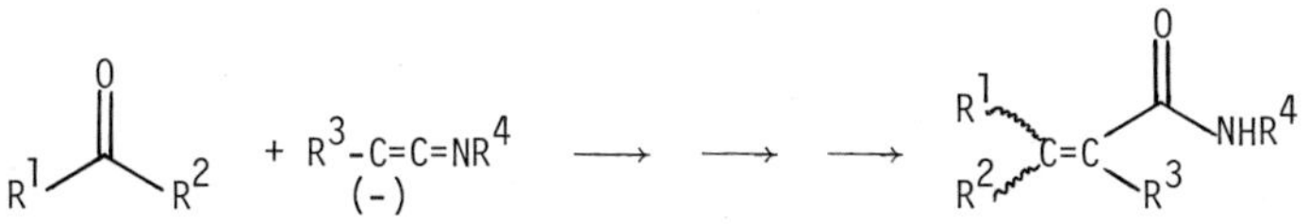

Liebigs Ann Chem (1974) 1655   60-70%

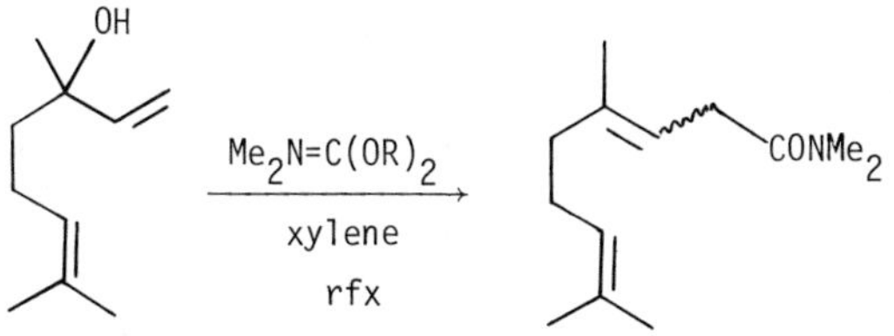

JACS (1974) 96 5563   75%

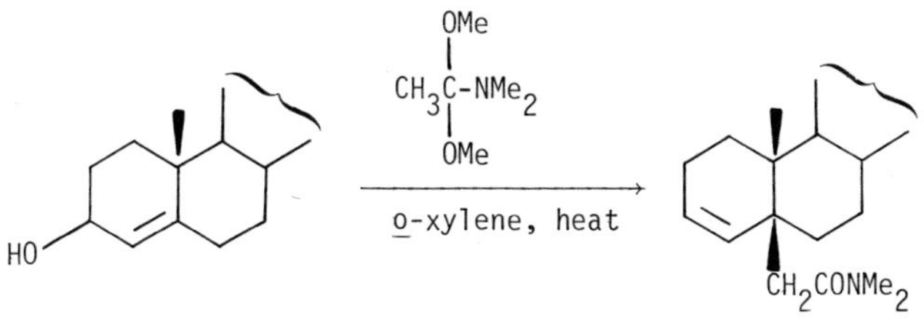

Org Synth (1974) 54 77   65%

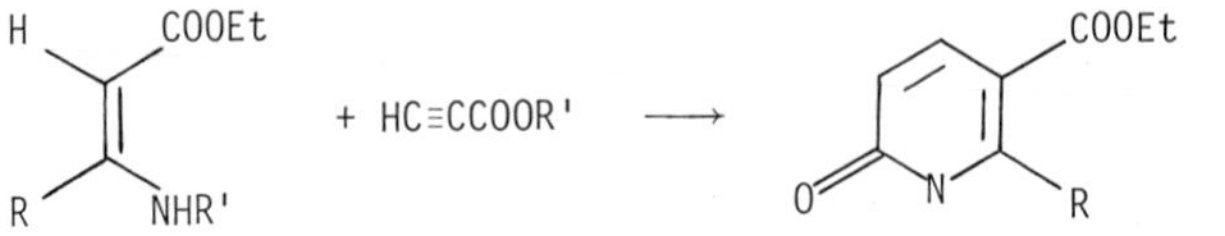

Tetr Lett (1974) 30 623

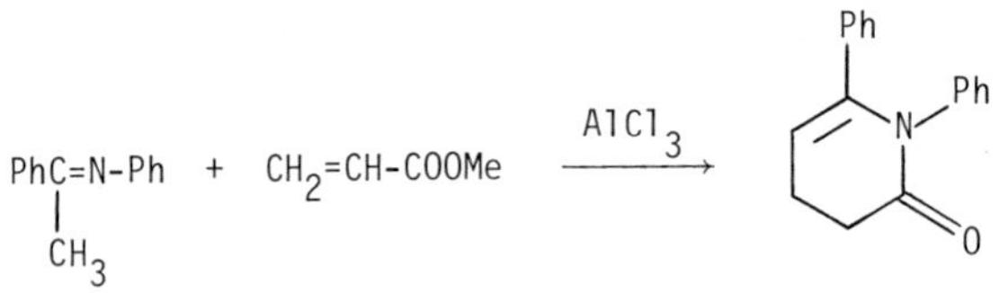

60-85%

Tetr Lett (1974) 977

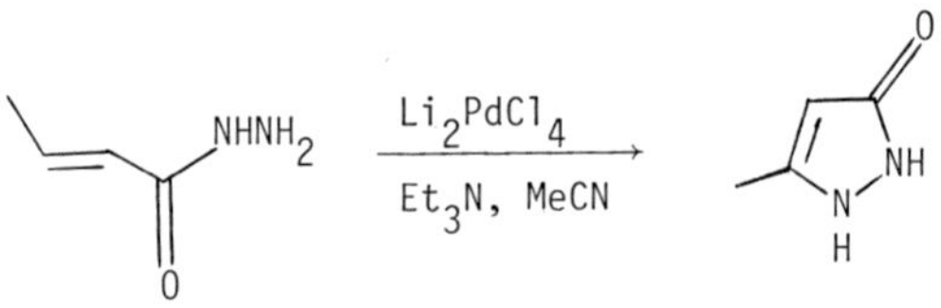

85%

Chem and Ind(1976) 1032

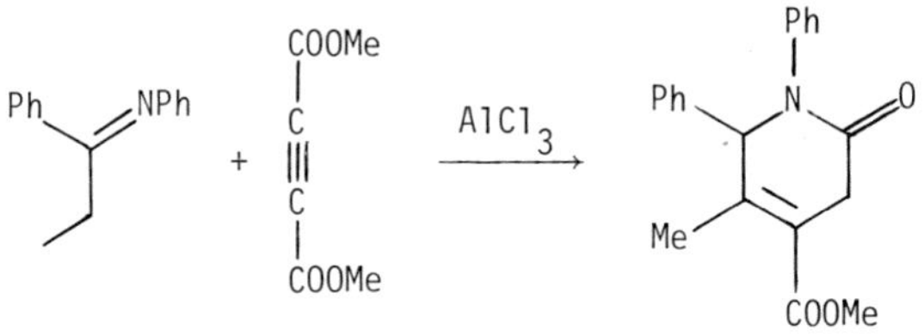

40%

Synthesis (1975) 643

81%

Also via: Olefinic acids (Section 322)

Section 350   Amine - Amine

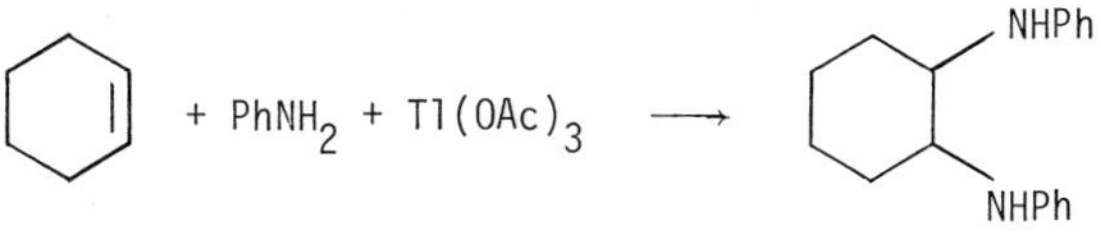

~70%

Synthesis (1974) 504

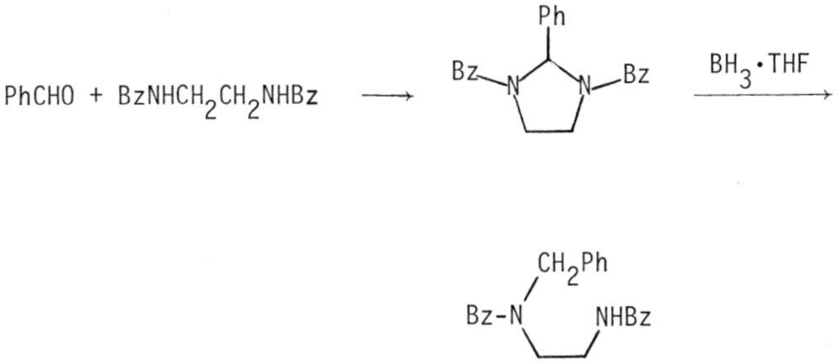

PhCHO + BzNHCH$_2$CH$_2$NHBz   →

BH$_3$·THF

57%

JOC (1975) 40 558

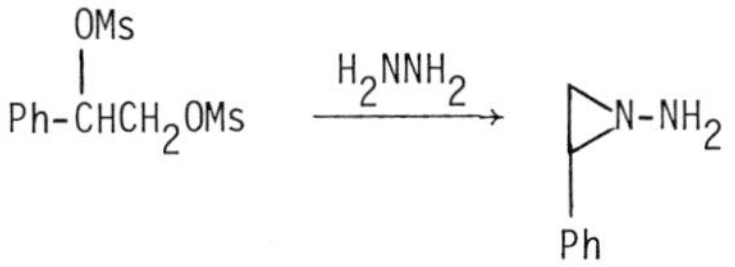

~80%

Org Synth (1976) 55 114

Section 351   Amine - Ester

Direct esterification of aminoacids and aminoacid halides. Other preparations
of esters of aminoacids. Esters of aminoalcohols.

1)

2) LDA

$NH_2CH_2CO_2\text{-}\underline{t}\text{-}Bu$

3) BzBr

4) H_3O^+

$H_2NCHCO_2\text{-}\underline{t}\text{-}Bu$ (with Bz substituent)

D 72% ee

79%

JCS Chem Comm (1976) 136

1) ^-OMe

2) HCl

3) NH_3

~60%

Angew Int Ed (1976) <u>15</u> 294

$MeC{\equiv}CNEt_2 + CH_2{=}C\begin{smallmatrix}Cl\\COOMe\end{smallmatrix}$ $\xrightarrow{Et_2O}$

89%

Tetrahedron (1974) <u>30</u> 3481
Tetr Lett (1974) 14<u>47</u> with cyclohexenone

+ $\begin{smallmatrix}CHCOOMe\\||\\C(COOMe)_2\end{smallmatrix}$ $\longrightarrow$

84%

Chem Comm (1974) 587

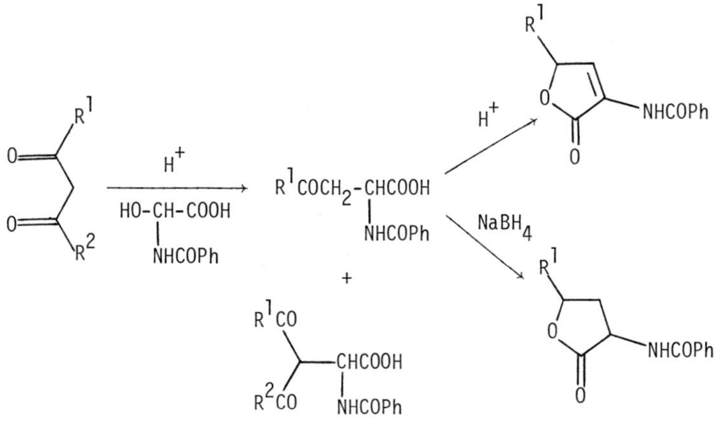

R's = Me, Ph, OMe

JCS Chem Comm (1975) 905

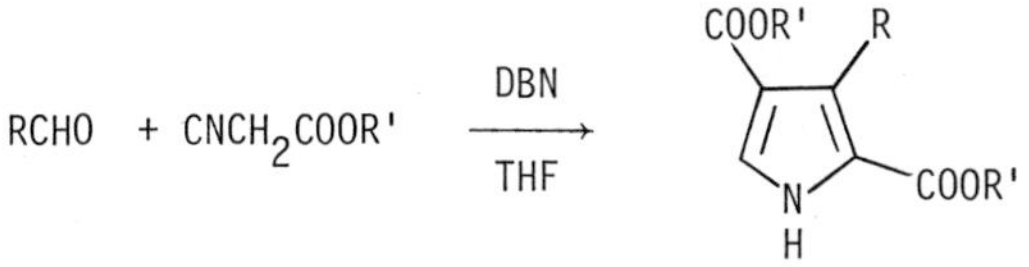

$$RCHO + CNCH_2COOR' \xrightarrow[THF]{DBN}$$

50-71%

JOC (1974) $\underline{39}$ 1980

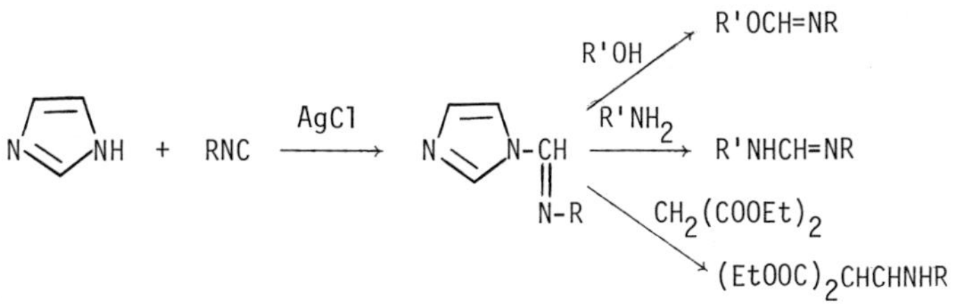

R'OCH=NR   65%

R'NHCH=NR   90%

$(EtOOC)_2CHCHNHR$ 80%

Tetr Lett (1974) 1283

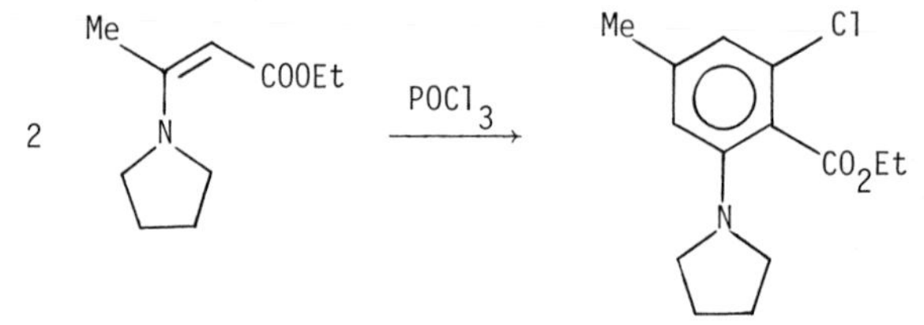

2

$$\text{POCl}_3$$

89%

Angew Int Ed (1976) 15 498

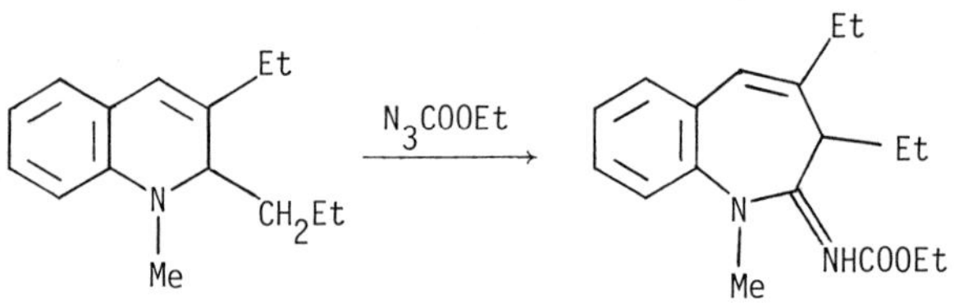

$$\text{N}_3\text{COOEt}$$

83%

JOC (1976) 41 195

Section 352   Amine - Ether

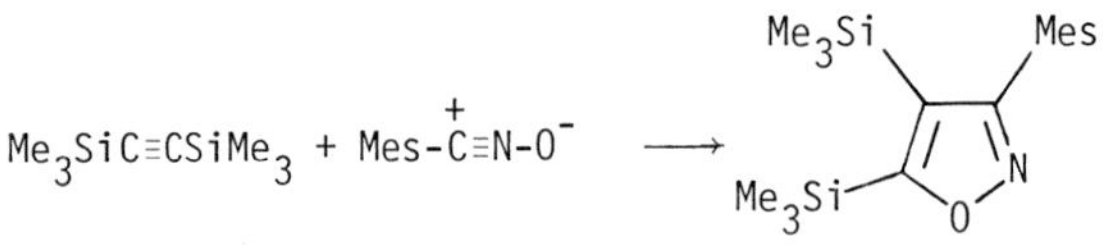

$$\text{Me}_3\text{SiC}\equiv\text{CSiMe}_3 + \text{Mes-}\overset{+}{\text{C}}\equiv\text{N-O}^- \longrightarrow$$

Chem Ber (1974) 107 3717

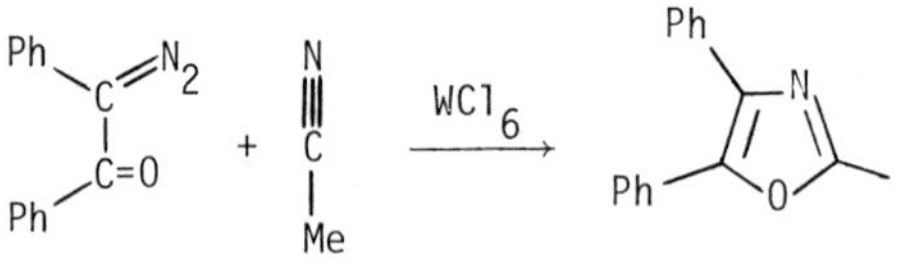

$$\text{WCl}_6$$

65%

Tetr Lett (1974) 1531

Section 353   Amine - Halide

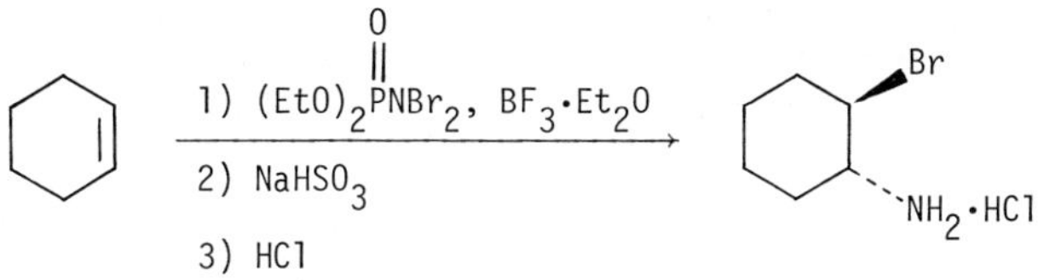

Angew Int Ed (1976) 15 302

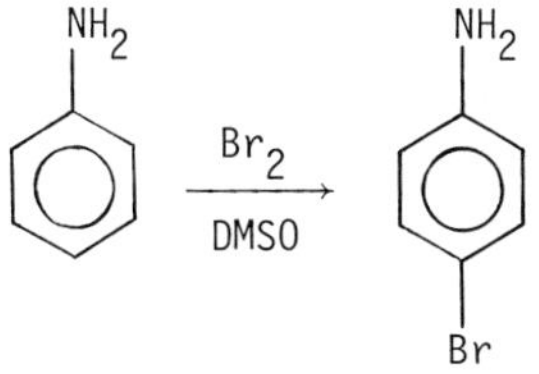

74%
(+6% ortho)

JOC (1975) 40 1867

$ArNH_2 \xrightarrow{Cl_2O} ArNHCl$

Aust J Chem (1976) 29 367

Section 354   Amine - Ketone

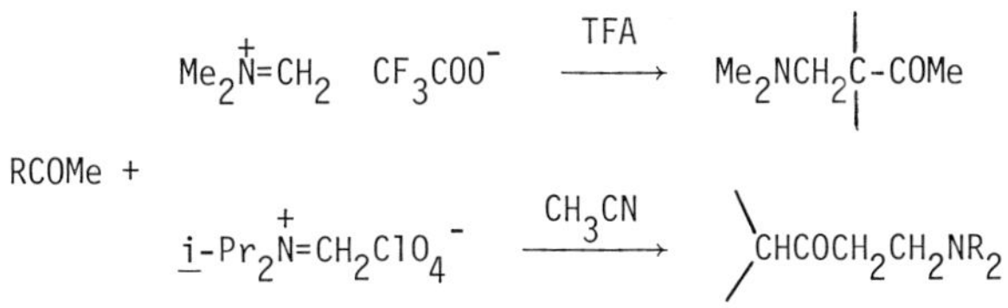

JCS Chem Comm (1974) 253

PhCCH=CHPh $\xrightarrow[\text{Cu}]{\text{PhN=S=O}}$ PhCCH$_2$CHPh

with O on first carbonyl, and NHPh substituent

JOC (1976) 41 3811

$$\text{Ph-}\underset{\text{O}}{\text{C}}\text{-CHMe}_2 \xrightarrow[\text{2) OH}^-]{\text{1) H}_2\text{C=NMe}_2{}^+ \text{ Cl}^-} \text{Ph-}\underset{\text{O}}{\text{C}}\text{-CMe}_2\text{-CH}_2\text{NMe}_2$$

78%

Angew Int Ed (1976) 15 239

$$\xrightarrow[\text{2) RX}]{\text{1) BuLi, -60°}}$$

Tetr Lett (1974) 3963

$$\xrightarrow[\text{W}_2 \text{ Raney Nickel}]{\text{CH}_3\text{OH, H}_2\text{CO}}$$

(aromatic ring with COCH$_3$, NO$_2$, OH substituents converted to COCH$_3$, NMe$_2$, OH)

~70%

Indian J Chem (1975) 13 33

$$\underset{\underset{\text{NH}_2}{|}}{\overset{\overset{\text{H}}{|}}{\text{RCCOOR'}}} + \underline{p}\text{-NO}_2\text{C}_6\text{H}_5\text{N}_2 \xrightarrow[\text{Et}_3\text{N}]{\text{DMF}} \text{R'O-}\underset{\text{O}}{\text{C}}\text{-}\underset{\text{R}}{\text{C}}\text{=N}_2$$

40-90%

JCS Chem Comm (1974) 558

$$CH_3COCH_2COCH_3 + tosN_3 \xrightarrow[\substack{\text{two phase} \\ \text{system}}]{R_4N^+X} CH_3CO\overset{\overset{\displaystyle N_2}{\|}}{C}\text{-}COCH_3$$

80%

Synthesis (1974) 347
Tetr Lett (1974) 1391 for polymer
   bound TsN$_3$

Section 355   Amine - Nitrile

PhCH=NCH$_2$Ph $\xrightarrow[\text{2) H}_2\text{O}]{\text{1) Me}_3\text{SiCN, ZnI}_2}$

$$\underset{\underset{\displaystyle NHCH_2Ph}{|}}{\overset{\overset{\displaystyle CN}{|}}{PhCH}}$$

98%

Chem Lett (1975) 331

NC(CH$_2$)$_5$CN $\xrightarrow[\text{KH, THF}]{}$

83%

Synthesis (1975) 326

Section 356   Amine - Olefin

$\xrightarrow[]{\text{NH}_4\text{Br}}$

89%

Synthesis (1976) 545

~50%

Synthesis (1976) 401

$$Me-\overset{O}{\underset{||}{C}}-\underset{\underset{Cl}{|}}{CH}-CN \xrightarrow{ArNH_2} Me-C=C-CN \\ \underset{ArNH}{} \underset{Cl}{}$$

~60%

Chem Ber (1976) 109 2908

$$R-\overset{O}{\underset{||}{C}}-CHCl-CHO \xrightarrow{HNR'_2} R-\overset{O}{\underset{||}{C}}-CCl=CHNR'_2$$

35-48%

R = Ph, OEt

R'$_2$ = Me$_2$, Et$_2$, (CH$_2$)$_5$

Comptes Rendus C (1976) 282 935

$$PhN\underset{\underset{Me}{|}}{CH_2}CH=CH_2 \xrightarrow[2)\ MeI]{1)\ BuLi} PhN\underset{\underset{Me}{|}}{-CH}=CHCH_3$$

75%

Synthesis (1974) 672

Tetr Lett (1976) 3919

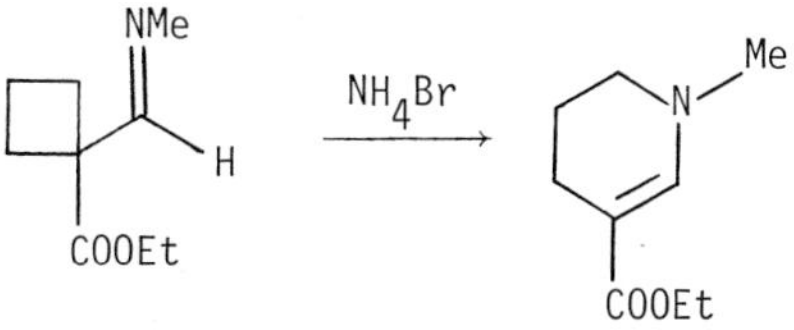

66%

JCS Chem Comm (1975) 682

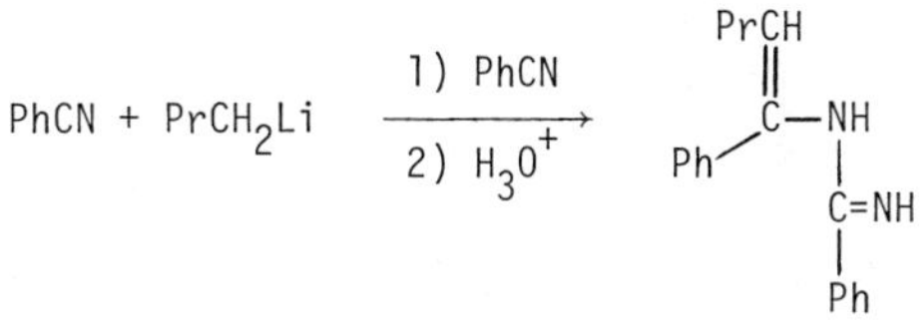

PhCN + PrCH$_2$Li $\xrightarrow{\text{1) PhCN} \atop \text{2) H}_3\text{O}^+}$

63%

Tetr Lett (1976) 147

Na[Et$_3$BC≡C-CH$_3$] + [Me$_2$NCH$_2$]Br$^-$

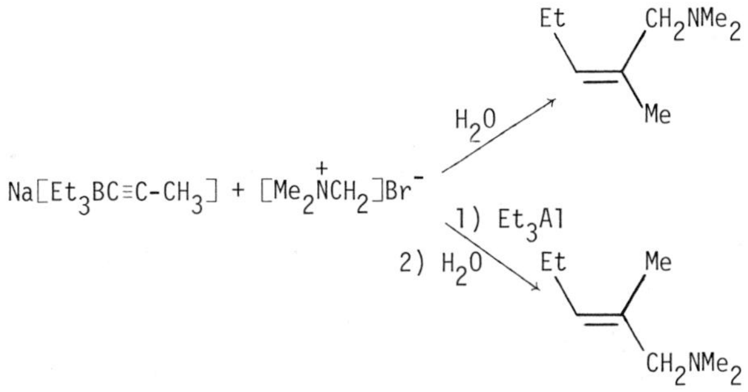

Chem Ber (1975) 108 395

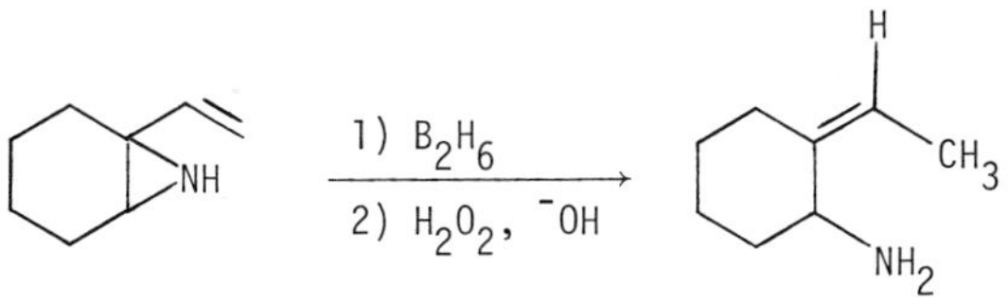

51%

Tetr Lett (1976) 757

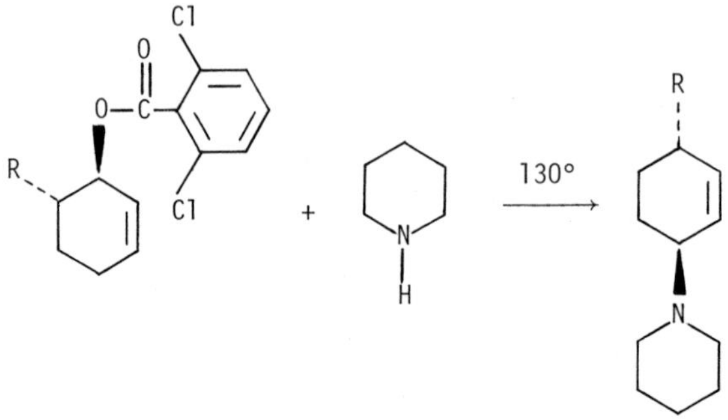

Cl
O
O—C
R
Cl
+
N
H
130°
R
N

60-73%

JACS (1976) 98 2901

Me
Me
+
Cl⁻
N
NaNH₂
N
(CH₂)₃

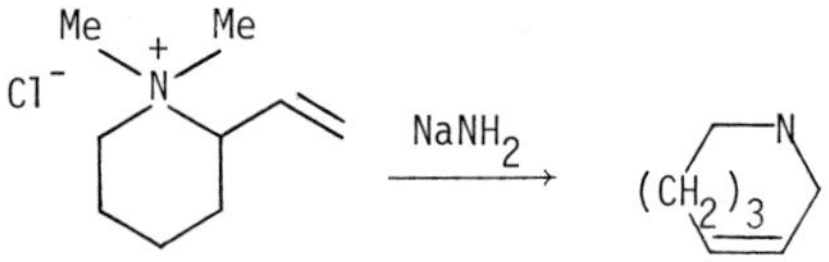

94%

Comptes Rendus C (1976) 282 1003

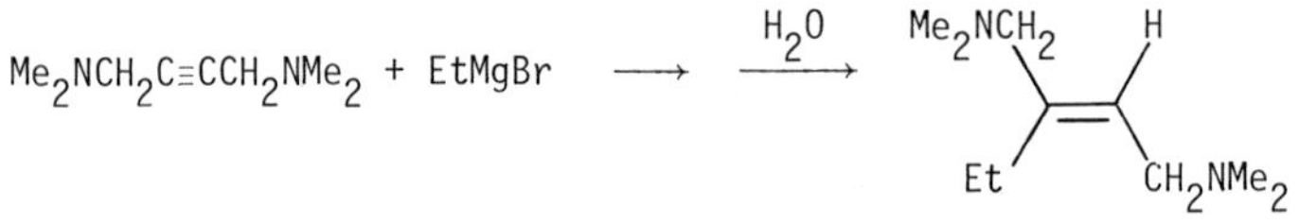

Me₂NCH₂C≡CCH₂NMe₂ + EtMgBr
H₂O
Me₂NCH₂
H
Et
CH₂NMe₂

73%

J Organometal Chem (1975) 86 297

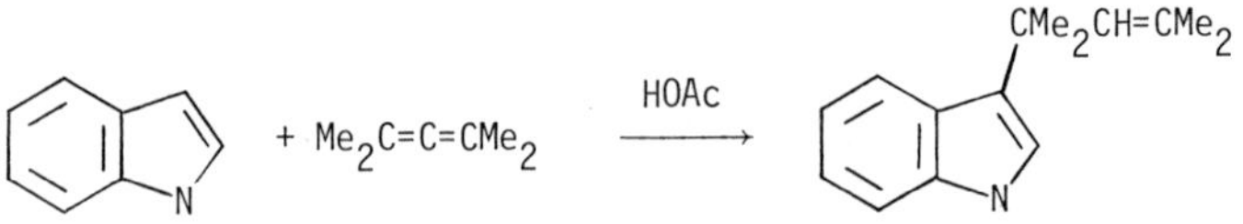

N
+ Me₂C=C=CMe₂
HOAc
CMe₂CH=CMe₂
N

100%

Indian J Chem (1975) 13 1124

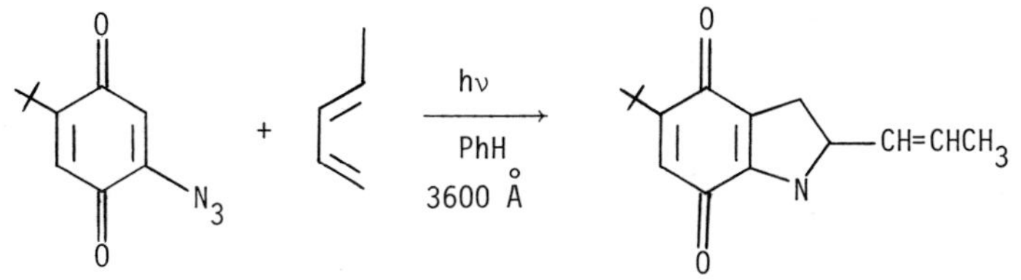

JOC (1974) <u>39</u> 781

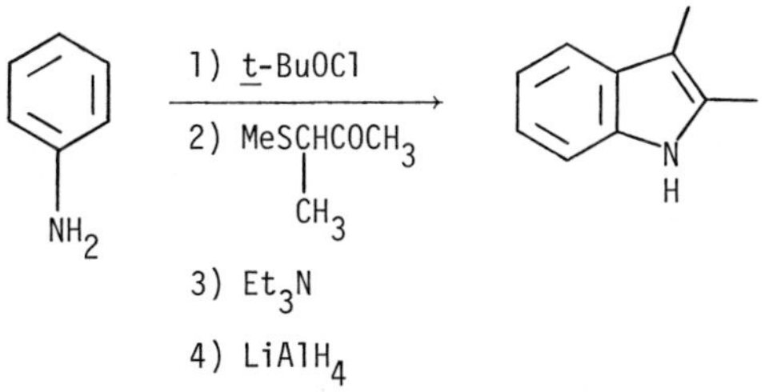

JCS Chem Comm (1974) 201

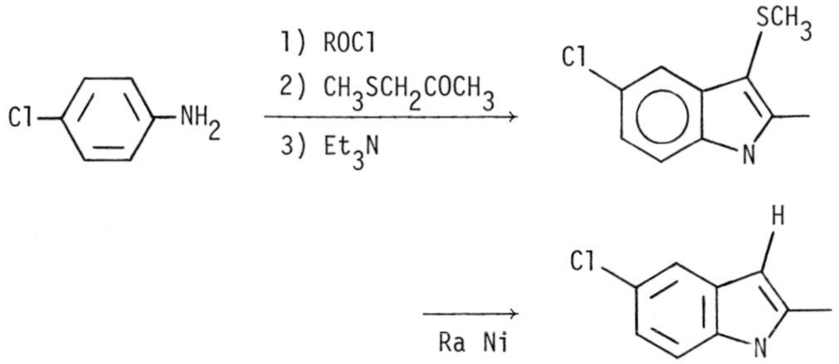

JACS (1974) <u>96</u> 5495
          <u>96</u> 5508 (oxindole)
          <u>96</u> 5512

Section 357   Ester - Ester

$$50\text{-}90\%$$

JACS (1974) 96 7091

$$Me_2CHCO_2Et \xrightarrow[\text{2) } I_2, \text{ THF}]{\text{1) LDA, THF}} \begin{array}{c} Me_2CCOOEt \\ | \\ Me_2C\text{-}COOEt \end{array}$$

85%

Synthesis (1975) 396
Chem Lett (1975) 621 (electrolytic
                          coupling)

Also via: Dicarboxylic acids (Section 312)
            Hydroxyesters (Section 327)
            Diols (Section 323)

Section 358   Ester - Ether, Epoxide

42%

Angew Int Ed (1976) 15 436

PhCHO + BrCH$_2$C(=O)-S-$\underline{t}$-Bu $\xrightarrow[\text{DMF}]{\text{NaH}}$ PhCH(—O—)C(=O)-S-$\underline{t}$-Bu   67%

JOC (1974) $\underline{39}$ 2938

Section 359   Ester - Halide

$\xrightarrow{\text{MgBr}_2,\ \text{H}_2\text{O}_2}$ 100%

Chem Pharm Bull (1976) $\underline{24}$ 820

1) Me$_3$SiC(SMe)$_2$Li
2) NBS, CH$_3$CN, CH$_3$OH   76%

Synthesis (1976) 121

+ CHBr$_2$CN $\xrightarrow[\text{i-PrOK}]{\text{i-PrOH}}$ 74%

Comptes Rendus (1974) $\underline{278}$ 77

$$\underset{\substack{|\\ Me}}{Me-CH=C-COOMe} \quad \xrightarrow[\text{2) PhHgCBrCl}_2]{\text{1) Et}_3\text{SiH, Rh(I)}} \quad \underset{\substack{|\\ Me}}{Me-CH_2-C=C(Cl)COOMe} \qquad 92\%$$

Chem Pharm Bull (1976) <u>24</u> 1957

70-95%

R = Me, Ar, CF$_3$, H

Tetr Lett (1976) 3661

high yield

JCS Perkin I (1974) 1858
JCS Perkin I (1974) 1864
(Iodolactonization)

$$\underset{\substack{|\\ OH}}{RCH-(CH_2)_nCH_2OH} + Ph_3PBr_2 \quad \xrightarrow[-20°]{DMF} \quad \underset{\substack{|\\ OCHO}}{RCH-(CH_2)_n-CH_2Br} \qquad 50\text{-}90\%$$

Differentiation between OH's in diols

1° → Br

2° → formate ester

Tetr Lett (1974) 913

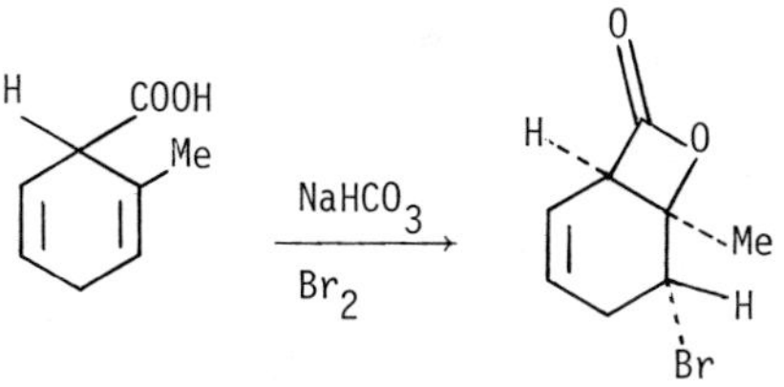

JOC (1975) 40 2843

Also via: Haloacids (Section 319)
           Halohydrins (Section 329)

Section 360   Ester - Ketone

$$RCOCH_2COR' \xrightarrow{SeO_2}$$

or

$$\xrightarrow[\text{2) DMSO}]{\text{1) Br}_2/\text{Py}} RCOCOCOR' \qquad 70\%$$

Helv Chim Acta (1974) 57 2201

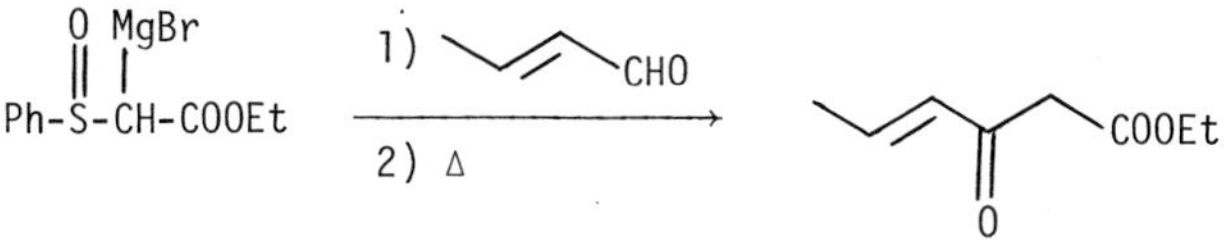

Tetr Lett (1975) 2841

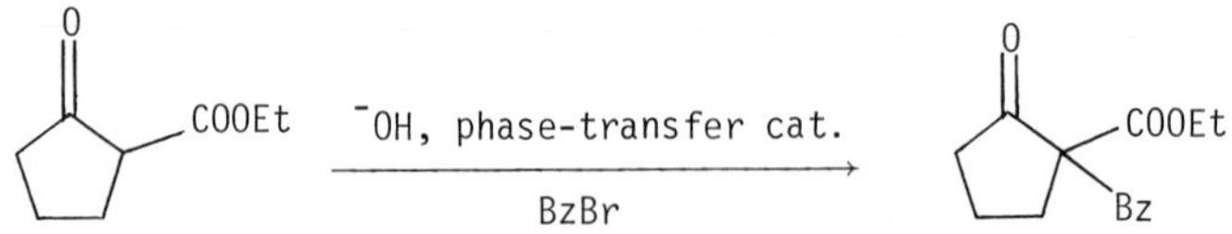

95%

Chem Lett (1976) 1259

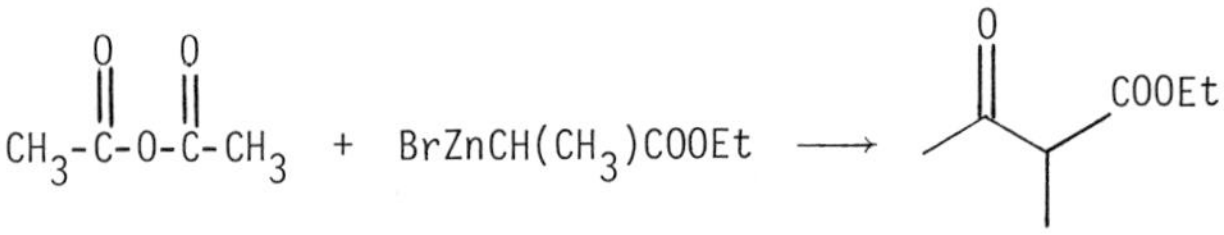

$CH_3-\overset{\overset{\displaystyle O}{\|}}{C}-O-\overset{\overset{\displaystyle O}{\|}}{C}-CH_3$ + BrZnCH(CH$_3$)COOEt $\longrightarrow$

87%

JOC USSR (1975) 11 2360

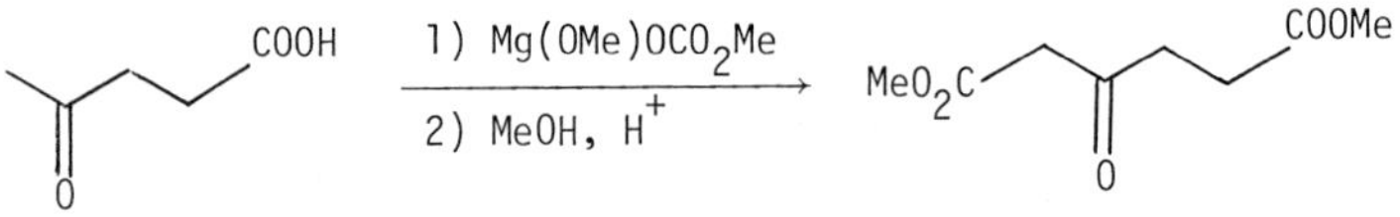

1) Mg(OMe)OCO$_2$Me
2) MeOH, H$^+$

92%

JOC (1974) 39 3144

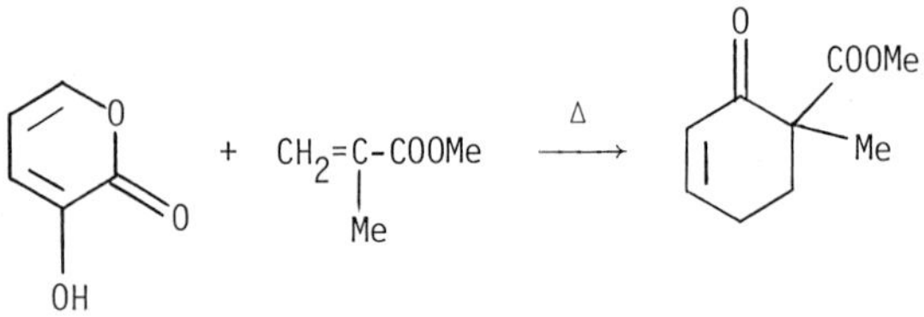

+ CH$_2$=C-COOMe
      |
      Me

$\overset{\Delta}{\longrightarrow}$

85%

Tetr Lett (1975) 2389

PhH
CH$_2$=CHCH$_2$Br
R$_3$NMe$^+$Cl$^-$

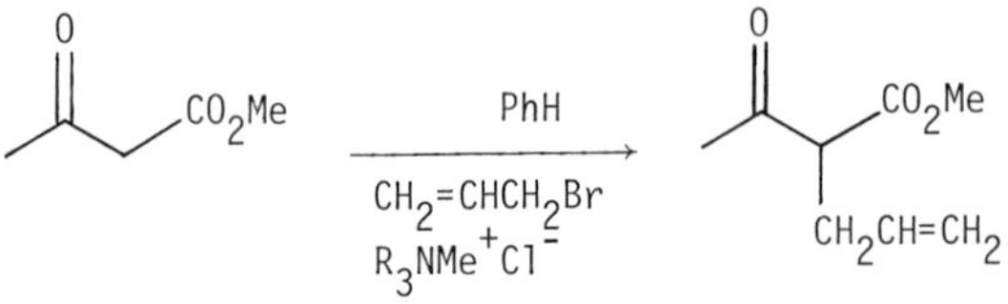

85%

JOC (1974) 39 3271

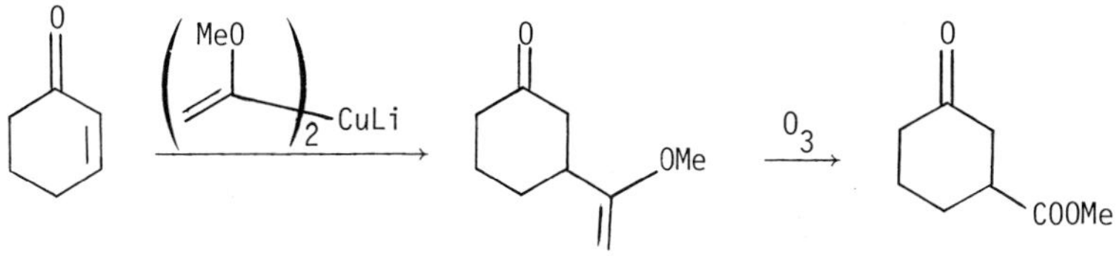

JACS (1975) 97 3822

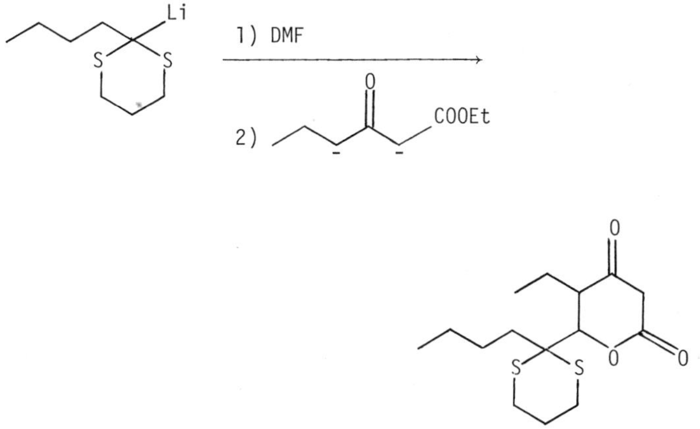

74%

Angew Int Ed (1974) 13 77

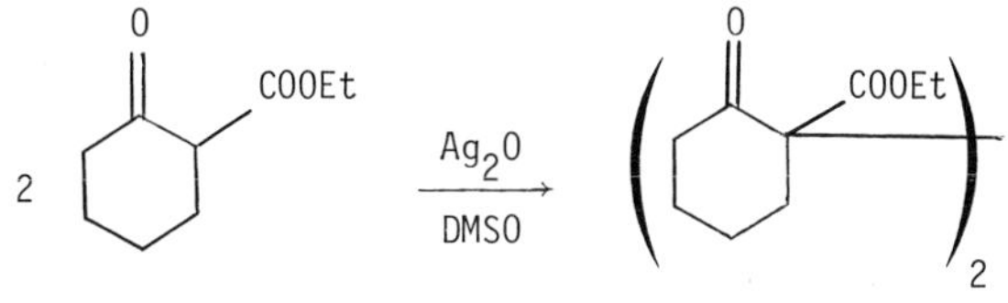

90%

Synth Comm (1976) 6 429

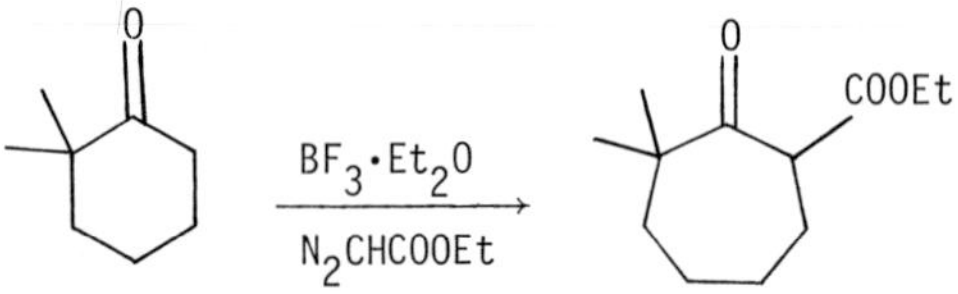

96%

Synth Comm (1975) 5 125

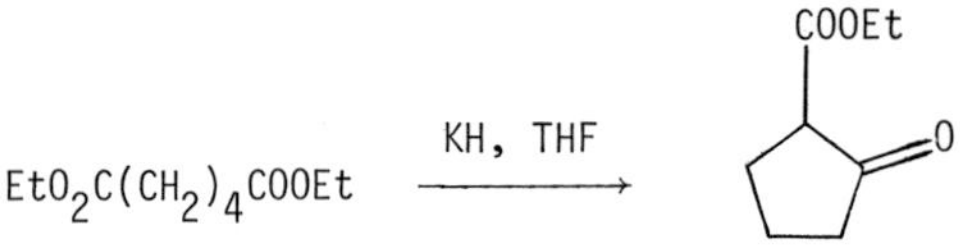

EtO$_2$C(CH$_2$)$_4$COOEt

KH, THF

95%

Synthesis (1975) 326

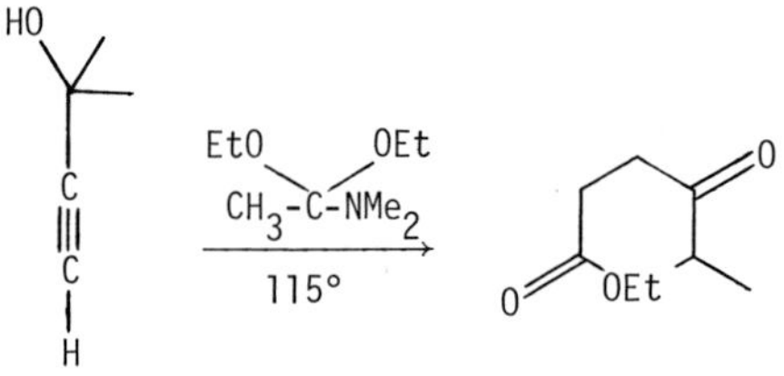

73%

Tetr Lett (1976) 341

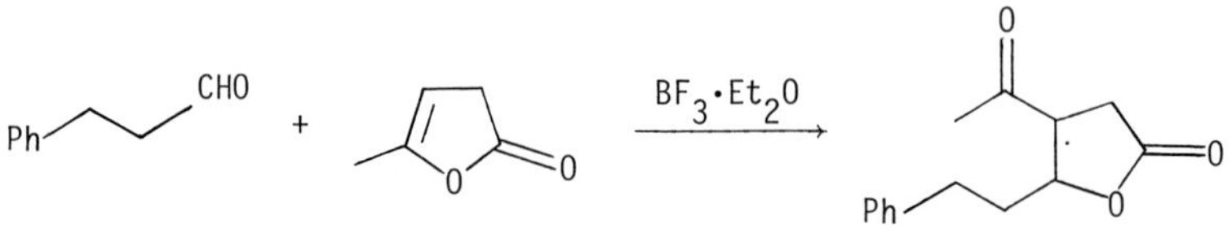

83%

Bull Chem Soc Japan (1976) 49 1055

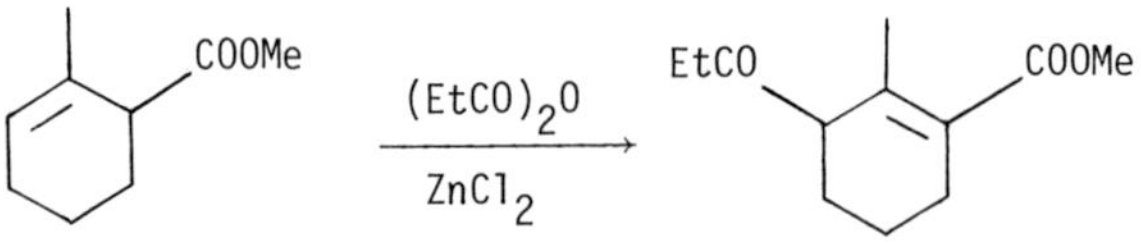

55%

Bull Soc Chim Fr (1975) 274

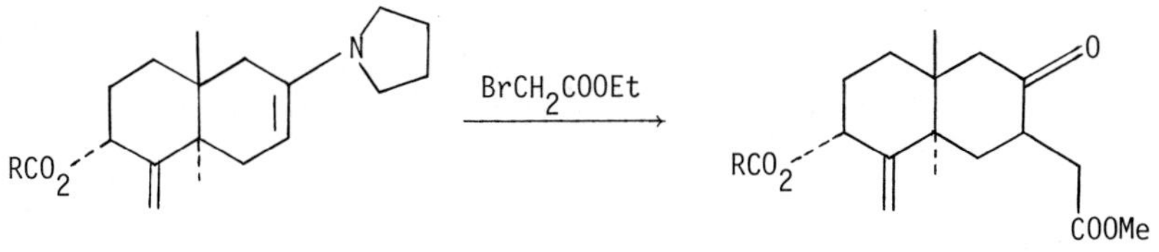

JACS (1974) 96 8102
JOC (1974) 39 1873

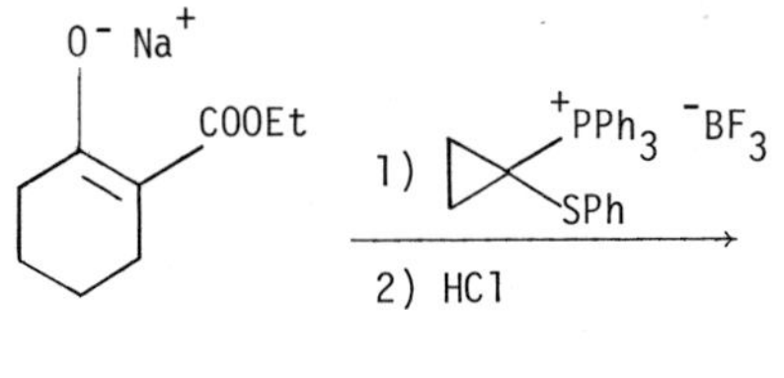

~70%

Tetr Lett (1975) 4531

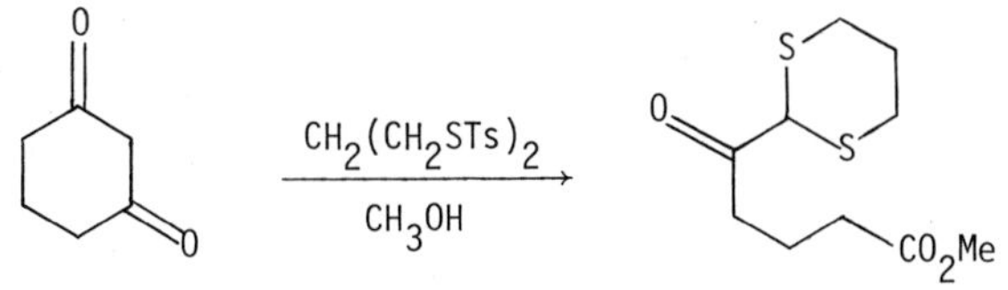

93%

Tetr Lett (1975) 3841

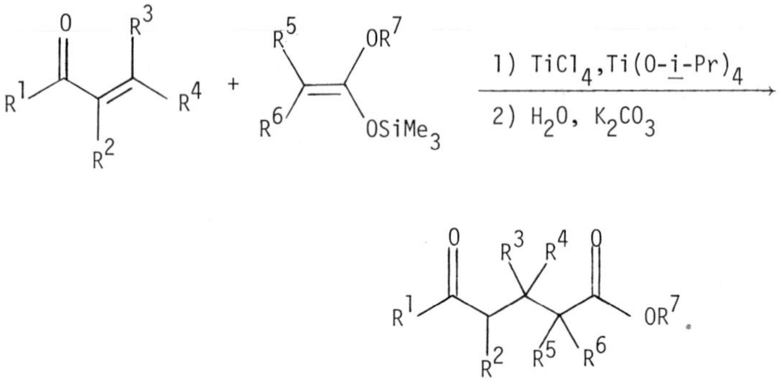

70-98%

R's = alkyl, Ph

Chem Lett (1976) 163

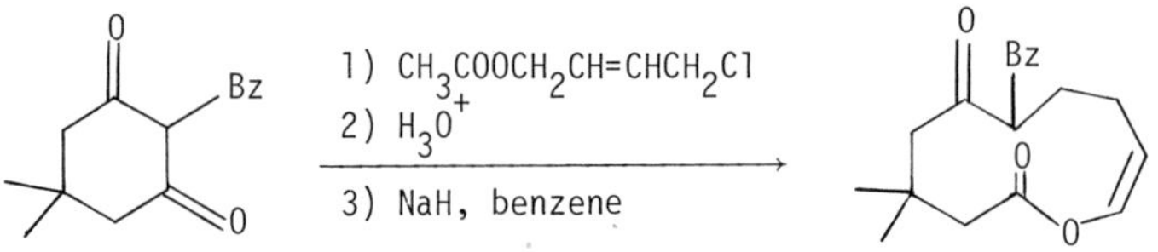

50%

Synthesis (1976) 110

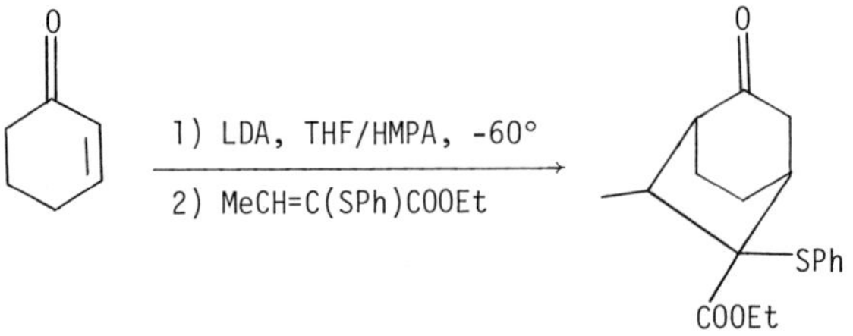

70%

Bull Chem Soc Japan (1975) 48 3769

$CH_2{=}CH(CH_2)_3COOMe$ +

$\xrightarrow{\text{ROOR}}$

70%

Coll Czech (1976) **41** 1698

OSiMe$_3$

$\xrightarrow[\text{2) Et}_3\text{NHF}]{\text{1) lead tetrabenzoate}}$

91%

JOC (1976) **41** 1673

$\xrightarrow[\text{2) py, (CH}_3\text{CO)}_2\text{O}]{\text{1) H}_2\text{NOH}}$

~80%

Bull Soc Chim France (1976) 642

OSiMe$_3$

$\xrightarrow[\text{2) H}_3\text{O}^+]{\text{1) Pb(OAc)}_4}$

95%

Synth Comm (1976) **6** 59

PhCHOCPh + MeCHO $\xrightarrow[\underline{t}\text{-BuOH}]{\text{K}_2\text{CO}_3}$ PhCOCHMe

|
CN

|
OC-Ph
||
O

59%

Tetr Lett (1975) 903

Also via: Ketoacids (Section 320)
           Hydroxyketones (Section 330)

Section 361   Ester - Nitrile

α, β and higher cyanoesters. Esters of cyanohydrins

$$Me_2C=C\begin{smallmatrix}CN\\COOEt\end{smallmatrix} \quad \xrightarrow[2)\ H_3O^+]{1)\ EtMgBr} \quad Me_2\overset{CN}{\underset{Et}{C}}-CHCOOEt$$

62%

Angew Int Ed (1975) <u>14</u> 629

$$NCCH_2COOEt + PhCH_2Br \quad \xrightarrow[160°]{LiCl,\ HMPA} \quad NCCH(CH_2Ph)COOEt + NCC(CH_2Ph)_2COOEt$$

30%   64%

Chem Lett (1975) 1149

$$\overset{COOEt}{\underset{}{MeCHN\equiv C}} \quad \xrightarrow[2)\ PhCH_2Br]{1)\ \underline{t}\text{-}BuOK} \quad \overset{COOEt}{\underset{CH_2Ph}{Me-C-N\equiv C}}$$

79%

Chem Ber (1975) <u>108</u> 1580

$$(CH_3C\!\!\!-\!\!\!\overset{O}{\overset{\|}{}}\!\!\!)_2O + Ph_3P=CHCOOEt \quad \xrightarrow{HCN} \quad$$

58%

cis-trans mixture

Coll Czech (1976) <u>41</u> 2040

85%

JCS Perkin I (1976) 1926

75%

Org Synth (1976) 55 57

Also via: Cyanoacids (Section 321)
            Hydroxynitriles (Section 331)

Section 362   Ester - Olefin

For allylic acetoxylation see section 116, Vol. 1 and 2 (Esters from Hydrides)

Review: "Methods for the Synthesis of α-Methylene Lactones"

                                Synthesis (1975) 67

Review: "Synthesis of α-Methylene-γ-Butyrolactones"

                                Synth Comm (1975) 5 245

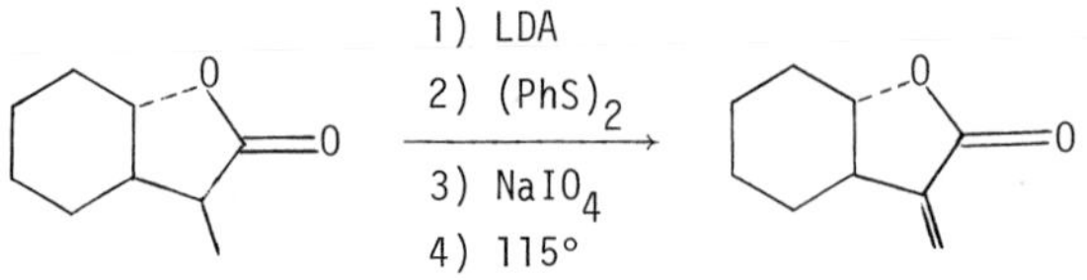

1) LDA
2) $(PhS)_2$
3) $NaIO_4$
4) 115°

72%

Tetr Lett (1974) 1100
JCS Chem Comm (1974) 135

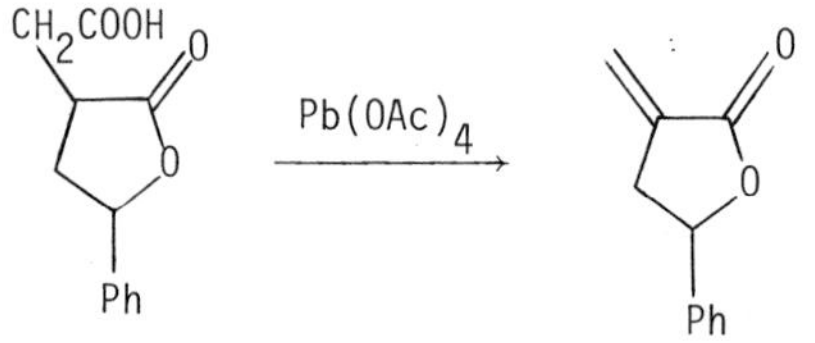

CH_2COOH
$Pb(OAc)_4$
Ph
Ph

50%

Tetr Lett (1974) 339
Synth Comm (1974) 4 133

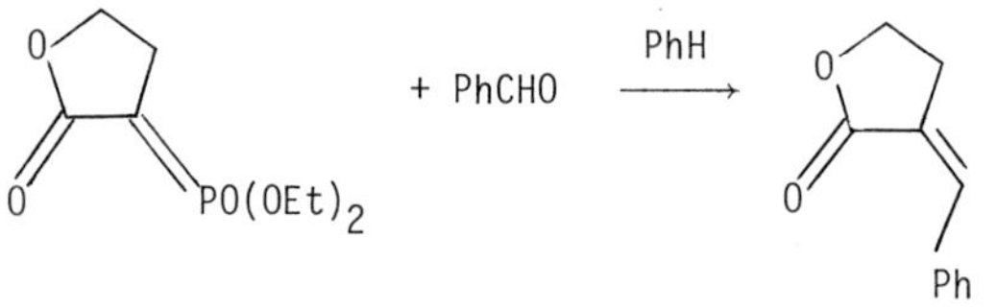

$PO(OEt)_2$
+ PhCHO
PhH
Ph

100%

JOC (1974) 39 3236

PPh_3
paraformaldehyde
THF

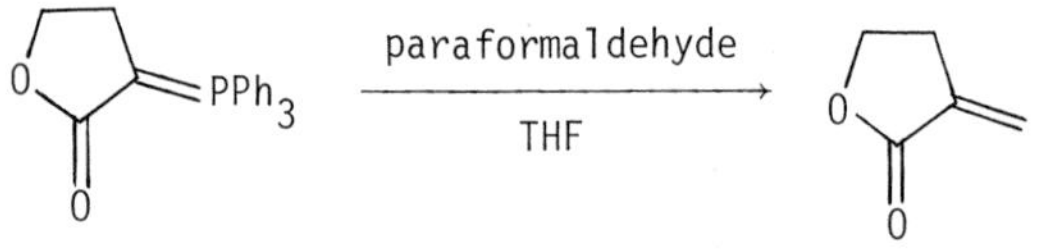

98%

JOC (1974) 39 1958

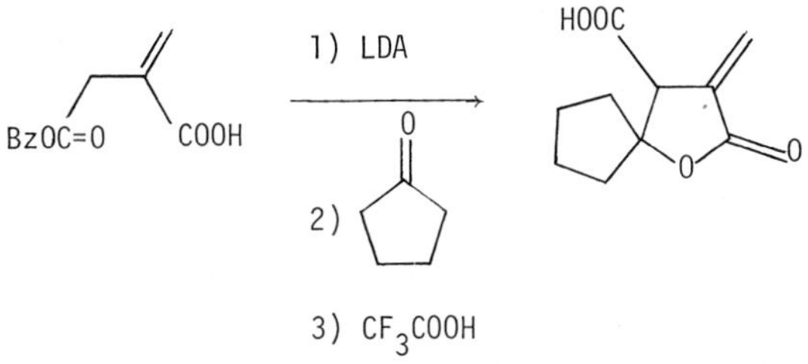

Tetr Lett (1975) 4099

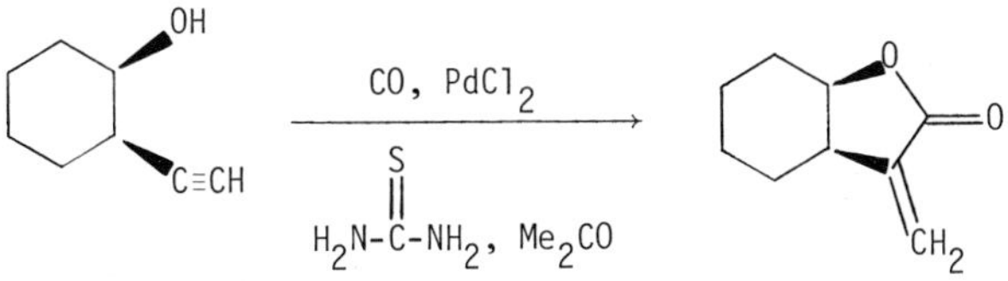

~50%

JCS Chem Comm (1976) 907

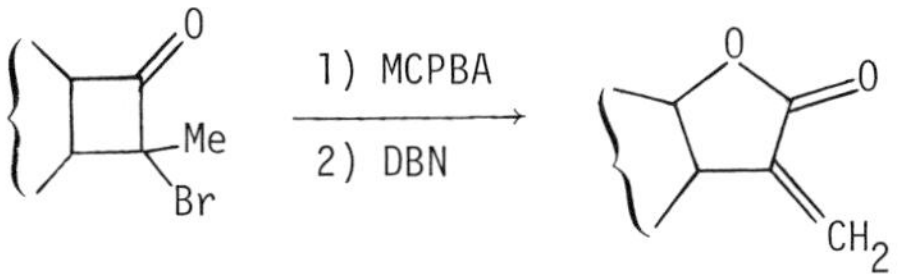

62%

JCS Chem Comm (1975) 887

60-80%

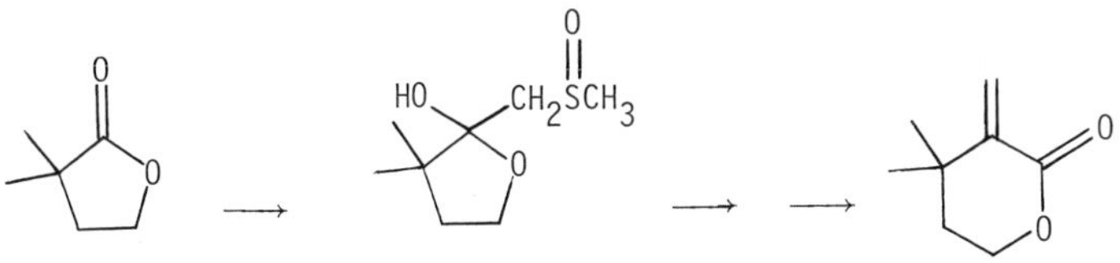

JACS (1975) 97 7182

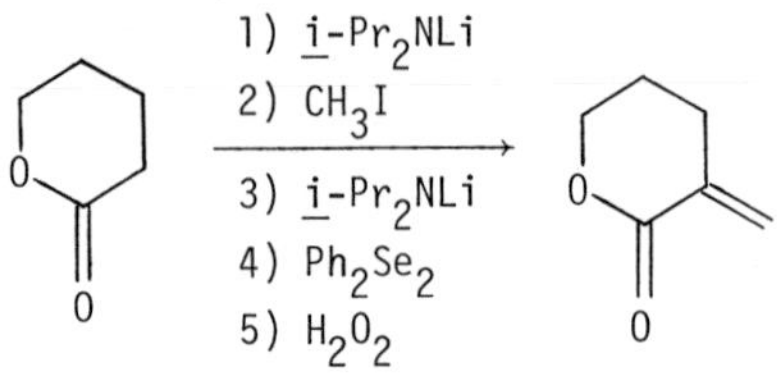

1) i-Pr$_2$NLi
2) CH$_3$I
3) i-Pr$_2$NLi
4) Ph$_2$Se$_2$
5) H$_2$O$_2$

~60%

JOC (1974) 39 121

H
N
Me
Et
Br
COOEt
Me
Et

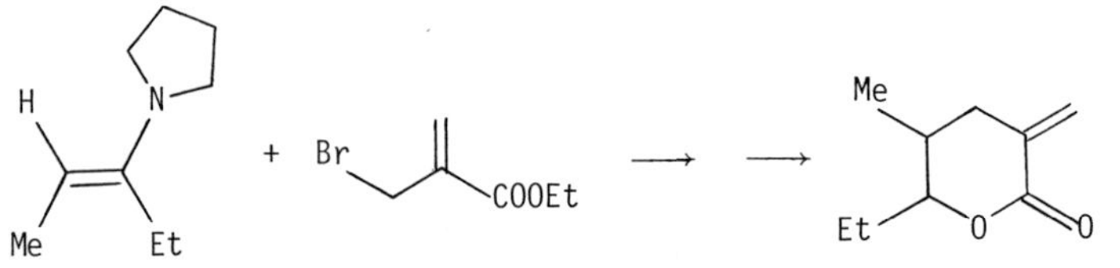

Chem Ber (1974) 107 2853

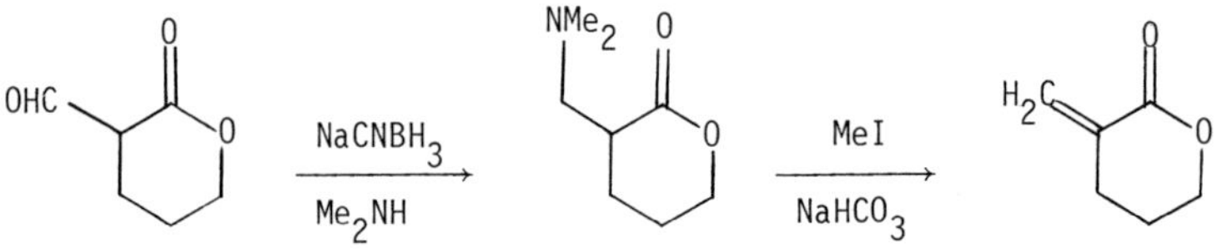

OHC
NaCNBH$_3$
Me$_2$NH
NMe$_2$
MeI
NaHCO$_3$
H$_2$C

52%

JOC (1975) 40 3474

COOEt
1) LDA
2) PhSSPh
COOEt
SPh
1) oxidation
2) Δ
COOEt

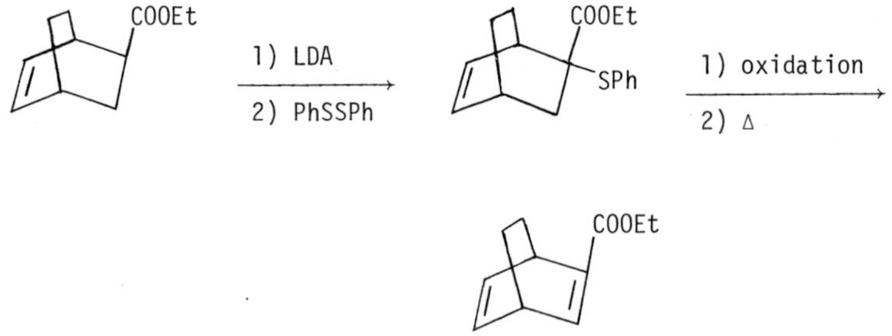

76%

JACS (1976) 98 4887

TMSCHCOOEt + [cyclohexanone] → [ethoxycarbonylmethylenecyclohexane]　　95%
|
Li

JACS (1974) 96 1620
Tetr Lett (1974) 1403

[R-CO-CH(R')-COOEt] $\xrightarrow[\text{2) }^{+PPh_3}\text{cyclopropylidene-COOEt}]{\text{1) NaH-THF}}$ [cyclopentene product with COOEt, R', R, COOEt]　　50-90%

JACS (1974) 96 1607

PhCHO + [MeO$_2$C, NHPh, H, COOR alkene] → [furanone with MeO$_2$C, NHPh, Ph, O]　　60%

Z Chem (1976) 16 13

[CH$_3$-CO-CH$_2$-COOMe] $\xrightarrow{CH_2Br_2}$ [cyclohexenone with COOMe, CH$_2$COOMe]　　62%

JACS (1974) 96 1082

HOCH$_2$C≡CH $\xrightarrow[\text{2) CO, Li}_2\text{PdCl}_4]{\text{1) HgCl}_2\text{, H}_2\text{O}}$ [chloro furanone]　　96%

Tetr Lett (1976) 4661

COOEt
1) base
2) PhSex
3) H_2O_2
COOEt

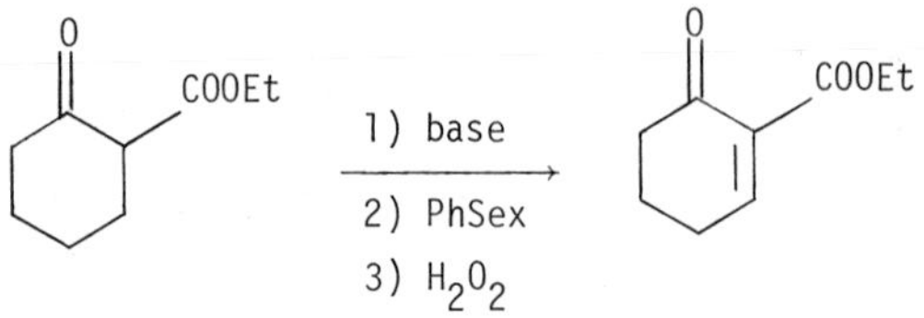

JOC (1974) 39 2135
39 120
Tetr Lett (1974) 2279

1) PhSeH
2) BuLi
3) CH_3OCOCl
PhSe
COOMe
H_2O_2
COOMe

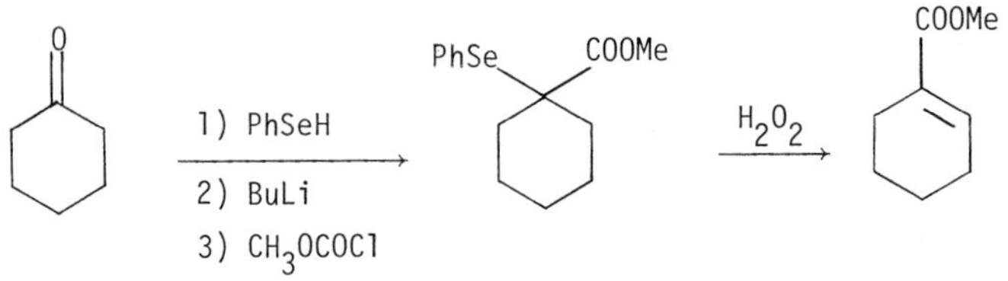

~50%

Tetr Lett (1976) 453

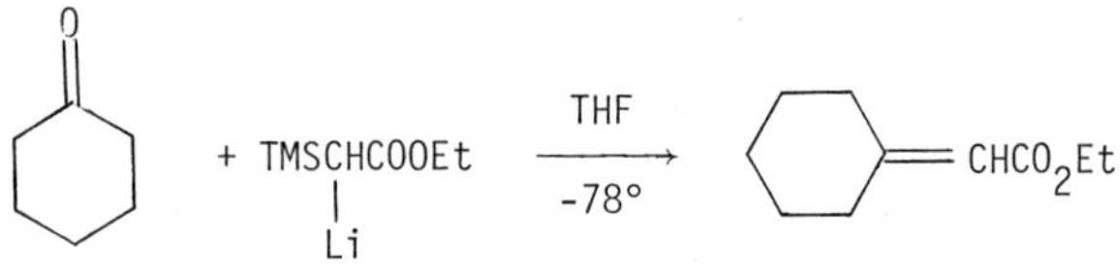

+ TMSCHCOOEt
Li
THF
-78°
$CHCO_2Et$

95%

Bull Chem Soc Jap (1974) 47 2529

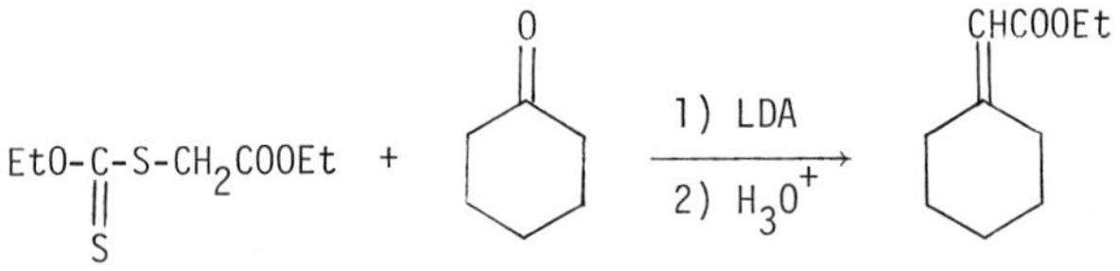

EtO-C-S-CH_2COOEt
S
+
1) LDA
2) H_3O^+
CHCOOEt

78%

Chem Lett (1976) 917

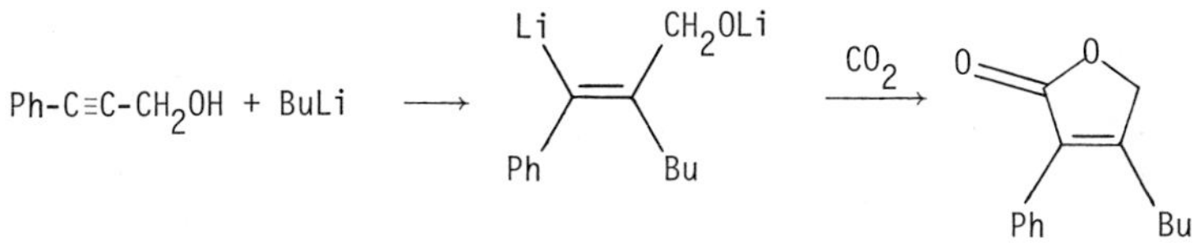

Ph-C≡C-CH$_2$OH + BuLi $\longrightarrow$

$\xrightarrow{CO_2}$ 77%

Acta Chem Scand B (1976) 30 521

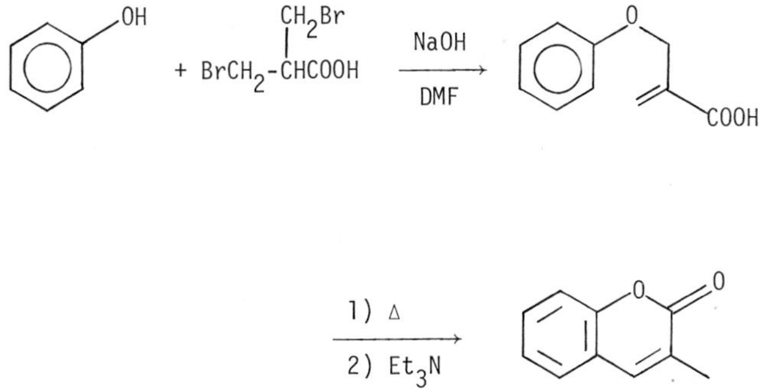

$\xrightarrow[\text{DMF}]{\text{NaOH}}$

37%

$\xrightarrow{\begin{array}{c}\text{1) } \Delta\\ \text{2) Et}_3\text{N}\end{array}}$

Synthesis (1975) 599

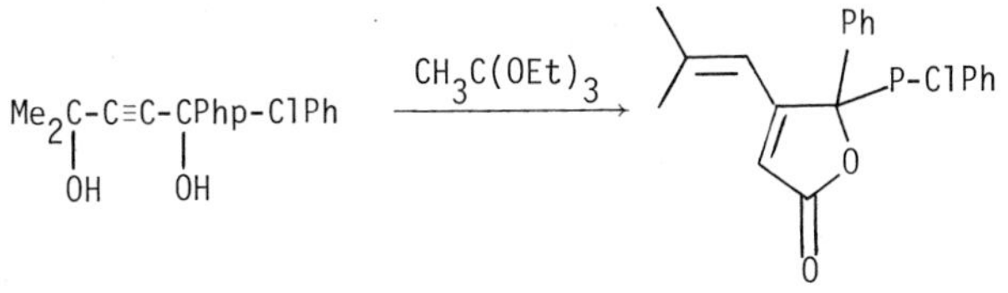

Me$_2$C-C≡C-CPhp-ClPh

$\xrightarrow{CH_3C(OEt)_3}$ 70%

Chem Lett (1975) 939

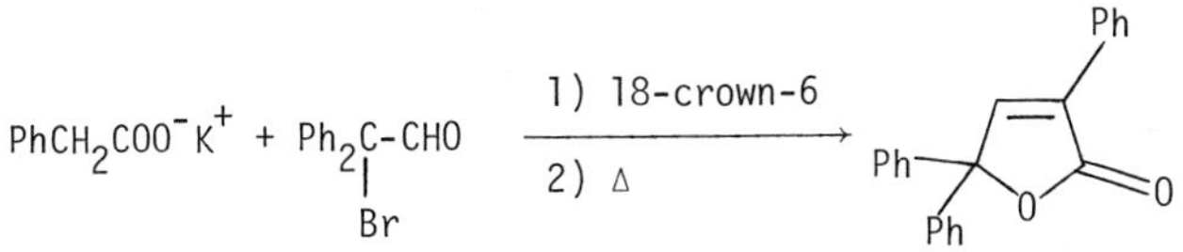

PhCH$_2$COO$^-$K$^+$ + Ph$_2$C-CHO

$\xrightarrow[\text{2) } \Delta]{\text{1) 18-crown-6}}$ 90%

JOC (1975) 40 3139

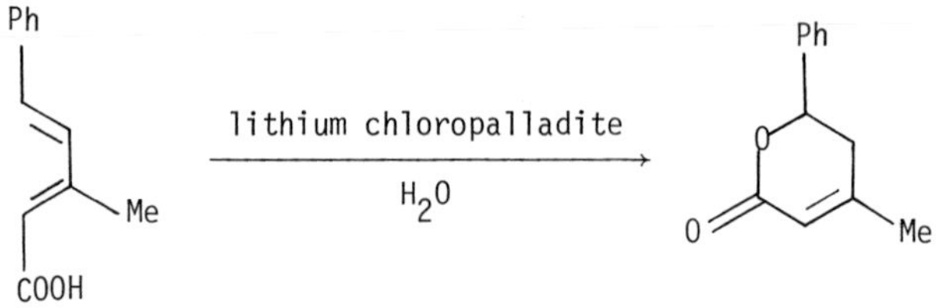

Ph
Me
COOH
lithium chloropalladite
H2O
Ph
O
O
Me

67%

Bull Chem Soc Japan (1975) 48 1673

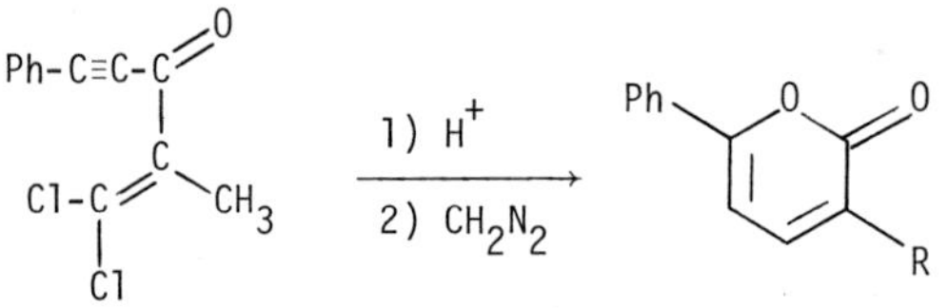

Ph-C≡C-C
O
C
Cl-C
CH3
Cl
1) H+
2) CH2N2
Ph
O
O
R

~40%

Bull Soc Chim France (1975) 751

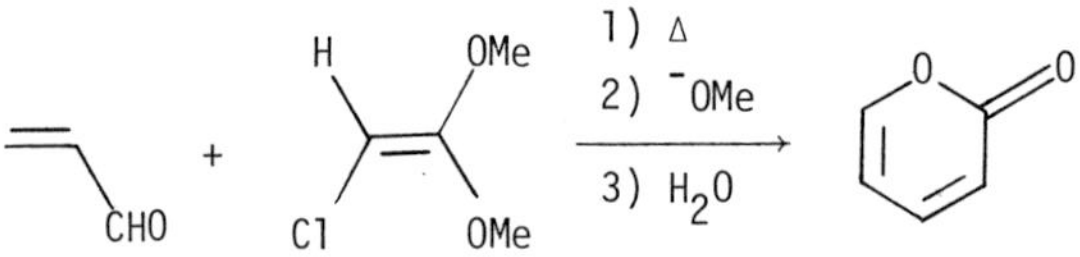

CHO
+
H
OMe
Cl
OMe
1) Δ
2) -OMe
3) H2O
O
O

~40%

Can J Chem (1975) 53 195

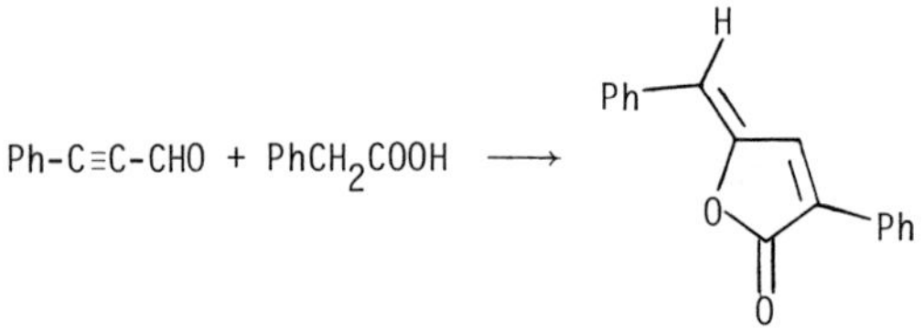

Ph-C≡C-CHO + PhCH2COOH
H
Ph
O
Ph
O

80%

Tetr Lett (1975) 1457

O
1) Me2CuLi
2) ClCOOMe
OCOMe
COOMe
Me

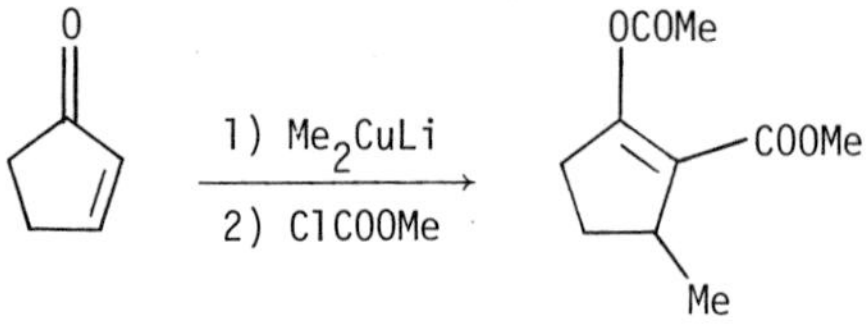

46%

JOC (1975) 40 1488

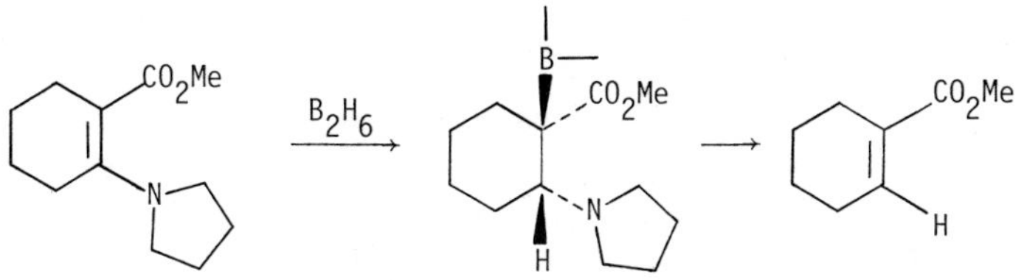

72%

Tetr Lett (1975) 1621

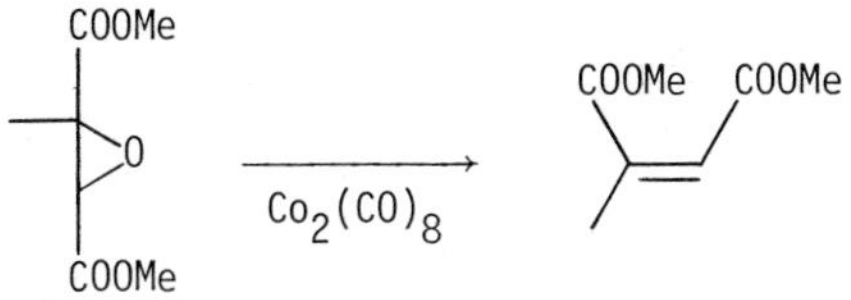

95%

JCS Chem Comm (1974) 384

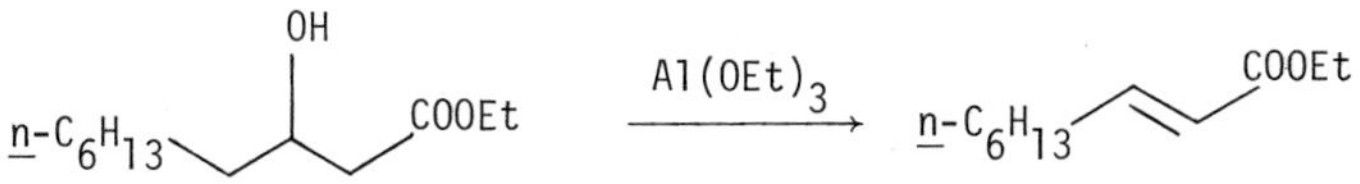

JACS (1974) 96 6153

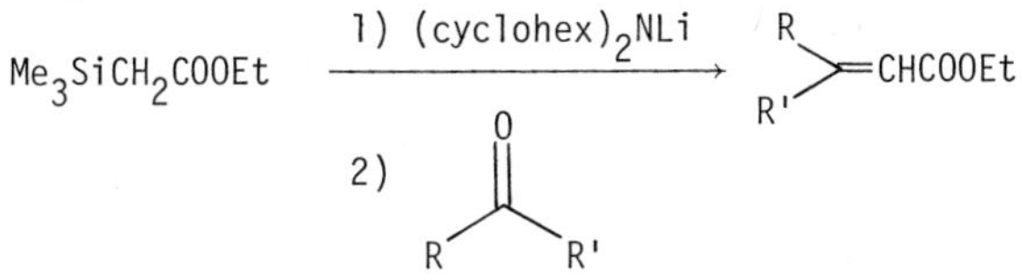

Me$_3$SiCH$_2$COOEt

1) (cyclohex)$_2$NLi

2) R—CO—R'

R,R'C=CHCOOEt

JACS (1974) 96 1620

TMSCH$_2$CO$_2$t-Bu

1) LDA
2) PhCHO
3) H$_3$O$^+$

PhCH=CHCOOtBu

75%

Tetr Lett (1974) 1403

$(EtO)_2POCHCO_2Et$ + $PhCOCH_3$ $\xrightarrow{\text{base}}$

Ph, Me
C=C
Me COOEt 65%

with Me on the lower left carbon

Synthesis (1974) 122
(1974) 869
Tetr Lett (1974) 711

$\underline{n}$-C_5H_{11}CHO + Ph$_3$$\overset{+}{P}CH_2$$\overset{\overset{\text{Me}}{|}}{C}$=CHCOOMe

$\xrightarrow{\text{CdI}_2\text{DBN}}$ $\underline{n}$-C_5H_{11}CH=CH-$\overset{\overset{\text{Me}}{|}}{C}$=CHCO$_2$Me 68%

$\xrightarrow{\underline{i}\text{-Pr}_2\text{NEt}}$ $\underline{n}$-C_5H_{11}CH=$\overset{\overset{}{|}}{C}$-COOMe 74%
MeC=CH$_2$

JOC (1974) $\underline{39}$ 821

BrCH$_2$CO$_2$Et +

PhSO$_2$
|
C$^-$
|
H

$\xrightarrow[\text{ethanol}]{\text{Na}_2\text{CO}_3}$

(branched diene structure with CO$_2$Et) 98%

Bull Soc Chim France (1976) 525

$\underline{t}$-Bu-C≡CH $\xrightarrow[\text{3) H}_3\text{O}^+]{\begin{array}{l}\text{1) DIBAL-H}\\\text{2) ClCOOEt}\end{array}}$

t-Bu
(alkene structure with COOEt) 72%

Synthesis (1976) 625

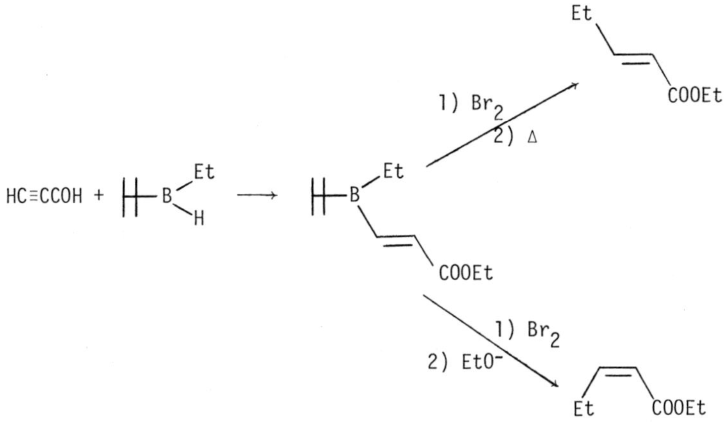

HC≡CCOH +
Et
B
H
Et
B
COOEt
1) Br2
2) Δ
Et
COOEt
1) Br2
2) EtO⁻
Et
COOEt

70%

76%

JOC (1974) 39 2321

[R'3B-C≡C-R]⁻Li⁺
1) AcCl
2) Jones reagent
R'
R
R'
COMe

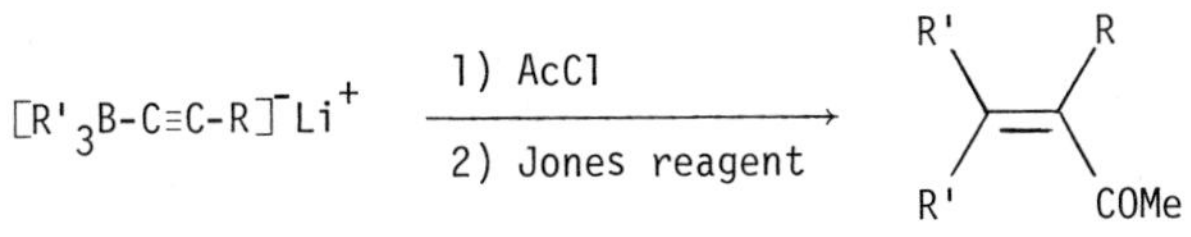

28-42%

Tetrahedron (1974) 30 835

COOEt
BuC≡CCu
+ CH2=CHCH2Br
Et2O
-78°
CH2=CHCH2
COOEt

70%

JACS (1974) 96 7138

(EtCO)2O + BrZnCHCOOEt
Me
EtC=CMeCOOEt
OCOEt

76%

JOC (USSR) (1975) 11 2360

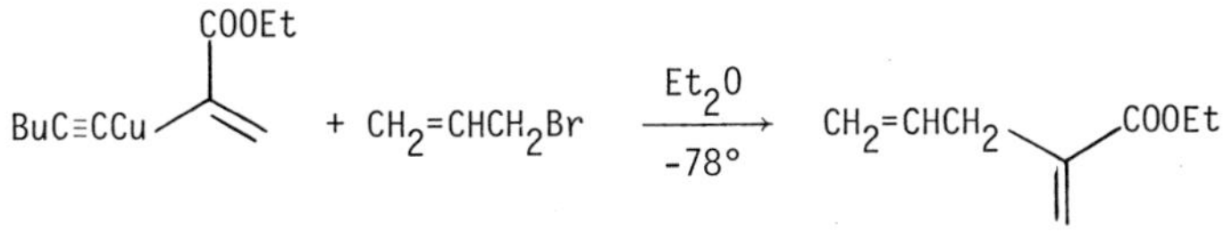

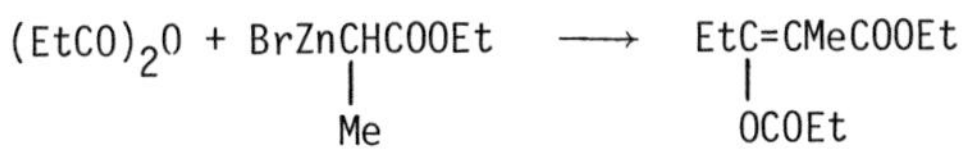

$CH_2=CHCH_3$ - - -> $\left(Pd\diagdown_2^{Cl}\right)$ + $PhSOCH=\overset{\overset{OLi}{|}}{C}CH_3$ $\xrightarrow{R_3P}$

40-90%

$\diagup\diagdown\diagup\diagdown_{COOMe}$

JACS (1974) 96 7165
JOC (1974) 39 737

+ $PhSO\overline{C}HCOOMe$ $\xrightarrow[\text{2) 80°}]{\text{1) HMPA, 25°}}$ 86%

JACS (1974) 96 7165

OH / OH (catechol) $\xrightarrow[\text{CH}_3\text{OH}]{O_2, \text{ CuCl}}$ COOMe / COOMe

JACS (1974) 96 7349

2 $\overset{CN}{\underset{COOEt}{CH_2}}$ $\xrightarrow[\text{THF, rfx}]{\text{SOCl}_2}$ $\overset{COOEt\quad CN}{\underset{NC\quad\quad COOEt}{C=C}}$ 64%

Synth Comm (1976) 6 185

$MeCH=C\overset{NMePh}{\underset{COOMe}{}}$ $\xrightarrow[\text{2) MeI}]{\text{1) i-Pr}_2\text{NLi}}$ $MeCH_2CH=C\overset{NMePh}{\underset{COOMe}{}}$ 80%

Synthesis (1975) 512

PhTl(TFA)$_2$ + HC≡CCOOMe $\xrightarrow[\text{CuCl}_2, \text{ MeOH}]{\text{Li}_2\text{PdCl}_4}$ PhCH=CHCOOMe   88%

J Organometal Chem (1975) 98 C8

$\overset{+}{P}Ph_3$

1) LDA

2) CH=CHCOOMe
   |
   Cl

3) PhCHO

$\xrightarrow{\hspace{2cm}}$ PhCH=CHCH=CH-CH=CHCOOEt   84%

Tetr Lett (1975) 1359

PhCH$_2$CCOOMe
     ‖
     N$_2$

NaOMe $\xrightarrow{\hspace{2cm}}$

72%

$\xrightarrow[\text{DCC}]{\text{BF}_3 \cdot \text{Et}_2\text{O}}$

80%

Chem Pharm Bull (1975) 23 229

CH$_2$(COMe)$_2$ + t-BuNC $\xrightarrow[\text{PhH}]{\text{Cu}_2\text{O}}$ (MeOC)$_2$C=CHNt-Bu   34%

J Organometal Chem (1975) 85 395

PhCH=CHHgCl $\xrightarrow[\text{PdCl}_2]{\text{CO, MeOH}}$ PhCH=CHCOOMe   100%

JOC (1975) 40 3237

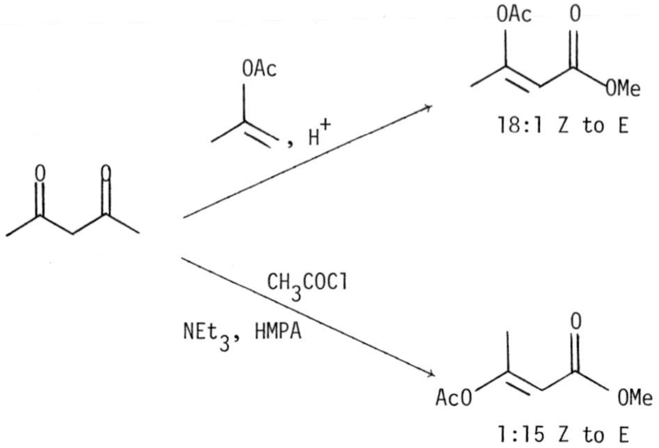

OAc
O
OMe
OAc
, H+
18:1 Z to E
CH3COCl
NEt3, HMPA
O
AcO
OMe
1:15 Z to E

Tetr Lett (1974) 925

EtCOOH + $HgOAc_2$ + vinyl versate 10 $\xrightarrow{H^+}$ $EtCOOCHCH_2$

Tetrahedron (1974) 30 4205

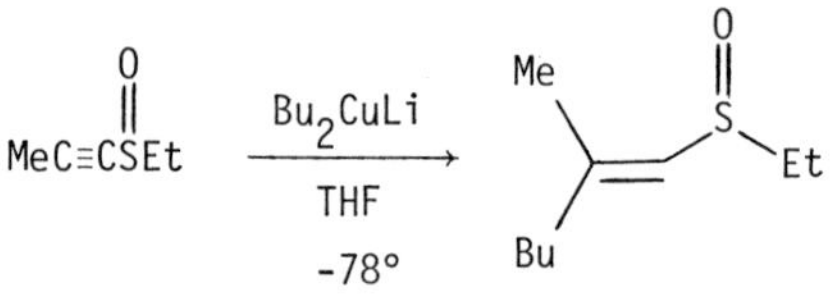

O
MeC≡CSEt
Bu2CuLi
THF
-78°
Me
O
S
Et
Bu

100%

JOC (1974) 39 3174

O
O
O
+ Ph3P-CHCOEt
O
O
H
EtOC
O

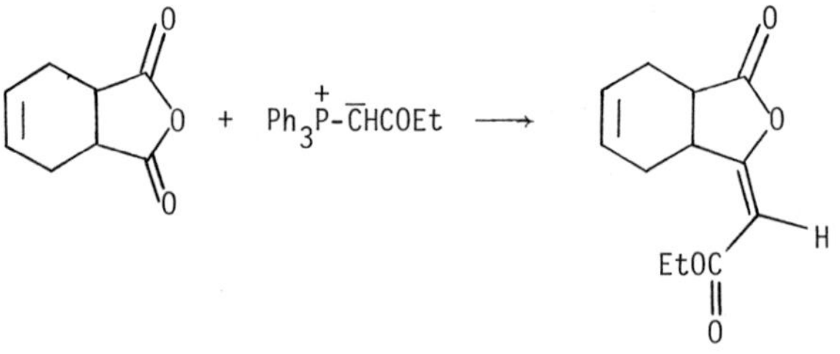

60%

Aust J Chem (1975) 28 2499

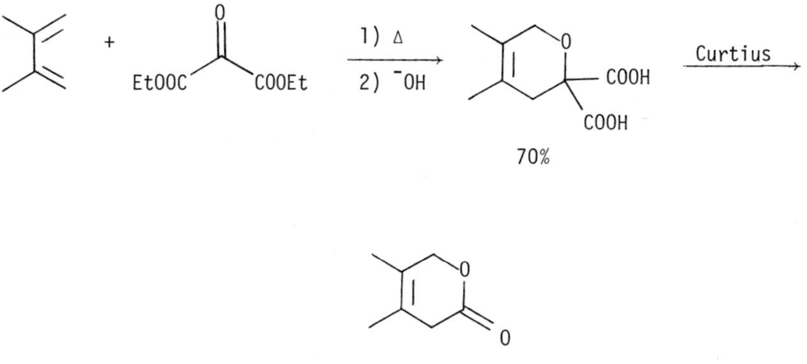

70%

72%

JACS (1975) 97 6892

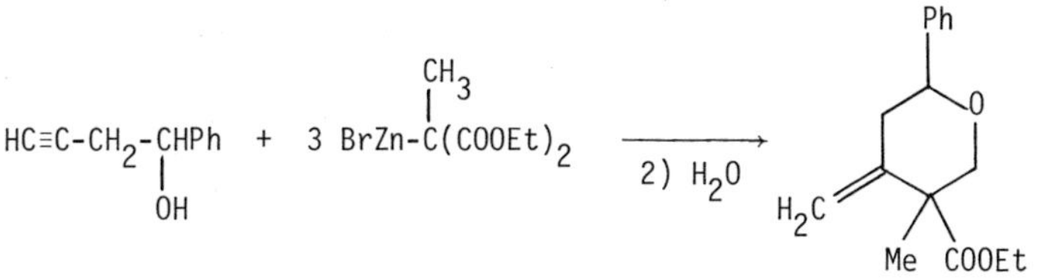

HC≡C-CH₂-CHPh   +   3 BrZn-C(COOEt)₂
  | |
  OH   CH₃

70%

Comptes Rendus C (1975) 280 999

MeCH=CHCH₂ZnBr + ClCO₂Et   —Et₂O→   MeCH=CHCH₂COOEt   70%
                            -15°

J Organometal Chem (1975) 96 163

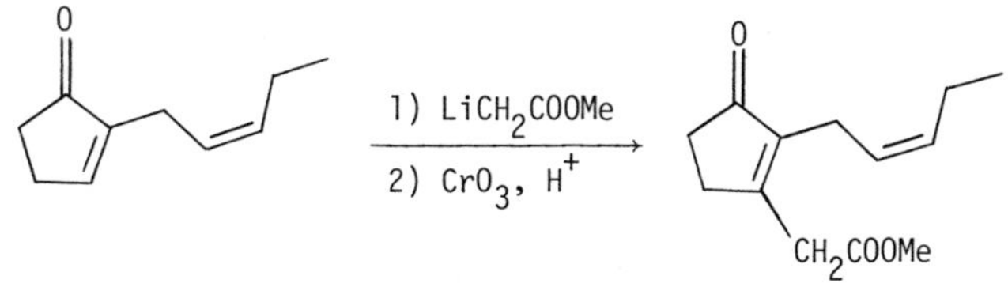

1) LiCH₂COOMe
2) CrO₃, H⁺

80%

JOC (1974) 39 2637

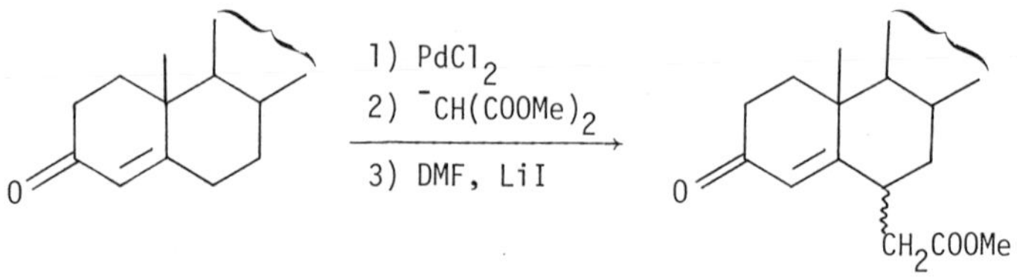

1) PdCl2
2) ⁻CH(COOMe)2
3) DMF, LiI
O
O
CH2COOMe

~50%

Tetr Lett (1976) 495

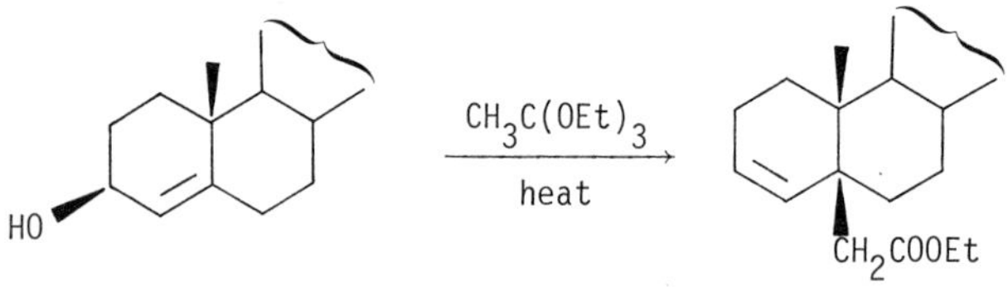

CH3C(OEt)3
heat
HO
CH2COOEt

~50%

Org Synth (1974) 54 74

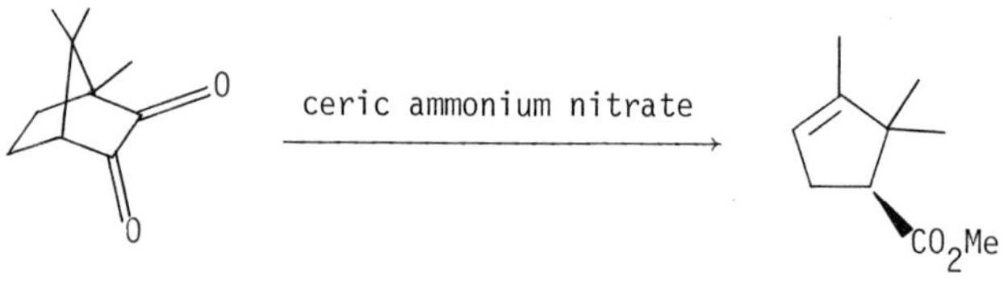

ceric ammonium nitrate
O
O
CO2Me

65%

Chem and Ind (1976) 565

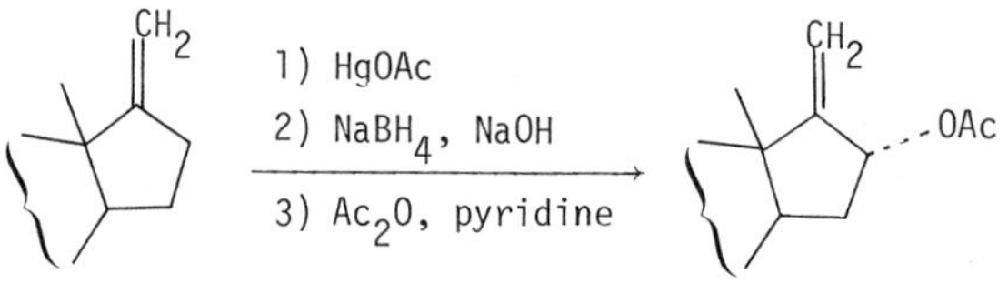

CH2
1) HgOAc
2) NaBH4, NaOH
3) Ac2O, pyridine
CH2
OAc

57%

Steroids (1976) 27 197

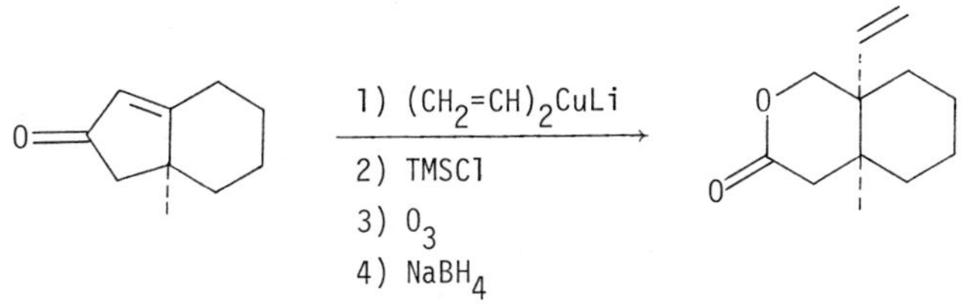

1) $(CH_2=CH)_2CuLi$
2) TMSCl
3) O_3
4) $NaBH_4$

65%

Tetr Lett (1974) 1713

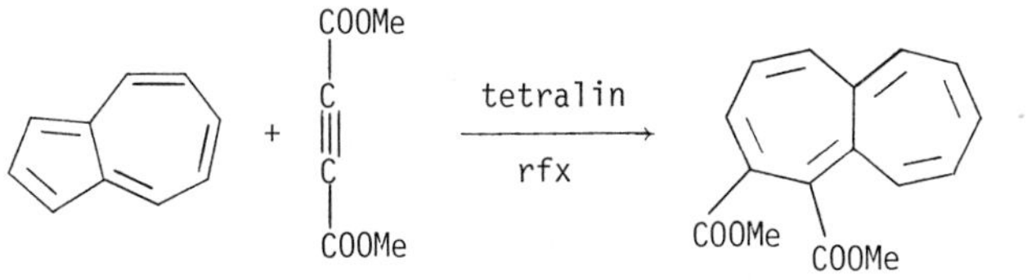

+

COOMe
|
C
‖
C
|
COOMe

tetralin
―――――→
rfx

15%

Angew Int Ed (1976) 15 104

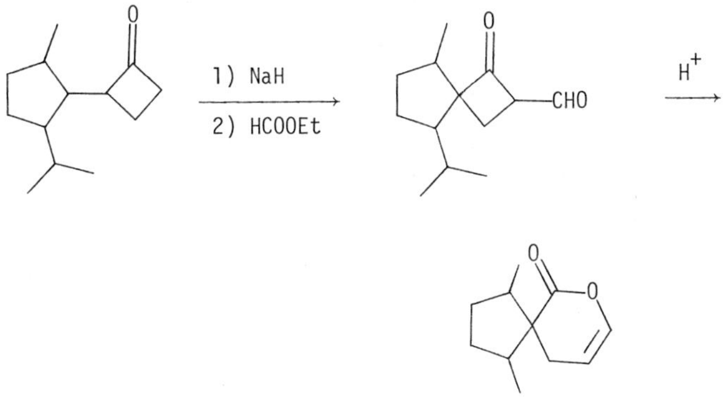

1) NaH
2) HCOOEt

―――→

H^+
―――→

62%

JACS (1975) 97 5873

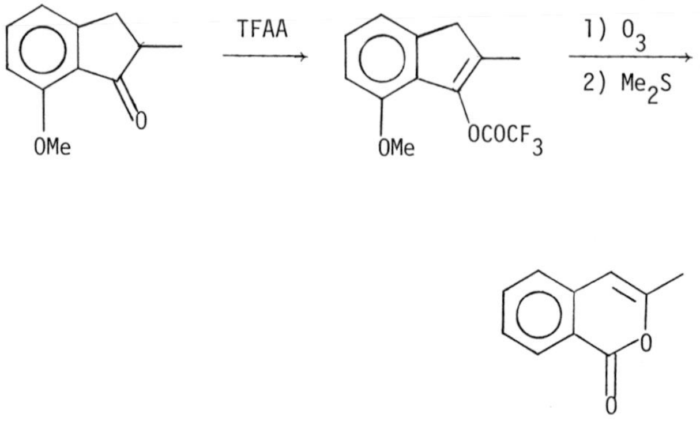

~40%

JCS Perkin I (1976) 1438

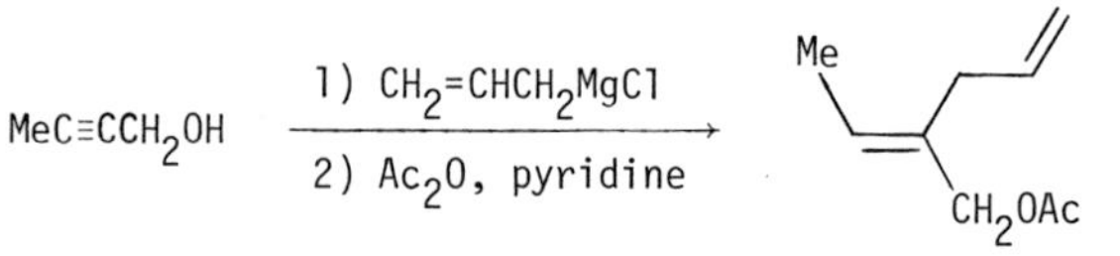

MeC≡CCH₂OH $\xrightarrow{\text{1) CH}_2\text{=CHCH}_2\text{MgCl}}{\text{2) Ac}_2\text{O, pyridine}}$

75%

Synth Comm (1976) 6 319

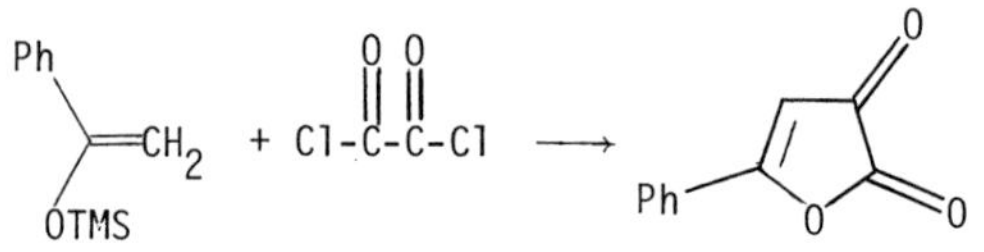

95%

Angew Int Ed (1975) 14 636

Also via: Acetylenic esters (Section 306)
           Olefinic acids (Section 322)
           β-Hydroxyesters (Section 327)

Section 363   Ether - Ether

$$Me(CH_2)_4CHO \xrightarrow[\text{DME}]{\text{Li, Me}_3\text{SiCl}} Me(CH_2)_4\overset{\overset{\displaystyle OTMS}{|}}{\underset{\underset{\displaystyle OTMS}{|}}{CH}}\text{-CH-}(CH_2)_4Me \qquad 60\%$$

Synthesis (1975) 787

$$\xrightarrow[\text{electrolysis}]{\text{CH}_3\text{OH}}$$

71%

Synthesis (1975) 717

Section 364   Ether - Halide

$$\text{EtO-CH=CH}_2 \quad + \quad CuCl_2 \quad + \quad KI \quad + \quad MeOH \quad \longrightarrow$$

moderate yield

Bull Chem Soc Jap (1974) 47 2818

$$\xrightarrow[\text{EtOCH}_2\text{Cl}]{BF_3 \cdot Et_2O}$$

$$\text{(P)}\text{-}\bigcirc\text{-OCH}_2\text{Cl}$$

Tetr Lett (1975) 4637

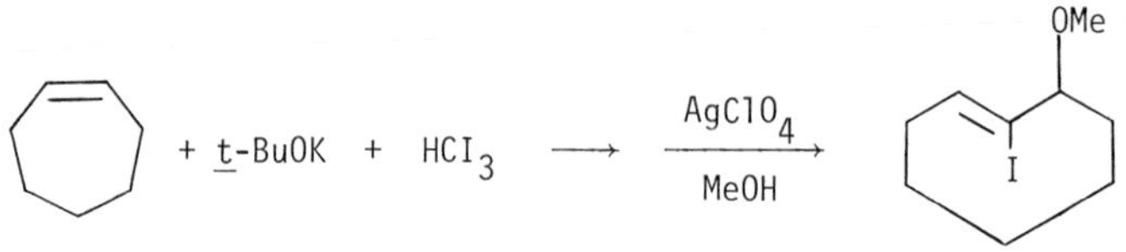

55%

JCS Chem Comm (1974) 196

Section 365   Ether, Epoxide - Ketone

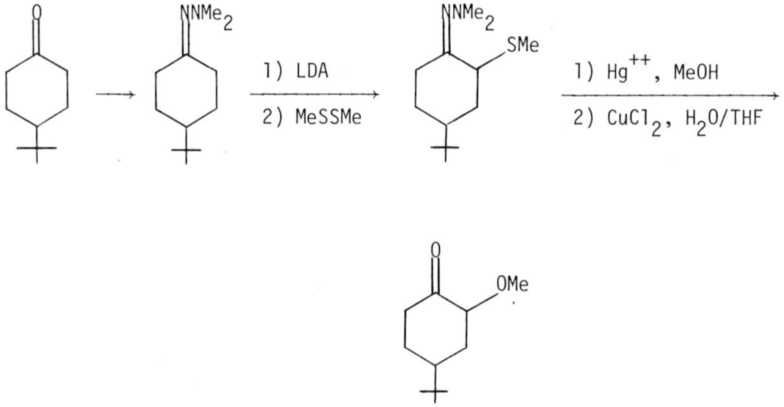

Individual yields >90%

Tetr Lett (1976) 4687

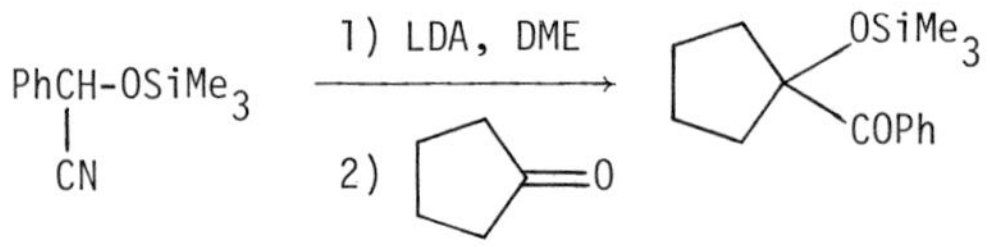

98%

Synthesis (1975) 391

$$RC \equiv CCH(OEt)_2 + 2R'_2BH \xrightarrow{\text{NaOH/H}_2\text{O}} RCH_2\overset{\overset{\displaystyle O}{\|}}{C}CH_2OEt$$

JACS (1974) 96 316

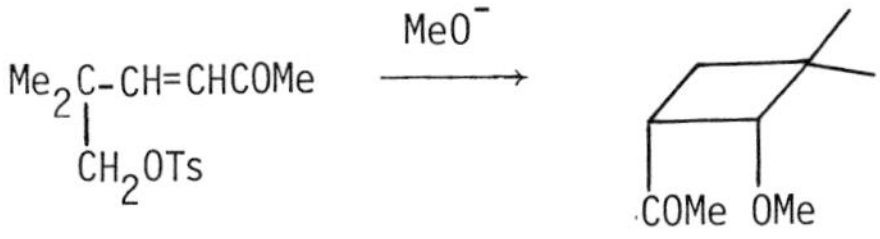

$$\underset{\underset{\displaystyle CH_2OTs}{|}}{Me_2C}-CH=CHCOMe \xrightarrow{\text{MeO}^-}$$

COMe OMe

Chem Ber (1974) 107 887

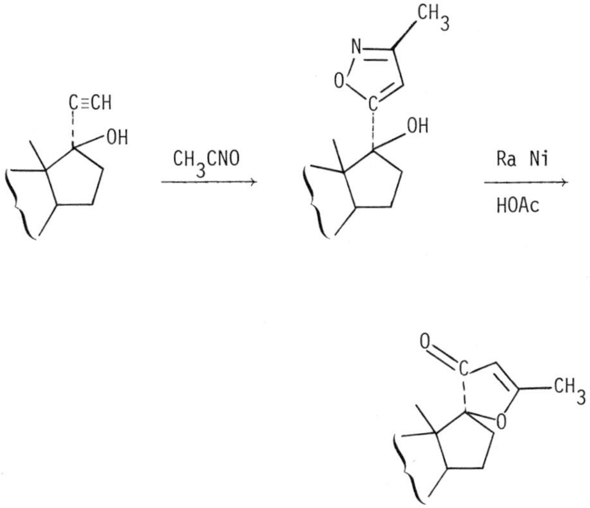

$$\xrightarrow{\text{CH}_3\text{CNO}} \qquad \xrightarrow[\text{HOAc}]{\text{Ra Ni}}$$

74%

J Gen Chem USSR (1975) 45 2534

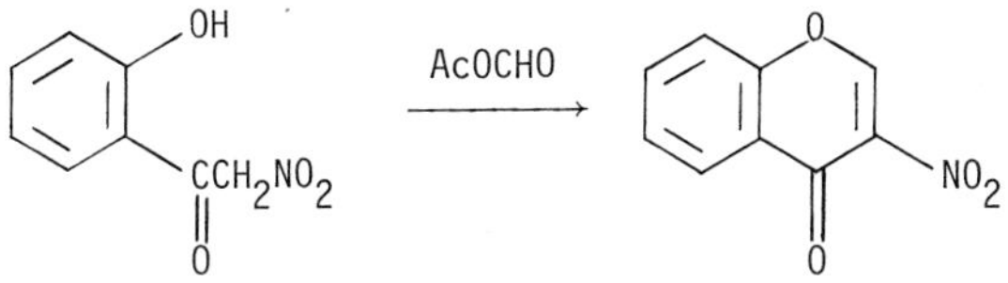

$$\xrightarrow{\text{AcOCHO}}$$

~100%

Tetr Lett (1976) 719

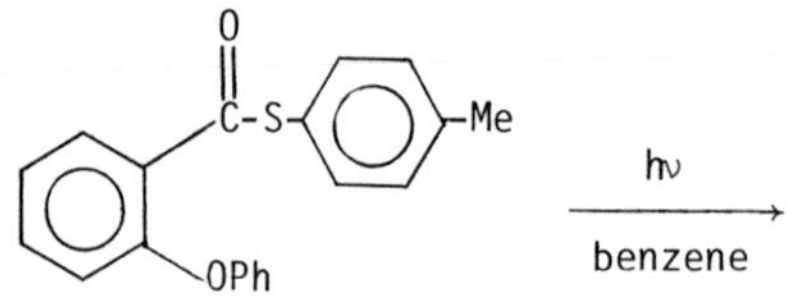

84%

Synthesis (1976) 532

Section 366 Ether, Epoxide - Nitrile

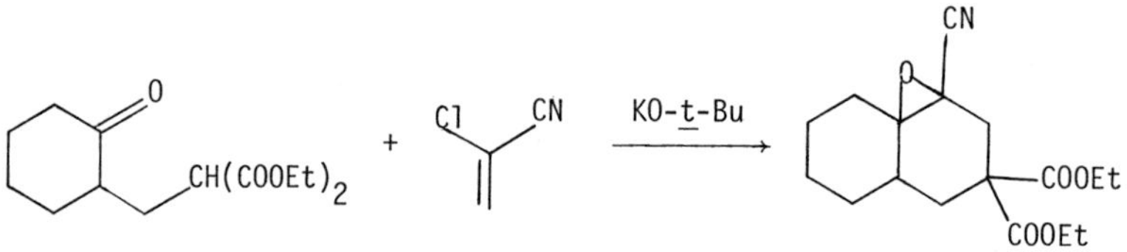

54%

JCS Chem Comm (1975) 95

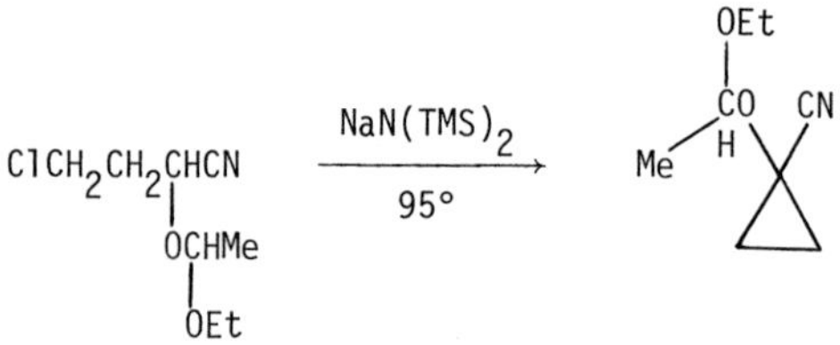

62%

Tetr Lett (1975) 389

Section 367 Ether - Olefin

$$R'\overset{O}{\overset{\|}{C}}CH_2R^2 + EtO_3CH \xrightarrow[\text{cat}]{\text{Amberlyst 15}} R'-\overset{OEt}{\overset{|}{C}}=CHR^2$$

70-100%

Synthesis (1974) 348

PhCH$_2$Cl $\xrightarrow[\text{2) HCOOEt}]{\text{1) Ph}_3\text{P}}$ Ph—CH=CH—OEt   90%

JOC (1976) <u>41</u> 1272

PhOH + Ph$_3\overset{+}{P}$-CH=CH-$\overset{+}{P}$Ph$_3$ $\xrightarrow[\text{2) NaOH/H}_2\text{O}]{\text{1) Et}_3\text{N}}$ Ph-O-CH=CH$_2$ 76%
(2Br$^-$)

Synthesis (1975) 736

CH$_2$=CH-CH$_2$OTHP $\xrightarrow[\text{2) MeI}]{\begin{array}{c}\text{1) }\underline{t}\text{-BuOK}\\\underline{\text{sec}}\text{-BuLi}\end{array}}$ CH$_3$CH=C$\begin{array}{l}\text{—OTHP}\\\text{Me}\end{array}$ 83%

Synthesis (1974) 888

$\xrightarrow[\text{PhLi, THF}]{\text{Ph}_3\overset{+}{P}\text{CH}_2\text{OTHP}}$ 85%

Tetrahedron (1975) <u>31</u> 89

PhOH + Ph-$\overset{\overset{\text{O}}{\|}}{\text{C}}$-C≡CH $\xrightarrow{\text{Et}_3\text{N}}$ 75%

Indian J Chem (1975) <u>13</u> 852

$$\text{cyclohexene-CHO} \quad \xrightarrow[\text{2) CH}_2\text{CHCH}_2\text{Br}]{\text{1) KNH}_2} \quad \text{cyclohexene=CHOCH}_2\text{CHCH}_2 \qquad 90\%$$

Tetr Lett (1974) 1653

$$\text{MeSC}\equiv\text{CCH}_2\text{OMe} \quad \xrightarrow[\text{2) EtBr}]{\text{1) i-Pr}_2\text{NLi}} \quad \underset{\underset{\text{Et}}{|}}{\text{MeSC}=\text{C}=\text{CHOMe}} \qquad 93\%$$

Tetr Lett (1975) 1741

$$\text{BrCH}=\text{CHOEt} \quad + \quad \text{(benzodioxole-MgBr)} \quad \xrightarrow{[\text{Ni(dpp)Cl}_2]}$$

77%

(benzodioxole-CH=CHOEt)

Chem Lett (1976) 1237

$$[\text{i-PrCuBr}]\text{MgCl} + \text{H}_2\text{C}=\text{C}=\text{CHOCH}_3 \quad \xrightarrow[\text{2) H}_3\text{O}^+]{} \quad \text{i-Pr-CH}_2 \diagdown \diagup \text{OMe} \qquad \begin{array}{c}\text{no}\\ \text{yield}\end{array}$$

Tetr Lett (1976) 947

$$\text{i-PrCOC(Me)}_2\text{CO}_2\text{TMS} \quad \xrightarrow{250°} \quad \underset{\underset{\text{OTMS}}{|}}{\text{i-PrC}=\text{CMe}_2} \qquad 80\%$$

JACS (1975) 97 1619

+ $[CH_2=C(OEt)]_2CuLi$ $\longrightarrow$

74%

JCS Chem Comm (1975) 519

$CH_2=CHCH(OEt)_2$ $\xrightarrow[\text{Et}_2O]{\text{n-BuMgX, CuBr}}$ $\underline{n}\text{-BuCH}_2\text{CH=CHOEt}$

83%

Tetr Lett (1975) 3833

$\xrightarrow[\text{HC(OMe)}_3]{\text{p-TsOH}}$

95%

Synthesis (1974) 38

$CH_2=CH-CH_2OSiEt_3$ $\xrightarrow[\text{2) MeI}]{\text{1) \underline{sec}-BuLi}}$ $\text{Me-CH}_2\text{CH=CHOSiEt}_3$

95%

JACS (1974) 96 5561

$\xrightarrow[\text{Me}_3\text{SiCl}]{\text{Et}_3\text{N}\cdot\text{ZnCl}_2}$

68%

JACS (1974) 96 7807

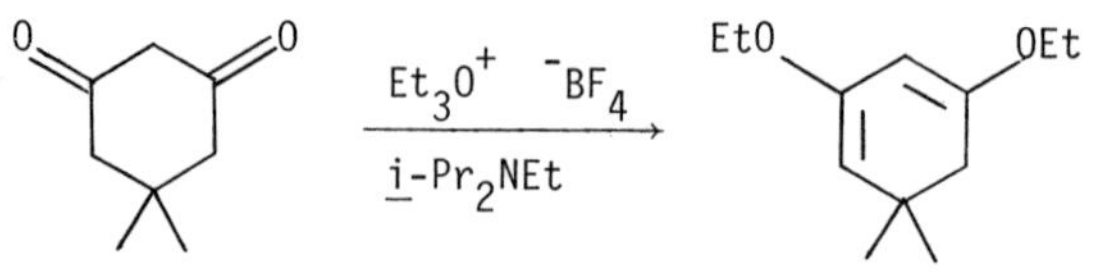

JCS Chem Comm (1975) 644

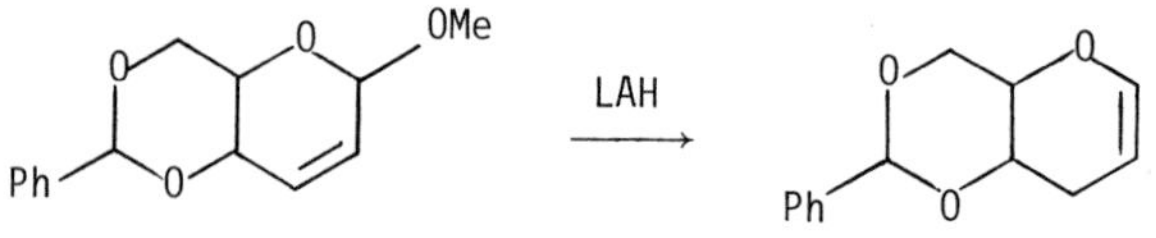

95%

Can J Chem (1975) 53 2005

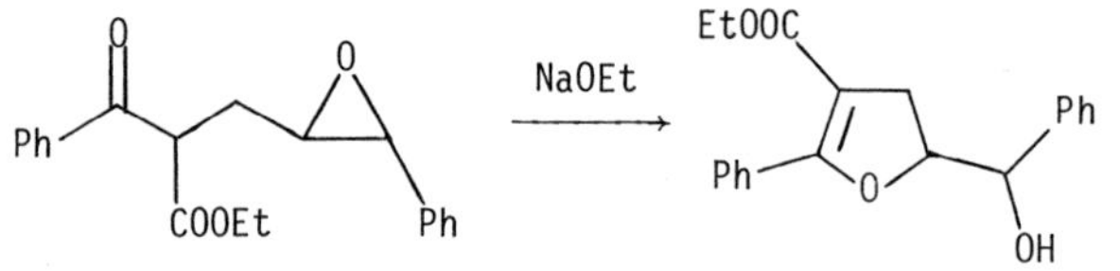

70%

Indian J Chem (1976) 14B 47

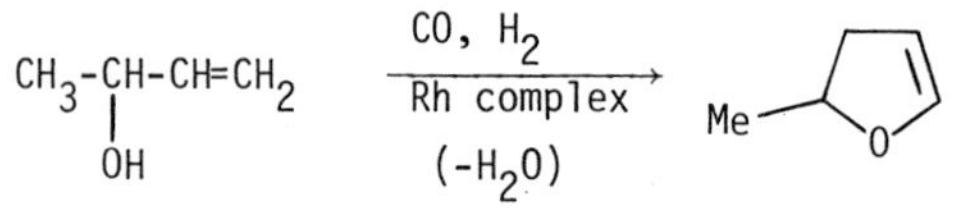

$$CH_3-CH-CH=CH_2 \quad \xrightarrow[\text{(-H}_2\text{O)}]{\substack{CO, H_2 \\ \text{Rh complex}}}$$

~50%

Gazz Chim Ital (1975) 105 233

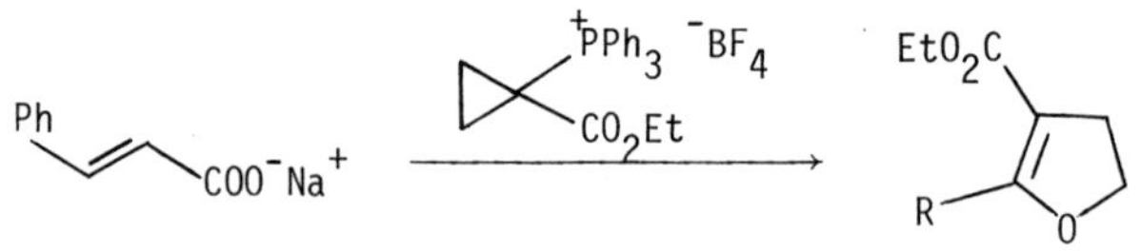

93%

Tetr Lett (1975) 4353

Me$_3$SiCH$_2$CH=CH$_2$ +

$\xrightarrow[\text{2) H}_2\text{O}]{\text{1) TiCl}_4}$

71%

Chem Lett (1976) 941

$\xrightarrow[\text{3) H}_3\text{O}^+]{\begin{array}{l}\text{1) DIBAL-H}\\\text{2) EtOCH}_2\text{Cl}\end{array}}$

72%

Synthesis (1976) 816

PhSex +

$\xrightarrow{\text{MeOH}}$ $\xrightarrow{\text{H}_2\text{O}_2}$

78%

JOC (1974) 39 429

$\xrightarrow[\text{3) } \underline{t}\text{-BuOK, DMSO}]{\begin{array}{l}\text{1) MeSeMe}_2\text{C}^- \text{ Li}^+\\\text{2) MeI}\end{array}}$

55%

Tetr Lett (1976) 457

Ph-C-C-Ph + CH$_3$C≡CSEt
   ‖ ‖
   O O

$\xrightarrow[\begin{array}{l}\text{2) SnCl}_2\text{/HCl}\\\text{HOAc}\end{array}]{\text{1) h}\nu}$

85%

Tetr Lett (1974) 4179

$$\xrightarrow[\text{MeSO}_2\text{Cl, HCONMe}_2]{\text{BF}_3 \cdot \text{Et}_2\text{O}}$$

82%

JCS Chem Comm (1976) 78

$$+ \quad \text{PhI} \quad \longrightarrow$$

81%

(86:14 E:Z)

Chem Lett (1975) 747

1) MeI
2) anion-exchange resin
3) Δ

$$\text{H}_2\text{C=CHOCHPh}_2$$

∿80%

Org Synth (1976) 55 3

$$\xrightarrow[\text{DMF}]{\text{KOH}}$$

90%

Liebig's Ann Chem (1974) 523

Section 368   Halide - Halide

$$\text{Ph}\diagdown\diagdown \quad \xrightarrow{\hspace{3cm}} \quad \begin{matrix}\text{Ph} \\ | \\ \text{CH}_3\end{matrix}\text{CH-CHF}_2$$

90%

JCS Chem Comm (1975) 715

$$PhCH_2CH_2NH_2 \xrightarrow[\text{CuBr}_2]{\underline{\text{t-BuOND}}} PhCH_2CHBr_2$$

70%

JCS Chem Comm (1976) 433

$$PhCH_2NH_2 \xrightarrow[\substack{2) \ (\text{CuBr}_2 \cdot \text{NO})_2, \\ \text{CH}_3\text{CN}}]{1) \ \text{CuBr}_2} RCHBr_2$$

JACS (1976) 98 1627

89%

Tetr Lett (1976) 943

$$PhN_2Cl \ + \ CH_2{=}CCl_2 \xrightarrow[\text{acetone}]{\text{CuCl}_2} PhCH_2CCl_3$$

79%

Org React (1976) 24 225

$$RCH_2CHCH_3 \ + \ SbCl_5 \xrightarrow{\text{CCl}_4} RCH{-}CH{-}CH_3$$
$$\overset{|}{\underset{Br}{}} \qquad\qquad \overset{|}{\underset{Cl}{}}\ \overset{|}{\underset{Br}{}}$$

70-90%

Tetr Lett (1974) 759

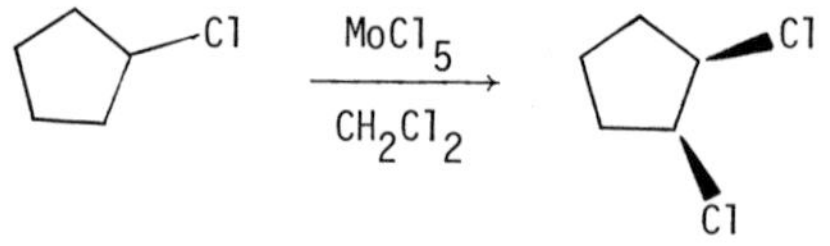

JOC (1975) <u>40</u> 3463

Section 369   Halide - Ketone

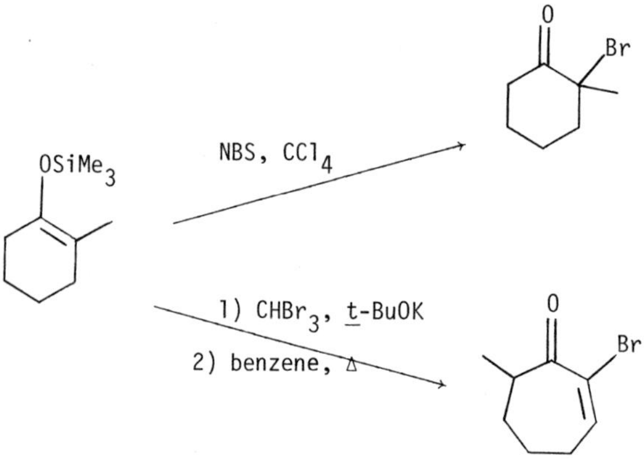

70%

16% overall

Synthesis (1976) 194 and 196

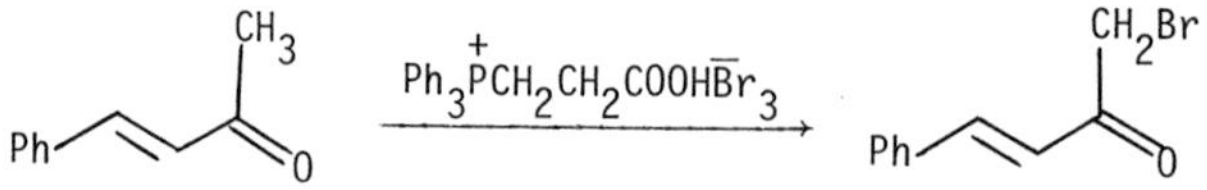

∿60%

Tetr Lett (1975) 373

$$Me_2CH\text{-}\overset{\displaystyle O}{\overset{\|}{C}}\text{-}CH_3 \xrightarrow[\text{CH}_3\text{OH}]{\text{Br}_2} Me_2CH\text{-}\overset{\displaystyle O}{\overset{\|}{C}}\text{-}CH_2Br$$

~70%

Org Synth (1976) <u>55</u> 24

MgBr$_2$ ether/CH$_2$Cl$_2$

60%

Tetr Lett (1976) 3677

$$Me_2N\text{-}\overset{\displaystyle O}{\overset{\|}{C}}\text{-}\overset{\displaystyle Br}{\overset{\|}{C}H}\text{-}CN$$

62%

Tetrahedron (1975) <u>31</u> 231

1) PCl$_5$
2) Br$_2$

80%

Experientia (1976) <u>32</u> 1491

$$R'CH=CH\text{-}\overset{\displaystyle O}{\overset{\|}{C}}\text{-}R \xrightarrow[\substack{\text{2) } \underline{t}\text{-BuOOH} \\ \text{3) Br}_2}]{\text{1) Hg(OAc)}_2} RCH(OO\text{-}\underline{t}\text{-Bu})CHBrCOR'$$

50-80%

JCS Perkin I (1974) 688

$$\underset{R-C-CHN_2}{\overset{O}{\overset{\|}{}}} \quad \xrightarrow[X = Cl, Br]{XI} \quad \underset{R-C-CHIX}{\overset{O}{\overset{\|}{}}} \qquad\qquad \sim 70\%$$

R = alkyl, aryl

JOC USSR (1976) $\underline{12}$ 228

$$\begin{array}{c} \overset{O}{\overset{\|}{Ph-C}} \\ CH_2 \\ \underset{Ph-C}{\underset{\overset{\|}{O}}{}} \end{array} \quad \xrightarrow[CH_3CO_3H]{I_2} \quad \begin{array}{c} \overset{O}{\overset{\|}{Ph-C}} \\ CHI \\ \underset{Ph-C}{\underset{\overset{\|}{O}}{}} \end{array} \qquad\qquad 92\%$$

JCS Perkin I (1975) 1285

$$RCOOEt + HCCl_2Li \quad \longrightarrow \quad \xrightarrow{H_3O^+} \quad RCOCHCl_2$$

moderate to
high yield

Comptes Rendus (1974) $\underline{278}$ 929

$$+ Br_2 \quad \longrightarrow \quad \underset{PhC-CH_2CH_2Br}{\overset{O}{\overset{\|}{}}}$$

quantitative

JCS Chem Comm (1974) 1032

Section 370   Halide - Nitrile

$$\begin{array}{c} Ph \\ \overset{(-)}{>}C-CN \\ CH_3 \end{array} \quad \xrightarrow{CCl_4} \quad \begin{array}{c} Ph \\ >C-CN \\ CH_3 \underset{Cl}{|} \end{array} \qquad\qquad 80\%$$

Tetrahedron (1975) $\underline{31}$ 1335

Ph$_2$CHCN + (CH$_2$Br)$_2$ $\xrightarrow[\text{crown ether}]{\text{NaOH}}$ Ph$_2$C$\overset{\text{CN}}{\underset{\text{CH}_2\text{CH}_2\text{Br}}{<}}$ 75%

Angew Int Ed (1974) 13 665

Section 371   Halide - Olefin

EtMgBr $\xrightarrow[\begin{array}{l}\text{3) HgBr}_2\\ \text{4) Br}_2,\ \text{py}\end{array}]{\begin{array}{l}\text{1) CuBr, -30°}\\ \text{2) MeC}\equiv\text{CH}\end{array}}$ $\overset{\text{Et}}{\underset{\text{Me}}{>}}=\overset{\text{Br}}{\underset{\text{H}}{<}}$ 77%

Synthesis (1974) 803

RC≡CSiMe$_3$ + HBr $\xrightarrow{\text{radical}}$ $\overset{\text{Br}}{\underset{\text{R}}{>}}=\text{CH}_2$ 60-95%

JOC (1974) 39 3307

Et-C≡C-Et $\xrightarrow{\text{PhICl}_2}$ $\overset{\text{Et}}{\underset{\text{Cl}}{>}}=\overset{\text{Cl}}{\underset{\text{Et}}{<}}$ 96%

Bull Soc Chim France (1975) 2493

PhC≡CH + PhCH$_2$Br $\xrightarrow[\text{CH}_2\text{Cl}_2,\ \text{rfx}]{\text{ZnBr}_2}$ $\overset{\text{PhCH}_2}{\underset{\text{H}}{>}}\text{C=C}\overset{\text{Ph}}{\underset{\text{Br}}{<}}$ 90%

Gazz Chim Ital (1975) 105 495

n-BuC≡CH $\xrightarrow[\text{2) Me}_3\text{SiCl}]{\text{1) EtMgBr}}$ n-BuC≡CSiMe$_3$ $\xrightarrow[\text{3) H}_2\text{O}_2]{\substack{\text{1) (cyclohex)}_2\text{BH} \\ \text{2) HOAc, rfx}}}$

Bu, I / H, H   60%

Bu, SiMe$_3$ / H, H

$\xrightarrow{\text{I}_2\text{, CH}_2\text{Cl}_2}$

$\xrightarrow[\text{2) KF·2H}_2\text{O, DMSO}]{\text{1) I}_2\text{, CF}_3\text{COOAg}}$

Bu, H / H, I   60%

Tetr Lett (1974) 543

Ph-C≡C-Me $\xrightarrow[\text{CH}_3\text{I}]{\text{CH}_3\text{IF}_2}$ Ph, Me / F, I (C=C)   70%

Synthesis (1976) 473

polymer-PCl$_2$ $\xrightarrow{\text{PhCOCH}_3}$ PhCCl=CH$_2$

JACS (1974) 96 6469

$\xrightarrow[\text{Et}_2\text{O}]{\text{HCF}_2\text{Cl}}$

~100%

Tetr Lett (1976) 895

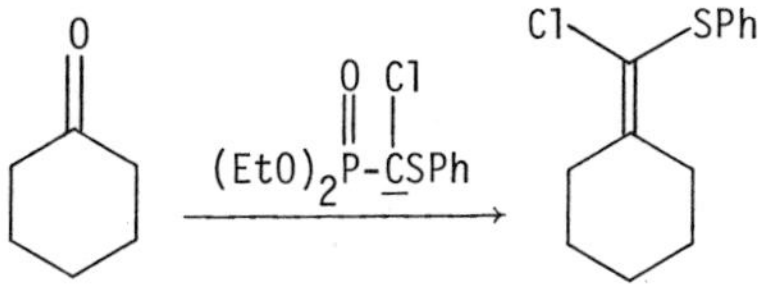

Synthesis (1976) 107

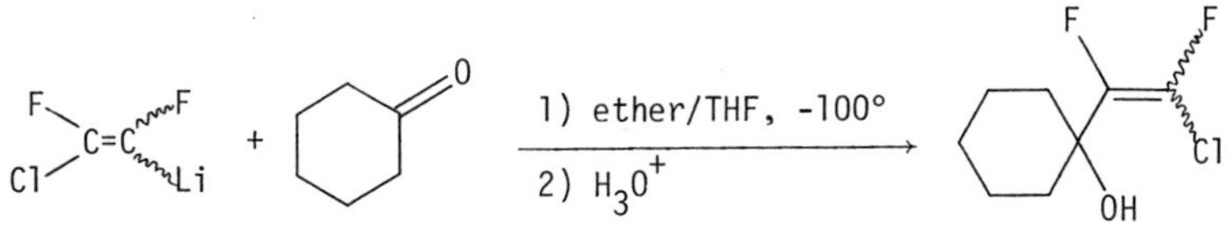

Synthesis (1976) 761

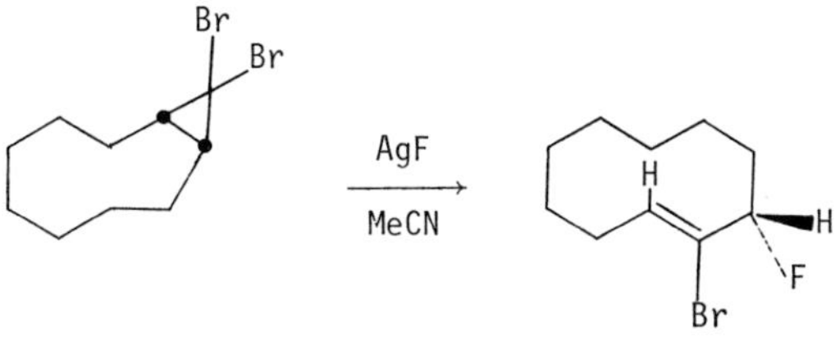

Rec Trav Chim (1976) 95 248

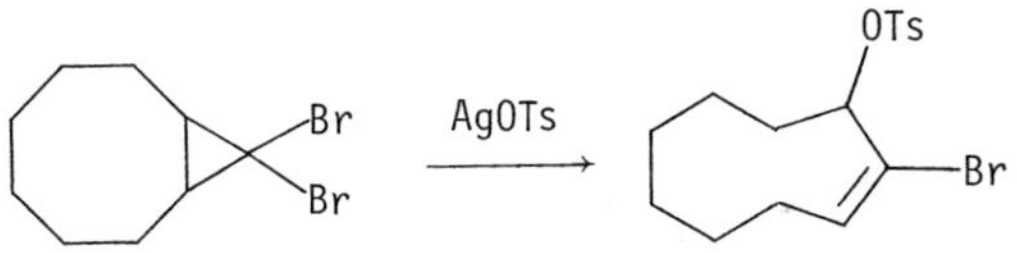

JOC (1976) 41 384

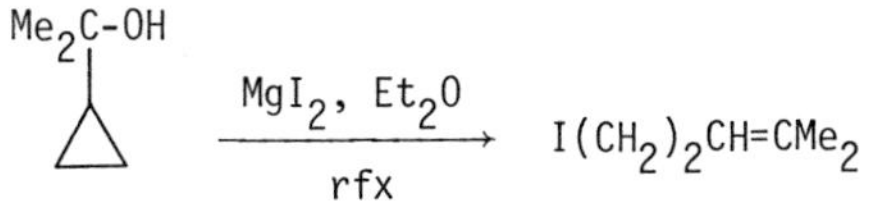

Me_2C-OH

$\xrightarrow[\text{rfx}]{MgI_2, Et_2O}$ $I(CH_2)_2CH=CMe_2$ 100%

JCS Chem Comm (1975) 303

$$PhCOCH_2COOEt + PCl_5 \xrightarrow{PhH} PhCCl=CHCOOH \qquad\qquad 50\%$$

JCS Chem Comm (1974) 288

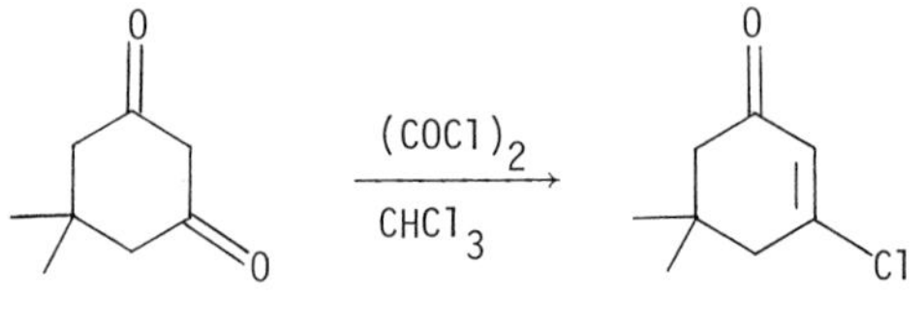

91%

JOC (1976) 41 636

$$Me_2CHCOCH_3 \xrightarrow[\text{pyridine}]{(CF_3SO_2)_2O} Me_2C=CMe + Me_2CHC=CH_2$$

with OSO_2CF_3 and OSO_2CF_3 substituents respectively

45%
total

Org Synth (1974) 54 79

Also via: Acetylenic halides (Section 308)

Section 372   Ketone - Ketone

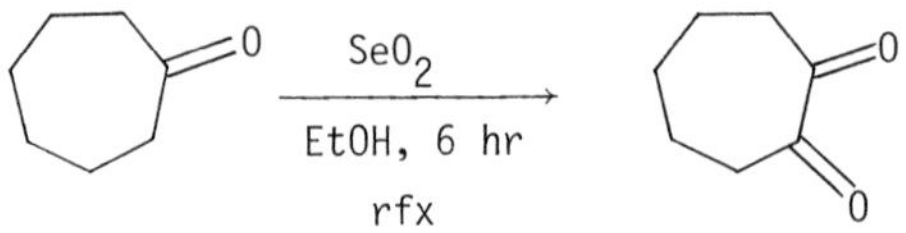

90%

Org React (1976) 24 261

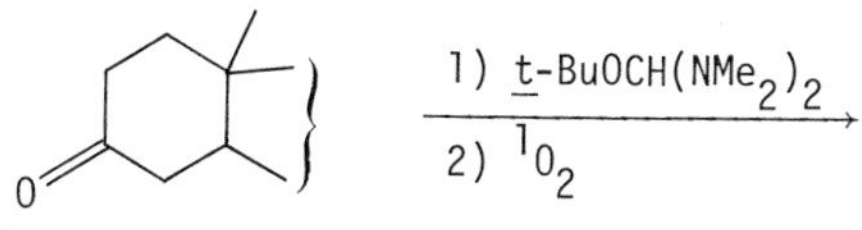

1) t-BuOCH(NMe$_2$)$_2$

2) $^1O_2$

87%

JACS (1976) 98 7868

OH O
| ||
Ph-CH-C-Ph

Yb(NO$_3$)$_3$

HCl, rfx

O O
|| ||
Ph-C-C-Ph

95%

Tetr Lett (1975) 4513

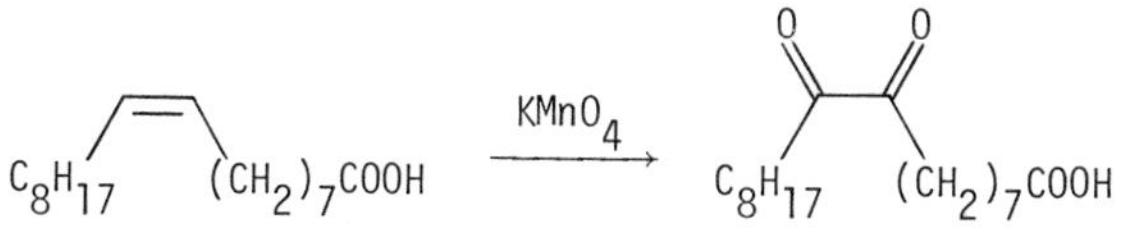

C$_8$H$_{17}$ (CH$_2$)$_7$COOH

KMnO$_4$

C$_8$H$_{17}$ (CH$_2$)$_7$COOH

46%

JOC (1974) 39 2314

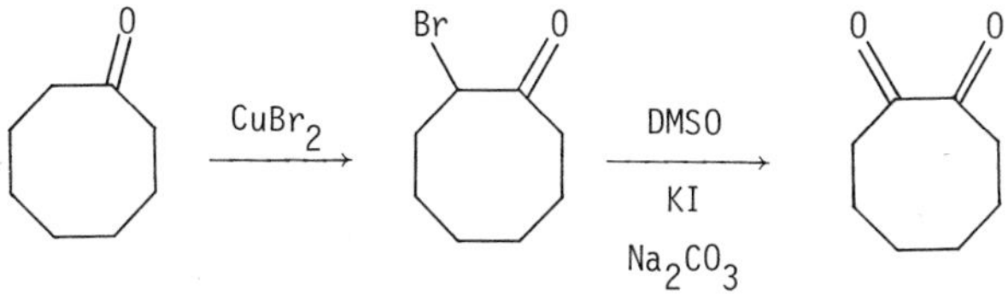

CuBr$_2$

DMSO

KI

Na$_2$CO$_3$

59%

JOC (1975) 40 1990

CHOH

CH$_2$STs$_2$
|
CH$_2$STs$_2$

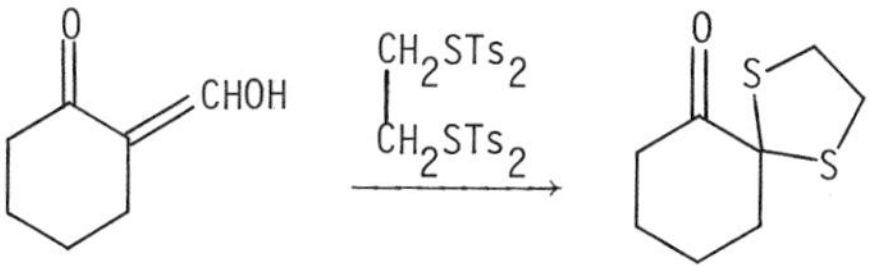

~50%

Org Synth (1974) 54 37

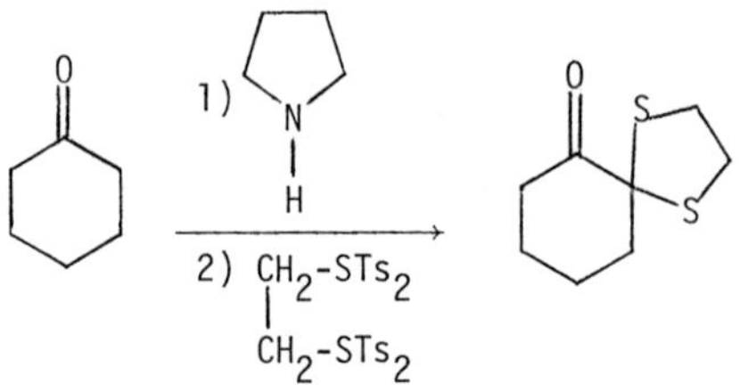

Org Synth (1974) <u>54</u> 39

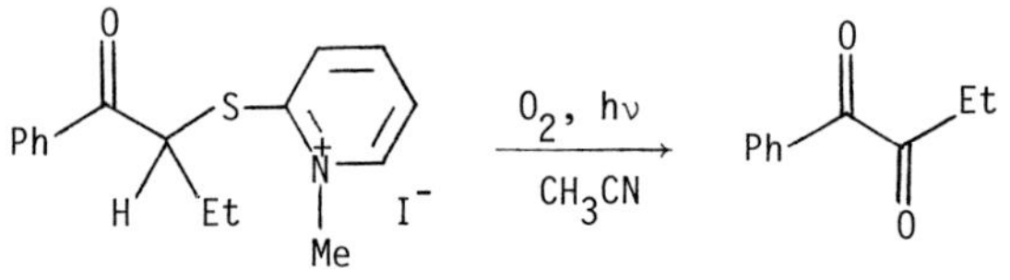

45%

Chem Lett (1975) 1033

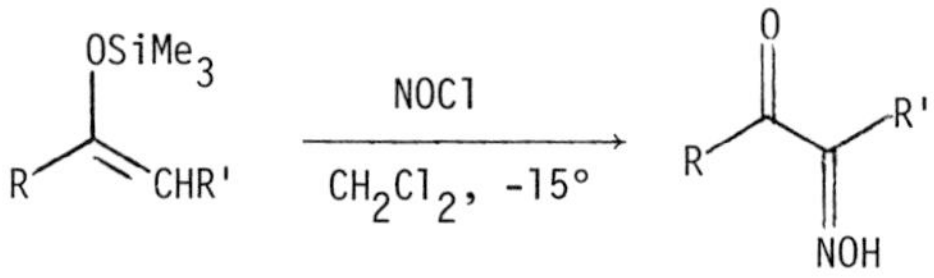

80%

JOC (1974) <u>39</u> 2558

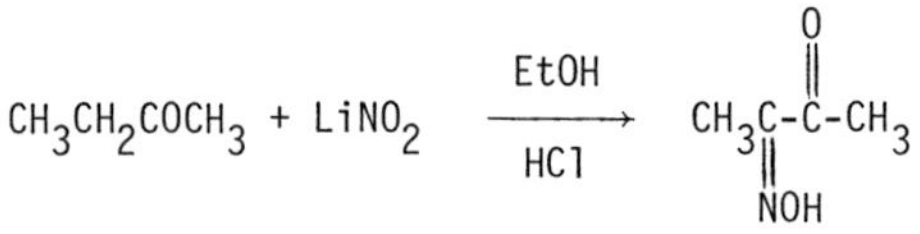

66-84%

$$CH_3CH_2COCH_3 + LiNO_2 \xrightarrow[HCl]{EtOH} CH_3C-\overset{O}{\underset{\underset{NOH}{\parallel}}{C}}-CH_3$$

Can J Chem (1974) <u>52</u> 1760

65%

$$Me_2C=C-NMe_2 \xrightarrow[\text{2) 10\% } H_2SO_4]{\text{1) PhLi}} Me_2CHCOCOPh$$
with CN substituent

JACS (1975) <u>97</u> 2276

90%

Cl
Cl
O
O
O
2) ⁻OH, H2O
O
O

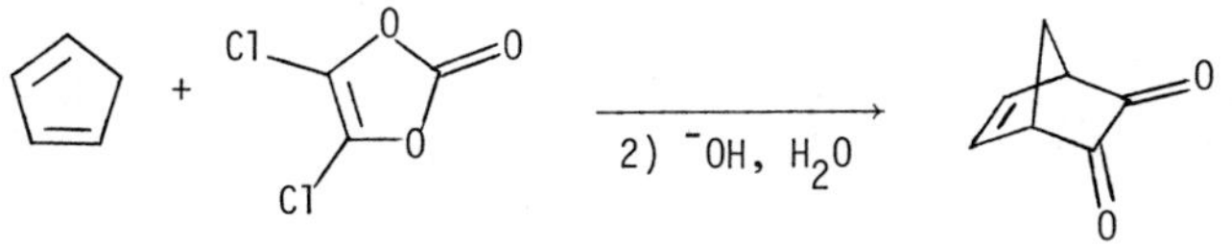

50%

Synthesis (1976) 256

Bz
N
S
Cl⁻
CHO
CHO
Me
Me
O
OH

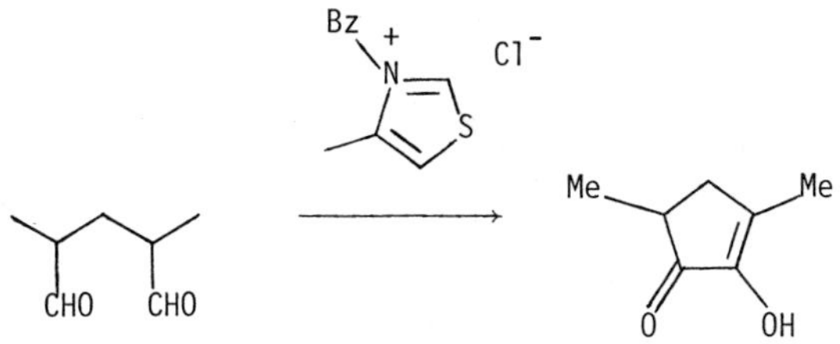

88%

JCS Chem Comm (1976) 804

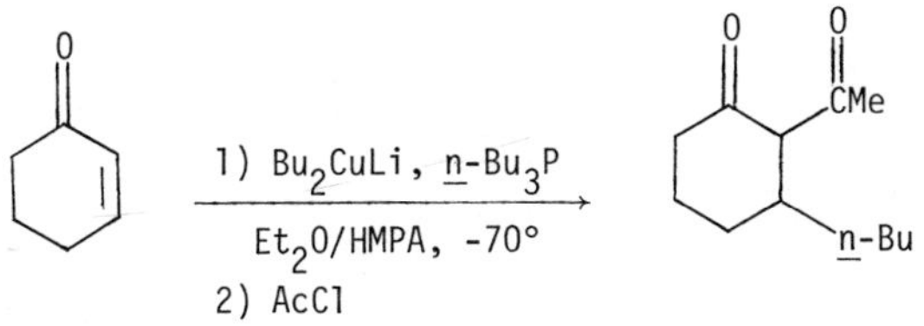

O
1) Bu2CuLi, n-Bu3P
Et2O/HMPA, -70°
2) AcCl
O
O
CMe
n-Bu

92%

Tetr Lett (1975) 1535

O
1) CH3CNO
2) H2/Pd
3) OH⁻
O
O
O

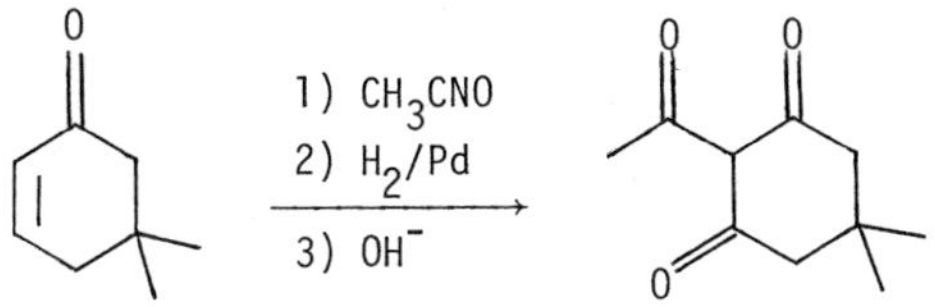

60%

Dokl Chem (1974) 216 404

MeCOCH$_2$CH$_2$COOEt $\xrightarrow[\text{xylene, rfx}]{\text{Ph}_3\text{COK}}$ [structure]   60%

Z Chem (1975) <u>15</u> 190

[structure] + R^{2}COCH$_2$COR^3 $\xrightarrow[\text{CuBr}]{\text{NaH}}$ [structure]

(Improved Hurtley reaction)

Tetrahedron (1975) <u>31</u> 2607
JCS Perkin I (1975) <u>1</u>267

[structure] + MeCOCH$_2$COMe $\xrightarrow[\text{CuBr}]{\text{NaH}}$ [structure]   98%

Angew Int Chem (1974) <u>13</u> 340

PhCH$_2$Br + Ni(acac)$_2$ $\xrightarrow{\text{DMF}}$ MeCOCH(CH$_2$Ph)COMe   69%

Tetr Lett (1975) 1727

[structure] $\xrightarrow[\text{2) EtOAc}]{\text{1) NaH}}$ [structure]   78%

Rec Trav Chim (1976) <u>95</u> 81

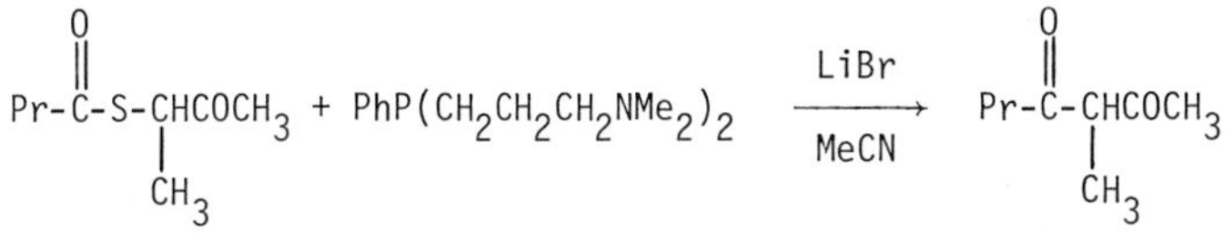

Pr-C-S-CHCOCH₃ + PhP(CH₂CH₂CH₂NMe₂)₂ →[LiBr][MeCN] Pr-C-CHCOCH₃   ~80%

Org Synth (1976) 55 127

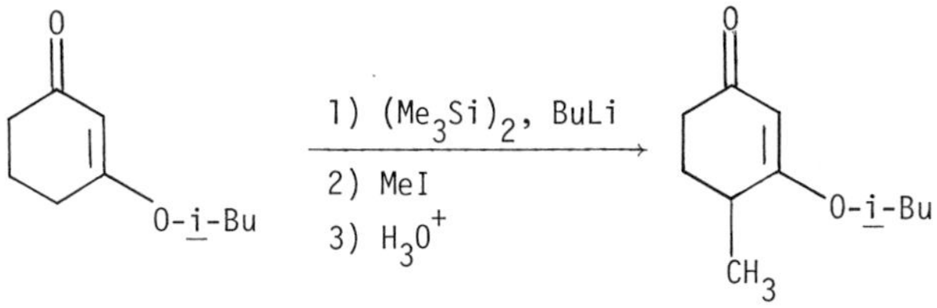

1) (Me₃Si)₂, BuLi
2) MeI
3) H₃O⁺

91%

Rec Trav Chim (1976) 95 81

Review: "The Synthesis of 1,4-Diketones"

Aust J Chem (1976) 29 339

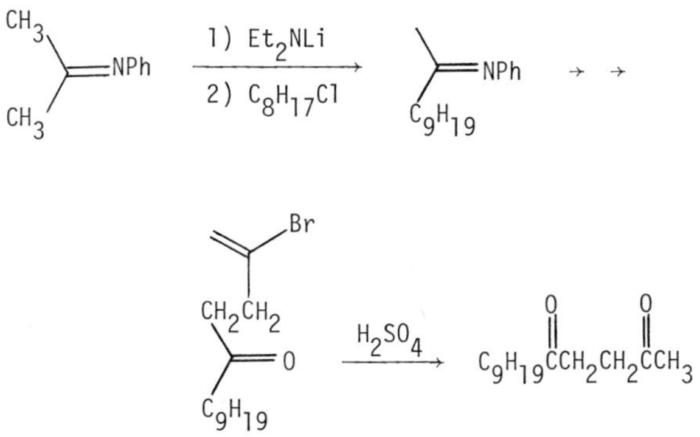

1) Et₂NLi
2) C₈H₁₇Cl

H₂SO₄

C₉H₁₉CCH₂CH₂CCH₃

Liebigs Ann Chem (1975) 719

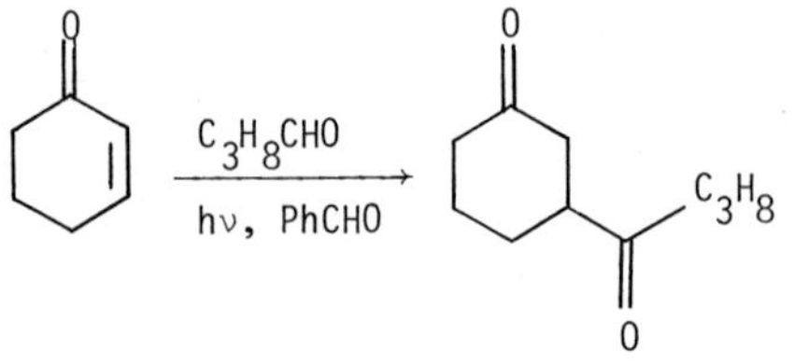

42%

Synth Comm (1976) 6 417

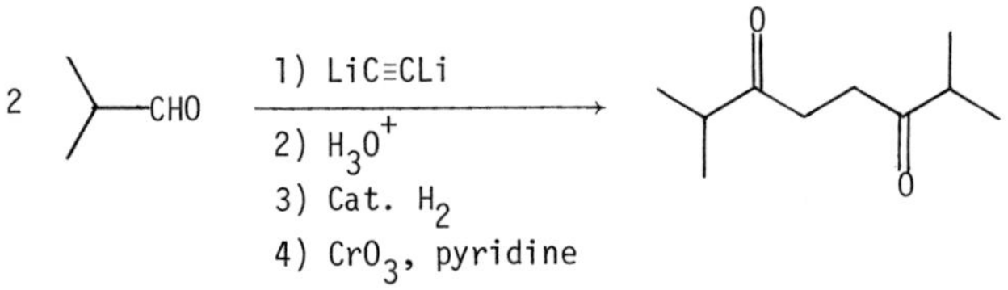

62%

JOC (1975) 40 1131

PhCHO + $H_2C=CHCOCH_3$ $\xrightarrow[\substack{DMF \\ \Delta}]{NaCN}$ $PhCOCH_2CH_2COCH_3$ 82%

Chem Ber (1974) 107 2453

Et-CH + $CH_2=CHCMe$ $\xrightarrow[\text{thiazolium salt}]{Et_3N}$ $EtCCH_2CH_2C-Me$ 66%

Tetr Lett (1974) 4505

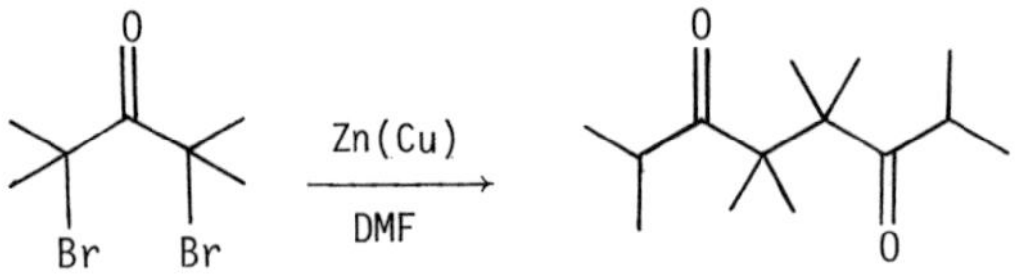

71%

JACS (1974) 96 606

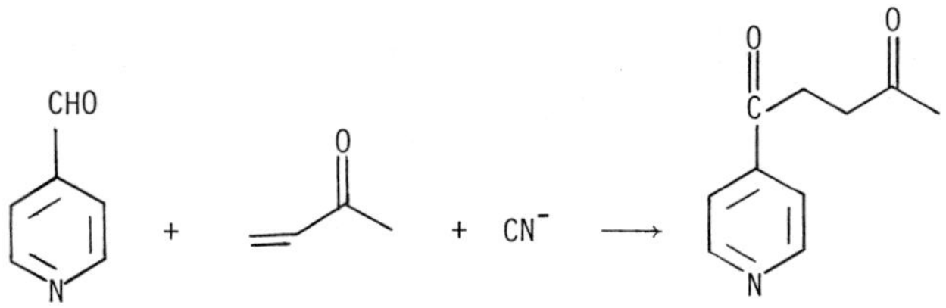

~75%

Chem Ber (1974) 107 2453

MeSO$_2$$\overline{C}$HCOPh + BrCH$_2$COPh $\longrightarrow$ $\xrightarrow{\text{Zn/HOAc}}$ PhCOCH$_2$CH$_2$COPh   76%

Coll Czech Chem Comm (1974) 39 192

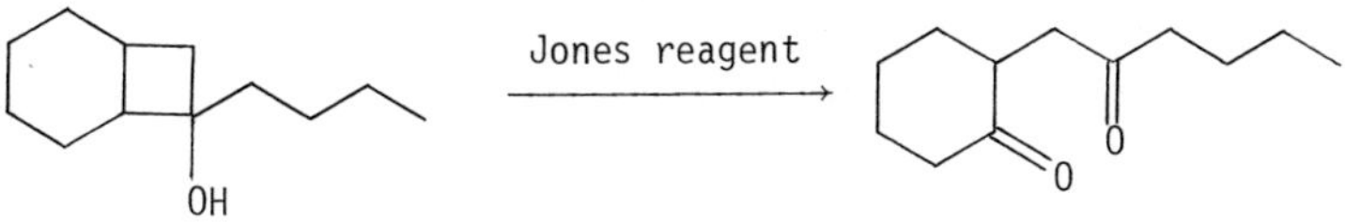

82%

Can J Chem (1976) 54 3113

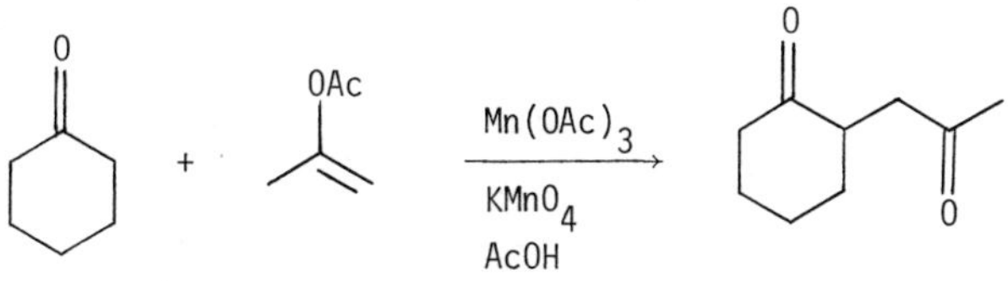

22%

JOC (1974) 39 3457

PhCOCH$_2$COMe $\xrightarrow{\begin{array}{l}\text{1) NaH, THF, 0°}\\ \text{2) \underline{n}-BuLi}\\ \text{3) CuCl cat, I}_2\end{array}}$ (PhCOCH$_2$COCH$_2$)$_2$ 75%

JOC (1975) 40 3887

t-BuCOCH$_3$ $\xrightarrow[\text{2) CuCl}_2\text{, DMF}]{\text{1) LDA, THF}}$ t-BuCOCH$_2$CH$_2$CO-t-Bu   95%

JACS (1975) <u>97</u> 2912

H$_2$C=C-Ph $\xrightarrow[\text{DMSO, }\Delta]{\text{Ag}_2\text{O}}$ PhCOCH$_2$CH$_2$COPh   73%
  |
 OTMS

JACS (1975) <u>97</u> 649

Et
   C=NR $\xrightarrow[\text{3) H}_3\text{O}^+]{\begin{array}{l}\text{1) LDA, THF}\\\text{2) I}_2\text{, THF}\end{array}}$ EtCOCH$_2$CH$_2$COEt   80%
Me

Synthesis (1975) 256

JACS (1974) <u>96</u> 5272   90%

Chem Lett (1975) 89   66%

$$CH_2=CHCOMe \quad + \quad \text{[OLi cyclohexenyl]} \quad \xrightarrow[-78°]{CH_3CN} \quad \text{[2-substituted cyclohexanone]} \quad CH_2-CHCOMe$$
$$\underset{Fp^+}{|} \qquad\qquad\qquad\qquad\qquad\qquad\qquad\qquad\qquad\qquad \underset{Fp}{|}$$

$Fp = \pi-C_5H_5Fe(CO)_2$

JOC (1975) <u>40</u> 3621

$$\overset{O}{\overset{||}{Cl C}}(CH_2)_8 COCl \quad \xrightarrow[\underset{-78°}{Cu^+}]{t\text{-BuMgCl}} \quad \underline{t}\text{-BuCO}(CH_2)_8 CO\underline{t}Bu \qquad\qquad 84\%$$

Aust J Chem (1974) <u>27</u> 2525

Review: "Acylation of Enamines"

Chem and Ind (1974) 731

Section 373   <u>Ketone - Nitrile</u>

$$PhCH_2CN \quad \xrightarrow[\text{2) } H_2O]{\text{1) } Me_2AlCl} \quad PhCH_2\overset{}{\underset{\underset{CN}{|}}{C}}OCH\text{-}Ph$$

Liebigs Ann Chem (1975) 636

70%

Tetr Lett (1976) 683

PhCHO + H$_2$C=CCN $\xrightarrow{\text{CN}^-}$ PhCOCH$_2$CH$_2$CN   80%

Chem Ber (1974) 107 210

86%

Chem Lett (1975) 237

Section 374   Ketone - Olefin

For the oxidation of allylic alcohols to olefinic ketones, see Section 168.
   Vol. 1 (Ketones from Alcohols and Phenols).

For the oxidation of allylic methylene groups (C=C-CH$_2$ → C=C-CO), see
   Section 170, Vol. 1 and 2 (Ketones from Alkyls and Methylenes).

For the alkylation of olefinic ketones, see also Section 177, Vol. 1 and 2
   (Ketones from Ketones) and Section 74 (Alkyls from Olefins), Vol. 3, for
   conjugate alkylations.

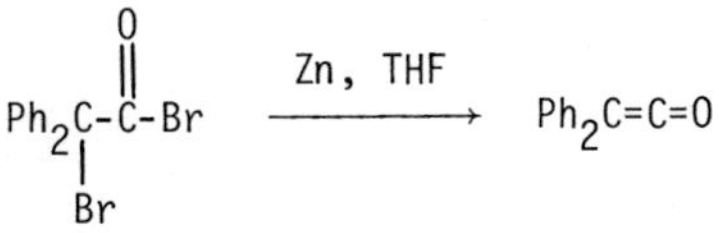

$$Ph_2C-\overset{\overset{\displaystyle O}{\|}}{C}-Br \xrightarrow{\text{Zn, THF}} Ph_2C=C=O \qquad \qquad 95\%$$
$$\underset{Br}{|}$$

JCS Perkin I (1975) 1600

$$Ac_2O + Bu_3Al\cdot NBu_3 \xrightarrow[\text{xylene}]{130°} CH_2=C=O \qquad \qquad 80\%$$

J Organometal Chem (1975) 97 21

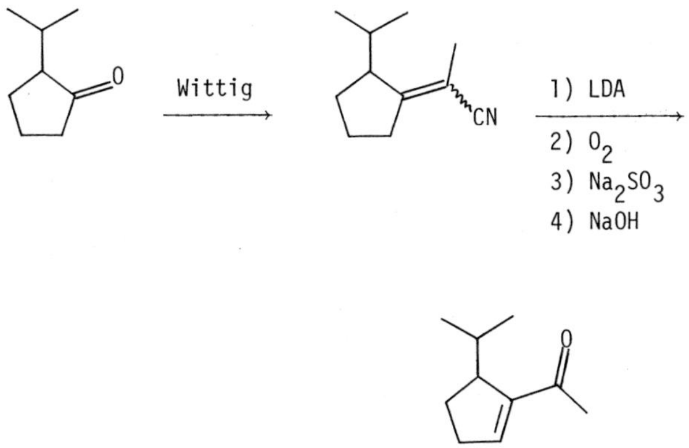

Wittig

1) LDA
2) O_2
3) Na_2SO_3
4) NaOH

55%

JOC (1976) 41 2939

1) t-BuO⁻
2) RCOR'
3) R"Li

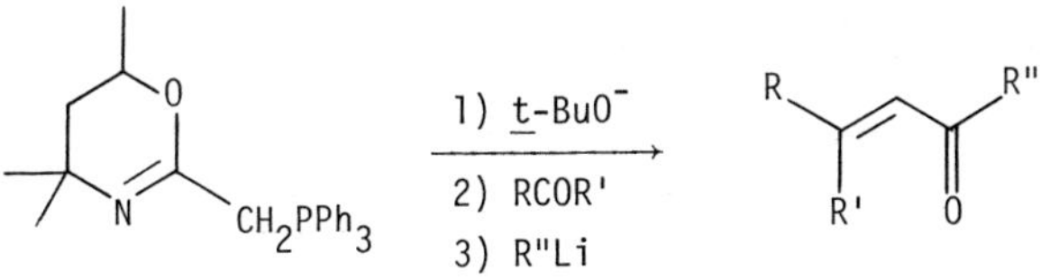

50-80%

JOC (1974) 39 629

Me$_2$C=CHCOMe + PhP=C=C(OLi)Ph $\xrightarrow[\text{HMPA}]{\text{PhH}}$

35%

Tetrahedron (1975) $\underline{31}$ 1331

$(EtO)_2\overset{\overset{\displaystyle Me}{|}}{\underset{\underset{\displaystyle O}{||}}{P}}-N-CH=C=CH_2$ $\xrightarrow[\text{2) BzBr}]{\text{1) BuLi}}$ PhCH$_2$$\underset{\underset{\displaystyle O}{||}}{C}$-CH=CH$_2$

65%

Tetr Lett (1976) 835

t-Bu-C≡CH $\begin{array}{l}\text{1) } \underline{i}\text{-Bu}_2\text{AlH} \\ \text{2) BuLi} \\ \text{3) } \end{array}$ ⟶

68%

CH$_2$$\underset{\underset{\displaystyle O}{||}}{C}$Me

JCS Chem Comm (1976) 17

$\xrightarrow[\text{2) PhCHO}]{\text{1) PCl}_5}$

80%

C=O
|
CH=CHPh

Experientia (1976) $\underline{32}$ 1491

+ MeCOCH=PPh$_3$ ⟶

92%

Gazz Chim Ital (1975) $\underline{105}$ 109

$Me_2NCH=CH-CH(OEt)_2$ + $PhCOCH_3$ $\longrightarrow$ $Me_2NCH=CHCH=CHCOPh$ 74%

Liebigs Ann Chem (1975) 874

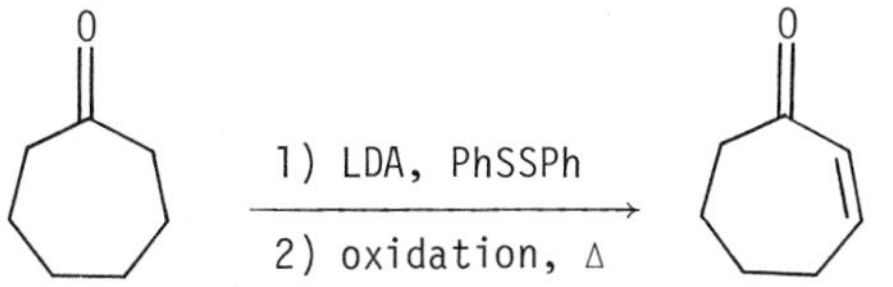

1) LDA, PhSSPh

2) oxidation, Δ

~70%

JACS (1976) 98 4887

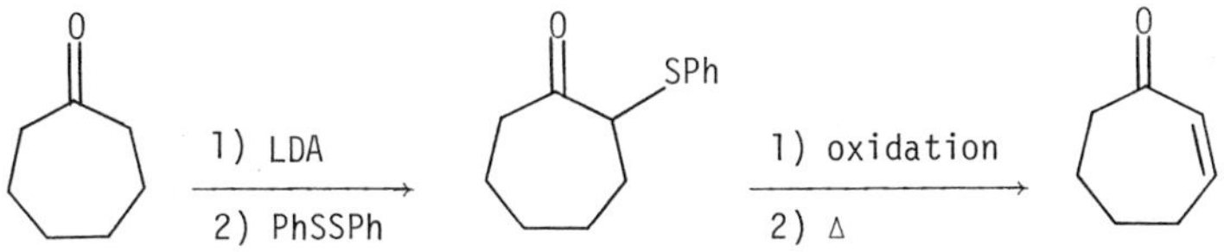

1) LDA

2) PhSSPh

1) oxidation

2) Δ

71%

JACS (1976) 98 4887

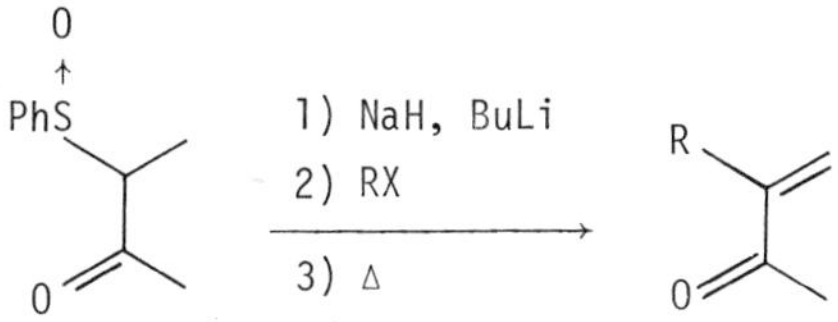

1) NaH, BuLi

2) RX

3) Δ

JCS Chem Comm (1974) 497

1) LDA

2) BzBr

3) Δ

55%

JCS Chem Comm (1975) 72

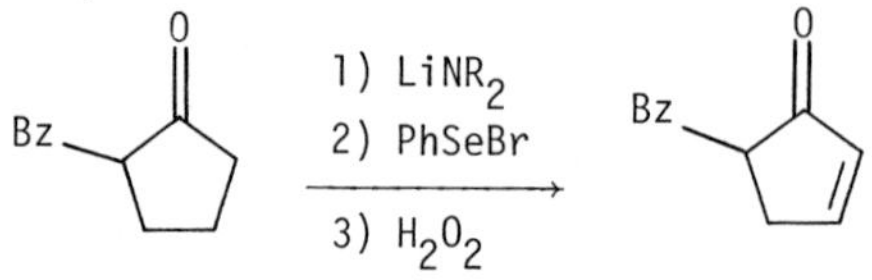

1) LiNR$_2$
2) PhSeBr
3) H$_2$O$_2$

66%

JACS (1975) 97 5434

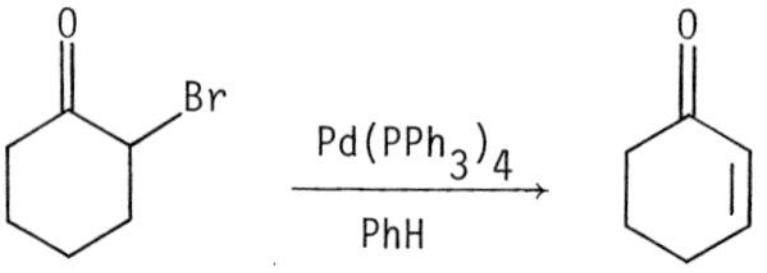

Pd(PPh$_3$)$_4$
PhH

70%

JOC (1975) 40 2976

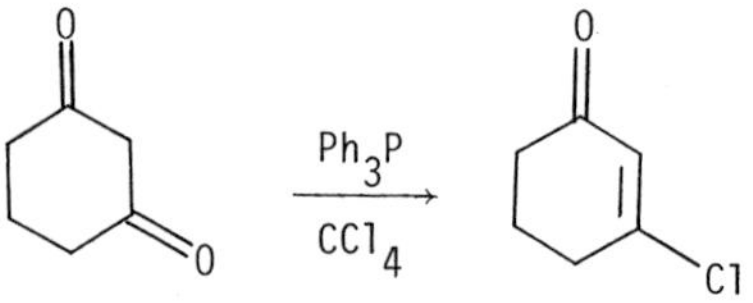

Ph$_3$P
CCl$_4$

Synthesis (1975) 708

Ph$_3$PCl$_2$
Et$_3$N, benzene

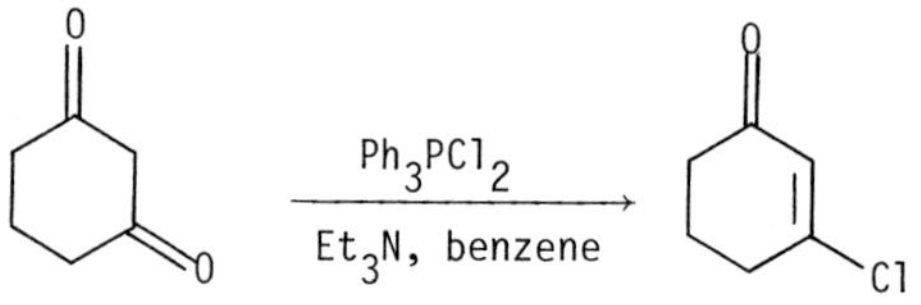

91%

Synth Comm (1975) 5 193

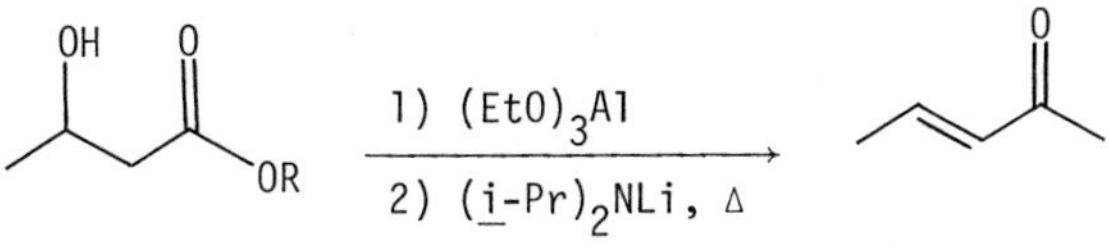

~50%

JACS (1974) <u>96</u> 6153

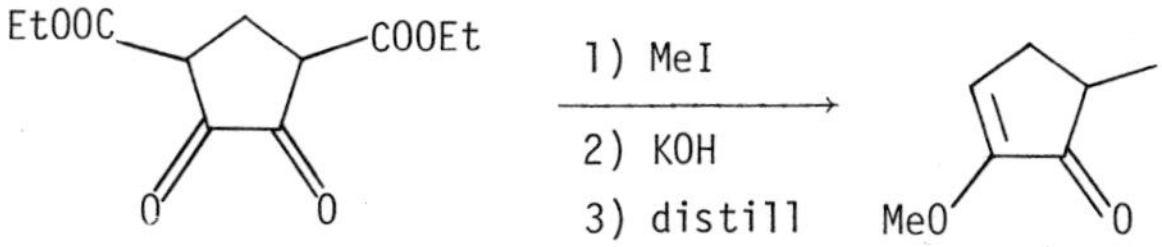

54%

Synthesis (1974) 33

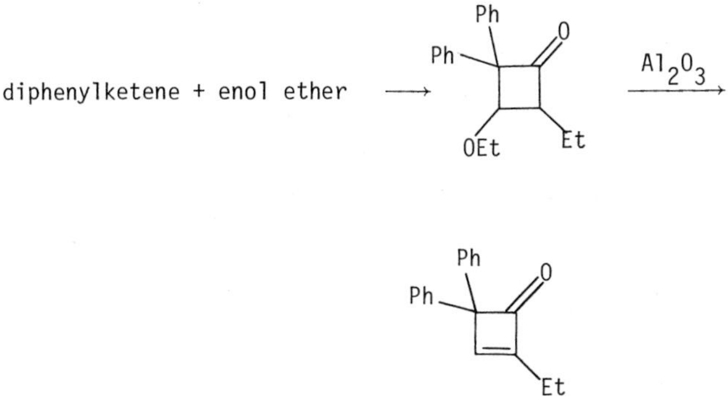

diphenylketene + enol ether ⟶

86%

Angew Int Ed (1975) <u>14</u> 499

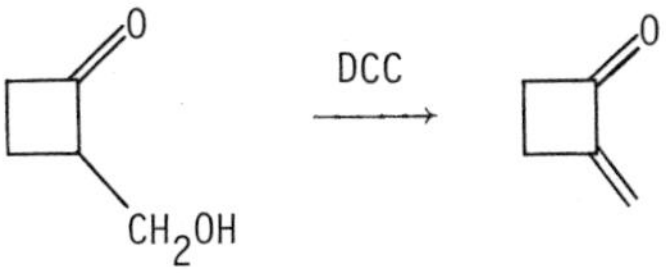

90%

Bull Soc Chim Fr (1974) 1015

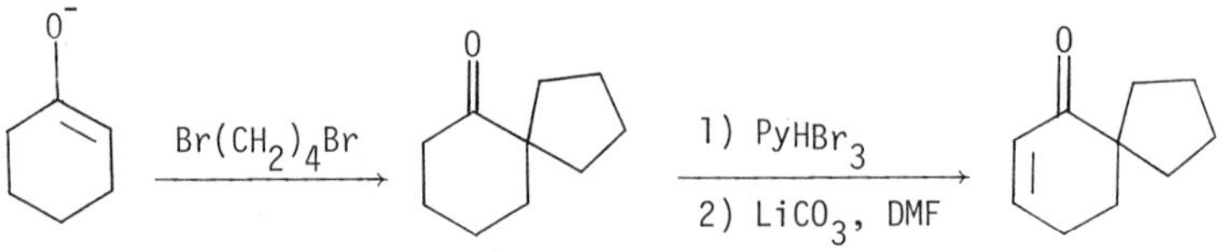

49%

50%

JCS Perkin I (1974) 964

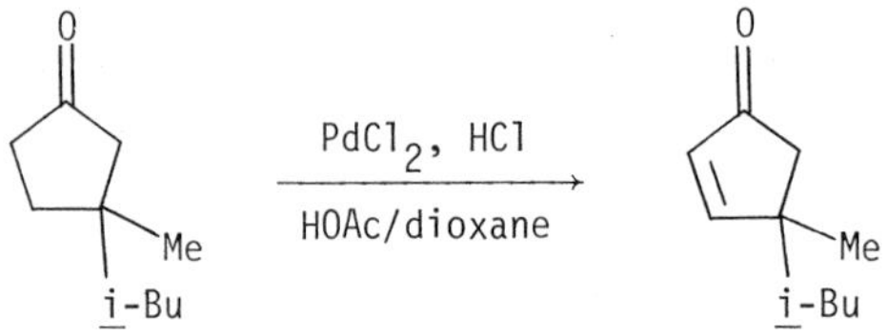

64%

Synthesis (1976) 240

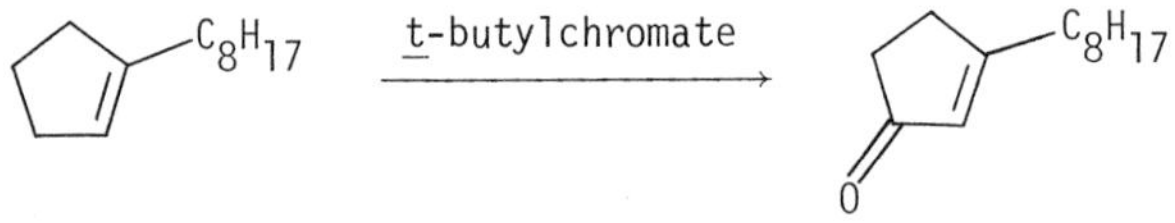

40%

Indian J Chem (1975) 13 29

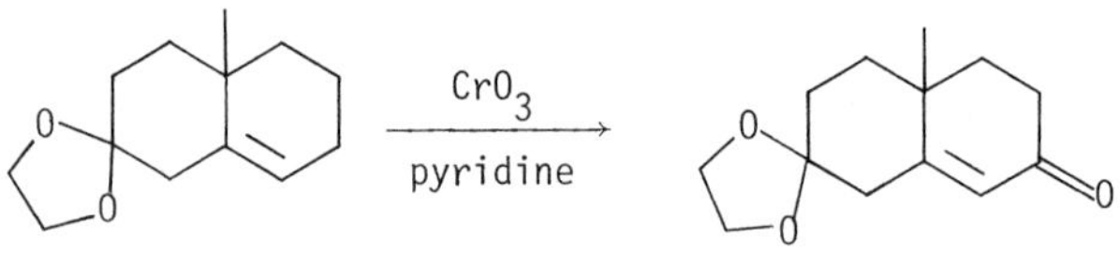

70%

Synth Comm (1976) 6 217

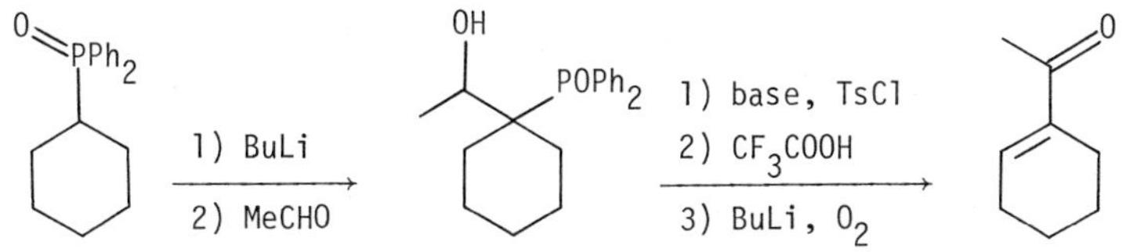

JCS Chem Comm (1975) 148

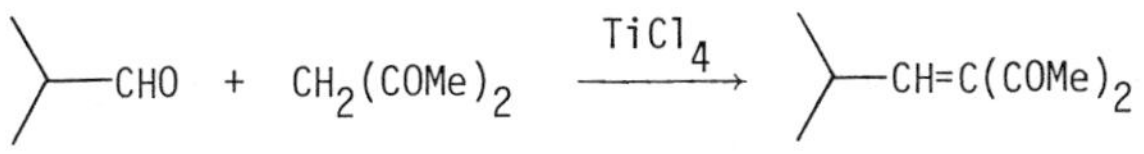

Synthesis (1974) 667

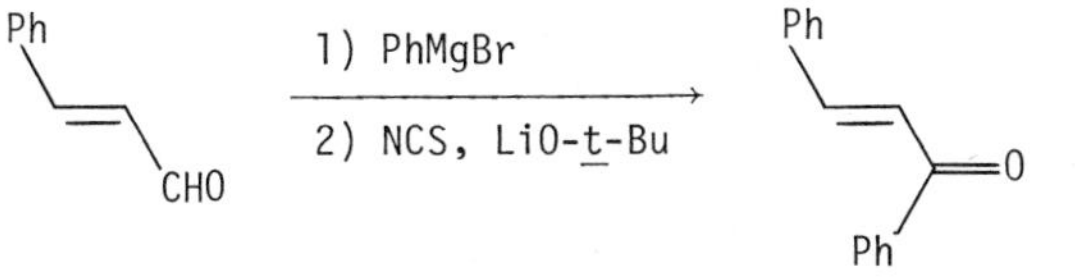

83%

Chem Lett (1976) 771

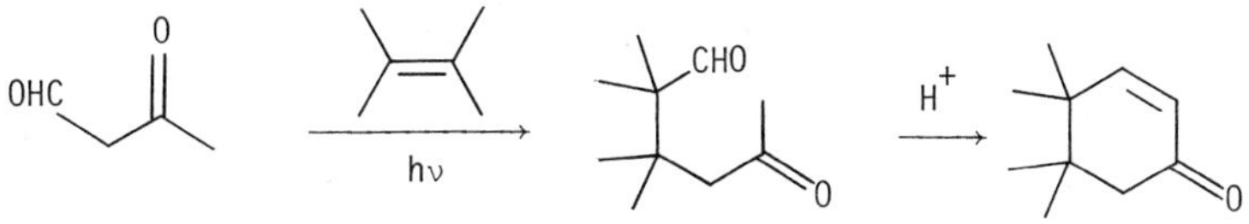

83%

JOC (1975) 40 1865

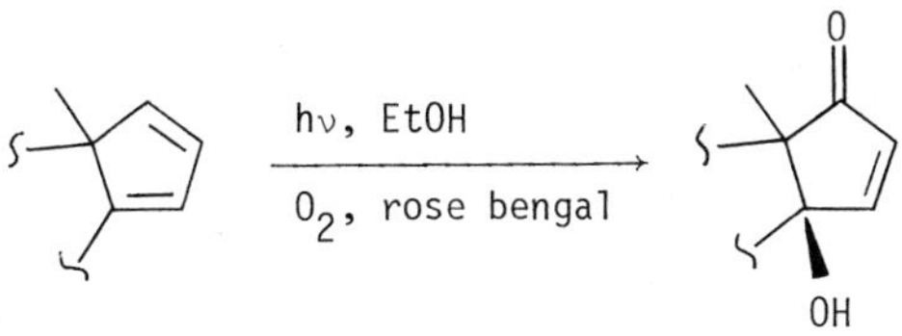

40-90%

Comptes Rendus C (1974) 279 347

30%

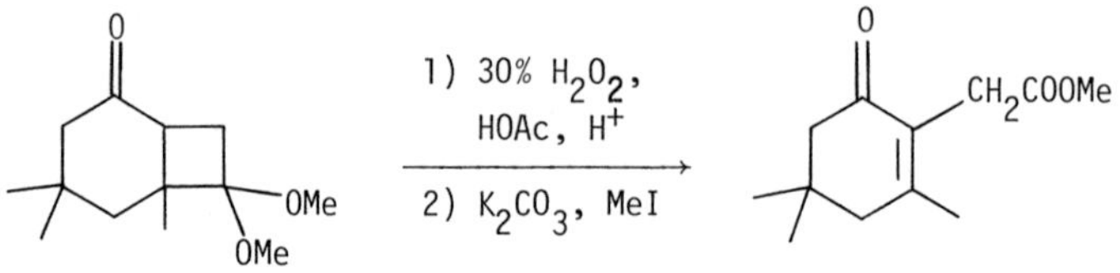

O
OMe
OMe
1) 30% H2O2,
HOAc, H+
2) K2CO3, MeI
O
CH2COOMe

42%

Synth Comm (1975) 5 161

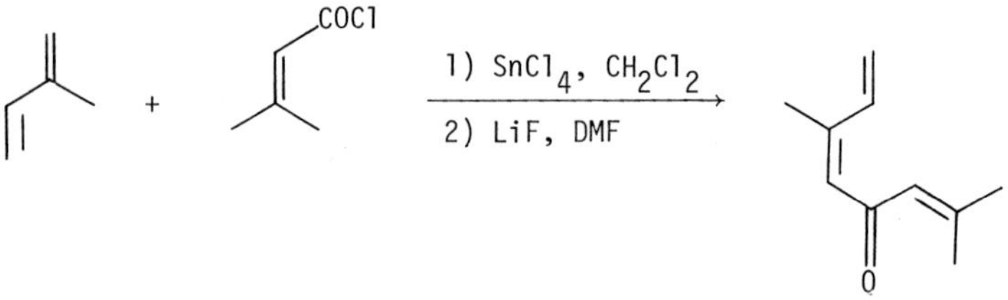

COCl
+
1) SnCl4, CH2Cl2
2) LiF, DMF
O

90%

Tetr Lett (1974) 3197

t-Bu
Me
C=C
H
HgCl
CH3COCl
AlCl3
t-Bu
Me
C=C
H
C-CH3
O

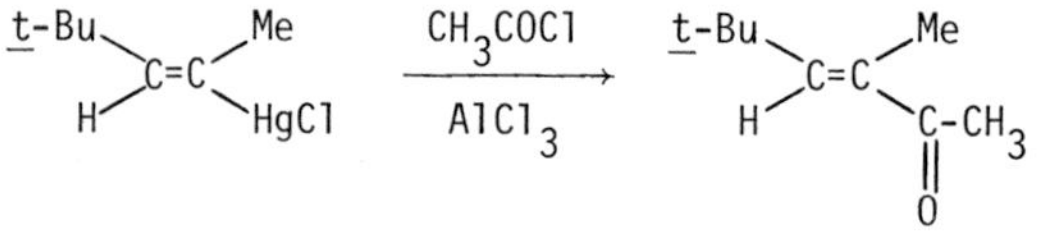

99%

Tetr Lett (1976) 3097

TMSCH=CHTMS + PhCOCl $\xrightarrow[-20°]{AlCl_3}$ TMSCH=CHCOPh 80%

Bull Soc Chim Fr (1975) 2143

$PhCH{=}CHSiMe_3$ + $PhCH_2COCl$ ⟶ $PhCH{=}CHCOCH_2Ph$ 86%

JCS Chem Comm (1975) 633

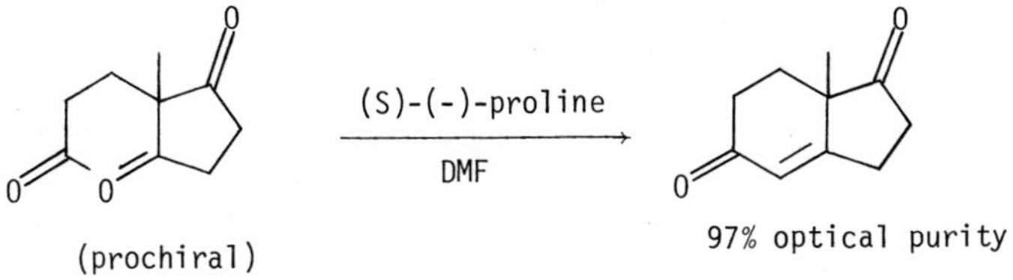

(S)-(-)-proline
DMF

(prochiral)

97% optical purity

97%

JOC (1974) 39 1615

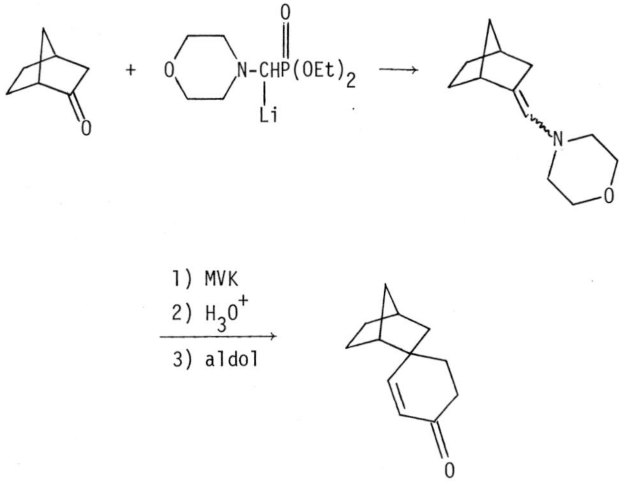

1) MVK
2) H₃O⁺
3) aldol

JOC (1976) 41 3337

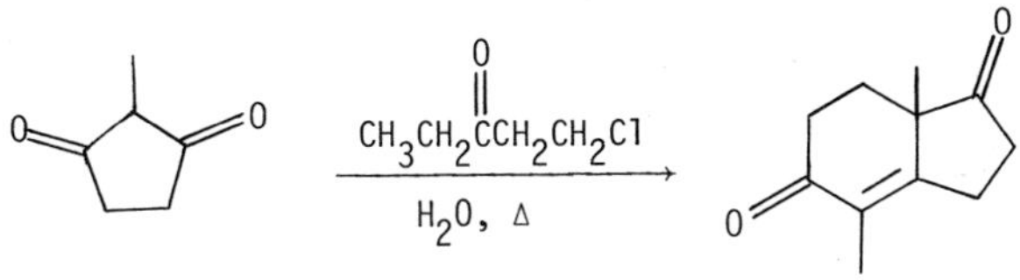

$CH_3CH_2CCH_2CH_2Cl$

H_2O, Δ

78%

JOC (1976) 41 3767

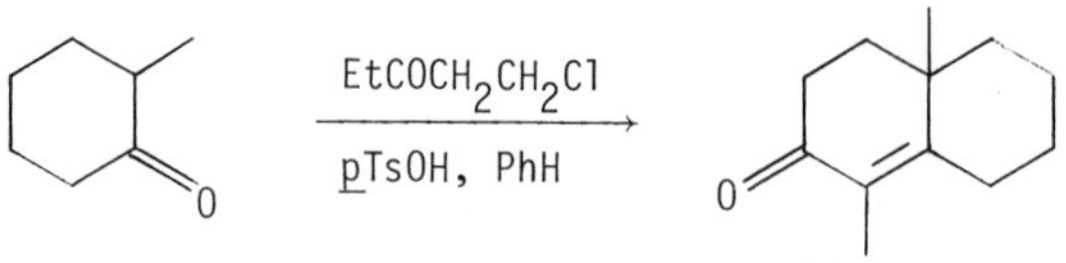

EtCOCH$_2$CH$_2$Cl
––––––––––––––––→
pTsOH, PhH

Tetr Lett (1975) 527

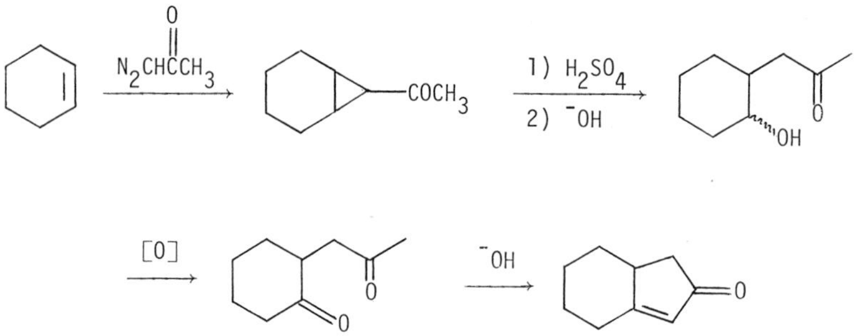

N$_2$CHCCH$_3$

—COCH$_3$

1) H$_2$SO$_4$
––––––––––→
2) $^-$OH

[0]
––––→

$^-$OH
––––→

Chem Lett (1976) 1025

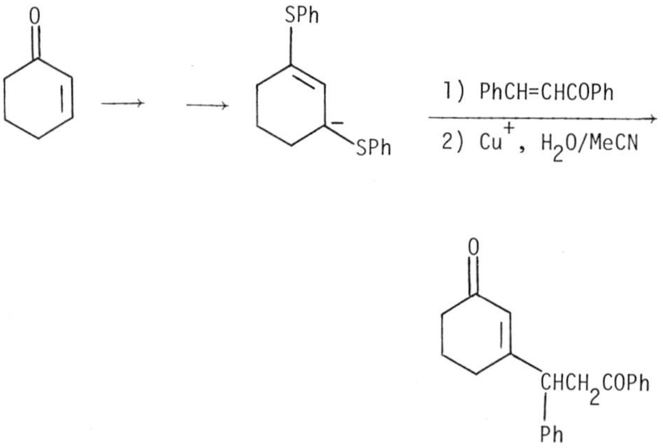

SPh

SPh

1) PhCH=CHCOPh
––––––––––––––––→
2) Cu$^+$, H$_2$O/MeCN

59%

CHCH$_2$COPh
|
Ph

JOC (1976) 41 2506

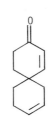

1) piperidine, -H$_2$O

2) MVK

3) HOAc, NaOAc, H$_2$O

45%

Synth Comm (1976) <u>6</u> 237

$$\xrightarrow[\text{L-phenylalanine}]{\text{HClO}_4}$$

82% yield

82% ee

$\longrightarrow$ estrone

JACS (1975) <u>97</u> 5282

$$\xrightarrow[\text{Et}_3\text{N}]{\text{Me}_3\text{SiCl}}$$

OTMS

1) CH$_2$I$_2$, Zn/Ag

2) $^-$OH

∿80%
overall

Pure and Appl Chem (1975) <u>48</u> 317

CH_3CH_2COOH $\xrightarrow[\text{2) } H_2O]{\text{1) } CH_2=CHLi}$ $CH_3CH_2COCH=CH_2$ 92%

Tetr Lett (1974) 2877

+ PhCHO $\xrightarrow{HCl}$ 89%

Can J Chem (1974) 52 2157

$\underline{n}$-PrCH$_2$CH= $\xrightarrow[\substack{\text{2) } \underline{n}\text{-hexyl I} \\ \text{3) hydrolysis}}]{\text{1) } \underline{i}\text{-Pr}_2\text{NLi}}$ $\underline{n}$-PrCH=CH-C-hexyl   81%

Tetr Lett (1975) 925

COOMe $\xrightarrow[\substack{\text{2) NaIO}_4 \\ \text{3) } \Delta}]{\text{1) PhSCH}_2\text{I}}$ $CH_2=CH-COCH_2COOMe$ 72%

JOC (1974) 39 2648

$\xrightarrow[\text{2) } H_3O^+]{\text{1) PhMgBr or PhCu}}$ 77%

JOC (1976) 41 2935; 2937

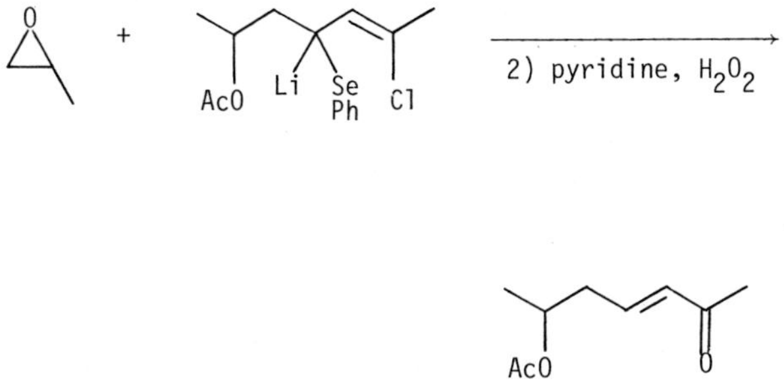

AcO
Li
Se
Ph
Cl
2) pyridine, H2O2
AcO

80%

JOC (1975) 40 2570

H
N
Et2N-C
O
Li+
+ PhCOOCH3
COPh

95%

Angew Int Ed (1976) 15 171

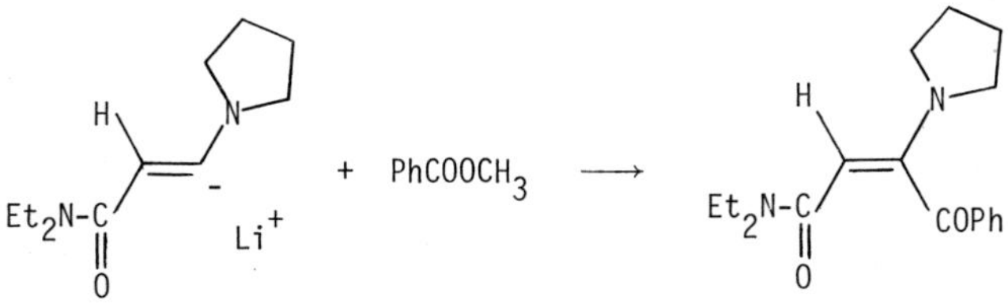

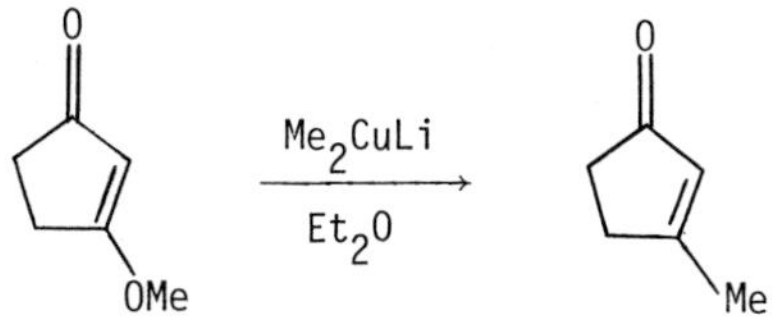

O
OMe
Me2CuLi
Et2O
Me

70%

Indian J Chem (1974) 12 325
Chem Comm (1974) 1005

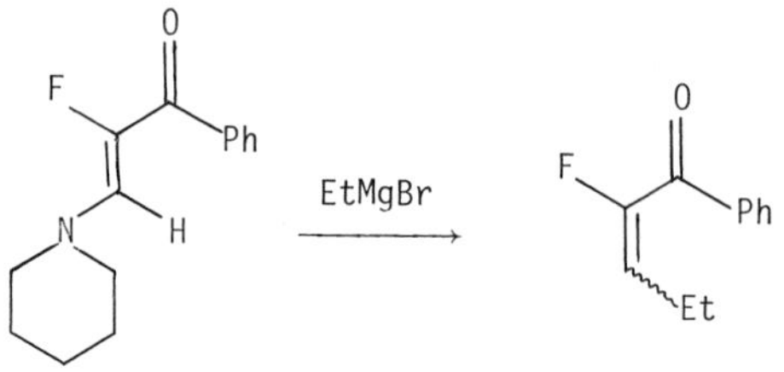

Bull Soc Chim France (1976) 439   61%

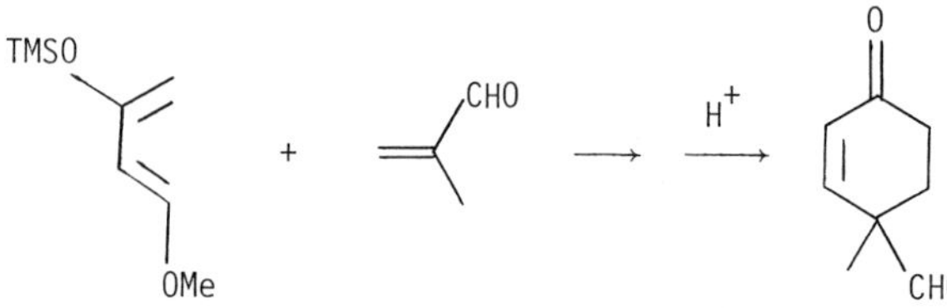

JACS (1974) <u>96</u> 7807   72%

Synth Comm (1975) <u>5</u> 27   ~50%

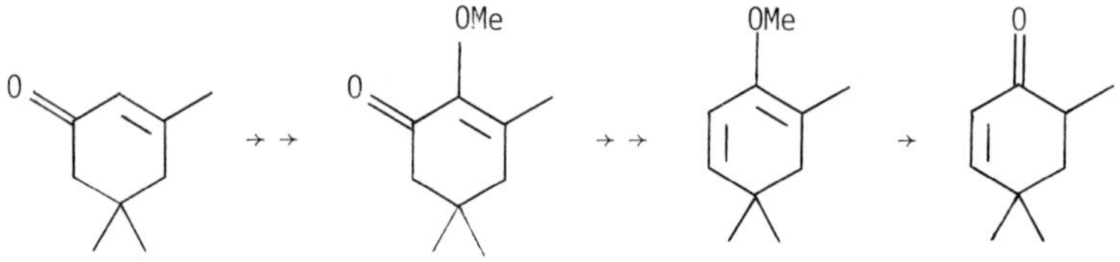

1) LDA, THF

2) TiCl$_3$, HCl, H$_2$O

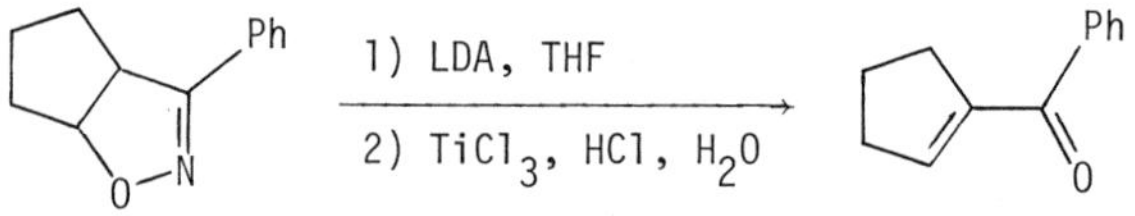

Angew Int Ed (1976) <u>15</u> 50   61%

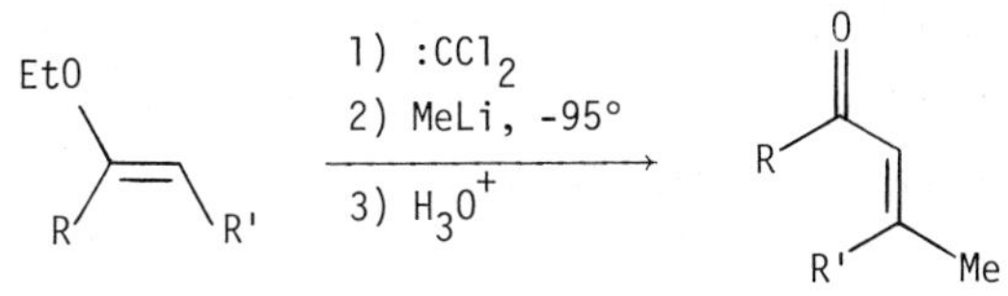

1) :CCl$_2$
2) MeLi, -95°
3) H$_3$O$^+$

40-50%

Tetr Lett (1974) 3297

MeCOCH$_2$Me

1) Me$_2$NCH=CMe (NO$_2$)

NaH

2) Si gel

→ MeCOC=CHCOMe (Me)

high yields

Chem Ber (1974) 107 1499

PhCH$_2$COPh + PhN=CHOEt → PhCCOPh (CHNHPh)

190°

61%

Z Chem (1975) 15 15

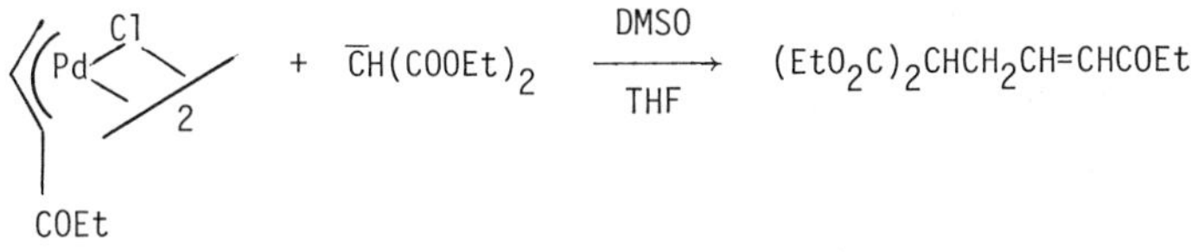

+ $\overline{C}$H(COOEt)$_2$

DMSO
THF

(EtO$_2$C)$_2$CHCH$_2$CH=CHCOEt

100%

Tetr Lett (1975) 2591

1) Li-HMPA-THF-EtOH
2) HCl

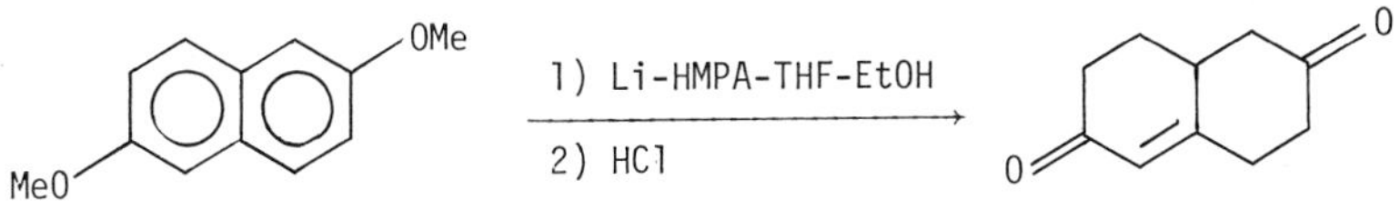

60%

JOC (1975) 40 2841

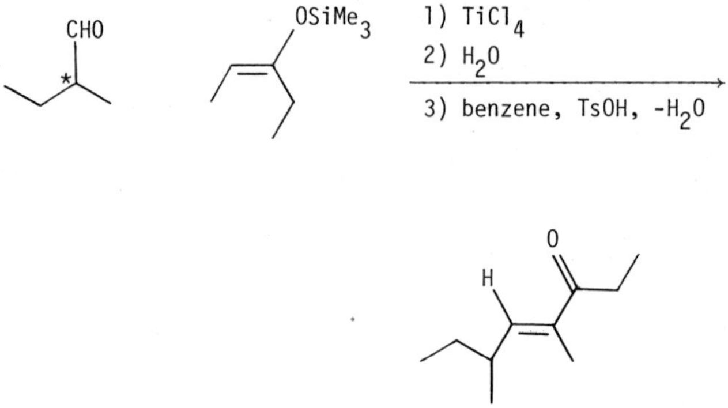

83%

97% ee

Chem Lett (1976) 279

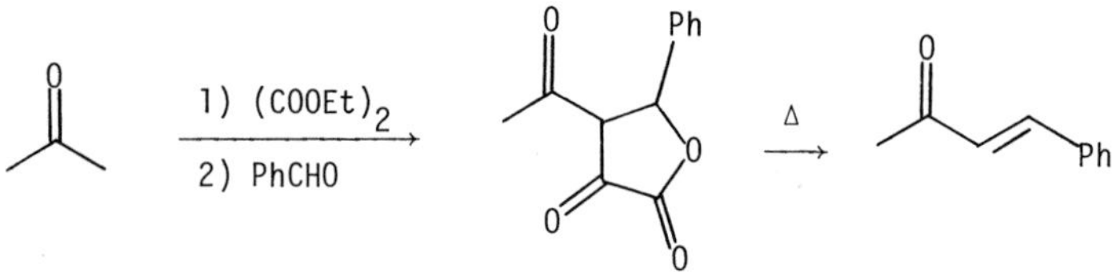

90%

Tetr Lett (1976) 4691

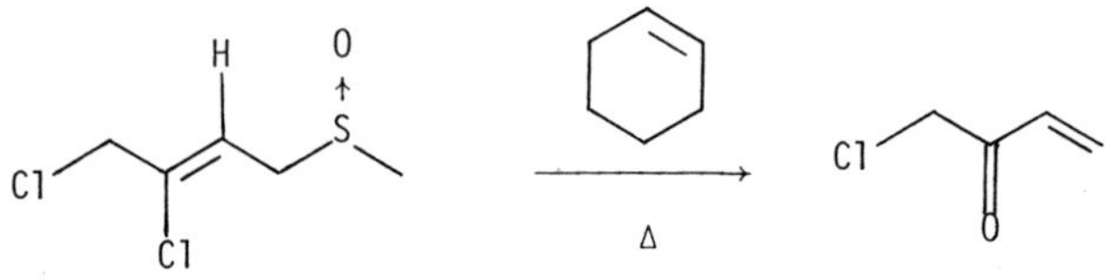

90%

JACS (1976) 98 4577

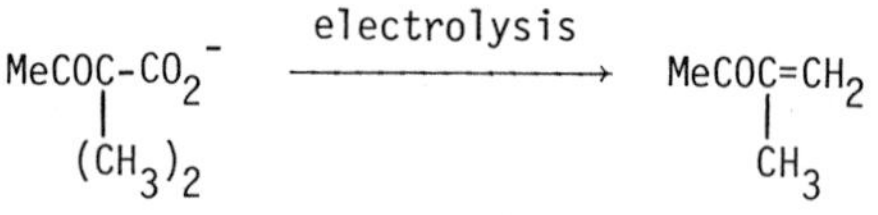

electrolysis

$MeCOC-CO_2^-$ $\longrightarrow$ $MeCOC=CH_2$ 90%

$(CH_3)_2$ CH_3

Comptes Rendus C (1975) 280 731

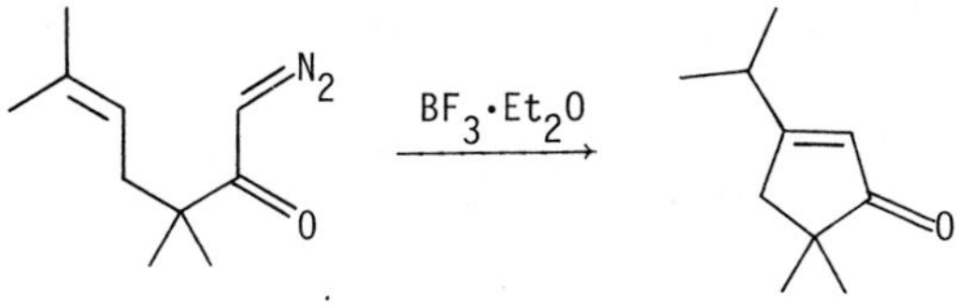

~50%

Comptes Rendus C (1976) 282 761

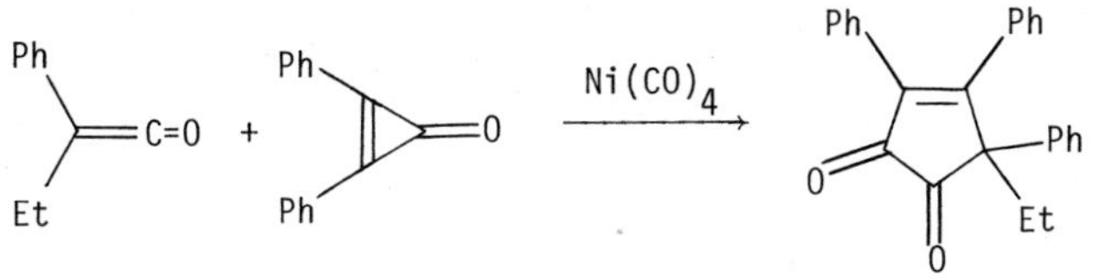

85%

J Organometal Chem (1976) 110 121

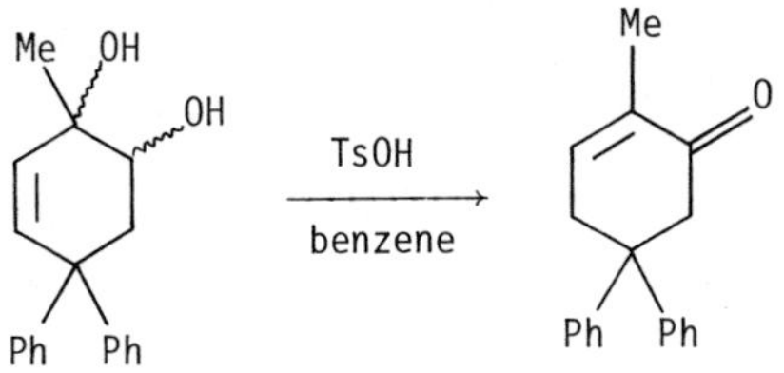

90%

Helv Chim Acta (1976) 59 2012

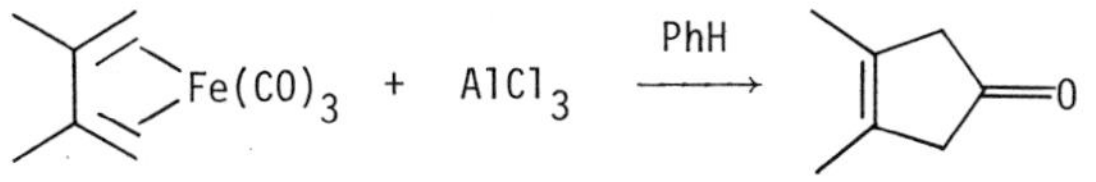

moderate yield

Tetr Lett (1974) 3789

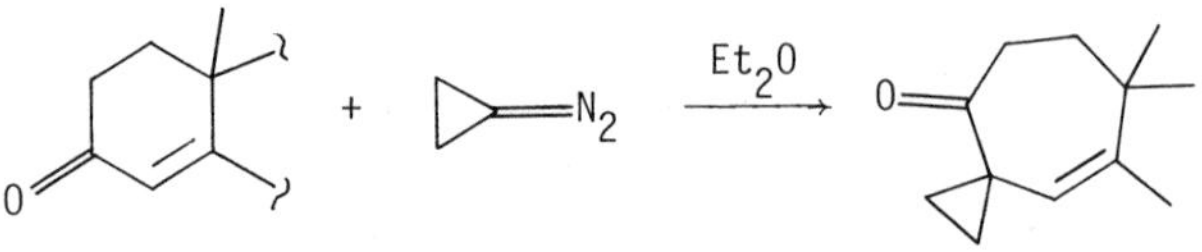

30%

JCS Perkin I (1974) 2240

$$TMSCH_2CH=CH_2 \; + \; PhCOCl \; \xrightarrow{\text{AlCl}_3} \; PhCOCH_2CH=CH_2$$

90%

J Organometal Chem (1975) 85 149

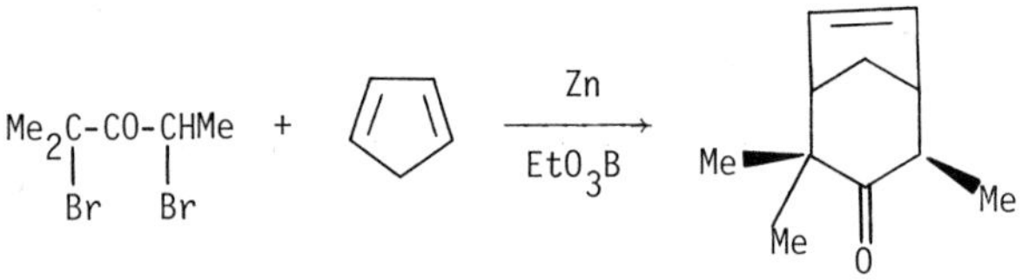

78%

Tetr Lett (1975) 4487

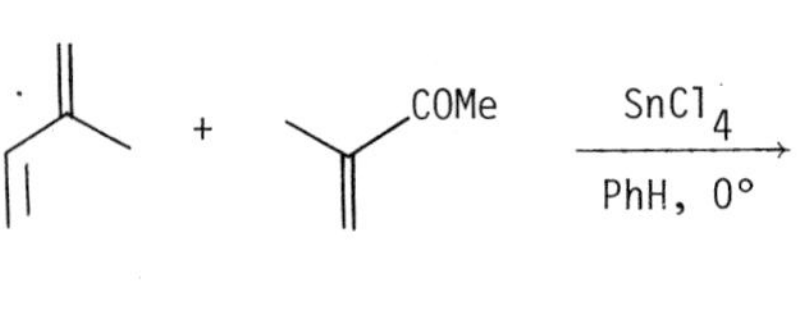

70%

Helv Chim Acta (1974) 57 164

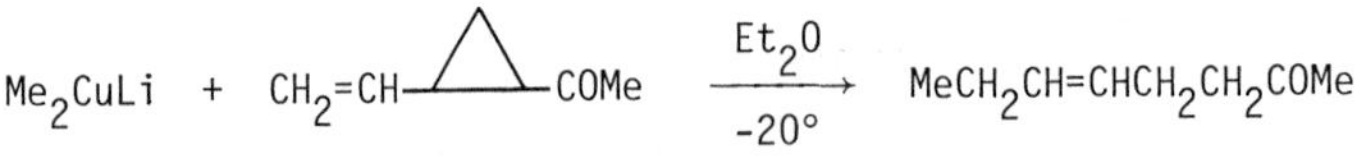

$$Me_2CuLi \; + \; CH_2=CH-\!\!\triangle\!\!-COMe \; \xrightarrow[-20°]{\text{Et}_2\text{O}} \; MeCH_2CH=CHCH_2CH_2COMe$$

81%

Synthesis (1975) 317

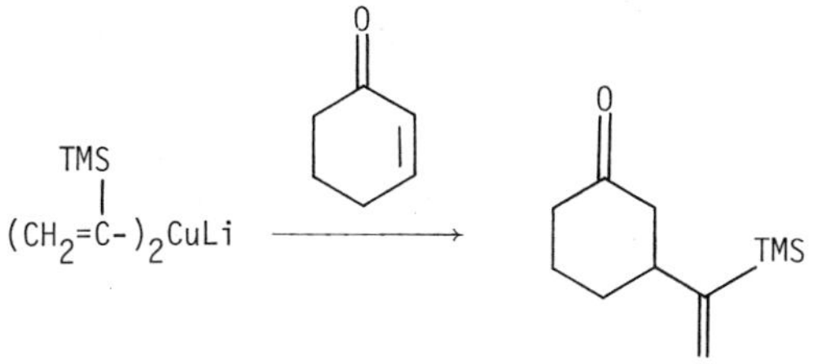

TMS
(CH2=C-)2CuLi
O
O
TMS

90%

Tetr Lett (1974) 3365

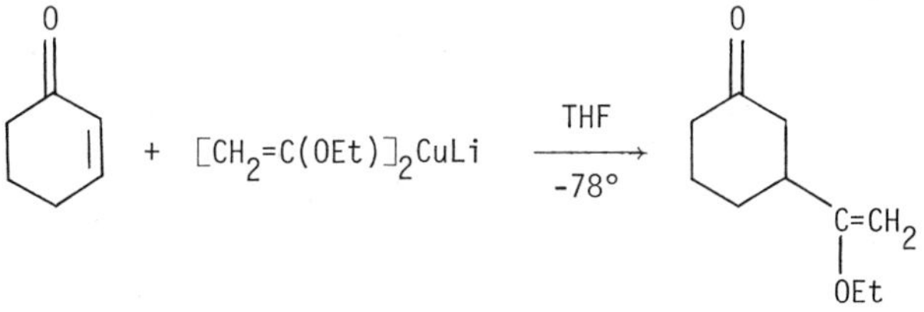

O
+ [CH2=C(OEt)]2CuLi
THF
-78°
O
C=CH2
OEt

84%

JCS Chem Comm (1975) 519
JACS (1975) 97 3822

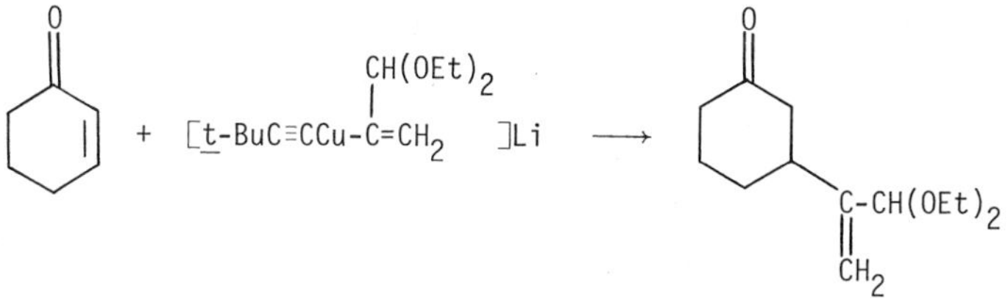

O
CH(OEt)2
+ [t-BuC≡CCu-C=CH2]Li
O
C-CH(OEt)2
CH2

96%

Tetr Lett (1975) 3901

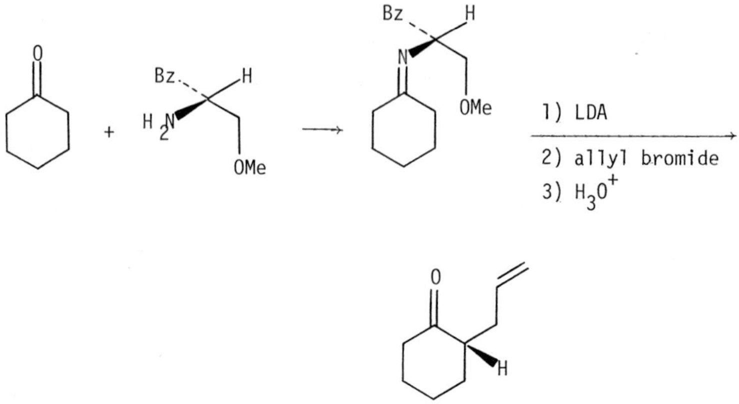

1) LDA

2) allyl bromide

3) H_3O^+

80%

>90% ee

JACS (1976) <u>98</u> 3032

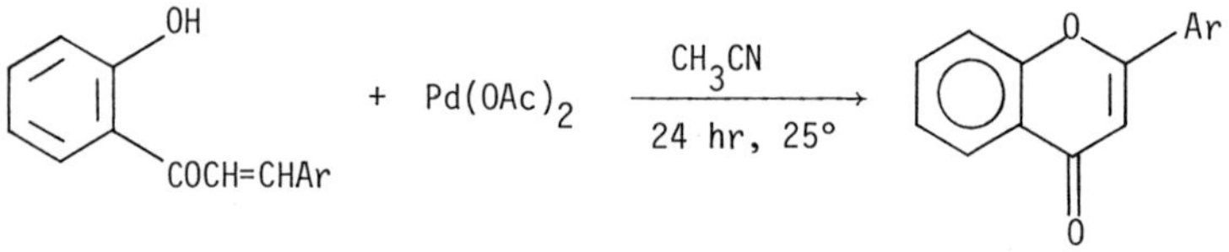

+ Pd(OAc)$_2$

$\xrightarrow[\text{24 hr, 25°}]{\text{CH}_3\text{CN}}$

80%

Bull Chem Soc Jap (1974) <u>47</u> 2526

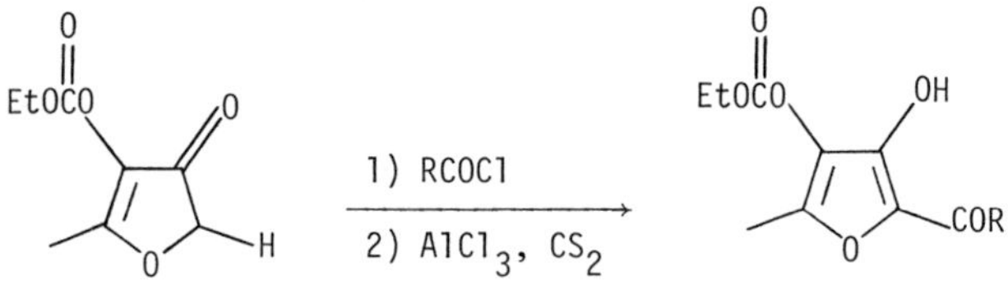

1) RCOCl

2) AlCl$_3$, CS$_2$

Comptes Rendus (1974) <u>278</u> 267

Also via: Acetylenic ketones (Section 309)
           β-Hydroxyketones (Section 330)

Section 375   **Nitrile - Nitrile**

90%

Chem Lett (1976) 147

Section 376   **Nitrile - Olefin**

RCH_2CH_2CN $\xrightarrow{\begin{array}{l}1)\ LDA,\ -75°\\ 2)\ (PhSe)_2\\ 3)\ H_2O_2\end{array}}$

90%

Tetr Lett (1974) 2279

$H_2C=C=O$ + TMSCN $\longrightarrow$ $H_2C=C\begin{array}{l}\diagup OTMS\\ \diagdown CN\end{array}$

96%

Angew Int Ed (1975) 14 179

MeĊ-CN + EtCOMe $\longrightarrow$

95%

Tetr Lett (1974) 4005

PhCHO + CH_3CN $\xrightarrow[\text{crown ether}]{\text{KOH}}$ PhCH=CHCN

82%

Tetr Lett (1976) 3495

PhCHO $\xrightarrow[\text{NaOH}]{\text{CH}_2(\text{CO}_2\text{H})\text{CN}}$ [structure: Ph-CH=C(COOH)(CN)] $\xrightarrow[\Delta]{\text{CuO}_2}$ [structure: Ph-CH=CH-CN]   ~40%

Tetr Lett (1975) 3843

[structure: isopropyl-C(CN)(COOEt)⁻ Na⁺] + [structure: 1-bromo-1-nitrocyclohexane] $\xrightarrow[\Delta]{\text{HMPT}}$ [structure: cyclohexylidene with NC and isopropyl group]   63%

Chem Lett (1976) 757

2 [structure: CH₂ with CN and COOEt] $\xrightarrow[\text{THF, rfx}]{\text{SOCl}_2}$ [structure: (COOEt)(NC)C=C(CN)(COOEt)]   64%

Synth Comm (1976) 6 185

PhC≡CH $\xrightarrow{\begin{array}{l}\text{1) }[\text{Ni}_2(\text{CN})_2(\text{DPB})_3]\\ \text{2) HCl}\end{array}}$ PhCH=CHCN   70%

Gazz Chim Ital (1974) 104 1279

[structure: PhCHCN with Et] PhCHCN / Et $\xrightarrow[\text{phase-transfer cat}]{\text{HC≡CH, KOH, DMSO}}$ [structure: Ph-C(CH=CH₂)(CN)(Et)]   60%

Org Synth (1976) 55 99

Section 377　　Olefin - Olefin

$$Me_2C=CCH_3 \quad \xrightarrow[100°]{\text{quinoline}} \quad Me_2C=C=CH_2 \qquad\qquad 70\%$$
$$\underset{OTf}{|}$$

JOC (1975) $\underline{40}$ 647

$$R'C\equiv CCH-OTs \quad \xrightarrow[\text{THF}]{R^3MgX,\ CuBr} \quad R^3R'C=C=CHR^2 \qquad\qquad 90\%$$
$$\underset{R^2}{\overset{|}{}}$$

Rec Trav Chim (1975) $\underline{94}$ 112

$$RC\equiv CSMe \quad \xrightarrow[\text{2) MeI, 0°}]{\text{1) R'}_2CuLi} \quad \underset{R'}{\overset{R}{>}}C=C=C=C\underset{R'}{\overset{R}{<}} \qquad 20\text{-}50\%$$

Tetr Lett (1975) 2923

$$Ph_2C=C=C\underset{Br}{\overset{Ph}{<}} \quad \xrightarrow[\text{DMF}]{CuCl} \quad Ph_2C=C=C\underset{\underset{Ph}{}}{\overset{Ph}{<}}C=C=CPh_2 \qquad 72\%$$

JCS Chem Comm (1975) 174

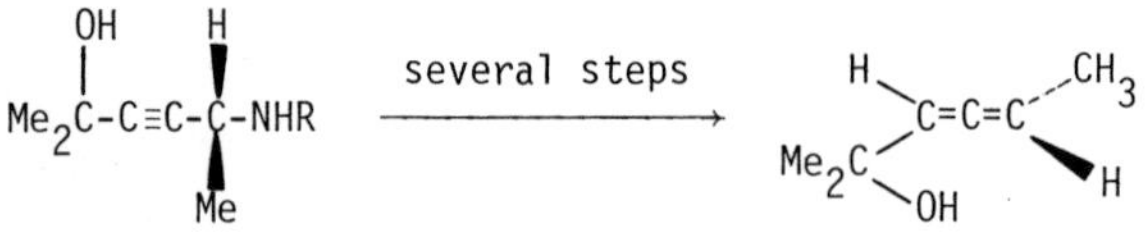

~30%

33% ee

JACS (1975) 97 2919

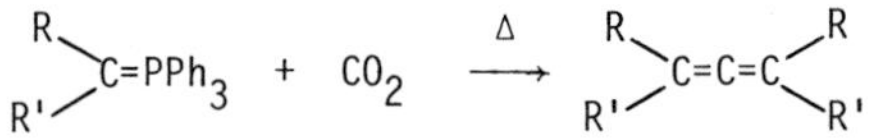

~30%

Tetr Lett (1974) 1275

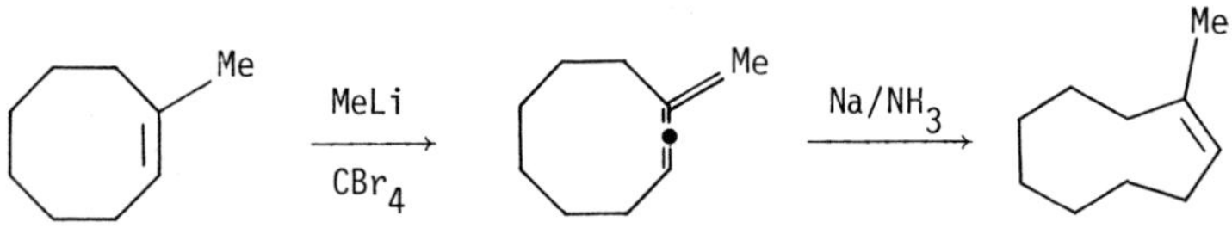

95%

65%

Synthesis (1975) 194

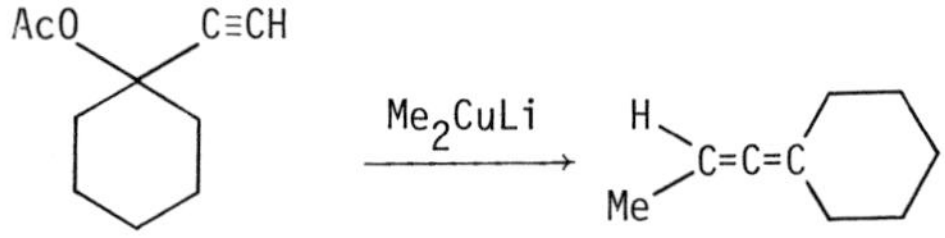

80%

JCS Chem Comm (1976) 183

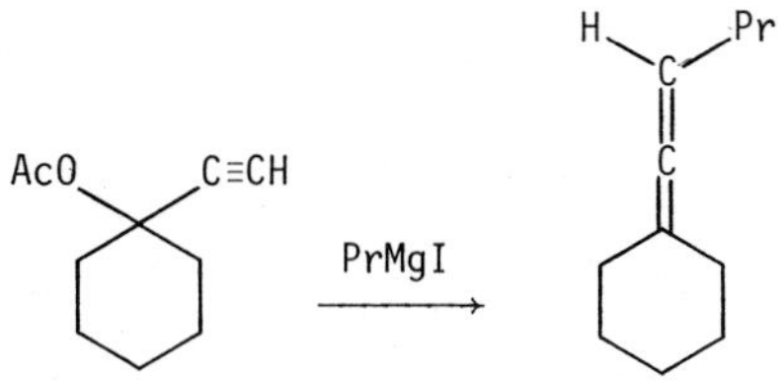

AcO   C≡CH

PrMgI
⟶

72%

Angew Int Ed (1976) 15 496

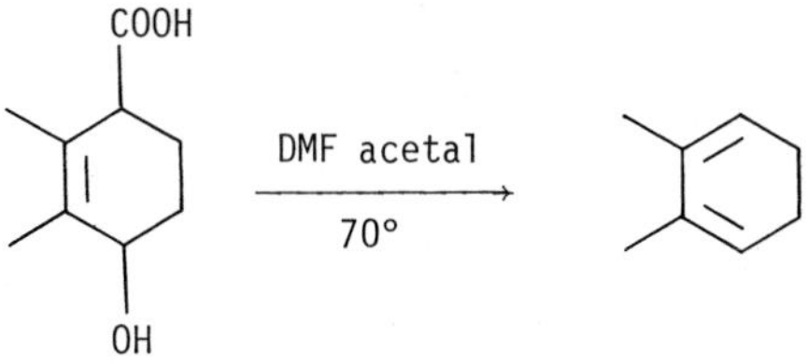

COOH

DMF acetal
⟶
70°

94%

OH

Helv Chim Acta (1975) 58 1450

Me₂C=CHCH₂SCHMeS(O)Me

1) LDA, -10°
⟶
2) MeI

SMe
|
CH₂=CHCMe₂C=CH₂

54%

Comptes Rendus (1975) 280 1231;1327

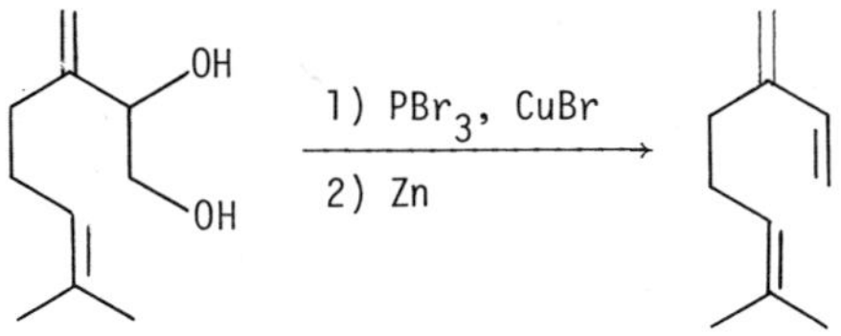

OH

OH

1) PBr₃, CuBr
⟶
2) Zn

58%

JACS (1975) 97 3252

$$\begin{array}{c} \text{OH} \\ \end{array} \quad \xrightarrow[\text{HMPT, 50°}]{[(PhO)_3PCH_3]^+ \ I^-} \quad \text{(benzene ring)} \qquad 78\%$$

Synthesis (1976) 108

$$R'R^2C=CHBr \ + \ R^3CH=CHR^4 \quad \xrightarrow[\text{L}_2Pd(OAc)_2, \ 100°]{Et_3N} \quad R'R^2C=CHCR^3=CHR^4$$

JOC (1975) 40 1083

$$CH_2=CHCH=CH_2 \quad \xrightarrow[\text{NaBH}_4, \text{ THF-EtOH, 100°}]{(Ph_3P)_2NiBr_2} \quad CH_2=CHCH=CHCH_2CH=CHMe \qquad 95\%$$

JACS (1975) 97 341

$$EtC\equiv CEt \quad \xrightarrow[\text{2) MeCu}]{1) \ H_2BCl}$$

Et   Et
 \   /
  C=C
 /   \
H   C=C
       / \
     Et   Et
          H   100%

JACS (1975) 97 5606

$$\begin{array}{c} \text{Ph} \\ \text{Ph} \end{array}C=CH_2 \quad \xrightarrow[\text{CH}_2Cl_2, \ h\nu]{\text{(P)}-C_6H_4-I}$$

Ph   Ph
  \   /
   C=C
  /   \
Ph   CH
          \
           C=C
          /   \
        Ph   Ph   80%

Synth Comm (1976) 6 309

JOC (1975) <u>40</u> 100

PhC≡CR + ⟨(Ni)⟩ —→ $\xrightarrow{\text{HCl}}$ 50%

J Organometal Chem (1975) <u>96</u> C19

PhSe(O)CH₃

1) LDA, THF, -78°
$\xrightarrow{\hspace{2cm}}$ PhCH=CHCH₂CH=CHPh   75%
2) PhCH=CHCH₂Br
3) HOAc
4) CH₂Cl₂, rfx

JACS (1975) <u>97</u> 3250

CH₂=CHCH=CH₂

1) <u>t</u>-BuCu, MgBr₂
$\xrightarrow{\hspace{2cm}}$ <u>t</u>-BuCH₂CH=CHCH₂   90%
2)   Br

J Organometal Chem (1975) <u>92</u> C28

<u>i</u>-PrSCH₂CH=CH₂

1) <u>s</u>-BuLi, -25°
$\xrightarrow{\hspace{2cm}}$ <u>i</u>-PrSCH=CHCH₂CH₂CH=CH₂   88%
2) CuI, -78°
3) CH₂=CHCH₂Br

Bull Chem Soc Japan (1975) <u>48</u> 1567